Lecture Notes in Computer Science 4932

Commenced Publication in 1973
Founding and Former Series Editors:
Gerhard Goos, Juris Hartmanis, and Jan van Leeuwen

T0223167

Sven Hartmann   Gabriele Kern-Isberner (Eds.)

Foundations of Information and Knowledge Systems

5th International Symposium, FoIKS 2008
Pisa, Italy, February 11-15, 2008
Proceedings

 Springer

Volume Editors

Sven Hartmann
Massey University, Information Science Research Centre
PN 311, Private Bag 11 222, Palmerston North 5301, New Zealand
E-mail: s.hartmann@massey.ac.nz

Gabriele Kern-Isberner
University of Dortmund, Department of Computer Science
44221 Dortmund, Germany
E-mail: gabriele.kern-isberner@cs.uni-dortmund.de

Library of Congress Control Number: 2007943324

CR Subject Classification (1998): H.2, H.3, H.5, I.2.3-4, F.3.2, G.2

LNCS Sublibrary: SL 3 – Information Systems and Application, incl. Internet/Web
and HCI

ISSN 0302-9743
ISBN 978-3-540-77683-3 Springer Berlin Heidelberg New York

Springer is a part of Springer Science+Business Media

springer.com

© Springer-Verlag Berlin Heidelberg 2008

Typesetting: Camera-ready by author, data conversion by Scientific Publishing Services, Chennai, India
Printed on acid-free paper   SPIN: 12216253   06/3180   5 4 3 2 1 0

Preface

This volume contains the papers presented at the Fifth International Symposium on Foundations of Information and Knowledge Systems (FoIKS 2008) which was held in Pisa, Italy during February 11–15, 2008. On behalf of the Program Committee we commend these papers to you and hope you find them useful.

The FoIKS symposia provide a biennial forum for presenting and discussing theoretical and applied research on information and knowledge systems. The goal is to bring together researchers with an interest in this subject, share research experiences, promote collaboration and identify new issues and directions for future research.

FoIKS 2008 solicited original contributions dealing with any foundational aspect of information and knowledge systems, including submissions from researchers working in fields such as discrete mathematics, logic and algebra, model theory, information theory, complexity theory, algorithmics and computation, geometry, analysis, statistics and optimization who are interested in applying their ideas, theories and methods to research on information and knowledge systems.

Previous FoIKS symposia were held in Budapest (Hungary) in 2006, Vienna (Austria) in 2004, Schloß Salzau near Kiel (Germany) in 2002, and Burg / Spreewald near Berlin (Germany) in 2000. FoIKS took up the tradition of the conference series Mathematical Fundamentals of Database Systems (MFDBS), which initiated East-West collaboration in the field of database theory. Former MFDBS conferences were held in Rostock (Germany) in 1991, Visegrad (Hungary) in 1989, and Dresden (Germany) in 1987.

The FoIKS symposia are a forum for intensive discussions. Speakers are given sufficient time to present their results, expound relevant background information and put their research into context. Furthermore, participants are asked in advance to prepare as correspondents to a contribution of another author.

Suggested topics for FoIKS 2008 included, but were not limited to:

- Database design: formal models, dependency theory, schema translations, desirable properties
- Dynamics of information and knowledge systems: models of transactions, models of interaction, updates, consistency preservation, concurrency control
- Information integration: heterogeneous data, views, schema dominance and equivalence
- Integrity and constraint management: verification, validation, and enforcement of consistency, triggers
- Intelligent agents: multi-agent systems, autonomous agents, foundations of software agents, cooperative agents
- Knowledge discovery and information retrieval: machine learning, data mining, text mining, information extraction

- Knowledge representation: planning, reasoning techniques, description logics, knowledge and belief, belief revision and update, non-monotonic formalisms, uncertainty
- Logic in databases and AI: non-classical logics, spatial and temporal logics, probabilistic logics, deontic logic, logic programming
- Mathematical foundations: discrete structures and algorithms, graphs, grammars, automata, abstract machines, finite model theory, information theory
- Security and risk management in information and knowledge systems: privacy, trust, cryptography, steganography, information hiding
- Semi-structured data and XML: data modelling, data processing, data compression, data exchange
- Social and collaborative computing: symbiotic intelligence, self-organization, knowledge flow, decision making
- The Semantic Web and knowledge management: languages, ontologies, agents, adaption, intelligent algorithms
- The WWW: models of Web databases, Web dynamics, Web services, Web transactions and negotiations

Following the call for papers which yielded 79 submissions, there was a rigorous refereeing process that saw each paper refereed by at least three international experts. The 13 papers judged best by the Program Committee were accepted for long presentation. Further nine papers were accepted for short presentation. This volume contains revised versions of all these papers that have been polished by their authors to address the comments provided in the reviews. After the conference, authors of a few selected papers were asked to prepare extended versions of their papers for publication in a special issue of the journal *Annals of Mathematics and Artificial Intelligence.*

We wish to thank all authors who submitted papers and all conference participants for the fruitful discussions. We are grateful to Egon Börger, Jürgen Dix and Gyula O.H. Katona, who presented invited talks at the conference. We would like to thank the members of the Program Committee and external referees for their timely expertise in carefully reviewing the submissions. A special thank you goes to Markus Kirchberg for his outstanding work as FoIKS Publicity Chair. Without him this volume would not have seen the light of day. We would also like to acknowledge the excellent work of Dagong Dong, who programmed and supported the MuCoMS conference management system. Finally, we wish to express our appreciation to Carlo Meghini and his team for being our hosts and for the wonderful days in Pisa.

February 2008   Sven Hartmann
                                                       Gabriele Kern-Isberner

Conference Organization

Program Committee Chairs

Sven Hartmann, Massey University, New Zealand
Gabriele Kern-Isberner, University of Dortmund, Germany

Program Committee

Rudolf Ahlswede, University of Bielefeld, Germany
Catriel Beeri, The Hebrew University of Jerusalem, Israel
Leopoldo Bertossi, Carleton University, Canada
Joachim Biskup, University of Dortmund, Germany
Stefan Brass, University of Halle, Germany
Cristian S. Calude, University of Auckland, New Zealand
John Cantwell, Royal Institute of Technology, Sweden
Samir Chopra, City University of New York, USA
James P. Delgrande, Simon Fraser University, Canada
Jürgen Dix, Clausthal University of Technology, Germany
Rod Downey, Victoria University of Wellington, New Zealand
Thomas Eiter, Vienna University of Technology, Austria
Lluís Godo, Institut d'Investigació en Intel·ligència Artificial, CSIC, Spain
Stephen J. Hegner, Umeå University, Sweden
Anthony Hunter, University College London, UK
Hyunchul Kang, Chung-Ang University Seoul, Korea
Odej Kao, Berlin University of Technology, Germany
Gyula O.H. Katona, Hungarian Academy of Sciences, Hungary
Hans-Joachim Klein, University of Kiel, Germany
Dexter Kozen, Cornell University, USA
Jerome Lang, Institut de Recherche en Informatique de Toulouse, CNRS, France
Uwe Leck, University of Wisconsin, USA
Mark Levene, Birbeck University of London, UK
Sebastian Link, Massey University, New Zealand
Yue Lu, East China Normal University Shanghai, China
Thomas Lukasiewicz, Sapienza Università di Roma, Italy
Carlo Meghini, Istituto di Scienza e Tecnologie dell'Informazione, Italy
Peter Mika, Yahoo! Research Barcelona, Spain
Wilfred Ng, Hong Kong University of Science and Technology, China
Beng Chin Ooi, National University of Singapore
Jeff B. Paris, University of Manchester, UK
Henri Prade, Université Paul Sabatier, France
Attila Sali, Hungarian Academy of Sciences, Hungary

Vladimir Sazonov, University of Liverpool, UK
Klaus-Dieter Schewe, Massey University, New Zealand
Karl Schlechta, Université de Provence, France
Dietmar Seipel, University of Würzburg, Germany
Guillermo R. Simari, Universidad Nacional del Sur, Argentina
Nicolas Spyratos, University of Paris-South, France
Ernest Teniente, Universitat Politècnica de Catalunya, Spain
Bernhard Thalheim, University of Kiel, Germany
Yannis Theodoridis, University of Piraeus, Greece
Miroslav Truszczynski, University of Kentucky, USA
José María Turull-Torres, Massey University Wellington, New Zealand
Dirk Van Gucht, Indiana University, USA
Marina de Vos, University of Bath, UK
Jef Wijsen, University of Mons-Hainaut, Belgium
Ian H. Witten, University of Waikato, New Zealand
Jeffrey Xu Yu, Chinese University of Hong Kong, China

External Referees

Alberto Abello
Martin Brain
Nils Bulling
Andrea Calì
Elena Calude
Carlos Chesñevar
Tiago de Lima
Michel de Rougemont
Flavio Ferrarotti
Michael Fink
Giorgos Flouris
Alejandro Garcia
Bhakhadyr Khoussainov
Henning Koehler
Dominique Laurent
Alexei Lisitsa

Emiliano Lorini
Enrico Marchioni
Pavle Mogin
Peter Novak
Julian Padget
Jordi Sabater-Mir
Mantas Simkus
Shiliang Sun
Henning Thielemann
Hans Tompits
Thu Trinh
Panayiotis Tsaparas
Wiebe van der Hoek
Geoff Whittle
Stefan Woltran
Rui Zhou

Local Arrangements Chair

Carlo Meghini, Istituto di Scienza e Tecnologie dell'Informazione, Italy

Publicity Chair

Markus Kirchberg, Massey University, New Zealand

Sponsored By

Association for Symbolic Logic (ASL)
European Association for Theoretical Computer Science (EATCS)

Table of Contents

Coupling Design and Verification in Software Product Lines

Egon Börger[1] and Don Batory[2]

[1] Università di Pisa, Dipartimento di Informatica, I-56125 Pisa, Italy
boerger@di.unipi.it
[2] Department of Computer Sciences, University of Texas at Austin, USA
batory@cs.utexas.edu

Abstract. We propose an ASM-based method to integrate into current feature-based software design practice modular verification techniques.

1 Motivation

Scaling verification to large programs is a long-standing problem which has recently received renewed interest in the Formal Methods community, see [13, 14, 15]. For scaling to occur and to become of practical relevance, verification—instead of being an afterthought—must be intimately integrated with software design [5], using compositional verification techniques that support modular software design methods. One such design technique is feature-based development of software product lines [16]. In this talk, which is based upon the forthcoming paper [1], we present an approach that shows how combining feature-based modular design and proof techniques may scale verification to software product lines (families of related programs) and integrate it into current software development practice.

We illustrate the approach by a case study taken from the Jbook [17], showing that the work done there can be viewed as a lock-step development of a product line for OO programming language constructs and for their provably correct implementation by Virtual Machine bytecode. From the theorems proved in the Jbook we choose here the Java-to-JVM compilation correctness proof, which is carried out in terms of a grammar generating Java programs, of an interpreter and a compilation scheme for them and of a bytecode (JVM) interpreter for the compiled code. The idea is to link feature modularization and composition to the structured way ASMs are used in Jbook to incrementally develop the involved components—Java grammar, interpreter, compiler, bytecode (JVM) interpreter (including a bytecode verifier)— and to accompany the component definition by an equally incremental development also of the statements and proofs of the theorems verifying the desired program properties (here compilation correctness).

S. Hartmann and G. Kern-Isberner (Eds.): FoIKS 2008, LNCS 4932, pp. 1–4, 2008.

2   Combined Stepwise Refinement of Design and Verification Features

The main observation is that the stepwise definition of the language grammar, interpreter and compiler components for Java/JVM in the Jbook can be associated with a stepwise definition of the statements of the properties of interest and of their proofs. The language levels (horizontal structure), introduced in the Jbook by stepwise ASM refinement, capture imperative (I), class module (C), object oriented (O), exception handling (E) and thread concurrency features (T) for expressions and statements. Their combination can be represented in the Gen-Voca model of product-lines [3] as functional composition where *Java* denotes a definition of the Java language; Exp_I denotes the definition of imperative Java expressions, Stm_I the definition of imperative Java statements, etc.:

$$Java = Stm_T.Stm_E.Exp_E.Exp_O.Stm_C.Exp_C.Stm_I.Exp_I$$

We explain here how theorem statements and correctness proofs can be assembled at each horizontal composition step simultaneously with the language and program constituents. One can define a vector that covers the vertical component structure to be considered in the given step, what in AHEAD [2,4] is exposed more generally as different program representations. For example, for the feature Exp_I of imperative expressions one has the vector

$$[G_{Exp_I}, I_{Exp_I}, C_{Exp_I}, J_{Exp_I}, T_{Exp_I}]$$

of the corresponding grammar (G), the interpreter (I) and the compiler (C) for the generated expressions, the interpreter for the compiled JVM-bytecode (J) and the compilation correctness theorem (T) relating the two interpreter runs via the compilation, where the theorem component itself is a vector $[S_{Exp_I}, P_{Exp_I}]$ of the statement and its proof. To illustrate the ability to refine representations by features consider adding the statement Stm_I feature to define the imperative component $Java_I$. It can be realized in the GenVoca framework as elementwise vector composition as follows:

$$\begin{aligned} Java_I &= Stm_I.Exp_I \\ &= [\Delta G_{Stm_I}, \Delta I_{Stm_I}, \Delta C_{Stm_I}, \Delta J_{Stm_I}, \Delta T_{Stm_I}] \cdot [G_{Exp_I}, I_{Exp_I}, C_{Exp_I}, J_{Exp_I}, T_{Exp_I}] \\ &= [\Delta G_{Stm_I}.G_{Exp_I}, \Delta I_{Stm_I}.I_{Exp_I}, \Delta C_{Stm_I}.C_{Exp_I}, \Delta J_{Stm_I}.J_{Exp_I}, \Delta T_{Stm_I}.T_{Exp_I}] \end{aligned}$$

That is, the grammar of the $Java_I$ language is the base grammar G_{Exp_I} composed with its refinement by ΔG_{Stm_I}—adding new rules for Java statements and tokens—resulting in $\Delta G_{Stm_I}.G_{Exp_I}$, the ASM definition of the $Java_I$ interpreter is the base definition composed with its refinement by ΔI_{Stm_I}—implementing the new statements—resulting in $\Delta I_{Stm_I}.I_{Exp_I}$, and so on. In general, the representations of a program are assembled by taking a GenVoca expression, replacing each term with its corresponding vector, and composing vectors.

The theorem statement component $T.S_{Exp_I}$ of the theorem vector T_{Exp_I} for Exp_I in the Jbook consists of three invariants called there (reg), (begE) and (exp). The first invariant (reg) expresses the equivalence of local variables in the language interpreter I_{Exp_I} and the associated registers in the JVM interpreter J_{Exp_I} when both are in what is defined as 'corresponding' states. The second invariant (begE) expresses that in corresponding states, when the language interpreter begins to execute an expression, the JVM interpreter begins to execute the compiled code for that expression and the computed intermediate values are equivalent. The third invariant (exp) expresses the same as (begE) for a value returning termination of an expression execution. The Stm_I feature refines the theorem statement component $T.S_{Exp_I}$ by adding $\Delta T.S_{Stm_I}$ consisting of three invariants—(begS), (stm) and (abr)—which deal with the normal and abrupted termination of statement executions.

The proof component $T.P_{Exp_I}$ of the theorem vector T_{Exp_I} for Exp_I is a case analysis using structural induction on the definition of expressions and of their compilation, showing that the invariants of the statement component $T.S_{Exp_I}$ are preserved in each case during (properly initialized) interpreter runs of I_{Exp_I} and J_{Exp_I}. For the composed interpreters $\Delta I_{Stm_I}.I_{Exp_I}$ and $\Delta J_{Stm_I}.J_{Exp_I}$ to satisfy the invariants of $\Delta T.S_{Stm_I} \cdot T.S_{Exp_I}$, $\Delta T.P_{Stm_I}$ adds proof cases to $T.P_{Exp_I}$ for each new feature in ΔG_{Stm_I}. These additional cases prove the invariants of $\Delta T.S_{Stm_I} \cdot T_{Exp_I}$ to be preserved in statement executing interpreter runs. Note that this induction on statements uses the proofs for the expression invariants as induction hypothesis.

3 Outlook

To conclude we explain that the method illustrated here can be reused for the other theorems proved in Jbook and also for other languages of the family of object-oriented languages. See for example the work done for C# and the .NET CLR (see [6,7,8,9,11,10]). Also the work in [12] on the development of UML and SDL profiles can be interpreted along these lines. The method is not restricted to the development of language patterns and their implementations, but seems to be applicable to the verification of properties of programs in arbitrary software product lines, at least wherever the development proceeds incrementally, adding features and new elements and extend existing elements. One of the major issues for the practicality of refining theorems about feature behavior consists in supporting the certification of assembled theorems by appropriate tools, in particular proof-checkers.

References

1. Batory, D., Börger, E.: Modularizing theorems for software product lines (2008)
2. Batory, D., Lofaso, B., Smaragdakis, Y.: Jts: Tools for implementing domain-specific languages. In: Proc. ICSR (1998)
3. Batory, D., O'Malley, S.: The design and implementation of hierarchical software systems with reusable components. ACM TOSEM (October 1992)
4. Batory, D., Sarvela, J., Rauschmayer, A.: Scaling step-wise refinement. IEEE TSE (June 2004)

5. Börger, E.: Construction and analysis of ground models and their refinements as a foundation for validating computer based systems. Formal Aspects of Computing 19, 225–241 (2007)
6. Börger, E., Fruja, G., Gervasi, V., Stärk, R.: A high-level modular definition of the semantics of C#. Theoretical Computer Science 336(2–3), 235–284 (2005)
7. Börger, E., Stärk, R.F.: Exploiting Abstraction for Specification Reuse. The Java/C# Case Study. In: de Boer, F.S., Bonsangue, M.M., Graf, S., de Roever, W.-P. (eds.) FMCO 2003. LNCS, vol. 3188, pp. 42–76. Springer, Heidelberg (2004)
8. Fruja, N.G.: The correctness of the definite assignment analysis in C#. Journal of Object Technology 3(9), 29–52 (2004)
9. Fruja, N.G.: A modular design for the Common Language Runtime (CLR) architecture. In: Beauquier, D., Börger, E., Slissenko, A. (eds.) Proc. ASM 2005, vol. 12, pp. 175–200. Université de Paris (2005)
10. Fruja, N.G.: Type Safety of C# and .NET CLR. PhD thesis, ETH Zürich (2006)
11. Fruja, N.G., Börger, E.: Modeling the .NET CLR Exception Handling Mechanism for a Mathematical Analysis. Journal of Object Technology 5(3), 5–34 (2006)
12. Grammes, R.: Syntactic and Semantic Modularization of Modelling Languages. PhD thesis, University of Kaiserslautern, Germany (2007)
13. Hoare, C.A.R.: The verifying compiler: A grand challenge for computing research. J. ACM 50(1), 63–69 (2003)
14. Hoare, T., Misra, J.: Verified software: theories, tools, experiments. Vision of a Grand Challenge project. In: Meyer, B. (ed.) Proc. IFIP WG Conference on Verified Software: Tools, Techniques, and Experiments. ETH, Zürich (October 2005), http://vstte.ethz.ch/papers.html
15. Jones, C., O'Hearn, P., Woodcock, J.: Verified software: A grand challenge. IEEE Computer (April 2006)
16. Pohl, K., Bockle, G., Linden, F.v.d.: Software Product Line Engineering: Foundations, Principles and Techniques. Springer, Heidelberg (2005)
17. Stärk, R.F., Schmid, J., Börger, E.: Java and the Java Virtual Machine: Definition, Verification, Validation. Springer, Heidelberg (2001)

Random Geometric Identification

Gyula O.H. Katona

Rényi Institute, Budapest, Hungary
ohkatona@renyi.hu

Abstract. The practical problem can be described in the following way. Physical objects (credit cards, important documents) should be identified using geometric labels. An optical device reads the label and a simple computation checks whether the label belongs to the given object or not. This could be done by "asking" an authority which stores certain data (e.g. the reading of the label) of the objects in question. This, however, supposes the existence of an online connection which could technically be difficult, on the other hand it would be a source of evedropping.

This is why the identification will be done using the label and a 0,1 sequence both placed on the object. The sequence can be determined from the label. This calculation is done first when the document is supplied with them, later each time when identification is needed the reading and calculation are repeated and it is checked whether the result is the 0,1 sequence written on the object or not.

Of course the method has practical values only when the label cannot be easily reproduced. Moreover we suppose that the labels are randomly generated.

More mathematically, the space S of all possible labels is known and a (deterministic) function f mapping S into $\{0,1\}^n$, that is, the set of 0,1 sequences of length n. This function is used first in the "factory" where the labels and the 0,1 sequences are placed on the objects, and later each time when the identification is necessary. The size $|S|$ of S is either infinite or finite, but even in the latter case $|S|$ is much larger than the number of 0,1 sequences, 2^n. The subset $A \subset S$ satisfying $A = \{a : f(a) = i\}$ for a given 0,1 sequence i is denoted by A_i. Of course, $A_i \cap A_{i'} = \emptyset$ must hold for $i \neq i'$. The situation however is even more serious. The reading device can read the label only with a certain error. Therefore we have to suppose that the space is endowed with a distance d. ($0 \leq d(a,b)(= d(b,a))$ is defined for all pairs $a, b \in S$, where $d(a,b) = 0$ iff $a = b$ and the triangle inequality $d(a,c) \leq d(a,b) + d(b,c)$ holds.) The reading device can read the label with an error at most $\varepsilon > 0$. Then if the label is within A_i, the reading device might sense it anywhere in the set $n(A_i, \varepsilon) = \{x \in S : d(A, x) \leq \varepsilon\}$. The function f has to be defined and must have the value i within $n(A_i, \varepsilon)$. Therefore these sets must be also disjoint.

If a label x is randomly generated it might fall outside of the set $\cup_{i=1}^{2^n} n(A_i, \varepsilon)$. Then x cannot be used, it is a waste. The proportion of the the waste should be low. Hence we have the condition $\mu(\cup_{i=1}^{2^n} n(A_i, \varepsilon)) \geq \alpha$ for some (not very little) $\alpha > 0$.

S. Hartmann and G. Kern-Isberner (Eds.): FoIKS 2008, LNCS 4932, pp. 5–6, 2008.

If $\mu(A_i)$ is too large for some i then the falsifier has a good chance to choose a point randomly which falls in A_i. Therefore $\mu(A_i) \leq \rho$ must hold for a rather small $0 < \rho$.

Now we are ready to define the *geometric identifying codes in S of size n with error tolerance ε, waste-rate α and security ρ* as a family of subsets $A_1, \ldots, A_{2^n}$ where $n(A_i, \varepsilon)$ are disjoint $(1 \leq i \leq 2^n)$, $\mu(\cup_{i=1}^{2^n} n(A_i, \varepsilon)) \geq \alpha$, and $\mu(A_i) \leq \rho$ holds for $1 \leq i \leq 2^n$. The paper investigates when these codes exist in the case of a general S.

A practical algorithm using these principles was worked out by the following team: L. Csirmaz, A. Haraszti, Gy. Katona, L. Marsovszky, D. Miklós and T. Nemetz. The theoretical investigations above were done jointly by Csirmaz and the present author.

In our practical case an element of the space S is a set of points where the points are in a (two-dimensional) rectangle, with coordinates of the form $\frac{j}{N}$ where N is an integer (because the coordinates can be determined only up to a certain exactness) and the number of points chosen is between given lower and upper bounds. The distance between two elements of S is defined by the distances of their member points and the set difference between them. (Some of the points can disappear during the reading. This is a practical experience.) The investigations for this special space lead to combinatorial problems related to the *shadow problem*: given m k-element subsets of an n-element set, what can be said about the minimum number of $k-1$-element subsets of these k-element subsets? Results in this direction are obtained jointly with P. Frankl.

Strategic Abilities of Agents

Jürgen Dix

Department of Computer Science, Clausthal University of Technology,
Julius-Albert-Str. 4, D-38678 Clausthal-Zellerfeld, Germany
`dix@tu-clausthal.de`

Abstract. In recent years, there has been a trend in the area of multi-agent systems to model and reason about complex systems by using appropriate logics.

In this talk, we focus on logics that have been defined to deal with describing and reasoning about groups of agents and their abilities: what are they able to bring about.

Starting from CTL, we discuss ATL and various extensions. We discuss complexity results and, in particular, focus on symbolic representations and their implication for the overall complexity.

We discuss several relatively new ideas to add plausibility operators to ATL, to combine ATL with argumentation theory and to combine branching time logic CTL* with Markov models: Markov Temporal Logic (due to Jamroga).

S. Hartmann and G. Kern-Isberner (Eds.): FoIKS 2008, LNCS 4932, p. 7, 2008.
© Springer-Verlag Berlin Heidelberg 2008

Aggregation of Attack Relations: A Social-Choice Theoretical Analysis of Defeasibility Criteria*

Fernando A. Tohmé, Gustavo A. Bodanza, and Guillermo R. Simari

Artificial Intelligence Research and Development Laboratory (LIDIA)
Universidad Nacional del Sur, Av.Alem 1253, (8000) Baha Blanca, Argentina
Consejo Nacional de Investigaciones Científicas y Técnicas (CONICET)
{ftohme,ccbodanza}@criba.edu.ar, grs@cs.uns.edu.ar

Abstract. This paper analyzes the aggregation of different abstract attack relations over a common set of arguments. Each of those attack relations can be considered as the representation of a *criterion* of warrant. It is well known in the field of Social Choice Theory that if some "fairness" conditions are imposed over an aggregation of preferences, it becomes impossible to yield a result. When the criteria lead to *acyclic* attack relations, a positive result may ensue under the same conditions, namely that if the class of winning coalitions in an aggregation process by voting is a *proper prefilter* an outcome will exist. This outcome may preserve some features of the competing attack relations, such as the highly desirable property of acyclicity which can be associated with the existence of a single extension of an argumentation system. The downside of this is that, in fact, the resulting attack relation must be a portion common to the "hidden dictators" in the system, that is, all the attack relations that belong to all the winning coalitions.

1 Introduction

Defeasible reasoning relies on the possibility of comparing conclusions in terms of their support. This support is often given by a set of arguments. Only those arguments (and consequently their conclusions) that remain undefeated in a series of comparisons are deemed warranted. While the literature contains alternative formalisms capturing this intuition [1,2], the groundbreaking work on Abstract Argumentation Frameworks reported in [3] presents a view according to which all the features that are not essential for the study of the attack relation in defeasible argumentation are eliminated. What remains is a system formed by a family of abstract arguments and a relation of *attack* among them. Several alternative semantics have been introduced, but the essential idea is that the set of arguments that survive all possible attacks of other arguments in the system constitute the so-called *extensions* of the system and capture its semantics.

One aspect that has received little attention in the literature[1] is the possibility of considering different relations of attack among the same arguments. In this scenario, the warrant of arguments cannot be established in an unambiguous way without first

---

* Partially supported by SeCyT - Universidad Nacional del Sur, CONICET, and ANPCYT.
[1] A remarkable exception being [4].

S. Hartmann and G. Kern-Isberner (Eds.): FoIKS 2008, LNCS 4932, pp. 8–23, 2008.

coalescing the multiple attack relation onto a single one acting over the family of arguments. Notice that this ensuing relation does not need to coincide with anyone of those defined over the arguments. On the other hand, nothing precludes this possibility.

In Economics, the process by which a single preference ordering is obtained given a class of individual preferences over the same alternatives, it is known as an *aggregation* of them [5]. Similarly, we can consider that each of the attack relations among arguments represents an individual *criterion* of warrant, since it defines which extensions should obtain. Then, the aggregation process weights up the different criteria and determines which extensions will actually appear, but instead of simply enumerating extensions, it yields an attack relation that supports them.

While there might exist many ways of doing this, a natural form is by means of pairwise voting [6]. That is, each alternative attack relation "votes" over pairs of arguments, and the winning relation over those two arguments is incorporated in the aggregate attack relation.

But such procedure has been shown to have, in certain contexts, serious shortcomings. It is widely known that it may fail to verify some required constraints over the aggregation process [7,8]. These constraints are actually desiderata for a fair aggregation process. Social Choice Theory (SCT) has been studying them for over fifty years and it seems natural to transfer its results to the problem of aggregating attack relations. Arrow's Impossibility Theorem [9] claims that four quite natural constraints, that capture abstractly the properties of a democratic aggregation process, cannot be simultaneously satisfied. That is true for the case of reflexive and transitive preference relations over the alternatives. Once those constraints become incorporated in the framework of argumentation, we could expect something like Arrow's theorem to ensue. But attack relations and preference relations are different in many respects. This point must be emphasized, since it involves the reason why an Arrow-like result may not be a necessary outcome for argumentation systems. This fact makes our purpose non trivial.

The difference between aggregating individual preferences and attack criteria originates from their corresponding order-theoretic characterizations. While in Economics preferences are usually assumed to be *weak orders* (i.e., reflexive, transitive and complete relations), attack relations are free to adopt any configuration. On the other hand, preference relations are expected to have maximal elements, while this is not the case for attack relations. If A attacks B and B attacks C, it is commonly accepted that not only A does not (necessarily) attack C, but that A "defends" C, which implies that A and C can be jointly warranted. So, while a preference relation can lead to the choice of its maximal elements, and attack relation can lead to the choice of a maximal (w.r.t. $\subseteq$) set of "defensible" arguments.

Viewed as criteria of acceptance, the choices should verify at least a minimal degree of rationality. In SCT that requirement is fulfilled by the condition that chosen options should not be transitively better than themselves, i.e. they should not be part of cycles of preference [10]. In the context of attack criteria this condition can be interpreted as that each of the arguments that will be deemed warranted under a criterion should be supported by chains of attacks that do not include themselves. A sufficient condition that ensures this is the *acyclicity* of the attack relations.[2]

---

[2] An argumentation framework in which the attack relation is acyclic is said *well-founded* [3].

Once we require the acyclicity of the attack relations we look for aggregation processes that have as inputs finite numbers of acyclic attack relations and output also acyclic relations. In that case, as we will show in this paper, under the same conditions of fairness as Arrow's Theorem, we can prove the existence of an aggregate attack relation. In fact, following [7] we show that the class of winning coalitions of attack criteria constitutes an algebraic structure called a *proper prefilter*.

As it has been discussed in the literature on Arrow's Theorem, a prefilter indicates the existence of a *collegium* of attack relations. Each member of the collegium belongs to a winning coalition, while the collegium itself does not need to be one. Each collegium member, by itself, cannot determine the outcome of the aggregation process, but can instead veto the behaviors that run contrary to its prescription. The final outcome can be seen as the agreement of the representatives of the different winning coalitions. In this sense it indicates a very basic consensus among the attack relations.

In a sense this means that even in the case of "equal opportunity" aggregation procedures there will exist some fragment of the individual attack relations that will become imposed on the aggregate one. But while in social context this seems rather undesirable (in the literature the members of the collegium are called *hidden dictators*), in the case of argument systems is far more reassuring, since it indicates that when the attack relations are minimally rational, a consensual outcome may arise.

2   Aggregating Attack Relations

Dung defines an argumentation framework as a pair $AF = \langle AR; \rightarrow \rangle$, where AR is a set of abstract entities called 'arguments' and $\rightarrow \subseteq AR \times AR$ denotes an attack relation among arguments. This relation determines which sets of arguments become "defended" from attacks. Different characterizations of the notion of defense yield alternative sets called *extensions* of AF. These extensions are seen as the semantics of the argumentation framework, i.e. the classes of arguments that can be deemed as the outcomes of the whole process of argumentation. Dung introduces the notions of *preferred*, *stable*, *complete*, and *grounded* extensions, each corresponding to different requirements on the attack relation.

Definition 1. (Dung ([3])). *In any argumentation framework AF an argument σ is said* acceptable *w.r.t. a subset S of arguments of AR, in case that for every argument τ such that $\tau \rightarrow \sigma$, there exists some argument $\rho \in S$ such that $\rho \rightarrow \tau$. A set of arguments S is said* admissible *if each $\sigma \in S$ is acceptable w.r.t. S, and is conflict-free, i.e., the attack relation does not hold for any pair of arguments belonging to S. A* preferred extension *is any maximally admissible set of arguments of AF. A* complete extension *of AF is any conflict-free subset of arguments which is a fixed point of $\Phi(\cdot)$, where $\Phi(S) = \{\sigma : \sigma$ is acceptable w.r.t. $S\}$, while the* grounded extension *is the least (w.r.t. $\subseteq$) complete extension. Moreover, a* stable extension *is a conflict-free set S of arguments which attacks every argument not belonging to S.*

Interestingly, if the attack relation is acyclic, the framework has only one extension that is grounded, preferred stable and complete (*cf.* [3], theorem 30, pp. 331). The main application of argumentation frameworks is the field of *defeasible reasoning*. Roughly, arguments are structures that support certain conclusions (claims). The extensions include the arguments, and more importantly their conclusions, that become warranted by a reasoning process that considers the attack relation.

We consider, instead, for a given n an *extended* argumentation framework $AF^n = \langle AR; \rightarrow_1, \ldots, \rightarrow_n \rangle$. Each $\rightarrow_i$ is a particular attack relation among the arguments in AR, representing different criteria according to which arguments are evaluated one against another. Such extended frameworks may arise naturally in the context of defeasible reasoning, since there might exist more than one criterion of defeat among arguments.

The determination of *preferred, complete* or *grounded* extensions in an argumentation framework is based upon the properties of the single attack relation. There are no equivalent notions for an extended argumentation framework, except for those corresponding to an *aggregate* argumentation framework $AF^* = \langle AR; \mathcal{F}(\rightarrow_1, \ldots, \rightarrow_n) \rangle$, where $\mathcal{F}(\rightarrow_1, \ldots, \rightarrow_n) = \rightarrow$, *i.e.*, $\mathcal{F}(\rightarrow_1, \ldots, \rightarrow_n)$ is the aggregated attack relation of AF^*. That is, AF^* is a an argumentation framework in which its attack relation arises as a function of the attack relations of AF^n. Notice that $\mathcal{F}$ may be applied over any extended argumentation framework with n attack relations. It embodies a method that yields a single attack relation up from n alternatives.

To postulate an aggregate relation addresses the problem of managing the diversity of criteria, by yielding a single approach. This is of course analogous to a social system, in which a unified criterion must by reached. While there exist many alternative ways to aggregate different criteria, most of them are based in some form of voting. In fact, the best known case of $\mathcal{F}$ is *majority voting*. Unlike political contests in which for each pair $A, B \in AR$ a majority selects either $A \rightarrow B$ or $B \rightarrow A$, we allow for a third alternative in which the majority votes for the absence of attacks between A and B. Formally:

- $A \rightarrow B$ if $|\{i : A \rightarrow_i B\}| > \max(|\{i : B \rightarrow_i A\}|, |\{i : B \not\rightarrow_i A \wedge A \not\rightarrow_i B\}|)$.
- $B \rightarrow A$ if $|\{i : B \rightarrow_i A\}| > \max(|\{i : A \rightarrow_i B\}|, |\{i : B \not\rightarrow_i A \wedge A \not\rightarrow_i B\}|)$.
- $(A, B) \notin \rightarrow$ (*i.e.*, A does not attack B, nor B does attack A in $\rightarrow$) if $|\{i : B \not\rightarrow_i A \wedge A \not\rightarrow_i B\}| > \max(|\{i : A \rightarrow_i B\}|, |\{i : B \rightarrow_i A\}|)$.

For instance, if out of 100 individual relations, 34 are such that A attacks B, while 33 verify that B attacks A and the rest that there is no attack relation between A and B, majority voting would yield that $A \rightarrow B$. That is, it only matters which alternative is verified by more individual relations than the other two.

Example 1. Consider the following framework in which $AR = \{A, B, C\}$ and the arguments are:

A : "Symptoms x, y and z suggest the presence of disease d_1, so we should apply therapy t_1";

B : "Symptoms x, w and z suggest the presence of disease d_2, so we should apply therapy t_2";

C : "Symptoms x and z suggest the presence of disease d_3, so we should apply therapy t_3".

Assume these are the main arguments discussed in a group of three agents (M.D.s), 1, 2 and 3, having to make a decision on which therapy should be applied to some patient. Suppose that each agent i, $i \in \{1, 2, 3\}$, proposes an attack relation $\rightarrow_i$ over the arguments as follows:

- $\rightarrow_1 = \{(A, B), (B, C)\}$ (agent 1 thinks that it is not convenient to make a joint application of therapies t_1 and t_2 or t_2 and t_3; moreover she thinks that B is more specific than C, hence B defeats C, and that, in the case at stake, symptom y is more clearly present than symptom w. Hence A defeats B),
- $\rightarrow_2 = \{(A, C), (B, C)\}$ (agent 2 thinks that it is not convenient to apply therapies t_1 together with t_3 or t_2 joint with t_3; moreover she thinks that symptoms y and w are equally present in the case at stake. Furthermore, both A and B are more specific than argument C, hence both A and B defeat C),
- $\rightarrow_3 = \{(A, C), (C, B)\}$ (agent 3 thinks that it is not convenient to apply t_1 together with t_3 or t_2 with t_3; moreover she thinks that symptom w is not clearly detectable, hence C defeats B, but A is more specific than C, hence A defeats C).

According to majority voting we obtain $\rightarrow$ over AR:

- $A \rightarrow C$ since A attacks C under $\rightarrow_2$ and $\rightarrow_3$.
- $B \rightarrow C$ since B attacks C under $\rightarrow_1$ and $\rightarrow_2$.
- $(A, B) \notin \rightarrow$ since $(A, B) \notin \rightarrow_2$ and $(A, B) \notin \rightarrow_3$.

In this example, majority voting picks out one of the individual attack relations, showing that $\rightarrow = \rightarrow_2$.[3]

On the other hand, majority voting may yield cycles of attacks up from acyclical individual relations:

Example 2. Consider the following three attack relations over the set $AR = \{A, B, C\}$: $C \rightarrow_1 B \rightarrow_1 A$, $A \rightarrow_2 C \rightarrow_2 B$, and $B \rightarrow_3 A \rightarrow_3 C$. We obtain $\rightarrow$ over AR as follows:

- $A \rightarrow C$ since A attacks C under $\rightarrow_2$ and $\rightarrow_3$.
- $B \rightarrow A$ since B attacks C under $\rightarrow_1$ and $\rightarrow_3$.
- $C \rightarrow B$ since C attacks B under $\rightarrow_1$ and $\rightarrow_2$.

Thus, $\rightarrow$ yields a cycle $A \rightarrow C \rightarrow B \rightarrow A$. This phenomenon is known in the literature on voting systems as the *Condorcet's Paradox* and it shows clearly that even the most natural aggregation procedures may have drawbacks.

Another way of aggregating attack relations is by restricting majority voting to a *qualified voting* aggregation function. It fixes a given class of relations as those that will have more weight in the aggregate. Then, the outcome of majority voting over a pair of arguments is imposed on the aggregate only if the fixed attack relations belong to the majority. Otherwise, in the attack relation none of the arguments attacks the other. That is, given a set $U \subset \{1, \ldots, n\}$:

---

[3] In any of the extension semantics introduced by [3], arguments A and B become justified under the aggregate attack relation, supporting the decision of applying both therapies t_1 and t_2.

- $A \rightarrow B$ iff $|\{i : A \rightarrow_i B\}| > \max(|\{i : B \rightarrow_i A\}|, |\{i : B \not\rightarrow_i A \wedge A \not\rightarrow_i B\}|)$ and $U \subseteq \{i : A \rightarrow_i B\}$.
- $B \rightarrow A$ iff $|\{i : B \rightarrow_i A\}| > \max(|\{i : A \rightarrow_i B\}|, |\{i : B \not\rightarrow_i A \wedge A \not\rightarrow_i B\}|)$ and $U \subseteq \{i : B \rightarrow_i A\}$.

$(A, B) \not\rightarrow$ (*i.e.*, A does not attack B, nor B does attack A in $\rightarrow$) can arise as follows:

- either $|\{i : B \not\rightarrow_i A \wedge A \not\rightarrow_i B\}| > \max(|\{i : A \rightarrow_i B\}|, |\{i : B \rightarrow_i A\}|)$ and $U \subseteq \{i : B \not\rightarrow_i A \wedge A \not\rightarrow_i B\}$,
- or if U is not a subset of either $\{i : A \rightarrow_i B\}$, $\{i : B \rightarrow_i A\}$ or $\{i : B \not\rightarrow_i A \wedge A \not\rightarrow_i B\}$.

Example 3. Consider again the individual attack relations in Example 1. If $U = \{2, 3\}$ we have that $A \rightarrow C$, since $A \rightarrow_2 C$ and $A \rightarrow_3 C$. Again $(A, B) \not\rightarrow$ because $(A, B) \not\rightarrow_2$ and $(A, B) \not\rightarrow_3$. But we have also that $(B, C) \not\rightarrow$ because although there exists a majority for B attacking C ($\{1, 2\}$), $C \rightarrow_3 B$, *i.e.*, there is no consensus among the members of U on B and C.

3   Arrow's Conditions on Aggregation Functions

While different schemes of aggregation of attack relations can be postulated, most of SCT, up from the seminal work of Kenneth Arrow [9] points towards a higher degree of abstraction. Instead of looking for particular functional forms, the goal is to set general constraints over aggregation processes and see if they can be jointly fulfilled. We carry out a similar exercise in the setting of extended argumentation frameworks, in order to investigate the features of aggregation processes that ensure that a few reasonable axioms are satisfied.

Social choice-theoretic analysis can be carried out in terms of an aggregation process that, up from a family of *weak orders* (complete, transitive and reflexive orderings), yields a weak order over the same set of alternatives. This is because both individual and social *preference* relations are represented as weak orders. But attack relations cannot be assimilated to *preference* orderings, since attacks do not verify necessarily any of the conditions that define a weak order.[4] Therefore, the difference of our setting with the usual Arrovian context is quite significant.

Let us begin with a few properties that, very much like in SCT, we would like to be verified in any aggregation function. Below, we will use the alternative notation $\rightarrow_{\mathcal{F}}$ instead of $\mathcal{F}(\rightarrow_1, \ldots, \rightarrow_n)$ when no confusion could arise.

- **Pareto condition.** For all $A, B \in AR$ if for every $i = 1, \ldots, n$, $A \rightarrow_i B$ then $A \rightarrow_{\mathcal{F}} B$.
- **Positive Responsiveness.** For all $A, B \in AR$, and two n-tuples of attack relations, $(\rightarrow_1, \ldots, \rightarrow_n)$, $(\rightarrow'_1, \ldots, \rightarrow'_n)$, if $\{i : A \rightarrow_i B\} \subseteq \{i : A \rightarrow'_i B\}$ and $A \rightarrow_{\mathcal{F}} B$, then $A \rightarrow'_{\mathcal{F}} B$, where $\rightarrow'_{\mathcal{F}} = \mathcal{F}(\rightarrow'_1, \ldots, \rightarrow'_n)$.

---

[4] So for instance, reflexivity in an attack relation would mean that each argument attacks itself. While isolated cases of self-attack may arise, this is not a general feature of attack relations. The same is true of transitivity that means that if, say $A \rightarrow B$ and $B \rightarrow C$ then $A \rightarrow C$. In fact, in many cases of interest, $A \rightarrow B \rightarrow C$ can be interpreted as indicating that A *defends* B. Finally, completeness is by no means a necessary feature of attacks, since there might exist at least two arguments A and B such that neither $A \rightarrow B$ and $B \rightarrow A$.

- **Independence of Irrelevant Alternatives.** For all $A, B \in AR$, and given two n-tuples of attack relations, $(\rightarrow_1, \ldots, \rightarrow_n)$, $(\rightarrow'_1, \ldots, \rightarrow'_n)$, if $\rightarrow_i = \rightarrow'_i$ for each i, over (A, B), then $\rightarrow_{\mathcal{F}} = \rightarrow'_{\mathcal{F}}$ over (A, B).
- **Non-dictatorship.** There does not exist i_0 such that for all $A, B \in AR$ and every $(\rightarrow_1, \ldots, \rightarrow_n)$, if $A \rightarrow_{i_0} B$ then $A \rightarrow_{\mathcal{F}} B$.

All these requirements were intended to represent the abstract features of a democratic collective decision-making system. While in our setting this does no longer apply, we still consider that an aggregation function should yield a fair representative of the whole class of attack relations. Let us see why these conditions imply the fairness of the aggregation process.[5]

The *Pareto condition* indicates that if all the attacks relations coincide over a pair of arguments, the aggregate attack should also agree with them. That is, if all the individual attack relations agree on some arguments, this agreement should translate into the aggregate attack relation.

The *positive responsiveness condition* just asks that the aggregation function should yield the same outcome over a pair of arguments if some attack relation previously dissident over them, now change towards an agreement with the others. It can be better understood in terms of political elections: if a candidate won an election, she should keep winning in an alternative context in which somebody who voted against her now turns to vote for her.

The axiom of *independence of irrelevant alternatives* just states that if there is an agreement over a pair of arguments among alternative n-tuples of attacks, this should be also be true for the aggregation function over both n-tuples. Again, some intuition from political elections may be useful. If the individual preferences over two candidates a and b remain the same when a third candidate c arises, the rank of a and b should be the same in elections with and without c. That is, the third party should be irrelevant to the other two.

Finally, the *non-dictatorship condition* just stipulates that no fixed entry in the n-tuples of attacks should become the outcome in every possible instance. That is, there is no 'dictator' among the individual attack relations. We have the following proposition:

Proposition 1
Both the majority and the qualified voting (with $|U| \geq 2$) aggregation functions verify trivially the four axioms.

PROOF
Majority voting:

- *(Pareto): if for all $A, B \in AR$ if for every $i = 1, \ldots, n$, $A \rightarrow_i B$ then trivially $|\{i : A \rightarrow_i B\}| > \max(|\{i : B \rightarrow_i A\}|, |\{i : B \not\rightarrow_i A \wedge A \not\rightarrow_i B\}|)$ which in turns implies that $A \rightarrow_{\mathcal{F}} B$.*

---

[5] Whether fairness is *exactly* captured by these requirements is still debated in the philosophy of Social Choice. Nevertheless, there exists a consensus on that they are desirable conditions for an aggregation function.

- *(Positive responsiveness): if for all $A, B \in AR$, and two n-tuples of attack relations, $(\rightarrow_1, \ldots, \rightarrow_n)$, $(\rightarrow'_1, \ldots, \rightarrow'_n)$, if $A \rightarrow_{\mathcal{F}} B$, this means that $|\{i : A \rightarrow_i B\}| > \max(|\{i : B \rightarrow_i A\}|, |\{i : B \not\rightarrow_i A \wedge A \not\rightarrow_i B\}|)$ and therefore, if $\{i : A \rightarrow_i B\} \subseteq \{i : A \rightarrow'_i B\}$ it follows that $|\{i : A \rightarrow'_i B\}| > \max(|\{i : B \rightarrow'_i A\}|, |\{i : B \not\rightarrow'_i A \wedge A \not\rightarrow'_i B\}|)$ which in turn implies that $A \rightarrow'_{\mathcal{F}} B$.*

- *(Independence of Irrelevant Alternatives): suppose that for any given $A, B \in AR$, and two n-tuples of attack relations, $(\rightarrow_1, \ldots, \rightarrow_n)$, $(\rightarrow'_1, \ldots, \rightarrow'_n)$, $\rightarrow_i = \rightarrow'_i$ for each i, over (A, B). Without loss of generality assume that $|\{i : A \rightarrow_i B\}| > \max(|\{i : B \rightarrow_i A\}|, |\{i : B \not\rightarrow_i A \wedge A \not\rightarrow_i B\}|)$ then, $A \rightarrow_{\mathcal{F}} B$. But then $|\{i : A \rightarrow'_i B\}| > \max(|\{i : B \rightarrow'_i A\}|, |\{i : B \not\rightarrow'_i A \wedge A \not\rightarrow'_i B\}|)$, which implies that $A \rightarrow'_{\mathcal{F}} B$. That is, $\rightarrow_{\mathcal{F}} = \rightarrow'_{\mathcal{F}}$ over (A, B).*

- *(Non-dictatorship): suppose there where a i_0 such that for all $A, B \in AR$ and every $(\rightarrow_1, \ldots, \rightarrow_n)$, if $A \rightarrow_{i_0} B$ then $A \rightarrow_{\mathcal{F}} B$. Consider in particular that $A \rightarrow_{i_0} B$ while $|\{i : B \rightarrow_i A\}| = n - 1$, i.e. except i_0 all other attack relations have B attacking A. But then $B \rightarrow_{\mathcal{F}} A$. Contradiction.*

The proof for qualified voting, when $|U| \geq 2$, is quite similar:

- *(Pareto): if for all $A, B \in AR$ if for every $i = 1, \ldots, n$, $A \rightarrow_i B$ then trivially $|\{i : A \rightarrow_i B\}| > \max(|\{i : B \rightarrow_i A\}|, |\{i : B \not\rightarrow_i A \wedge A \not\rightarrow_i B\}|)$ and $U \subseteq \{i : A \rightarrow_i B\}$ which implies that $A \rightarrow_{\mathcal{F}} B$.*

- *(Positive responsiveness): if for all $A, B \in AR$, and two n-tuples of attack relations, $(\rightarrow_1, \ldots, \rightarrow_n)$, $(\rightarrow'_1, \ldots, \rightarrow'_n)$, if $A \rightarrow_{\mathcal{F}} B$, this means that $|\{i : A \rightarrow_i B\}| > \max(|\{i : B \rightarrow_i A\}|, |\{i : B \not\rightarrow_i A \wedge A \not\rightarrow_i B\}|)$ and $U \subseteq \{i : A \rightarrow_i B\}$. Therefore, if $\{i : A \rightarrow_i B\} \subseteq \{i : A \rightarrow'_i B\}$ it follows that $|\{i : A \rightarrow'_i B\}| > \max(|\{i : B \rightarrow'_i A\}|, |\{i : B \not\rightarrow'_i A \wedge A \not\rightarrow'_i B\}|)$ and $U \subseteq \{i : A \rightarrow'^i B\}$ which in turn implies that $A \rightarrow'_{\mathcal{F}} B$.*

- *(Independence of Irrelevant Alternatives): suppose that for any given $A, B \in AR$, and two n-tuples of attack relations, $(\rightarrow_1, \ldots, \rightarrow_n)$, $(\rightarrow'_1, \ldots, \rightarrow'_n)$, $\rightarrow_i = \rightarrow'_i$ for each i, over (A, B). Without loss of generality assume that $|\{i : A \rightarrow_i B\}| > \max(|\{i : B \rightarrow_i A\}|, |\{i : B \not\rightarrow_i A \wedge A \not\rightarrow_i B\}|)$ and $U \subseteq \{i : A \rightarrow_i B\}$ then, $A \rightarrow_{\mathcal{F}} B$. But then $|\{i : A \rightarrow'_i B\}| > \max(|\{i : B \rightarrow'_i A\}|, |\{i : B \not\rightarrow'_i A \wedge A \not\rightarrow'_i B\}|)$ and also $U \subseteq \{i : A \rightarrow'_i B\}$ which implies that $A \rightarrow'_{\mathcal{F}} B$. That is, $\rightarrow_{\mathcal{F}} = \rightarrow'_{\mathcal{F}}$ over (A, B).*

- *(Non-dictatorship): suppose there where a i_0 such that for all $A, B \in AR$ and every $(\rightarrow_1, \ldots, \rightarrow_n)$, if $A \rightarrow_{i_0} B$ then $A \rightarrow_{\mathcal{F}} B$. Consider in particular that $A \rightarrow_{i_0} B$ while $|\{i : B \rightarrow_i A\}| = n - 1$, i.e. except i_0 all other attack relations have B attacking A. If $i_0 \notin U$, $B \rightarrow_{\mathcal{F}} A$, while if $i_0 \in U$, $(A, B) \notin \rightarrow_{\mathcal{F}}$. In either case we have a contradiction.* $\qquad\qquad\square$

While qualified voting seems in certain sense less fair than majority voting it can be shown that it is not prone to phenomena like Condorcet's Paradox:

Proposition 2

If F is a qualified voting aggregation function and each $\rightarrow_i$ is acyclic, then $\rightarrow_{\mathcal{F}}$ is acyclic.

PROOF

Suppose that $\to_{\mathcal{F}}$ has a cycle of attacks, say $A^0 \to_{\mathcal{F}} A^1 \to_{\mathcal{F}} \ldots \to_{\mathcal{F}} A^k \to_{\mathcal{F}} A^0$. By definition of qualified voting, for $j = 0 \ldots k - 1$ $A^j \to_{\mathcal{F}} A^{j+1}$ if and only if $U \subseteq \{i : A^j \to_i A^{j+1}\}$. By the same token, $A^k \to_{\mathcal{F}} A^0$ iff

$$U \subseteq \{i : A^k \to_i A^0\}.$$

That is, for each $i \in U$, $A^0 \to_i A^1 \to_i \ldots \to_i A^k \to_i A^0$. But this contradicts that each individual attack relation is acyclic. □

4   Decisive Sets of Attack Relations

The aggregation function determines a class of *decisive* sets (*i.e.*, winning coalitions) of attack relations. Interestingly, the structure of this class exhibits (in relevant cases) clear algebraic features that shed light on the behavior of the aggregation function. Formally: $\Omega \subset \{1, \ldots, n\}$ be a *decisive* set if for every possible n-tuple $(\to_1, \ldots, \to_n)$ and every $A, B \in AR$, if $A \to_i B$, for *every* $i \in \Omega$, then $A \to_{\mathcal{F}} B$ (*i.e.*, $A \, \mathcal{F}(\to_1, \ldots, \to_n) \, B$). As we have already seen in the case of qualified voting aggregation functions, if not every member of a decisive set agrees with the others over a pair of arguments, the aggregate attack relation should not include the pair. But this is so unless any other decisive set can force the pair of arguments into the aggregate attack relation.[6]

Example 4. In Example 1, each of $\{1, 2\}, \{2, 3\}$ is a decisive set, since they include more than half of the agents that coincide with pairs of attacks in the aggregate attack relation. On the other hand, for the qualified voting function of Example 3, $U = \{2, 3\}$ is decisive, but not $\{1, 2\}$ or $\{1, 3\}$.

If we recall that the U is a decisive set for qualified voting, we can conjecture that there might exist a close relation between the characterization of an aggregation function and the class of its decision sets. Furthermore, if a function verifies Arrow's axioms and yields an acyclic attack relation up from acyclic individual attack relations, it can be completely characterized in terms of the class of its decision sets:

Proposition 3
Consider an aggregate attack relation $\mathcal{F}$ that for every n-tuple $(\to_1, \ldots, \to_n)$ of acyclic attack relations yields an acyclic $\to_{\mathcal{F}}$. It verifies the Pareto condition, Positive Responsiveness, Independence of Irrelevant Alternatives, and Non-Dictatorship if and only if its class of decisive sets $\bar{\Omega} = \{\Omega^j\}_{j \in J}$ verifies the following properties:

- $\{1, \ldots, n\} \in \bar{\Omega}$.
- *If $O \in \bar{\Omega}$ and $O \subseteq O'$ then $O' \in \bar{\Omega}$.*
- *Given $\bar{\Omega} = \{\Omega^j\}_{j \in J}$, where $J = |\bar{\Omega}|$, $\cap \bar{\Omega} = \bigcap_{j=1}^{J} \Omega^j \neq \emptyset$.*
- *No $O \in \bar{\Omega}$ is such that $|O| = 1$.*

---

[6] Of course, in a qualified voting function U is always a decisive set.

PROOF

$(\Rightarrow)$

We will begin our proof noticing that $|\bar{\Omega}| \leq 2^n$. That is, it includes only a finite number of decisive sets. Then, by Pareto, the grand coalition $\{1, \ldots, n\}$ must be decisive. On the other hand, by Positive Responsiveness, if a set O is decisive and $O \subseteq O'$, if the attack relations in $O' \setminus O$ agree with those in O, the result will be the same, and therefore O' becomes decisive too.

By Independence of Irrelevant Alternatives, if over a pair of arguments A, B, the attack relations remain the same then the aggregate attack relation will be the same over A, B. We will prove that this implies that $\cap \bar{\Omega} \neq \emptyset$. First, consider the case where at least two decisive sets $O, W \in \bar{\Omega}$ are such that $O \cap W = \emptyset$. Suppose furthermore that O determines $\rightarrow_{\mathcal{F}}$ up from $\{\rightarrow_i\}_{i=1}^n$ while W defines $\rightarrow_{\mathcal{F}}'$ up from $\{\rightarrow_i'\}_{i=1}^n$. Then, if over $A, B \rightarrow_i = \rightarrow_i'$ then $\rightarrow_{\mathcal{F}} = \rightarrow_{\mathcal{F}}'$ over A, B. But then, since there is no element common to O and W, the choice over A, B will differ from $\rightarrow_{\mathcal{F}}$ to $\rightarrow_{\mathcal{F}}'$. Contradiction. Furthermore, if $\cap \bar{\Omega} = \emptyset$, then there is no $\bar{i}$ such that $\rightarrow_{\mathcal{F}} \subseteq \rightarrow_{\bar{i}}$. But then, $\rightarrow_{\mathcal{F}}$ includes other attacks than those in each individual attack relation. Without loss of generality, consider an extended argument framework over n arguments and a profile in which the attack relations over them is such that each of them constitutes a linear chain of attacks:

- *$A^1 \rightarrow_1 A^2 \ldots \rightarrow_1 A^n$,*
- *$A^2 \rightarrow_2 \ldots A^n \rightarrow_2 A^1$,*
- *$\ldots$,*
- *$A^n \rightarrow_n A^1 \ldots \rightarrow_n A^{n-1}$.*

Then, over each pair A^j, A^k, $\rightarrow_{\mathcal{F}}$ has to coincide with some of the individual attack relations. In particular for each pair of arguments A^j, A^{j+1}. But also on A^n, A^1. But then, $\rightarrow_{\mathcal{F}}$ yields a cycle (see Example 2): $A^1 \rightarrow_{\mathcal{F}} A^2 \ldots \rightarrow_{\mathcal{F}} A^n \rightarrow_{\mathcal{F}} A^1$ But this contradicts the assumption that $\rightarrow_{\mathcal{F}}$ is acyclic. Then, $\cap \bar{\Omega} \neq \emptyset$.

Finally, a dictator i_0 is such that $\{i_0\} \in \bar{\Omega}$. Therefore, non-dictatorship implies that there is no $O \in \bar{\Omega}$ such that $|O| = 1$.

$(\Leftarrow)$

The Pareto condition follows from the fact that $\{1, \ldots, n\} \in \bar{\Omega}$. That is, if for a given pair $A, B \in AR$, $A \rightarrow_i B$ for every $i = 1, \ldots, n$, since $\{1, \ldots, n\}$ is decisive, it follows that $A \rightarrow_{\mathcal{F}} B$.

Positive Responsiveness follows from the fact that if O is decisive and $O \subseteq O'$, O' is also decisive. This is so since, given any $A, B \in AR$, and two n-tuples of attack relations, $(\rightarrow_1, \ldots, \rightarrow_n)$, $(\rightarrow_1', \ldots, \rightarrow_n')$, if $\{i : A \rightarrow_i B\} \subseteq \{i : A \rightarrow_i' B\}$ and $A \rightarrow_{\mathcal{F}} B$, then $\{i : A \rightarrow_i B\}$ is decisive, and therefore $\{i : A \rightarrow_i' B\}$ is also decisive, and then $A \rightarrow_{\mathcal{F}}' B$.

Independence of Irrelevant Alternatives obtains from the fact that $\cap \bar{\Omega} \neq \emptyset$. Suppose this were not the case. That is, there exists a pair $A, B \in AR$, and two n-tuples of attack relations, $(\rightarrow_1, \ldots, \rightarrow_n)$, $(\rightarrow_1', \ldots, \rightarrow_n')$, such that $\rightarrow_i = \rightarrow_i'$ over (A, B), but $\rightarrow_{\mathcal{F}} \neq \rightarrow_{\mathcal{F}}'$ over (A, B). Consider $\bar{i} \in \cap \bar{\Omega} \neq \emptyset$. That is $\bar{i}$ belongs to every decisive set. Then if, without loss of generality, $A \rightarrow_{\bar{i}} B$ then $A \rightarrow_{\mathcal{F}} B$, but also, since $A \rightarrow_{\bar{i}}' B$, we have that $A \rightarrow_{\mathcal{F}}' B$. Contradiction.

Non-dictatorship *follows from the fact that no set with a single criterion is decisive and therefore, no single attack relation can be imposed over the aggregate for every profile of attack relations.*

Finally, notice that since there exists $\bar{i} \in \cap \bar{\Omega}$ over each pair of arguments A, B, $\rightarrow_{\mathcal{F}}$ either coincides with $\rightarrow_{\bar{i}}$ or $(A, B) \notin \rightarrow_{\mathcal{F}}$. Since $\rightarrow_{\bar{i}}$ has no cycles of attacks, $\rightarrow_{\mathcal{F}}$ will also be acyclic. □

When $\bar{\Omega}$ satisfies the properties described in Proposition 3, we say that $\bar{\Omega}$ is a *proper prefilter* over $\{1, \ldots, n\}$ [7]. Moreover, if the class of decision sets for an aggregation function has this structure, it aggregates acyclic attack relations into an acyclic relation, verifying Arrow's conditions.

Example 5. Over $\{\rightarrow_1, \rightarrow_2, \rightarrow_3\}$ (or $\{1, 2, 3\}$, for short), the only possible proper prefilters are:

- $\bar{\Omega}^I = \{\{1, 2\}, \{1, 2, 3\}\}$.
- $\bar{\Omega}^{II} = \{\{1, 3\}, \{1, 2, 3\}\}$.
- $\bar{\Omega}^{III} = \{\{2, 3\}, \{1, 2, 3\}\}$.
- $\bar{\Omega}^{IV} = \{\{1, 2\}, \{2, 3\}, \{1, 2, 3\}\}$.
- $\bar{\Omega}^V = \{\{1, 3\}, \{2, 3\}, \{1, 2, 3\}\}$.
- $\bar{\Omega}^{VI} = \{\{1, 2\}, \{1, 3\}, \{1, 2, 3\}\}$.

Notice that the corresponding aggregation functions $\mathcal{F}^I$, $\mathcal{F}^{II}$ and $\mathcal{F}^{III}$ are qualified voting functions. To see how the other three functions act, just consider $\mathcal{F}^{IV}$ over $A \rightarrow_1 B \rightarrow_1 C, A \rightarrow_2 C$, $B \rightarrow_2 C$, and $A \rightarrow_3 C \rightarrow_3 B$. Then, $\rightarrow_{\mathcal{F}} = \mathcal{F}^{IV}(\rightarrow_1, \rightarrow_2, \rightarrow_3)$ is defined as follows:

- $A \rightarrow_{\mathcal{F}} C$ since while there is no agreement in $\{1, 2\}$, $\{2, 3\}$ agree in that A attacks C.
- $B \rightarrow_{\mathcal{F}} C$ since B attacks C in $\rightarrow_1$ and $\rightarrow_2$, but there is no agreement in $\{2, 3\}$.
- $(A, B) \notin \rightarrow_{\mathcal{F}}$ since $(A, B) \notin \rightarrow_2$ and $(A, B) \notin \rightarrow_{\mathcal{F}} 3$, but there is no agreement in $\{1, 2\}$.

That means that $\mathcal{F}^{IV}$ behaves in the same manner as a majority function over the following profile: $(\rightarrow_1, \rightarrow_2, \rightarrow_3)$. The same conclusion can be drawn for $\mathcal{F}^V$ and $\mathcal{F}^{VI}$.

Notice that, $\mathcal{F}^{IV}$ is actually a majority function only in the case that $\rightarrow_{\mathcal{F}}$ is acyclic. That is, it behaves like the majority function in well-behaved cases. Instead, for the individual attack relations in Example 2 it yields an acyclic order $A \rightarrow_{\mathcal{F}} C \rightarrow_{\mathcal{F}} B$, which is *not* the outcome of the majority function. Therefore, we should actually say that $\mathcal{F}^{IV}$, $\mathcal{F}^V$ and $\mathcal{F}^{VI}$ are *acyclic majority* functions. Notice that any $\bar{i} \in \cap \bar{\Omega}$ is kind of a "hidden dictator", in the sense made precise in the following result:

Proposition 4
If $\bar{i} \in \cap \bar{\Omega}$, and $\mathcal{F}_{\bar{\Omega}}$ is the aggregation function characterized by the prefilter then $\bar{\Omega}$, $\rightarrow_{\mathcal{F}_{\bar{\Omega}}} \subseteq \rightarrow_{\bar{i}}$.

PROOF

Suppose that given $A, B \in AR$, we have, without loss of generality, that $A \rightarrow_{\bar{i}} B$. Let us consider two cases:

- *There exists a decisive set $O \in \bar{\Omega}$ such that for every $i \in O$ (by definition $\bar{i} \in O$), $A \rightarrow_i B$. Then $A \rightarrow_{\mathcal{F}_{\bar{\Omega}}} B$, and therefore $\rightarrow_{\mathcal{F}_{\bar{\Omega}}}$ coincides with $\rightarrow_{\bar{i}}$ over (A, B).*
- *There does not exist any decisive set O in which for every $i \in O$, $A \rightarrow_i B$. Then, neither $A \rightarrow_{\mathcal{F}_{\bar{\Omega}}} B$ nor $B \rightarrow_{\mathcal{F}_{\bar{\Omega}}} A$ can obtain. Therefore $\bar{i}$ vetoes $B \rightarrow_i A$, although it cannot imposes $A \rightarrow B$. In this case $\rightarrow_{\mathcal{F}_{\bar{\Omega}}} \subset \rightarrow_{\bar{i}}$ over (A, B).*

□

We will consider now the question of *Aggregation and Cycles of Attack*. The analysis of argumentation systems is usually carried out in terms of their *extensions*. The existence and properties of the extensions can be ascertained according to the properties of the attack relation. In the case that several alternative attack relations compete over the same class of arguments, the class of extensions may vary from one to another. The structure of extensions of such an argumentation system should not be seen as just the enumeration of the classes corresponding to each attack relation but should arise from the same aggregation process we have discussed previously. That is, it should follow from the properties of the aggregate attack relation.

In particular, since our main results concern the aggregation of acyclic attack relations into an acyclic aggregate one, we will focus on the case of well-founded argumentation frameworks (*cf.* [3], p. 10). We can say, roughly, that their main feature is the absence of cycles of attack among their arguments. For them, all the types of extensions described by Dung coincide. Furthermore, they all yield a single set of arguments (*cf.* [3], theorem 30, p. 331). To see how such a single extension of an argument system over a family of individual attack relations may obtain, let us recall that if for each i, $\rightarrow_i$ has no cycles of attack, an aggregate relation $\rightarrow_{\mathcal{F}_{\bar{\Omega}}}$, obtained through an aggregation function F with a prefilter of decisive sets $\bar{\Omega}$, is acyclic as well. The following result is an immediate consequence of this claim.

Proposition 5

Consider an aggregate argument framework $AF^* = \langle AR; \ \mathcal{F}(\rightarrow_1, \ldots, \rightarrow_n) \rangle$. If each $\rightarrow_i$ $(i = 1, \ldots, n)$ is acyclic and $\mathcal{F}$ is such that its corresponding class of decisive sets $\bar{\Omega}$ is a prefilter, then $\rightarrow_{\mathcal{F}} = \mathcal{F}(\rightarrow_1, \ldots, \rightarrow_n)$ is acyclic and AF^* has a single extension which is grounded, preferred and stable.

Furthermore:

Corollary 1. *If $AF^* = \langle AR; \ \mathcal{F}(\rightarrow_1, \ldots, \rightarrow_n) \rangle$ has a single extension when each $\rightarrow_i$ $(i = 1, \ldots, n)$ is acyclic, then if $\mathcal{F}$ is such that its corresponding class of decisive sets $\bar{\Omega}$ is a prefilter, it also verifies the Pareto condition, Positive Responsiveness, Independence of Irrelevant Alternatives, and Non-Dictatorship.*

PROOF

Immediate. If AF^* has a single extension and $\mathcal{F}$ is such that its corresponding class of decisive sets $\bar{\Omega}$ is a prefilter, then by Proposition 5 the aggregate attack $\rightarrow_{\mathcal{F}}$ is acyclic. Then, the claim follows from Proposition 3.

□

5 Discussion

As indicated by Brown in [11], the fact that the class of decisive sets constitutes a prefilter is an indication of the existence of a *collegium*. In SCT this means a kind of "shadow" decisive set, being its members interspersed among all the actual decisive sets. Their actual power comes not from being able to enforce outcomes but from their ability to veto alternatives that are not desirable for them. In the current application, the existence of a collegium means that there exists a class of attack relations that by themselves cannot determine the resulting attack relation, but can instead block (veto) alternatives.

This feature still leaves many possibilities open, but as our examples intended to show, there are few aggregation functions that may adopt this form, while at the same time verifying the conditions postulated by Arrow for fair aggregation functions. The main instance is constituted by the acyclic majority function, but qualified voting functions yield also fair outcomes. The difference is that with the acyclic majority functions one of the several attack relations in AF^n is selected, while with qualified voting functions, new attack relations may arise. But these new attack relations just combine those of the winning coalitions, and therefore can be seen as resulting from the application of generalized variants of the majority function. That is, if two or more rules belong to all the decisive sets, their common fragments plus the non-conflicting ones add up to constitute the aggregate attack relation. In a way or another, the attack relations that are always decisive end up acting as hidden dictators in the aggregation process.

Explicit dictators arise in aggregation processes in other branches of non-monotonic reasoning. Doyle and Wellman [12], in particular, suggested to translate Reiter's defaults into total preorders of autoepistemic formulas, representing preferences over worlds. Reasoning with different defaults implies to find, first, the aggregation of the different preorders. These authors show that it is an immediate consequence of Arrow's theorem that no aggregation function can fulfill all the properties that characterize fairness (*i.e.*, the equivalents of the Pareto condition, Positive Responsiveness, Independence of Irrelevant Alternatives, and Non-Dictatorship). In terms of decisive sets, it means that $\hat{\Omega}$ constitutes a *principal ultrafilter*.[7] Since the number of defaults is assumed to be finite, it follows that there exists one of these default rules, say R^* that belongs to each $U \in \hat{\Omega}$. Of course, the existence of R^* violates the Non-Dictatorship condition, and consequently the actual class of formulas that arise in the aggregation are determined by R^*.

It can be said that Doyle and Wellman's analysis is concerned with the *generation* of a class of arguments arising from different default rules while we, instead, concentrate on the *comparison* among arguments in an abstract argumentation framework. But Dung [3] has shown that Reiter's system can be rewritten as an argumentation framework, and therefore both approaches can be made compatible. In this sense, Doyle and Wellman's result can be now interpreted as indicating that $\bar{\Omega}$ (over attack relations) is **not** a *proper prefilter* over $\{1, \ldots, n\}$ (where these indexes range over the attack relations determined each by a corresponding default rule). Instead, as said, it constitutes a

---

[7] Notice that $\hat{\Omega}$ denotes the decisive set over *default rules* and therefore should not be confused with $\bar{\Omega}$, the class of decisive sets over *attack relations*.

principal ultrafilter and therefore it implies the existence of a "dictatorial" attack relation, that is imposed over the framework.

Finally, the approach most related to ours is Coste-Marquis et al. [4], which already presented some early ideas on how to merge Dung's argumentation frameworks. The authors' aim is to find a set of arguments collectively warranted, addressing two main problems. One is the individual problem faced by each agent while considering a set of arguments different to that of other agents. The other is the aggregate problem of getting the collectively supported extension. The second one is the most clearly related to our approach, but differs in that the authors postulate a specific way of merging the individual frameworks.

The approach is based on a notion of *distance* between partial argumentation frameworks (PAFs, each one representing one agent's criterion) over a common set of arguments A. Each partial argumentation framework is defined by three binary relations R, I and N over A: R is the attack relation sanctioned by the agent, I includes the pairs of arguments about which the agent cannot establish any attacks, and $N = (A \times A) \setminus R \cup I$. A pseudo-distance d between PAFs over A is a mapping that associates a real number to each pair of PAFs over A and satisfies the properties of symmetry $(d(x, y) = d(y, x))$ and minimality $(d(x, y) = 0$ iff $x = y)$. d is a distance if it satisfies also the triangular inequality $(d(x, y) \leq d(x, y) + d(y, z))$. These mappings give a way of measuring how "close" is a collective framework from a given profile.

In a further step the authors define an aggregation function as a mapping from $(R+)^n$ to $(R+)$ that satisfies non-decreasingness (if $x_i \geq x_i'$), then $\otimes(x_1, \ldots, x_i, \ldots, x_n) \geq \otimes(x_1, \ldots, x_i', \ldots, x_n)$, minimality $(\otimes(x_1, \ldots, x_n) = 0$ if $\forall i \ x_i = 0)$, and identity $(\otimes(x) = x)$. The idea is that merging a profile of AFs is a two-step process: first, to compute an *expansion* of each AF_i over the profile; and second, a *fusion* in which the AFs over A that are selected as result of the merging are the ones that are the "closest" to the profile. We are currently trying to establish a formal relation between this approach and our results about decisive sets of agents, in particular to determine whether Coste-Marquis et al.'s aggregation procedure satisfies the Arrovian properties.

6 Further Work

A relevant question arises from our analysis of the semantics of aggregate frameworks when the attack relations are acyclic. Namely, whether there exist a sensible notion of aggregation of extensions that could correspond to the aggregation of attack criteria. In this paper we have focused on the path that goes from several attack relations to a single aggregate one and from it to its corresponding extension.

An open question is whether it is possible to go through the alternative path from several attack relations to their corresponding extensions and from there on to a single family of extensions. If, furthermore, these two alternative paths commute (in category-theoretic terms), the aggregation of attack relations would be preferable in applications since it is simpler to aggregate orderings than families of sets.

Nevertheless, there are reasons to be pessimistic, due to the similarities between this problem and the aggregation of judgments for which List and Pettit have found negative results [13]. They consider the so-called "discursive dilemma" in which several judgments (*i.e.*, pairs of the form $\langle premises, conclusion \rangle$), are aggregated component-wise, that is, a pair formed by an aggregated premises set and an aggregated conclusion is obtained, but it does not constitute an acceptable judgment. It could happen that similar problems may arise while trying to match aggregated attack criteria and aggregated extensions.

Pigozzi [14] postulates a solution to the discursive dilemma based on the use of operators for merging belief bases in AI [15]. To pose a *merge* operation as an *aggregation* one involves to incorporate a series of trade-offs among the several alternatives that hardly will respect Arrow's conditions, as it is well known in the literature on political systems (see [16]). In relation to this issue, a wider point that we plan to address is to systematize the *non*-fair aggregation procedures that could be applied to the aggregation of attack relations in argument frameworks. The idea would be to lesser the demands on the aggregation function and to see which features arise in the aggregate. It seems sensible to think that depending on the goals of the aggregation process, one or another function should be chosen.

Another question to investigate is the connections between the correspondence of the aggregate attack relation among arguments and the relation of *dominance* among alternatives ([17]). From a SCT view, alternative A *dominates* alternative B iff the number of individuals for which A is preferred to B is larger than the number of individuals for which B is preferred to A. This implies that the dominance relation is asymmetric. Although it is not commonly assumed in the literature that attack relations are asymmetric, it follows from our definition of majority voting over pairs of arguments that the resulting attack relations will have this property (even when cycles of order > 2 may occur). Dominance relations lead to the choice of stable sets. A stable set is such that none of its elements dominate another, and every alternative outside the set is dominated by some of its elements. The correspondence between stable semantics in argumentation frameworks and stable sets was previously studied by Dung ([3]). It is natural, so, to inquire about the relationship between our majoritarian voting aggregation mechanism on attack relations and stable sets in argumentation frameworks.

References

1. Chesñevar, C., Maguitman, A., Loui, R.: Logical Models of Argument. ACM Computing Surveys 32, 337–383 (2000)
2. Prakken, H., Vreeswijk, G.: Logical systems for defeasible argumentation. In: Gabbay, D. (ed.) Handbook of Philosophical Logic, 2nd edn. Kluwer Academic Pub., Dordrecht (2002)
3. Dung, P.M.: On the acceptability of arguments and its fundamental role in nonmonotonic reasoning, logic programming and n-person games. Artificial Intelligence 77, 321–358 (1995)
4. Coste-Marquis, S., Devred, C., Konieczny, S., Lagasquie-Schiex, M.C., Marquis, P.: Merging argumentation systems. In: Veloso, M.M., Kambhampati, S. (eds.) Proc. of the 20th National Conference on Artificial Intelligence and the 17th Innovative Applications of Artificial Intelligence Conference, pp. 614–619. AAAI Press, USA (2005)
5. Moulin, H.: Social choice. In: Aumann, R., Hart, S. (eds.) Handbook of Game Theory, vol. 2. North-Holland, Amsterdam (1994)
6. Black, D.: The Theory of Committees and Elections. Cambridge University Press, Cambridge (1958)

7. Brown, D.J.: Aggregation of preferences. Quarterly Journal of Economics 89, 456–469 (1975)
8. Blair, D., Pollack, R.: Acyclic collective choice rules. Econometrica 50, 931–943 (1982)
9. Arrow, K.J.: Social Choice and Individual Values, 2nd edn. Yale University Press, London (1970)
10. Sen, A.: Maximization and the act of choice. Econometrica 65, 745–779 (1997)
11. Brown, D.J.: An approximate solution to arrow's problem. Journal of Economic Theory 9, 375–383 (1974)
12. Doyle, J., Wellman, M.: Impediments to universal preference-based default theories. Artificial Intelligence 49, 97–128 (1991)
13. List, C., Pettit, P.: Aggregating sets of judgments. Two impossibility results compared. Synthese 140, 207–235
14. Pigozzi, G.: Belief merging and the discursive dilemma: An argument-based account to paradoxes of judgment aggregation. Synthese 152, 285–298 (2006)
15. Konieczny, S., Pino-Pérez: Propositional belief base merging or how to merge beliefs/goals coming from several sources and some links with social choice theory. European Journal of Operational Research 160, 785–802 (2005)
16. Austen-Smith, D., Banks, J.S.: Positive Political Theory I: Collective Preference. In: Michigan Studies in Political Analysis. University of Michigan Press, Ann Arbor, Michigan (2000)
17. Brandt, F., Fischer, F., Harrenstein, P.: The Computational Complexity of Choice Sets. In: Samet, D. (ed.) TARK. Proceedings of the 11th Conference on Theoretical Aspects of Rationality and Knowledge, pp. 82–91. Presses Universitaires de Louvain (2007)

Alternative Characterizations for Program Equivalence under Answer-Set Semantics Based on Unfounded Sets[*]

Martin Gebser[1], Torsten Schaub[1], Hans Tompits[2], and Stefan Woltran[3]

[1] Institut für Informatik, Universität Potsdam,
August-Bebel-Straße 89, D-14482 Potsdam, Germany
{gebser,torsten}@cs.uni-potsdam.de
[2] Institut für Informationssysteme 184/3, Technische Universität Wien,
Favoritenstraße 9-11, A-1040 Vienna, Austria
tompits@kr.tuwien.ac.at
[3] Institut für Informationssysteme 184/2, Technische Universität Wien,
Favoritenstraße 9-11, A-1040 Vienna, Austria
woltran@dbai.tuwien.ac.at

Abstract. Logic programs under answer-set semantics constitute an important tool for declarative problem solving. In recent years, two research issues received growing attention. On the one hand, concepts like loops and elementary sets have been proposed in order to extend Clark's completion for computing answer sets of logic programs by means of propositional logic. On the other hand, different concepts of program equivalence, like strong and uniform equivalence, have been studied in the context of program optimization and modular programming. In this paper, we bring these two lines of research together and provide alternative characterizations for different conceptions of equivalence in terms of unfounded sets, along with the related concepts of loops and elementary sets. Our results yield new insights into the model theory of equivalence checking. We further exploit these characterizations to develop novel encodings of program equivalence in terms of standard and quantified propositional logic, respectively.

1 Introduction

Among the plethora of semantics that emerged during the nineties of the twentieth century for giving meaning to logic programs with nonmonotonic negation, two still play a major role today: firstly, the *answer-set semantics*, due to Gelfond and Lifschitz [1], and secondly, the *well-founded semantics*, due to Van Gelder, Ross, and Schlipf [2]. While the answer-set semantics adheres to a multiple intended models approach, representing the canonical instance of the *answer-set programming* (ASP) paradigm [3], the well-founded semantics is geared toward efficient query answering and can be seen as a skeptical approximation of the answer-set semantics. In this paper, our interest lies with the answer-set semantics of nonmonotonic logic programs. The results developed here can informally be described as linking two research issues in the context of the answer-set semantics by way of the central constituents of the well-founded semantics, viz. *unfounded sets* [2,4]. Let us explain this in more detail.

---

[*] This work was partially supported by the Austrian Science Fund (FWF) under grant P18019.

S. Hartmann and G. Kern-Isberner (Eds.): FoIKS 2008, LNCS 4932, pp. 24–41, 2008.
© Springer-Verlag Berlin Heidelberg 2008

An important concept in logic programming is *Clark's completion* [5], which associates logic programs with theories of classical logic. While every answer set of a logic program P is also a model of the completion of P, the converse does not hold in general. As shown by Fages [6], a one-to-one correspondence is obtained if P satisfies certain syntactic restrictions. In recent years, a large body of work was devoted to extensions of Fages' characterization in which the syntactic proviso is dropped at the expense of introducing additional formulas to Clark's completion, referred to as *loop formulas* [7]. Although exponentially many such loop formulas must be added in the worst case [8], implementations for the answer-set semantics based on this technique, like ASSAT [7] and Cmodels [9], exploiting solvers for classical logic as back-end inference engines, behave surprisingly well compared to dedicated answer-set tools like DLV [10] and Smodels [11]. While DLV exploits unfounded sets as an approximation technique [4], recent work also reveals relations between unfounded sets and the semantical concepts underlying loop formulas [12,13].

Another issue extensively studied in the context of answer-set semantics are different notions of program equivalence. The main reason for dealing with varying forms of equivalence for logic programs is that ordinary equivalence, in the sense that two programs are equivalent if they have the same answer sets, does not satisfy a substitution principle similar to that of classical logic. That is to say, replacing a subprogram Q of an overall program P by an equivalent program R does not, in general, yield a program that is equivalent to P. This is of course undesirable for modular programming or program optimization when submodules should be replaced by other, more efficient ones. This led to the introduction of more robust notions of equivalence, notably of *strong equivalence* [14] and *uniform equivalence* [15], defined as follows: two programs, P and Q, are strongly equivalent iff, for every program R, $P \cup R$ and $Q \cup R$ have the same answer sets; and P and Q are uniformly equivalent iff the former condition holds for every set R of facts.

The interesting fact about strong equivalence is that it can be reduced to equivalence in the nonclassical logic of *here-and-there* (also known as *Gödel's three-valued logic*) [14], which is basically intuitionistic logic restricted to two worlds, "here" and "there". This characterization was subsequently adapted by Turner [16] by introducing *SE-models*: an SE-model of a program P is a pair (X, Y), where X, Y are interpretations such that $X \subseteq Y$, $Y \models P$, and $X \models P^Y$, where P^Y is the usual Gelfond-Lifschitz reduct [1] of P relative to Y. Two programs are then strongly equivalent iff they possess the same SE-models. Uniform equivalence, in turn, can be captured by certain *maximal* SE-models, termed *UE-models* [15].

We provide a new perspective on SE- and UE-models by relating them to unfounded sets. As it turns out, an explicit reference to the reduct is not required; for a model Y of a program, unfounded sets U with respect to Y allow us to characterize and distinguish SE- and UE-models of the form $(Y \setminus U, Y)$.[1] While UE-models are certain maximal SE-models, our new characterization of UE-models involves *minimal* unfounded sets.

---

[1] A similar relationship has been established by Eiter, Leone, and Pearce [17] with respect to the logic $N2$; a logic that later has been used as a first characterization of strong equivalence. Thus, certain connections between unfounded sets and strong equivalence already appear in their work, but only in an implicit manner.

Using our characterization of SE- and UE-models, we also derive novel characterizations of program equivalence in terms of unfounded sets. These can in turn be linked to loop formulas, and consequently to classical logic. Similar to the observation that expressing answer sets in terms of loop formulas yields an exponential blow-up in the worst case, our reductions into standard propositional logic are likewise exponentially sized. However, we can avoid this exponential increase by switching to *quantified propositional logic* as the target language, which extends standard propositional logic by admitting quantifications over atomic formulas. Our encodings not only provide us with new theoretical insights, but the availability of practicably efficient solvers for quantified propositional logic also gives an easy means to build implementations for equivalence checking in a straightforward way. Indeed, other axiomatizations of strong equivalence in terms of propositional logic and of ordinary and uniform equivalence in terms of quantified propositional logic already appeared in the literature [18,19,20] with that purpose in mind, but these differ significantly from ours as they were based upon different approaches. Finally, all of our encodings are adequate in the sense that the evaluation problems obtained are of the same complexity as the encoded equivalence problems.

The outline of this paper is as follows. In Section 2, we introduce the formal background, and in Section 3, we develop characterizations of models substantial for program equivalence, viz. answer sets, SE-models, and UE-models, based on unfounded sets. We further exploit these characterizations in Section 4 for providing novel specifications of program equivalence as well as encodings in standard and quantified propositional logic. Finally, we discuss our results in Section 5.

2   Background

A propositional *disjunctive logic program* is a finite set of rules of the form

$$a_1 \vee \cdots \vee a_k \leftarrow a_{k+1}, \ldots, a_m, not\, a_{m+1}, \ldots, not\, a_n, \tag{1}$$

where $1 \leq k \leq m \leq n$, every a_i ($1 \leq i \leq n$) is a propositional atom from some universe $\mathcal{U}$, and *not* denotes default negation. A rule r of form (1) is called a *fact* if $k = n = 1$, and *positive* if $m = n$. Furthermore, $H(r) = \{a_1, \ldots, a_k\}$ is the *head* of r, $B(r) = \{a_{k+1}, \ldots, a_m, not\, a_{m+1}, \ldots, not\, a_n\}$ is the *body* of r, $B^+(r) = \{a_{k+1}, \ldots, a_m\}$ is the *positive body* of r, and $B^-(r) = \{a_{m+1}, \ldots, a_n\}$ is the *negative body* of r. We sometimes denote a rule r by $H(r) \leftarrow B(r)$.

The (*positive*) *dependency graph* of a program P is the pair

$$(\mathcal{U}, \{(a, b) \mid r \in P, a \in H(r), b \in B^+(r)\}).$$

A nonempty set $U \subseteq \mathcal{U}$ is a *loop* of P if the subgraph of the dependency graph of P induced by U is strongly connected. Following Lee [12], we consider every singleton over $\mathcal{U}$ as a loop. A program P is *tight* [6,21] if every loop of P is a singleton.

As usual, an interpretation Y is a set of atoms over $\mathcal{U}$. For a rule r, we write $Y \models r$ iff $H(r) \cap Y \neq \emptyset$, $B^+(r) \not\subseteq Y$, or $B^-(r) \cap Y \neq \emptyset$. An interpretation Y is a *model* of a program P, denoted by $Y \models P$, iff $Y \models r$ for every $r \in P$. The *reduct* of P with respect to Y is $P^Y = \{H(r) \leftarrow B^+(r) \mid r \in P, B^-(r) \cap Y = \emptyset\}$; Y is an *answer set* of P iff Y is a minimal model of P^Y.

Two programs, P and Q, are *ordinarily equivalent* iff their answer sets coincide. Furthermore, P and Q are *strongly equivalent* [14] (resp., *uniformly equivalent* [15]) iff, for every program (resp., set of facts) R, $P \cup R$ and $Q \cup R$ have the same answer sets. For interpretations X, Y, the pair (X, Y) is an *SE-interpretation* iff $X \subseteq Y$. An SE-interpretation (X, Y) is an *SE-model* [16] of a program P iff $Y \models P$ and $X \models P^Y$. The pair (X, Y) is a *UE-model* [15] of P iff it is an SE-model of P and there is no SE-model (Z, Y) of P such that $X \subset Z \subset Y$. The set of all SE-models (resp., UE-models) of P is denoted by $SE(P)$ (resp., $UE(P)$). Two programs, P and Q, are strongly (resp., uniformly) equivalent iff $SE(P) = SE(Q)$ (resp., $UE(P) = UE(Q)$) [16,15].

Example 1. Consider $P = \{a \vee b \leftarrow\}$ and $Q = \{a \leftarrow not\, b;\ b \leftarrow not\, a\}$. Clearly, both programs are ordinarily equivalent as $\{a\}$ and $\{b\}$ are their respective answer sets. However, they are not strongly equivalent. Indeed, since P is positive, we have that $SE(P) = \{(a, a), (b, b), (ab, ab), (a, ab), (b, ab)\}$.[2] For Q, we have to take the reduct into account. In particular, we have $Q^{\{a,b\}} = \emptyset$, and so any interpretation is a model of $Q^{\{a,b\}}$. Hence, each pair (X, ab) with $X \subseteq \{a, b\}$ is an SE-model of Q. We thus have $SE(Q) = \{(a, a), (b, b), (ab, ab), (a, ab), (b, ab), (\emptyset, ab)\}$. That is, $SE(P) \neq SE(Q)$, so P and Q are not strongly equivalent. A witness for this is $R = \{a \leftarrow b;\ b \leftarrow a\}$, as $P \cup R$ has $\{a, b\}$ as its (single) answer set, while $Q \cup R$ has no answer set.

Concerning uniform equivalence, observe first that $UE(P) = SE(P)$. This is not the case for Q, where the SE-model $(\emptyset, ab)$ is not a UE-model since there exist further SE-models (Z, ab) of Q with $\emptyset \subset Z \subset \{a, b\}$, viz. (a, ab) and (b, ab). One can check that $(\emptyset, ab)$ is in fact the only pair in $SE(Q)$ that is not a UE-model of Q. So, $UE(Q) = SE(Q) \setminus \{(\emptyset, ab)\} = SE(P) = UE(P)$. Thus, P and Q are uniformly equivalent. $\lozenge$

We conclude this section with the following known properties. First, for any program P and any interpretation Y, the following statements are equivalent: (i) $Y \models P$; (ii) $Y \models P^Y$; (iii) $(Y, Y) \in SE(P)$; and (iv) $(Y, Y) \in UE(P)$. Second, if $Y \models P$, Y is an answer set of P iff, for each SE-model (resp., UE-model) (X, Y) of P, $X = Y$.

3   Model-Theoretic Characterizations by Unfounded Sets

In this section, we exploit the notion of an unfounded set [2,4] and provide alternative characterizations of models for logic programs and program equivalence. Roughly speaking, the aim of unfounded sets is to collect atoms that cannot be derived from a program with respect to a fixed interpretation. Given the closed-world reasoning flavor of answer sets, such atoms are considered to be false. However, we shall relate here unfounded sets also to SE- and UE-models, and thus to concepts that do not fall under the closed-world assumption (since they implicitly deal with program extensions). For

---

[2] Whenever convenient, we use strings like ab as a shorthand for $\{a, b\}$. As a convention, we let universe $\mathcal{U}$ be the set of atoms occurring in the programs under consideration.

the case of uniform equivalence, we shall also employ the recent concept of elementarily unfounded sets [13], which via elementary sets decouple the idea of (minimal) unfounded sets from fixed interpretations. Finally, we shall link our results to loops.

Given a program P and an interpretation Y, a set $U \subseteq \mathcal{U}$ is *unfounded* for P with respect to Y if, for each $r \in P$, at least one of the following conditions holds:

1. $H(r) \cap U = \emptyset$,
2. $H(r) \cap (Y \setminus U) \neq \emptyset$,
3. $B^+(r) \not\subseteq Y$ or $B^-(r) \cap Y \neq \emptyset$, or
4. $B^+(r) \cap U \neq \emptyset$.

Note that the empty set is unfounded for any program P with respect to any interpretation since the first condition, $H(r) \cap \emptyset = \emptyset$, holds for all $r \in P$.

Example 2. Consider the following program:

$$P = \left\{ \begin{array}{lll} r_1: & a \vee b \leftarrow & r_3: \quad c \leftarrow a & r_5: \quad c \leftarrow b, d \\ r_2: & b \vee c \leftarrow & r_4: \quad d \leftarrow not\ b & r_6: \quad d \leftarrow c, not\ a \end{array} \right\}.$$

Let $U = \{c, d\}$. We have $H(r_1) \cap U = \{a, b\} \cap \{c, d\} = \emptyset$, that is, r_1 satisfies Condition 1. For r_5 and r_6, $B^+(r_5) \cap U = \{b, d\} \cap \{c, d\} \neq \emptyset$ and $B^+(r_6) \cap U = \{c\} \cap \{c, d\} \neq \emptyset$. Hence, both rules satisfy Condition 4. Furthermore, consider the interpretation $Y = \{b, c, d\}$. We have $H(r_2) \cap (Y \setminus U) = \{b, c\} \cap \{b\} \neq \emptyset$. Thus, r_2 satisfies Condition 2. Finally, for r_3 and r_4, $B^+(r_3) = \{a\} \not\subseteq \{b, c, d\} = Y$ and $B^-(r_4) \cap Y = \{b\} \cap \{b, c, d\} \neq \emptyset$, that is, both rules satisfy Condition 3. From the fact that each rule in P satisfies at least one of the unfoundedness conditions, we conclude that $U = \{c, d\}$ is unfounded for P with respect to $Y = \{b, c, d\}$.   ◇

The basic relation between unfounded sets and answer sets is as follows.

Proposition 1 ([4,17]). *Let P be a program and Y an interpretation. Then, Y is an answer set of P iff $Y \models P$ and no nonempty subset of Y is unfounded for P with respect to Y.*

Example 3. Program P in Example 2 has two answer sets: $\{a, c, d\}$ and $\{b\}$. For the latter, we just have to check that $\{b\}$ is not unfounded for P with respect to $\{b\}$ itself, which holds in view of either rule r_1 or r_2. To verify via unfounded sets that $Y = \{a, c, d\}$ is an answer set of P, we have to check all nonempty subsets of Y. For instance, take $U = \{c, d\}$. We have already seen that r_1, r_5, and r_6 satisfy Condition 1 or 4, respectively; but the remaining rules r_2, r_3, and r_4 violate all four unfoundedness conditions for U with respect to Y. Hence, $U = \{c, d\}$ is not unfounded for P with respect to $Y = \{a, c, d\}$.   ◇

We next detail the relationship between unfounded sets and models of logic programs as well as of their reducts. First, we have the following relationships between models and unfounded sets.

Lemma 1. *Let P be a program and Y an interpretation. Then, the following statements are equivalent:*

(a) $Y \models P$;

(b) every set $U \subseteq \mathcal{U} \setminus Y$ is unfounded for P with respect to Y; and

(c) every singleton $U \subseteq \mathcal{U} \setminus Y$ is unfounded for P with respect to Y.

Proof. $(a) \Rightarrow (b)$: Assume that some set $U \subseteq \mathcal{U} \setminus Y$ is not unfounded for P with respect to Y. Then, for some rule $r \in P$, we have: (α) $H(r) \cap U \neq \emptyset$; (β) $H(r) \cap (Y \setminus U) = \emptyset$; (γ) $B^+(r) \subseteq Y$ and $B^-(r) \cap Y = \emptyset$; and (δ) $B^+(r) \cap U = \emptyset$. Since $U \cap Y = \emptyset$ by hypothesis, we conclude from (β) that $H(r) \cap Y = \emptyset$. Since (γ) holds in addition, we have $Y \not\models r$ and thus $Y \not\models P$.

$(b) \Rightarrow (c)$: Trivial.

$(c) \Rightarrow (a)$: Assume $Y \not\models P$. Then, there is a rule $r \in P$ such that $Y \not\models r$, that is, $H(r) \cap Y = \emptyset$ and (γ) hold. By the definition of rules, $H(r) \neq \emptyset$. So, consider any $a \in H(r)$ and the singleton $U = \{a\}$. Clearly, (α) holds for r, and (β) holds by $H(r) \cap Y = \emptyset$. Finally, since $B^+(r) \subseteq Y$ and $a \notin Y$, (δ) holds as well. That is, there is a singleton $U \subseteq \mathcal{U} \setminus Y$ that is not unfounded for P with respect to Y. □

We further describe the models of a program's reduct by unfounded sets.

Lemma 2. *Let P be a program, Y an interpretation such that $Y \models P$, and $U \subseteq \mathcal{U}$. Then, $(Y \setminus U) \models P^Y$ iff U is unfounded for P with respect to Y.*

Proof. $(\Rightarrow)$ Assume that U is not unfounded for P with respect to Y. Then, for some rule $r \in P$, (α)–(δ) from the proof of Lemma 1 hold. Clearly, $B^-(r) \cap Y = \emptyset$ implies $(H(r) \leftarrow B^+(r)) \in P^Y$. From $B^+(r) \subseteq Y$ and (δ), we conclude $B^+(r) \subseteq (Y \setminus U)$. Together with (β), we obtain $(Y \setminus U) \not\models (H(r) \leftarrow B^+(r))$ and thus $(Y \setminus U) \not\models P^Y$.

$(\Leftarrow)$ Assume $(Y \setminus U) \not\models P^Y$. Then, there is a rule $r \in P$ such that $(Y \setminus U) \not\models \{r\}^Y$. We conclude that r satisfies (β), $B^+(r) \subseteq (Y \setminus U)$, and $B^-(r) \cap Y = \emptyset$. Since $B^+(r) \subseteq (Y \setminus U)$ immediately implies $B^+(r) \subseteq Y$, (γ) holds. Moreover, $B^+(r) \subseteq (Y \setminus U)$ also implies (δ). It remains to show (α). From (γ) and $Y \models r$ (which holds by the assumption $Y \models P$), we conclude $H(r) \cap Y \neq \emptyset$. Together with (β), this implies (α). Since (α), (β), (γ), and (δ) jointly hold for some rule $r \in P$, we have that U is not unfounded for P with respect to Y. □

Example 4. For illustration, reconsider P from Example 2 and $Y = \{b, c, d\}$. For singleton $\{a\}$ and r_1, we have $H(r_1) \cap (Y \setminus \{a\}) = \{a, b\} \cap \{b, c, d\} \neq \emptyset$. Furthermore, $a \notin H(r)$ for all $r \in \{r_2, \ldots, r_6\}$. That is, $\{a\}$ is unfounded for P with respect to Y. From this, we can conclude by Lemma 1 that Y is a model of P, i.e., $Y \models P$.

As we have already seen in Example 2, $U = \{c, d\}$ is unfounded for P with respect to Y. Lemma 2 now tells us that $(Y \setminus U) = \{b\}$ is a model of $P^Y = \{r_1, r_2, r_3, r_5, (H(r_6) \leftarrow B^+(r_6))\}$. Moreover, one can check that $\{a, c, d\}$ is as well unfounded for P with respect to Y. ◇

The last observation in Example 4 stems from a more general side-effect of Lemma 2: for any program P, any interpretation Y such that $Y \models P$, and $U \subseteq \mathcal{U}$, U is unfounded for P with respect to Y iff $(U \cap Y)$ is unfounded for P with respect to Y. For models Y, this allows us to restrict our attention to unfounded sets $U \subseteq Y$.

We now are in a position to state an alternative characterization of SE-models.

Theorem 1. *Let P be a program, Y an interpretation such that $Y \models P$, and $U \subseteq \mathcal{U}$. Then, $(Y \setminus U, Y)$ is an SE-model of P iff $(U \cap Y)$ is unfounded for P with respect to Y.*

The following result reformulates the definition of UE-models in view of Theorem 1.

Corollary 1. *Let P be a program, Y an interpretation such that $Y \models P$, and $U \subseteq \mathcal{U}$ such that $(U \cap Y)$ is unfounded for P with respect to Y. Then, $(Y \setminus U, Y)$ is a UE-model of P iff, for each V with $(Y \setminus U) \subset (Y \setminus V) \subset Y$, $(V \cap Y)$ is not unfounded for P with respect to Y.*

A simple reformulation of this result provides us with the following novel characterization of UE-models.

Theorem 2. *Let P be a program, Y an interpretation such that $Y \models P$, and $U \subseteq \mathcal{U}$. Then, $(Y \setminus U, Y)$ is a UE-model of P iff $(U \cap Y)$ is unfounded for P with respect to Y and no nonempty proper subset of $(U \cap Y)$ is unfounded for P with respect to Y.*

Note that the inherent maximality criterion of UE-models is now reflected by a *minimality condition* on (nonempty) unfounded sets. Theorems 1 and 2 allow us to characterize strong and uniform equivalence in terms of unfounded sets, avoiding an explicit use of programs' reducts. This will be detailed in Section 4.

Example 5. Recall programs $P = \{a \vee b \leftarrow\}$ and $Q = \{a \leftarrow not\, b;\ b \leftarrow not\, a\}$ from Example 1. We have seen that the only difference in their SE-models is the pair $(\emptyset, ab)$, which is an SE-model of Q, but not of P. Clearly, $Y = \{a, b\}$ is a classical model of P and of Q, and in view of Theorem 1, we expect that Y is unfounded for Q with respect to Y, but not for P with respect to Y. The latter is easily checked since the rule $r = (a \vee b \leftarrow)$ yields (1) $H(r) \cap Y \neq \emptyset$; (2) $H(r) \cap (Y \setminus Y) = \emptyset$; (3) $B^+(r) \subseteq Y$ and $B^-(r) \cap Y = \emptyset$; and (4) $B^+(r) \cap Y = \emptyset$. Thus, none of the four unfoundedness conditions is met. However, for $r_1 = (a \leftarrow not\, b)$ and $r_2 = (b \leftarrow not\, a)$, we have $B^-(r_i) \cap Y \neq \emptyset$, for $i \in \{1, 2\}$, and thus Y is unfounded for Q with respect to Y.

Recall that $(\emptyset, ab)$ is not a UE-model of Q. In view of Theorem 2, we thus expect that $Y = \{a, b\}$ is not a minimal nonempty unfounded set. As one can check, both nonempty proper subsets $\{a\}$ and $\{b\}$ are in fact unfounded for Q with respect to Y. $\Diamond$

In the remainder of this section, we provide a further characterization of UE-models that makes use of elementary sets [13]. For a UE-model (X, Y), this not only gives us a more intrinsic characterization of the difference $U = (Y \setminus X)$ than stated in Theorem 2, but it also yields a further direct relation to loops. We make use of this fact and provide a new result for the UE-models of tight programs.

We define a nonempty set $U \subseteq \mathcal{U}$ as *elementary* for a program P if, for each V such that $\emptyset \subset V \subset U$, there is some $r \in P$ jointly satisfying

1. $H(r) \cap V \neq \emptyset$,
2. $H(r) \cap (U \setminus V) = \emptyset$,
3. $B^+(r) \cap V = \emptyset$, and
4. $B^+(r) \cap (U \setminus V) \neq \emptyset$.

Due to Conditions 1 and 4, every elementary set is also a loop of P, but the converse does not hold in general. Similar to loops, however, elementary sets can be used to characterize answer sets in terms of propositional logic (cf. Proposition 3).

Example 6. Consider the following program:

$$P = \begin{cases} r_1: & a \vee d \leftarrow c \\ r_2: & a \leftarrow b, d \end{cases} \quad \begin{matrix} r_3: & b \leftarrow a, c \\ r_4: & c \leftarrow a \end{matrix} \quad \begin{matrix} r_5: & b \vee c \leftarrow d, e \\ r_6: & d \vee e \leftarrow \end{matrix} \Bigg\}.$$

As one can check, the subgraph of the dependency graph of P induced by $\{a, c, d\}$ is strongly connected, hence, $\{a, c, d\}$ is a loop of P. However, looking at subset $\{d\}$, we can verify that there is no rule in P satisfying all four elementary set conditions: only for $r \in \{r_1, r_6\}$, we have $H(r) \cap \{d\} \neq \emptyset$, but $H(r_1) \cap (\{a, c, d\} \setminus \{d\}) = \{a, d\} \cap \{a, c\} \neq \emptyset$ and $B^+(r_6) \cap (\{a, c, d\} \setminus \{d\}) = \emptyset \cap \{a, c\} = \emptyset$. This shows that the loop $\{a, c, d\}$ is not an elementary set of P.

Next, consider $U = \{a, b, c\}$, which is again a loop of P. In contrast to $\{a, c, d\}$, U is as well elementary for P. For instance, taking $V = \{a, b\}$ and r_1, we have $H(r_1) \cap V = \{a, d\} \cap \{a, b\} \neq \emptyset$, $H(r_1) \cap (U \setminus V) = \{a, d\} \cap \{c\} = \emptyset$, $B^+(r_1) \cap V = \{c\} \cap \{a, b\} = \emptyset$, and $B^+(r_1) \cap (U \setminus V) = \{c\} \cap \{c\} \neq \emptyset$. ◇

To link elementary sets and unfounded sets together, for a program P, an interpretation Y, and $U \subseteq \mathcal{U}$, we define

$$P_{Y,U} = \{r \in P \mid H(r) \cap (Y \setminus U) = \emptyset, B^+(r) \subseteq Y, B^-(r) \cap Y = \emptyset\}.$$

Provided that $H(r) \cap U \neq \emptyset$, a rule $r \in P_{Y,U}$ supports U with respect to Y, while no rule in $(P \setminus P_{Y,U})$ supports U with respect to Y. Analogously to Gebser, Lee, and Lierler [13], we say that U is *elementarily unfounded* for P with respect to Y iff (i) U is unfounded for P with respect to Y and (ii) U is elementary for $P_{Y,U}$. Any elementarily unfounded set of P with respect to Y is also elementary for P, but an elementary set U that is unfounded for P with respect to Y is not necessarily elementarily unfounded because U might not be elementary for $P_{Y,U}$ [13].

Elementarily unfounded sets coincide with minimal nonempty unfounded sets.

Proposition 2 ([13]). *Let P be a program, Y an interpretation, and $U \subseteq \mathcal{U}$. Then, U is a minimal nonempty unfounded set of P with respect to Y iff U is elementarily unfounded for P with respect to Y.*

Example 7. Recall that $U = \{a, b, c\}$ is elementary for Program P from Example 6. For $Y = \{a, b, c, d\}$, we have $P_{Y,U} = \{r_2, r_3, r_4\}$ because $H(r_1) \cap (Y \setminus U) = H(r_6) \cap (Y \setminus U) = \{d\} \neq \emptyset$ and $B^+(r_5) = \{d, e\} \not\subseteq \{a, b, c, d\} = Y$. Since $B^+(r_i) \cap U \neq \emptyset$, for $i \in \{2, 3, 4\}$, we further conclude that U is unfounded for P with respect to Y. Although U is an elementary set of P and unfounded for P with respect to Y, it is not elementary for $P_{Y,U}$ and thus not an elementarily unfounded set of P with respect to Y. This is verified by taking $V = \{a, b\}$, where we have $B^+(r_i) \cap V \neq \emptyset$, for $i \in \{2, 3, 4\}$. As one can check, V itself is unfounded for P with respect to Y and elementary for

$P_{Y,V} = \{r_2, r_3\}$. Therefore, $V = \{a, b\}$ is elementarily unfounded for P with respect to Y. Finally, observe that neither $\{a\}$ nor $\{b\}$ is unfounded for P with respect to Y. Thus, V is indeed a minimal nonempty unfounded set of P with respect to Y.   ◊

The fact that every nonempty unfounded set contains some elementarily unfounded set, which by definition is an elementary set, allows us to derive some properties of the difference $U = (Y \setminus X)$ for SE-interpretations (X, Y). For instance, we can make use of the fact that every elementary set is also a loop, while it is not that obvious to conclude the same for minimal nonempty unfounded sets, only defined with respect to interpretations.

Formally, we derive the following properties for UE-models (resp., SE-models).

Corollary 2. *Let P be a program and (X, Y) a UE-model (resp., SE-model) of P. If $X \neq Y$, then $(Y \setminus X)$ is (resp., contains) (a) an elementarily unfounded set of P with respect to Y; (b) an elementary set of P; and (c) a loop of P.*

For tight programs, i.e., programs such that every loop is a singleton, we obtain the following property.

Corollary 3. *Let P be a tight program and (X, Y) an SE-model of P. Then, (X, Y) is a UE-model of P iff $X = Y$ or $(Y \setminus X)$ is a singleton that is unfounded for P with respect to Y.*

Proof. Given that P is tight, every loop of P is a singleton. Thus, the fact that any elementarily unfounded set is a loop of P implies that only singletons can be elementarily unfounded. From Theorem 2 and Proposition 2, we conclude that (X, Y) is a UE-model of P iff $X = Y$ or $Y \setminus X$ is a singleton that is unfounded for P with respect to Y.   □

Example 8. Recall the SE-model $(\emptyset, ab)$ of $Q = \{a \leftarrow not\, b;\ b \leftarrow not\, a\}$. The loops of Q are $\{a\}$ and $\{b\}$, thus, Q is tight. This allows us to immediately conclude that $(\emptyset, ab)$ is not a UE-model of Q, without looking for any further SE-model to rebut it. ◊

Corollary 3 shows that, for tight programs, the structure of UE-models is particularly simple, i.e., they are always of the form (Y, Y) or $(Y \setminus \{a\}, Y)$, for some $a \in Y$. As we will see in the next section, this also allows for simplified encodings.

4   Characterizations for Program Equivalence

In this section, we further exploit unfounded sets to characterize different notions of program equivalence. We start by comparing two programs, P and Q, regarding their unfounded sets for deriving conditions under which P and Q are ordinarily, strongly, and uniformly equivalent, respectively. Based on these conditions, we then provide novel encodings in standard and quantified propositional logic.

4.1   Characterizations Based on Unfounded Sets

Two programs are ordinarily equivalent if they possess the same answer sets. As Proposition 1 shows, answer sets are precisely the models of a program that do not contain any nonempty unfounded set. Hence, ordinary equivalence can be described as follows.

Theorem 3. *Let P and Q be programs. Then, P and Q are ordinarily equivalent iff, for every interpretation Y, the following statements are equivalent:*

(a) $Y \models P$ and no nonempty subset of Y is unfounded for P with respect to Y; and
(b) $Y \models Q$ and no nonempty subset of Y is unfounded for Q with respect to Y.

Note that ordinarily equivalent programs are not necessarily classically equivalent, as is for instance witnessed by programs $P = \{a \vee b \leftarrow\}$ and $Q = \{a \vee b \leftarrow; \ a \leftarrow c\}$ possessing the same answer sets: $\{a\}$ and $\{b\}$. However, $\{b, c\}$ is a model of P, but not of Q. In turn, for strong and uniform equivalence, classical equivalence is a necessary (but, in general, not a sufficient) condition. This follows from the fact that every model of a program participates in at least one SE-model (resp., UE-model) and is thus relevant for testing strong (resp., uniform) equivalence. Therefore, the following characterization of strong equivalence considers all classical models.

Theorem 4. *Let P and Q be programs. Then, P and Q are strongly equivalent iff, for every interpretation Y such that $Y \models P$ or $Y \models Q$, P and Q possess the same unfounded sets with respect to Y.*

Proof. ($\Rightarrow$) Assume that P and Q are strongly equivalent. Fix any interpretation Y such that $Y \models P$ (or $Y \models Q$). Then, (Y, Y) is an SE-model of P (or Q), and since P and Q are strongly equivalent, (Y, Y) is also an SE-model of Q (or P). That is, both $Y \models P$ and $Y \models Q$ hold. Fix any set $U \subseteq \mathcal{U}$. By Lemma 2, U is unfounded for P with respect to Y iff $(Y \setminus U, Y)$ is an SE-model of P. Since P and Q are strongly equivalent, the latter holds iff $(Y \setminus U, Y)$ is an SE-model of Q, which in turn holds iff U is unfounded for Q with respect to Y.

($\Leftarrow$) Assume that P and Q are not strongly equivalent. Then, without loss of generality, there is an SE-model (X, Y) of P that is not an SE-model of Q (the other case is symmetric). By the definition of SE-models, we have $Y \models P$, and by Lemma 2, $(Y \setminus X)$ is unfounded for P with respect to Y, but either $Y \not\models Q$ or $(Y \setminus X)$ is not unfounded for Q with respect to Y. If $(Y \setminus X)$ is not unfounded for Q with respect to Y, then P and Q do not possess the same unfounded sets with respect to Y. Otherwise, if $Y \not\models Q$, by Lemma 1, there is a set $U \subseteq \mathcal{U} \setminus Y$ that is not unfounded for Q with respect to Y, but U is unfounded for P with respect to Y. $\qquad\square$

Theorem 4 shows that strong equivalence focuses primarily on the unfounded sets admitted by the compared programs. In the setting of uniform equivalence, the consideration of unfounded sets is further restricted to minimal ones (cf. Theorem 2), and by Proposition 2, these are exactly the elementarily unfounded sets.

Theorem 5. *Let P and Q be programs. Then, P and Q are uniformly equivalent iff, for every interpretation Y such that $Y \models P$ or $Y \models Q$, P and Q possess the same elementarily unfounded sets with respect to Y.*

Proof. ($\Rightarrow$) Assume that P and Q are uniformly equivalent. Fix any interpretation Y such that $Y \models P$ (or $Y \models Q$). Then, (Y, Y) is a UE-model of P (or Q), and since P and Q are uniformly equivalent, (Y, Y) is also a UE-model of Q (or P). That is, both $Y \models P$ and $Y \models Q$ hold. Fix any elementarily unfounded set U of P (or Q) with respect to Y. If $U \subseteq \mathcal{U} \setminus Y$, by Lemma 1 and Proposition 2, U is a singleton that is unfounded for both P and Q with respect to Y, which implies that U is elementarily unfounded for Q (or P) with respect to Y. Otherwise, if $U \cap Y \neq \emptyset$, then Lemma 1 and Proposition 2 imply $U \subseteq Y$. By Theorem 2 and Proposition 2, $(Y \setminus U, Y)$ is a UE-model of P (or Q), and since P and Q are uniformly equivalent, $(Y \setminus U, Y)$ is as well a UE-model of Q (or P). Since $\emptyset \neq U \subseteq Y$, by Theorem 2 and Proposition 2, we conclude that U is elementarily unfounded for Q (or P) with respect to Y.

($\Leftarrow$) Assume that P and Q are not uniformly equivalent. Then, without loss of generality, there is a UE-model (X, Y) of P that is not a UE-model of Q (the other case is symmetric). Since (X, Y) is also an SE-model of P, we have $Y \models P$. If $Y \not\models Q$, by Lemma 1, there is a singleton $U \subseteq \mathcal{U} \setminus Y$ that is not unfounded for Q with respect to Y, but U is unfounded for P with respect to Y. That is, U is elementarily unfounded for P with respect to Y, but not for Q with respect to Y. Otherwise, if $Y \models Q$, (Y, Y) is a UE-model both of P and of Q. Hence, $X \subset Y$, and by Theorem 2 and Proposition 2, $(Y \setminus X)$ is elementarily unfounded for P with respect to Y. Furthermore, the fact that (X, Y) is not a UE-model of Q, by Theorem 2 and Proposition 2, implies that $(Y \setminus X)$ is not elementarily unfounded for Q with respect to Y.   □

In contrast to arbitrary unfounded sets, elementarily unfounded sets exhibit a certain structure as they are loops or, even more accurately, elementary sets (cf. Corollary 2). Theorem 5 tells us that such structures alone are material to uniform equivalence.

4.2   Characterizations in (Standard) Propositional Logic

We now exploit the above results on unfounded sets to encode program equivalence in propositional logic. For ordinary equivalence, we use the well-known concept of loop formulas, while for strong and uniform equivalence, we refer directly to unfounded sets.

In what follows, we write for a set of default literals, like $B(r)$, and a set of atoms, like $H(r)$, $B(r) \rightarrow H(r)$ as a shorthand for

$$\left(\bigwedge_{a \in B^+(r)} a \wedge \bigwedge_{a \in B^-(r)} \neg a \right) \rightarrow \bigvee_{a \in H(r)} a,$$

where, as usual, empty conjunctions (resp., disjunctions) are understood as $\top$ (resp., $\bot$). For instance, for a rule r of form (1), $B(r) \rightarrow H(r)$ stands for

$$a_{k+1} \wedge \cdots \wedge a_m \wedge \neg a_{m+1} \wedge \cdots \wedge \neg a_n \rightarrow a_1 \vee \cdots \vee a_k.$$

Furthermore, within the subsequent encodings, an occurrence of a program P is understood as $\bigwedge_{r \in P}(B(r) \rightarrow H(r))$.

As a basis for the encodings, we use the following concept. Following Lee [12], for a program P and $U \subseteq \mathcal{U}$, the *external support formula* of U for P is

$$ES_P(U) = \bigvee_{r \in P, H(r) \cap U \neq \emptyset, B^+(r) \cap U = \emptyset} \neg \big(B(r) \rightarrow (H(r) \setminus U)\big). \tag{2}$$

Intuitively, the models of $ES_P(U)$ are those interpretations Y such that U is externally supported by P with respect to Y. That is, there is a rule $r \in P$ that *supports* U

with respect to Y, i.e., $H(r) \cap U \neq \emptyset$, $H(r) \cap (Y \setminus U) = \emptyset$, $B^+(r) \subseteq Y$, and $B^-(r) \cap Y = \emptyset$ jointly hold. In addition, to make U *externally* supported, one requires $B^+(r) \cap U = \emptyset$, expressing that the support comes from "outside" U.

The relationship between unfounded sets and external support formulas is as follows.

Lemma 3. *Let P be a program, Y an interpretation, and $U \subseteq \mathcal{U}$. Then, U is unfounded for P with respect to Y iff $Y \not\models ES_P(U)$.*

Proof. ($\Rightarrow$) Assume $Y \models ES_P(U)$. Then, there is a rule $r \in P$ such that (α) $H(r) \cap U \neq \emptyset$; ($\beta$) $B^+(r) \cap U = \emptyset$; ($\gamma$) $B^+(r) \subseteq Y$ and $B^-(r) \cap Y = \emptyset$; and ($\delta$) $(H(r) \setminus U) \cap Y = H(r) \cap (Y \setminus U) = \emptyset$. That is, U is not unfounded for P with respect to Y.

($\Leftarrow$) Assume that U is not unfounded for P with respect to Y. Then, there is a rule $r \in P$ for which (α), (β), (γ), and (δ) hold. From (γ) and (δ), we conclude $Y \models \neg(B(r) \rightarrow (H(r) \setminus U))$, which together with ($\alpha$) and ($\beta$) implies $Y \models ES_P(U)$. $\square$

For a program P and $U \subseteq \mathcal{U}$, the (conjunctive) *loop formula* [12] of U for P is

$$LF_P(U) = \left(\bigwedge_{p \in U} p\right) \rightarrow ES_P(U). \qquad (3)$$

With respect to an interpretation Y, the loop formula of U is violated if Y contains U as an unfounded set, otherwise, the loop formula of U is satisfied.

Proposition 3 ([12,13]). *Let P be a program and Y an interpretation such that $Y \models P$. Then, the following statements are equivalent:*

(a) *Y is an answer set of P;*
(b) *$Y \models LF_P(U)$ for every nonempty subset U of $\mathcal{U}$;*
(c) *$Y \models LF_P(U)$ for every loop U of P;*
(d) *$Y \models LF_P(U)$ for every elementary set U of P.*

For ordinary equivalence, the following encodings (as well as different combinations thereof) can thus be obtained.

Theorem 6. *Let P and Q be programs, and let $\mathcal{L}$ and $\mathcal{E}$ denote the set of all loops and elementary sets, respectively, of P and Q. Then, the following statements are equivalent:*

(a) *P and Q are ordinarily equivalent;*
(b) *$\left(P \wedge \bigwedge_{\emptyset \neq U \subseteq \mathcal{U}} LF_P(U)\right) \leftrightarrow \left(Q \wedge \bigwedge_{\emptyset \neq U \subseteq \mathcal{U}} LF_Q(U)\right)$ is a tautology;*
(c) *$\left(P \wedge \bigwedge_{U \in \mathcal{L}} LF_P(U)\right) \leftrightarrow \left(Q \wedge \bigwedge_{U \in \mathcal{L}} LF_Q(U)\right)$ is a tautology;*
(d) *$\left(P \wedge \bigwedge_{U \in \mathcal{E}} LF_P(U)\right) \leftrightarrow \left(Q \wedge \bigwedge_{U \in \mathcal{E}} LF_Q(U)\right)$ is a tautology.*

Recall that, for tight programs, each loop (and thus each elementary set) is a singleton. In this case, the encodings in (c) and (d) are therefore polynomial in the size of the compared programs. Moreover, one can verify that they amount to checking whether the completions [5] of the compared programs are equivalent in classical logic.

For strong and uniform equivalence between P and Q, the models of P and Q along with the corresponding unfounded sets are compared, as Theorems 4 and 5 show. We thus directly consider external support formulas, rather than loop formulas.

Theorem 4 and Lemma 3 yield the following encoding for strong equivalence.

Theorem 7. *Let P and Q be programs. Then, P and Q are strongly equivalent iff $(P \vee Q) \rightarrow (\bigwedge_{U \subseteq \mathcal{U}} (ES_P(U) \leftrightarrow ES_Q(U)))$ is a tautology.*

Proof. By Theorem 4, P and Q are strongly equivalent iff, for every interpretation Y such that $Y \models P$ or $Y \models Q$, P and Q possess the same unfounded sets with respect to Y. By Lemma 3, a set $U \subseteq \mathcal{U}$ is unfounded for P (resp., Q) with respect to Y iff $Y \not\models ES_P(U)$ (resp., $Y \not\models ES_Q(U)$). From this, the statement follows. $\quad\square$

In order to encode also uniform equivalence, we have to single out elementarily unfounded sets. To this end, we modify the definition of the external support formula, $ES_P(U)$, and further encode the case that U is (not) a minimal nonempty unfounded set. For a program P and $U \subseteq \mathcal{U}$, we define the *minimality* external support formula as

$$ES_P^\star(U) = ES_P(U) \vee \neg\big(\bigwedge_{\emptyset \subset V \subset U} ES_P(V)\big). \tag{4}$$

Similar to external support formulas and unfounded sets, minimality external support formulas correspond to elementarily unfounded sets as follows.

Lemma 4. *Let P be a program, Y an interpretation, and $\emptyset \subset U \subseteq \mathcal{U}$. Then, U is elementarily unfounded for P with respect to Y iff $Y \not\models ES_P^\star(U)$.*

Proof. ($\Rightarrow$) Assume $Y \models ES_P^\star(U)$. We have two cases. First, $Y \models ES_P(U)$: By Lemma 3, U is not unfounded for P with respect to Y, which implies that U is not elementarily unfounded for P with respect to Y. Second, $Y \not\models (\bigwedge_{\emptyset \subset V \subset U} ES_P(V))$: For some V such that $\emptyset \subset V \subset U$, we have $Y \not\models ES_P(V)$. By Lemma 3, V is unfounded for P with respect to Y. We conclude that U is not a minimal nonempty unfounded set of P with respect to Y, and by Proposition 2, U is not elementarily unfounded for P with respect to Y.

($\Leftarrow$) Assume $Y \not\models ES_P^\star(U)$. Then, $Y \not\models ES_P(U)$, and by Lemma 3, U is unfounded for P with respect to Y. Furthermore, $Y \models (\bigwedge_{\emptyset \subset V \subset U} ES_P(V))$, and thus no set V such that $\emptyset \subset V \subset U$ is unfounded for P with respect to Y (again by Lemma 3). That is, U is a minimal nonempty unfounded set of P with respect to Y, and by Proposition 2, U is elementarily unfounded for P with respect to Y. $\quad\square$

Theorem 5 and Lemma 4 allow us to encode uniform equivalence as follows.

Theorem 8. *Let P and Q be programs, and let $\mathcal{L}$ and $\mathcal{E}$ denote the set of all loops and elementary sets, respectively, of P and Q. Then, the following statements are equivalent:*

(a) P and Q are uniformly equivalent;
(b) $(P \vee Q) \rightarrow (\bigwedge_{U \subseteq \mathcal{U}} (ES_P^\star(U) \leftrightarrow ES_Q^\star(U)))$ is a tautology;
(c) $(P \vee Q) \rightarrow (\bigwedge_{U \in \mathcal{L}} (ES_P^\star(U) \leftrightarrow ES_Q^\star(U)))$ is a tautology;
(d) $(P \vee Q) \rightarrow (\bigwedge_{U \in \mathcal{E}} (ES_P^\star(U) \leftrightarrow ES_Q^\star(U)))$ is a tautology.

Proof. By Theorem 5, P and Q are uniformly equivalent iff, for every interpretation Y such that $Y \models P$ or $Y \models Q$, P and Q possess the same elementarily unfounded sets with respect to Y. Clearly, any elementarily unfounded set of P or Q belongs to the set $\mathcal{E}$ of all elementary sets of P and Q, which is a subset of the set $\mathcal{L}$ of all loops of P and Q, and every element of $\mathcal{L}$ is a subset of $\mathcal{U}$. By Lemma 4, a set $\emptyset \subset U \subseteq \mathcal{U}$ is elementarily unfounded for P (resp., Q) with respect to Y iff $Y \not\models ES_P^\star(U)$ (resp., $Y \not\models ES_Q^\star(U)$). Finally, we have $ES_P^\star(\emptyset) = ES_Q^\star(\emptyset) = \bot \vee \neg\top \equiv \bot$, so that $Y \models \left(ES_P^\star(\emptyset) \leftrightarrow ES_Q^\star(\emptyset) \right)$ for any interpretation Y. From this, the statement follows. $\square$

Again, we exploit the fact that, for tight programs, all loops and elementary sets are singletons. It is thus sufficient to consider only the external support formulas of singletons. To the best of our knowledge, this provides a novel technique to decide uniform equivalence between tight programs. In fact, the following result is an immediate consequence of (c), or likewise (d), in Theorem 8.

Corollary 4. *Let P and Q be tight programs. Then, P and Q are uniformly equivalent iff $(P \vee Q) \rightarrow \left(\bigwedge_{a \in \mathcal{U}} \left(ES_P(\{a\}) \leftrightarrow ES_Q(\{a\}) \right) \right)$ is a tautology.*

Indeed, for singletons $\{a\}$, $\neg(\bigwedge_{\emptyset \subset V \subset \{a\}} ES_P(V))$ (resp., $\neg(\bigwedge_{\emptyset \subset V \subset \{a\}} ES_Q(V))$) can be dropped from $ES_P^\star(\{a\})$ (resp., $ES_Q^\star(\{a\})$) because it is equivalent to $\bot$.

Except for ordinary and uniform equivalence between tight programs, all of the above encodings are of exponential size. As with the well-known encodings for answer sets (cf. Proposition 3), we do not suggest to a priori reduce the problem of deciding program equivalence to propositional logic. Rather, our encodings provide an alternative view on the conditions underlying program equivalence; similar characterizations have already been successfully exploited in answer-set solving [7,9].

4.3   Characterizations in Quantified Propositional Logic

We now provide encodings that avoid the exponential blow-ups encountered above. Except for strong equivalence, this requires encodings to be stated in quantified propositional logic rather than in (standard) propositional logic.

We briefly recall the basic elements of quantified propositional logic. Syntactically, quantified propositional logic extends standard propositional logic by permitting quantifications over propositional variables. Formulas of quantified propositional logic are also referred to as *quantified Boolean formulas* (QBFs). An occurrence of an atom p is *free* in a QBF Φ if it is not in the scope of a quantifier $\mathbf{Q}p$, for $\mathbf{Q} \in \{\forall, \exists\}$. Given a finite set $P = \{p_1, \ldots, p_n\}$ of atoms, $\mathbf{Q}P\,\Psi$ stands for any QBF of the form $\mathbf{Q}p_1 \ldots \mathbf{Q}p_n\Psi$. For an atom p and a formula Ψ, $\Phi[p/\Psi]$ denotes the QBF resulting from Φ by replacing each free occurrence of p in Φ by Ψ. For (indexed) sets $P = \{p_1, \ldots, p_n\}$ and $S = \{\Psi_1, \ldots, \Psi_n\}$, $\Phi[P/S]$ is a shorthand for $(\cdots(\Phi[p_1/\Psi_1])\cdots)[p_n/\Psi_n]$.

For an interpretation I and a QBF Φ, the relation $I \models \Phi$ is defined analogously to standard propositional logic, with the additional cases that, if $\Phi = \forall p\Psi$, then $I \models \Phi$ iff $I \models \Psi[p/\top]$ and $I \models \Psi[p/\bot]$, and if $\Phi = \exists p\Psi$, then $I \models \Phi$ iff $I \models \Psi[p/\top]$ or $I \models \Psi[p/\bot]$. We say that a QBF Φ is *valid* if $I \models \Phi$ for every interpretation I.

Given a universe $\mathcal{U}$, we use further copies $\mathcal{U}' = \{p' \mid p \in \mathcal{U}\}, \mathcal{U}'' = \{p'' \mid p \in \mathcal{U}\}$, etc. of mutually disjoint sets of new atoms. Moreover, we adopt the following abbreviations to compare interpretations over $\mathcal{U}$ (via copies):

$\mathcal{U}' \leq \mathcal{U}$ stands for $\bigwedge_{p \in \mathcal{U}}(p' \to p)$; and

$\mathcal{U}' < \mathcal{U}$ stands for $\bigwedge_{p \in \mathcal{U}}(p' \to p) \wedge \neg(\bigwedge_{p \in \mathcal{U}}(p \to p'))$.

These building blocks allow us to check containedness between $Y, U \subseteq \mathcal{U}$ as follows. Let I be an interpretation over $\mathcal{U} \cup \mathcal{U}'$, $Y = I \cap \mathcal{U}$, and $U = \{p \mid p' \in (I \cap \mathcal{U}')\}$. Then, $I \models (\mathcal{U}' \leq \mathcal{U})$ iff $U \subseteq Y$, and $I \models (\mathcal{U}' < \mathcal{U})$ iff $U \subset Y$.

Making use of quantifiers, we next provide polynomial encodings for program equivalence. First, we introduce a module representing $ES_P(U)$, as given in (2), but without explicitly referring to certain sets U. Rather, a particular U is determined by the true atoms from a copy $\mathcal{U}'$ of $\mathcal{U}$. Given a rule r, we denote by r' the copy of r obtained by replacing each atom p with p', i.e., $r' = r[\mathcal{U}/\mathcal{U}']$.

Given the above notation, for a program P, we define

$$ES_P = \bigvee_{r \in P}\left(H(r') \wedge \bigwedge_{p \in H(r)}(p \to p') \wedge B(r) \wedge \bigwedge_{p \in B^+(r)}\neg p'\right).$$

For an interpretation Y over $\mathcal{U}$ and $U \subseteq \mathcal{U}$, we have $(Y \cup \{p' \mid p \in U\}) \models ES_P$ iff U is not unfounded for P with respect to Y. Regarding $ES_P(U)$, the following holds.

Lemma 5. *Let P be a program over $\mathcal{U}$, I an interpretation over $\mathcal{U} \cup \mathcal{U}'$, $Y = I \cap \mathcal{U}$, and $U = \{p \mid p' \in (I \cap \mathcal{U}')\}$. Then, $I \models ES_P$ iff $Y \models ES_P(U)$.*

Similar to $ES_P(U)$ and ES_P, we reformulate loop formulas $LF_P(U)$, as given in (3), without explicit reference to U:

$$LF_P = \forall \mathcal{U}'\left(\left(\left(\bigvee_{p \in \mathcal{U}}p'\right) \wedge (\mathcal{U}' \leq \mathcal{U})\right) \to ES_P\right).$$

We obtain the following counterpart of Proposition 3 in quantified propositional logic and afterwards reformulate the test for ordinary equivalence (as given by Theorem 6).

Lemma 6. *Let P be a program over $\mathcal{U}$ and Y an interpretation over $\mathcal{U}$ such that $Y \models P$. Then, Y is an answer set of P iff $Y \models LF_P$.*

Theorem 9. *Let P and Q be programs over $\mathcal{U}$. Then, P and Q are ordinarily equivalent iff $\forall \mathcal{U}\left((P \wedge LF_P) \leftrightarrow (Q \wedge LF_Q)\right)$ is valid.*

Modifying the encoding in Theorem 7 by using ES_P and ES_Q yields that deciding strong equivalence between P and Q amounts to checking whether

$$\forall \mathcal{U}\left((P \vee Q) \to \forall \mathcal{U}'(ES_P \leftrightarrow ES_Q)\right)$$

is valid. However, $\forall \mathcal{U}'$ can safely be placed in front of the formula, yielding

$$\forall \mathcal{U} \forall \mathcal{U}'\left((P \vee Q) \to (ES_P \leftrightarrow ES_Q)\right). \tag{5}$$

Verifying the validity of (5) can be done by checking whether the quantifier-free part in (5) is a tautology in standard propositional logic. We can thus state the following.

Theorem 10. *Let P and Q be programs over $\mathcal{U}$. Then, P and Q are strongly equivalent iff $(P \vee Q) \to (ES_P \leftrightarrow ES_Q)$ is a tautology.*

Finally, we consider uniform equivalence, where we first reformulate $ES_P^\star(U)$, as given in (4), in the same manner as we did above for $ES_P(U)$. To this end, define

$$ES_P^\star = ES_P \vee \exists \mathcal{U}'' \big((\textstyle\bigvee_{p \in \mathcal{U}} p'') \wedge (\mathcal{U}'' < \mathcal{U}') \wedge \neg ES_P[\mathcal{U}'/\mathcal{U}''] \big).$$

Similar to Lemma 5 above, we have the following.

Lemma 7. *Let P be a program over $\mathcal{U}$, I an interpretation over $\mathcal{U} \cup \mathcal{U}'$, $Y = I \cap \mathcal{U}$, and $U = \{p \mid p' \in (I \cap \mathcal{U}')\}$. Then, $I \models ES_P^\star$ iff $Y \models ES_P^\star(U)$.*

By replacing $ES_P^\star(U)$ with $ES_P^\star$, we obtain the following counterpart of Theorem 8.

Theorem 11. *Let P and Q be programs over $\mathcal{U}$. Then, P and Q are uniformly equivalent iff $\forall \mathcal{U} \forall \mathcal{U}' \big((P \vee Q) \to (ES_P^\star \leftrightarrow ES_Q^\star) \big)$ is valid.*

This encoding is very similar in shape to the one for strong equivalence in (5). In contrast to ES_P and ES_Q, $ES_P^\star$ and $ES_Q^\star$, however, contain additional quantifiers, which are needed to keep the size of the encoding polynomial with respect to P and Q. In fact, all of the above QBFs are constructible in polynomial time from P and Q. To the best of our knowledge, they are novel and differ from the encodings proposed in previous work [18,19,20].

5 Discussion

We have provided novel characterizations for program equivalence in terms of unfounded sets and related concepts, like loops and elementary sets. This allowed us to identify close relationships between these central concepts. While answer sets, and thus ordinary equivalence, rely on the absence of (nonempty) unfounded sets, we have shown that potential extensions of programs, captured by SE- and UE-models, can also be characterized directly by appeal to unfounded sets, thereby avoiding any reference to reducts of programs.

We have seen that uniform equivalence is located between ordinary and strong equivalence, in the sense that it considers all models, similar to strong equivalence, but only minimal unfounded sets, which also are sufficient to decide whether a model is an answer set. This allowed us to develop particularly simple characterizations for uniform equivalence in the case of tight programs. In fact, our results offer novel encodings for program equivalence in terms of (quantified) propositional logic, which are different from the ones found in the literature [18,19,20]. The respective encodings reflect the complexity of checking program equivalence: while checking strong equivalence is in coNP, in general, the checks for ordinary and uniform equivalence are Π_2^P-hard. For tight programs, however, our simpler characterization of uniform equivalence is reflected by the corresponding encoding (in standard propositional logic) and makes the complexity of checks drop into coNP, as with ordinary equivalence.

The relationship between (exponential size) propositional encodings and (polynomial size) QBF encodings has also (implicitly) been investigated by Ferraris, Lee, and

Lifschitz [22], where they transform a QBF encoding [20] for answer sets into a propositional formula by eliminating quantifiers, in order to develop a general theory of loop formulas. In fact, Ferraris, Lee, and Lifschitz [22] address answer sets (in a more general setting than ours), but neither strong nor uniform equivalence.

References

1. Gelfond, M., Lifschitz, V.: Classical negation in logic programs and disjunctive databases. New Generation Computing 9(3–4), 365–385 (1991)
2. Van Gelder, A., Ross, K., Schlipf, J.: The well-founded semantics for general logic programs. Journal of the ACM 38(3), 620–650 (1991)
3. Baral, C.: Knowledge Representation, Reasoning and Declarative Problem Solving. Cambridge University Press, Cambridge (2003)
4. Leone, N., Rullo, P., Scarcello, F.: Disjunctive stable models: Unfounded sets, fixpoint semantics, and computation. Information and Computation 135(2), 69–112 (1997)
5. Clark, K.: Negation as failure. In: Gallaire, H., Minker, J. (eds.) Logic and Data Bases, pp. 293–322. Plenum Press, New York (1978)
6. Fages, F.: Consistency of Clark's completion and the existence of stable models. Journal of Methods of Logic in Computer Science 1, 51–60 (1994)
7. Lin, F., Zhao, Y.: ASSAT: Computing answer sets of a logic program by SAT solvers. Artificial Intelligence 157(1–2), 115–137 (2004)
8. Lifschitz, V., Razborov, A.: Why are there so many loop formulas? ACM Transactions on Computational Logic 7(2), 261–268 (2006)
9. Giunchiglia, E., Lierler, Y., Maratea, M.: Answer set programming based on propositional satisfiability. Journal of Automated Reasoning 36(4), 345–377 (2006)
10. Leone, N., Pfeifer, G., Faber, W., Eiter, T., Gottlob, G., Perri, S., Scarcello, F.: The DLV system for knowledge representation and reasoning. ACM Transactions on Computational Logic 7(3), 499–562 (2006)
11. Simons, P., Niemelä, I., Soininen, T.: Extending and implementing the stable model semantics. Artificial Intelligence 138(1–2), 181–234 (2002)
12. Lee, J.: A model-theoretic counterpart of loop formulas. In: Kaelbling, L., Saffiotti, A. (eds.) IJCAI 2005. Proceedings of the 19th International Joint Conference on Artificial Intelligence, pp. 503–508. Professional Book Center (2005)
13. Gebser, M., Lee, J., Lierler, Y.: Elementary sets for logic programs. In: Gil, Y., Mooney, R. (eds.) AAAI 2006. Proceedings of the 21st National Conference on Artificial Intelligence. AAAI Press, California (2006)
14. Lifschitz, V., Pearce, D., Valverde, A.: Strongly equivalent logic programs. ACM Transactions on Computational Logic 2(4), 526–541 (2001)
15. Eiter, T., Fink, M.: Uniform equivalence of logic programs under the stable model semantics. In: Palamidessi, C. (ed.) ICLP 2003. LNCS, vol. 2916, pp. 224–238. Springer, Heidelberg (2003)
16. Turner, H.: Strong equivalence made easy: Nested expressions and weight constraints. Theory and Practice of Logic Programming 3(4–5), 602–622 (2003)
17. Eiter, T., Leone, N., Pearce, D.: Assumption sets for extended logic programs. In: Gerbrandy, J., Marx, M., de Rijke, M., Venema, Y. (eds.) JFAK. Essays Dedicated to Johan van Benthem on the Occasion of his 50th Birthday. Amsterdam University Press (1999)
18. Lin, F.: Reducing strong equivalence of logic programs to entailment in classical propositional logic. In: Fensel, D., Giunchiglia, F., McGuinness, D., Williams, M. (eds.) KR 2002. Proceedings of the 8th International Conference on Principles of Knowledge Representation and Reasoning, pp. 170–176. Morgan Kaufmann, San Francisco (2002)

19. Pearce, D., Tompits, H., Woltran, S.: Encodings for equilibrium logic and logic programs with nested expressions. In: Brazdil, P.B., Jorge, A.M. (eds.) EPIA 2001. LNCS (LNAI), vol. 2258, pp. 306–320. Springer, Heidelberg (2001)
20. Pearce, D., Tompits, H., Woltran, S.: Characterising equilibrium logic and nested logic programs: Reductions and complexity. Technical Report GIA-TR-2007-12-01, Universidad Rey Juan Carlos (2007)
21. Erdem, E., Lifschitz, V.: Tight logic programs. Theory and Practice of Logic Programming 3(4–5), 499–518 (2003)
22. Ferraris, P., Lee, J., Lifschitz, V.: A generalization of the Lin-Zhao theorem. Annals of Mathematics and Artificial Intelligence 47(1–2), 79–101 (2006)

An Alternative Foundation for DeLP: Defeating Relations and Truth Values

Ignacio D. Viglizzo[1,2,5], Fernando A. Tohmé[1,3,5], and Guillermo R. Simari[1,4]

[1] Artificial Intelligence Research and Development Laboratory
[2] Department of Mathematics
viglizzo@criba.edu.ar
[3] Department of Economics
ftohme@criba.edu.ar
[4] Department of Computer Science and Engineering
Universidad Nacional del Sur – Av. Alem 1253 - B8000CPB Bahía Blanca - Argentina
grs@cs.uns.edu.ar
[5] Consejo Nacional de Investigaciones Científicas y Técnicas (CONICET)

Abstract. In this paper we recast the formalism of argumentation formalism known as DeLP (Defeasible Logic Programming) in game-theoretic terms. By considering a game between a *Proponent* and an *Opponent*, in which they present arguments for and against each literal we obtain a bigger gamut of truth values for those literals and their negations as they are defended and attacked. An important role in the determination of warranted literals is assigned to a *defeating* relation among arguments. We consider first an unrestricted version in which these games may be infinite and then we analyze the underlying assumptions commonly used to make them finite. Under these restrictions the games are always *determined* -one of the players has a winning strategy. We show how varying the defeating relation may alter the set of truth values reachable under this formalism. We also show how alternative characterizations of the defeating relation may lead to different assignations of truth values to the literals in a DeLP program.

1 Introduction and Motivation

The development of *defeasible reasoning* in the last decades [Pol87, SL92, Nut94, Pol95, CML00, PV02], provided the foundations of an alternative form of declarative programming, Defeasible Logic Programming (DeLP) [GS04]. This formalism blends Logic Programming with Defeasible Argumentation, allowing the representation of tentative knowledge and leaving for the inference mechanism the task of finding the conclusions that the knowledge base warrants [CDSS03].

DeLP inherits from Logic Programming (LP) the formal characterization of *programs* as sets of rules. The difference is that in DeLP two kinds of rules are considered. On one hand, *strict rules*, which are assumed to represent sound knowledge and are handled as the rules in LP. On the other hand, *defeasible rules* represent tentative knowledge that may be defeated by other information.

S. Hartmann and G. Kern-Isberner (Eds.): FoIKS 2008, LNCS 4932, pp. 42–57, 2008.
© Springer-Verlag Berlin Heidelberg 2008

Again as in LP, DeLP operates by resolving the status of pieces of knowledge that we may deem as queries. A query l succeeds if there exists a warranted argument for l. Arguments are constructed using both types of rules and facts (which can be seen as special cases of strict rules).

The core of DeLP resides in the characterization of the *warrant procedure*. Defeasible Argumentation has provided a solid foundation over which the standard formalization of this procedure has been constructed. A key element in the warrant procedure is the set of criteria according to which two contradicting arguments are compared and eventually one of them deemed as *defeating* the other. While pure syntactic criteria like *specificity* constitute the main choice in the design of the warrant procedure [Poo85, SL92, SGCS03], the notion of warrant can be abstracted away from the details of the order relation among arguments.

The inference mechanism generates all the arguments that either support or contradict l. Then, a *warrant procedure* is applied determining which arguments end up undefeated. If there exists at least one argument warranting l, it yields a positive answer.

Various papers have been published by the LIDIA group in Argentina, developing the theory, providing prototype implementations and exploring a variety of applications. Some other groups have also been publishing papers that extend the theory and use it for applications. But while this framework has well-established and interesting features, we find that many aspects of the inference mechanism of DeLP as well as its corresponding semantics, have a strong game-theoretic flavor. This is inherited from the dialogical schema on which DeLP bases the warrant of queries [GS04].

As it is well known, one of the salient ways in which rational agents try to establish the appropriateness of claims is by means of discussions. The idea is that more appropriate contentions should be supported by the more cogent *arguments* in a given discussion. When discussions are seen as foundations for logic and reasoning systems, the goal is not to determine *who* wins the discussion, but instead which conclusions are supported by the better arguments. Furthermore, for the purpose of determining the relevant conclusions, it does not matter either whether a discussion has a non-cooperative or cooperative nature. The former case is known as a *debate*, in which some agents win and others lose. In a cooperative discussion, instead, agents collaborate in order to find the best answer to some question. But even if the goals of the discussants are non-conflicting, arguments for and against potential answers have to be weighted up to find out which one is more appropriate. It seems natural then, to resort to the idea of a game in which two fictitious players make their moves, that is, they select arguments to be uttered in a discussion. The winning strategies end up leading to a win by either one of the two players.

Although game-theoretic approaches to logic systems are well-known [Hin73, LL78, HS97, vB02], our goals here are somewhat different. We intend to provide an alternative foundation for DeLP closer to the developments in dialogue systems ([Ham70], [Ham71]), particularly in AI ([Bre01], [Lou98]) with applications in law ([LAP00], [MP02], [PWA03]). Even closer to our goals are developments like the explicit or implicit use of game-theoretic notions for *argumentation* [BH01, AC02, Dun95, Pra05].

The whole idea in that literature is that there exists a relation of defeat among arguments (which are basically derivations of claims). This relation allows to determine the arguments that are either undefeated or with defeated defeaters. The claims supported by these arguments are said warranted. The shape of the class of either these claims or of their supporting arguments has been assessed in many settings. But it has to be said that the use of the results of game-theoretic results (instead of just its terminology) in those analysis has been rather scarce.

What we intend to do here is to introduce an alternative, game-theoretic, presentation of DeLP. In particular, we use the notion of *winning strategy* as a central component of the inference mechanism, that yields answers to queries, and consequently for the determination of their truth values. We find that this alternative does preserve DeLP central features, in particular the class of warranted claims, and this is why we claim that the formalism presented here is an alternative foundation, instead of an entirely new formalism. On the other hand, our use of game-theoretic notions can be easily extended to the whole gamut of argumentation formalisms, yielding an alternative semantics to which strategic concepts can be applied.

As said, this paper takes some basic ideas from game theory and uses them for an alternative definition for DeLP. The approach is more abstract. It is interesting because it clarifies some of the notions in DeLP, and it suggests some alternative definitions for aspects of DeLP.

Perhaps more importantly the paper suggests a more general framework for viewing a variety of other argumentation systems in a common game-theoretic way. If so, this paper could be the first step of a very important line of research for understanding the foundations of argumentation. The goal of this paper is to provide an alternative specification of DeLP. It allows to express the main ideas in the formalism in terms of the sets of literals that are warranted. It allows a game-theoretic representation of the warrant procedure and yields a graded truth valuation of literals. We will show how this depends on the properties of the *defeating* relation among arguments.

2   The Basics of DeLP

In order to discuss the set-theoretical properties of DeLP, we have to present the basics of this formalism.[1]

Each *Defeasible Logic Program* $\mathbb{P}$ is a finite set of facts, strict rules, and defeasible rules $\mathbb{P} = \langle \Pi, \Delta \rangle$, where Π denotes the set of facts and strict rules, while Δ denotes the set of defeasible rules. The set Π is the disjoint union of the sets Π_F of facts and Π_R of strict rules.

Facts and rules are defined in terms of atoms. More precisely, let At be the set of atoms that occur in a given program $\mathbb{P}$. Given a set $X \subseteq$ At of atoms, $\sim X$

---

[1] We follow very closely the presentation in [GS04].

is the set $\{\sim x : x \in X\}$. Then, the set Lit is the set of all literals in At$\cup \sim$At. The complement $\bar{l}$ of a literal $l \in$ Lit is $\sim x$ if l is an atom x and x if l is a negated atom $\sim x$. This indicates the strong syntactic bent of DeLP and makes the formalism purely propositional.

Then, the main components of a program $\mathbb{P}$ are:

Π_F: *Facts*, which are ground literals in Lit.

Π_R: *Strict Rules* of the form $l_0 \leftarrow l_1, \ldots, l_n$, where l_0 is the *head* and $\{l_i\}_{i>0}$ is the *body*. Each l_i in the body or the head is in Lit.

Δ: *Defeasible Rules* of the form $l_0 \prec l_1, \ldots, l_n$, where l_0 is the *head* and $\{l_i\}_{i>0}$ is the *body*. Again, each l_i in the body or the head is a literal.

Example 1. Let's consider a simple program $\mathbb{P}_1$ with $\Pi_F = \{b, c\}$; $\Pi_R = \{d \leftarrow a\}$ and $\Delta = \{a \prec b, \sim a \prec c\}$.

Rules, both strict and defeasible act on facts, allowing to derive literals. More precisely, a *defeasible derivation* of l up from $X \subseteq$ Lit, $\mathcal{R} \subseteq \Pi_R$ and $\mathcal{A} \subseteq \Delta$ is a finite sequence $l_1, \ldots, l_n = l$ of literals in Lit such that each l_i is either in X or there exists a rule in $\mathcal{R}$ with l_i as its head, and every literal b_j in its body is such that $b_j \in \{l_k\}_{k<i}$.

Then:

Definition 1. *Given sets $X \subseteq$ Lit, $\mathcal{R} \subseteq \Pi_R$ and $\mathcal{A} \subseteq \Delta$, $C(X, \mathcal{R}, \mathcal{A})$ is the set of all literals defeasibly derivable from $X \cup \mathcal{R} \cup \mathcal{A}$. The set of strict consequences of $X \subseteq$ Lit is $C_s(X) = C(X, \Pi_R, \emptyset)$. Finally, we are going to use often $C(\mathcal{A}) = C(\Pi_F, \Pi_R, \mathcal{A})$ for sets $\mathcal{A} \subseteq \Delta$.*

Definition 2. *Given a set $X \subseteq$ Lit, let $X^+ = X \cap$ At and $X^- = \{a \in$ At $: \sim a \in X\}$. X is said to be contradictory if $X^+ \cap X^- \neq \emptyset$. A set of defeasible rules $\mathcal{A}$ is contradictory if $C(\mathcal{A})$ is contradictory.*

Example 2. For $\mathbb{P}_1$ of example 1, we have $C(\{b\}, \emptyset, \{a \prec b\}) = \{a, b\}$; $C_s(\{a\}) = \{a, d\}$; $C_s(\Pi_F) = \Pi_F$ and $C(\Delta) = \{a, b, c, d, \sim a\}$, a contradictory set.

We assume that for all programs, the set Π of facts and strict rules is not contradictory. A fundamental relation in DeLP is that of *disagreement* between literals:

Definition 3. *Two literals $h, q \in$ Lit are said to disagree (for a given program $\mathbb{P}$) if $C_s(\Pi_F \cup \{h, q\}) = C(\Pi_F \cup \{h, q\}, \Pi_R, \emptyset)$ is contradictory. We define a binary relation $D \subseteq$ Lit $\times$ Lit to record which pairs of literals disagree. This relation is clearly symmetric.*

This relation matters for the comparison of arguments. In order to get to that, let us define the fundamental concept of argument. We will use $\mathcal{P}$ to denote the powerset construction.

Definition 4. *An argument is a pair* $\langle \mathcal{A}, h \rangle \in \mathcal{P}(\Delta) \times \mathrm{Lit}$ *that satisfies:*

1. $h \in C(\mathcal{A})$
2. $\mathcal{A}$ *is not contradictory.*
3. *If* $h \in C(\mathcal{A}')$, *then* $\mathcal{A}' \not\subset \mathcal{A}$, *that is,* $\mathcal{A}'$ *is not a proper subset of* $\mathcal{A}$.

Definition 5. *Let* $\mathrm{Arg}(\mathbb{P})$ *be the set of all arguments of a program* $\mathbb{P}$. *The argument* $\langle \mathcal{A}_1, h \rangle$ *is a subargument of* $\langle \mathcal{A}_2, h' \rangle$ *iff* $\mathcal{A}_1 \subseteq \mathcal{A}_2$. *We denote this with* $\langle \mathcal{A}_1, h \rangle \preceq \langle \mathcal{A}_2, h' \rangle$.

Example 3. With the program $\mathbb{P}_1$ of Example 1, we let $\mathcal{A}_1 = \{a \prec b\}$ and $\mathcal{A}_2 = \{\sim a \prec c\}$. Then the arguments are:

$$\mathrm{Arg}(\mathbb{P}_1) = \{\langle \emptyset, b \rangle, \langle \emptyset, c \rangle, \langle \mathcal{A}_1, a \rangle, \langle \mathcal{A}_1, d \rangle, \langle \mathcal{A}_2, \sim a \rangle\}.$$

The *disagreeing relation D* is formed by the set $\{(a, \sim a), (d, \sim a)\}$ together with its symmetric pairs.

Note that the subargument relation $\preceq$ is a preorder over $\mathrm{Arg}(\mathbb{P})$. Furthermore, if $\langle \mathcal{A}_1, h \rangle \preceq \langle \mathcal{A}_2, h' \rangle$ and $\langle \mathcal{A}_2, h' \rangle \preceq \langle \mathcal{A}_1, h \rangle$, then $\mathcal{A}_1 = \mathcal{A}_2$. This means that if we identify arguments with the same first component, $\preceq$ is simply a restriction of the inclusion relation over $\mathcal{P}(\Delta)$.

Definition 6. *For each literal h we define the binary relation* R_h *on* $\mathrm{Arg}(\mathbb{P})$ *by* $\langle \mathcal{A}_1, h_1 \rangle R_h \langle \mathcal{A}_2, h_2 \rangle$ *iff there exists* $\langle \mathcal{A}, h \rangle \in \mathrm{Arg}(\mathbb{P})$ *such that* $\mathcal{A} \subseteq \mathcal{A}_1$ *and* $(h, h_2) \in D$. *We say in this case that the argument* $\langle \mathcal{A}_1, h_1 \rangle$ *is attacked by* $\langle \mathcal{A}_2, h_2 \rangle$ *at* h *or that* $\langle \mathcal{A}_2, h_2 \rangle$ *attacks or rebutts* $\langle \mathcal{A}_1, h_1 \rangle$ *at* h. *We also say that* $\langle \mathcal{A}_2, h_2 \rangle$ *is a* counter-argument *of* $\langle \mathcal{A}_1, h_1 \rangle$.

Example 4. Following again the analysis of $\mathbb{P}_1$, we have that, for example, $\langle \mathcal{A}_1, a \rangle R_a \langle \mathcal{A}_2, \sim a \rangle$ and also $\langle \mathcal{A}_1, a \rangle R_d \langle \mathcal{A}_2, \sim a \rangle$. On the other hand, we have that $\langle \mathcal{A}_2, \sim a \rangle R_{\sim a} \langle \mathcal{A}_1, a \rangle$ and $\langle \mathcal{A}_2, \sim a \rangle R_{\sim a} \langle \mathcal{A}_1, d \rangle$.

We have the following easy consequences of the definition:

Proposition 7. *1. If* $\langle \mathcal{A}_1, h_1 \rangle R_h \langle \mathcal{A}_2, h_2 \rangle$, *then for some* $\mathcal{A} \subseteq \mathcal{A}_1, \langle \mathcal{A}, h \rangle$ *is an argument and* $\langle \mathcal{A}_2, h_2 \rangle R_{h_2} \langle \mathcal{A}, h \rangle$.
2. *If* $\langle \mathcal{A}, l \rangle R_l \langle \mathcal{B}, p \rangle$, *then* $\langle \mathcal{B}, p \rangle R_p \langle \mathcal{A}, l \rangle$.
3. *If* $q \in C_s(\Pi_F)$, *then* $\langle \emptyset, q \rangle$ *is an argument and it has no counter-arguments.*
4. *Furthermore, an argument* $\langle \emptyset, q \rangle$ *cannot be a counterargument of any argument.*

Proof. 1. From the definition of R_h, we know that for some $\langle \mathcal{A}, h \rangle \in \mathrm{Arg}(\mathbb{P})$ we have that $\mathcal{A} \subseteq \mathcal{A}_1$ and $(h, h_2) \in D$. Therefore, there exists $\mathcal{A}_2 \subseteq \mathcal{A}_2$ and $(h, h_2) \in D$ so we can claim that $\langle \mathcal{A}_2, h_2 \rangle R_h \langle \mathcal{A}, h \rangle$.
2. Using the definition of R_l, this simply means that l and p disagree, and this is enough to justify that $\langle \mathcal{B}, p \rangle R_p \langle \mathcal{A}, l \rangle$, since the arguments are attacked precisely at the literals they support, so no subarguments need be considered. Recall also the minimality of the sets that form the argument.

3. It is easy to check that since $q \in C(\emptyset)$ while $\emptyset$ is not contradictory and has no proper subset, so $\langle \emptyset, q \rangle$ is an argument. Now assume it has a counterargument $\langle \mathcal{A}_2, h' \rangle$ that attacks $\langle \emptyset, q \rangle$ at h. Then there exists an argument $\langle \mathcal{A}, h \rangle$ with $\mathcal{A} \subseteq \emptyset$, so $h \in C(\emptyset)$, and $(h, h') \in D$. This means that $C_s(\Pi_F \cup \{h, h'\})$ is contradictory, but since $C(\emptyset) \subseteq C(\mathcal{A}_2)$, $\{h, h'\} \subseteq C(\mathcal{A}_2)$, so $\mathcal{A}_2$ is contradictory, and therefore $\langle \mathcal{A}_2, h' \rangle$ cannot be an argument.

4. If $\langle \mathcal{A}, h \rangle$ is an argument and $\langle \mathcal{A}, h \rangle R_p \langle \emptyset, q \rangle$, then there is an argument $\langle \mathcal{A}', p \rangle$ with $(p, q) \in D$ but we have seen in the previous proof that this contradicts the fact that $\mathcal{A}'$ is not contradictory.

Example 5. Consider the program $\mathbb{P}_2$ with $\Pi_F = \{c\}$ and $\Delta = \{a \prec b, b \prec c, \sim b \prec c\}$. Letting $\mathcal{A}_1$ be $\{a \prec b, b \prec c\}$ and $\mathcal{A}_2 = \{\sim b \prec c\}$ we have that $\langle \mathcal{A}_1, a \rangle R_b \langle \mathcal{A}_2, \sim b \rangle$. This follows from the fact that $\langle \mathcal{A}_2, \sim b \rangle$ attacks the subargument $\langle \{b \prec c\}, b \rangle$ of $\langle \mathcal{A}_1, a \rangle$. In symbols, this is $\langle \{b \prec c\}, b \rangle R_b \langle \mathcal{A}_2, \sim b \rangle$. Applying the previous proposition, part 2, we get $\langle \mathcal{A}_2, \sim b \rangle R_{\sim b} \langle \{b \prec c\}, b \rangle$.

Now we have the set $\mathrm{Arg}(\mathbb{P})$ with the relations R_h over it. We want to have a method to decide, given a literal l, whether it's supported by the program $\mathbb{P}$ or not. Clearly we want all literals in $C(\emptyset)$ to be supported or warranted. Which other literals should be supported? If there is a defeasible derivation of a literal l while $\bar{l}$ is not derivable, we want l to be warranted as well.

But what about the cases in which the derivation of l yields a contradictory set of literals, or there are arguments that support the negation of l as well?

We have seen that if the relation R_h holds between two arguments, there is also some attack on the attacking argument (Lemma 7, 1). We need to be able to tell which of these two arguments (if any) 'wins' the discussion.

There are several ways to do this. Our choice is to assume a binary relation $\leq$ contained in $R = \bigcup_{h \in \mathrm{Lit}} R_h$ (it will be of no consequence for us if $\leq$ holds in other cases, so we may as well just concentrate on subsets of R). The notation for this relation is somewhat misleading since we do not assume for the moment that $\leq$ is a partial order or even a preorder. It will just be an arbitrary way of deciding which one of two arguments, if any, is stronger than the other. This information will be used to construct a game as described in the following section.

We call $\leq$ the *defeating* relation. We define proper and blocking defeaters as follows:

Definition 8. $\langle \mathcal{A}_1, h_1 \rangle$ *is a* proper defeater *of* $\langle \mathcal{A}_2, h_2 \rangle$ *iff* $\langle \mathcal{A}_2, h_2 \rangle \leq \langle \mathcal{A}_1, h_1 \rangle$ *and it is not the case that* $\langle \mathcal{A}_1, h_1 \rangle \leq \langle \mathcal{A}_2, h_2 \rangle$.

$\langle \mathcal{A}_1, h_1 \rangle$ *is a* blocking defeater *of* $\langle \mathcal{A}_2, h_2 \rangle$ *iff* $\langle \mathcal{A}_2, h_2 \rangle \leq \langle \mathcal{A}_1, h_1 \rangle$ *and* $\langle \mathcal{A}_1, h_1 \rangle \leq \langle \mathcal{A}_2, h_2 \rangle$. *We denote this by* $\langle \mathcal{A}_1, h_1 \rangle \approx \langle \mathcal{A}_2, h_2 \rangle$.

Example 6. The defeating relation can be chosen arbitrarily as a subset of R, but this alone provides certain constraints. Continuing our ongoing example of $\mathbb{P}_1$ and its arguments, we observe that no defeating relation can compare the arguments $\langle \mathcal{A}_1, a \rangle$ and $\langle \mathcal{A}_1, d \rangle$. So if we have a defeating relation $\leq_1$ such that $\langle \mathcal{A}_1, d \rangle \leq_1 \langle \mathcal{A}_2, \sim a \rangle \leq_1 \langle \mathcal{A}_1, a \rangle$, we cannot expect it to be transitive. No defeating relation for this program could be a total order, either.

We can also have $\langle \mathcal{A}_1, a \rangle \leq_2 \langle \mathcal{A}_2, \sim a \rangle$, $\langle \mathcal{A}_2, \sim a \rangle \leq_2 \langle \mathcal{A}_1, a \rangle$ and $\langle \mathcal{A}_2, \sim a \rangle \leq_2 \langle \mathcal{A}_1, d \rangle$ so under this relation, $\langle \mathcal{A}_2, \sim a \rangle$ has $\langle \mathcal{A}_1, a \rangle$ as a blocking defeater and $\langle \mathcal{A}_1, d \rangle$ as a proper one.

Note that in [GS04], proper and blocking defeaters are defined in terms of an underlying comparison criterion $\prec$. We can always take as defeating relation the one stemming out of this prior relation, and thus subsume the previous work in our framework. Our definition, however, allows for the case in which two arguments attacking each other are not related under the defeating relation. In [GS04], these two would be blocking defeaters of each other, even if they were unrelated under the comparison criterion.

3 Warrant Games

Given a literal l we want to find out what the program $\mathbb{P}$ has to say about it. What if there are more than one argument supporting l? What if one of them is undefeated and the other is defeated? We will answer these questions through a slightly generalized version of the mechanism of warrant of DeLP [GS04], using the language of game theory, which turns out to be quite natural for framing the dialectical process that leads to the warrant of a literal.

First, we will introduce the definition of extensive games with perfect information. Then we will define *warrant games*, a class of games tailored to our needs, and finally we will show how to use them for answering queries to the program.

Definition 9. *(following [OR94]) An* extensive game with perfect information $G = \langle N, H, P, (U_i)_{i \in N} \rangle$ *consists of:*

- *A set N, the set of* players.
- *A set H of sequences (finite or infinite) that satisfies the following three properties:*
  - *The empty sequence $\emptyset$ is in H.*
  - *If $(a_k)_{k=1,\ldots,K} \in H$ (where K may be infinite) and $L < K$ then $(a_k)_{k=1,\ldots,L} \in H$.*
  - *If an infinite sequence $(a_k)_{k=1}^{\infty}$ satisfies $(a_k)_{k=1,\ldots,L} \in H$ for every positive integer L, then $(a_k)_{k=1}^{\infty} \in H$.*

  The members of H are called histories. *Each component a_k of a history is* an action *taken by a player. A history $(a_k)_{k=1,\ldots,K} \in H$ is terminal if it is infinite or there is no a_{K+1} such that $(a_k)_{k=1,\ldots,K+1} \in H$. The set of terminal histories is denoted with Z.*
- *A function $P : H \setminus Z \to N$, that indicates for each history in H which one of the players takes an action after the history.*
- *Functions $U_i : Z \to \mathbb{R}$ for $i \in N$ that give for each terminal history and each player, the* payoff *of that player after that history.*

The set H can be seen as a tree with root $\emptyset$, with its nodes labeled by the function P, and the leaves labeled by the functions U_n. We identify the elements a^k with edges of the tree. Therefore, each particular branch from the root is a history, in which the edges are the consequent actions chosen by the players. We call the game *finite* if H is finite. After any nonterminal history h player $P(h)$ chooses an action from the set $A(h) = \{a : (h, a) \in H\}$.

A *warrant game* for a literal l is an extensive game with perfect information with two players. We will call these two players *Proponent* and *Opponent*. We define the game as follows:

- $P(\emptyset) = Proponent.$
- The actions that the proponent can take at the root of the tree are all the arguments of the form $\langle \mathcal{A}, l \rangle$.
- The actions after a nonterminal history h are the arguments $\langle \mathcal{A}', q \rangle$ such that $\langle \mathcal{A}, p \rangle \leq \langle \mathcal{A}', q \rangle$, where $\langle \mathcal{A}, p \rangle$ is the last component in h.
- The utility for the proponent assumes the value 1(win) at a history $h \in Z$ if the length of h is odd, and -1 otherwise. The utility for the opponent is -1 times the utility of the proponent.

If we have that the relation $\leq$ is simply the relation $R = \bigcup_{h \in \text{Lit}} R_h$, and a literal l has some argument which can be attacked, then we will have an infinite tree.

Example 7. We build the warrant game for the literal a in the program $\mathbb{P}_1$ from example 1, assuming that $\leq = R$.

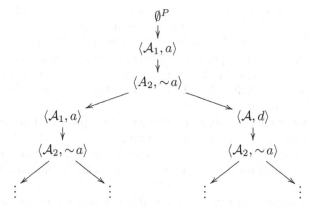

Notice that in the diagram we indicate the nodes by a single argument. The histories can be reconstructed by tracing the path in the tree up to the root. The superscript P or O on the root indicates the player who moves first.

If we consider instead the relation $\leq$ to favor arguments based on $\mathcal{A}_2$ to those based on $\mathcal{A}_1$, the game gets reduced to

$$\emptyset^P$$
$$\downarrow$$
$$\langle \mathcal{A}_1, a \rangle$$
$$\downarrow$$
$$\langle \mathcal{A}_2, \sim a \rangle$$

$$(-1, 1)$$

In this tree we have a terminal history and we have written below the payoffs for the players *Proponent* and *Opponent*, respectively.

A plan for a given player, in which she has a response for every possible contingency of the game is called a *strategy*.

Definition 10. *[OR94] A strategy for a player $i \in N$ in an extensive game with perfect information $\langle N, H, P, U_n \rangle$ is a function that assigns an action in $A(h)$ to each nonterminal history $h \in H \setminus Z$ for which $P(h) = i$.*

Since players are rational, they will seek to get the highest utility. In order to do that each one will seek the best possible strategy. The joint strategy profiles of all players yield a single history. In the case of warrant games, since each terminal history pays either 1 or -1, we can in principle define a *winning strategy* for a player:

Definition 11. *A winning strategy for one of the players in a warrant game is a strategy that yields a terminal history $z \in Z$ such that her utility is 1, no matter what the other player's actions are.*

It turns out that in some warrant games (for example those without terminal histories), none of the players has a winning strategy.

Now we pose a *query* to $\mathbb{P}$. The query is a literal l. Then we analyze two associated games. In the first place, we look at the warrant game for the literal l, and then the warrant game for the complement literal $\bar{l}$ in which the *Proponent* and *Opponent* change their roles. That is, the *Opponent* starts the game by choosing an argument for $\bar{l}$.

This presents a slight departure from the formalism presented in [GS04], where the existence of a winning strategy for the proponent of a literal l is enough to yield the answer "yes" to a query. On the other hand, in that paper, to yield the answer "no" to l, the status of $\bar{l}$ is analyzed, which clearly suggests the road we have taken here.

The following table summarizes the possible outcomes (meaning which of the players has a winning strategy) of both games and how they jointly yield an answer to the query:

Warrant game for l	Warrant game for $\bar{l}$	Answer to the query
Proponent	*Proponent*	YES
Proponent	*Opponent*	Undecided
Proponent	None	yes
Opponent	*Proponent*	Undecided
Opponent	*Opponent*	NO
Opponent	None	no
None	*Proponent*	yes
None	*Opponent*	no
None	None	Undecided

Thus we have a system in which a literal l can take different truth values which can be displayed in a partially ordered set:

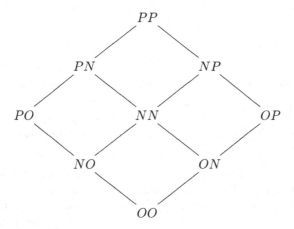

where each pair of letters indicates first who has a winning strategy in the game for l and then who has a winning strategy in the game for $\bar{l}$. P corresponds to the *Proponent*, O to the *Opponent* and N, in turn, indicates that none of them has a winning strategy.

We need an interpretation for each of these outcomes. We have marked in the table with YES and NO the cases in which the arguments are clearly settled for or against the literal l. If for l the *Proponent* has a winning strategy and nor she or the *Opponent* have one for the warrant game on $\bar{l}$, we want to give a positive answer for the literal l, but not as strong one as we would for the case in which the *Proponent* has a winning strategy for both games. This presents a slight departure from the formalism presented in [GS04].

Example 8. We look now at the warrant game initiated by the *Opponent* for the literal $\sim a$ in the conditions we established in Example 7. If $\leq = R$, the game is

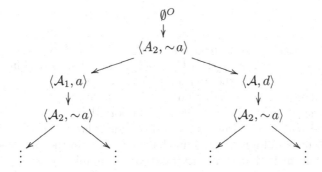

so the outcome for the query a is NN.

On the other hand, if the relation $\leq$ is empty, we just get

$$\emptyset^O$$
$$\downarrow$$
$$\langle A_2, \sim a \rangle$$

$$(-1, 1)$$

So here we get the result OO, and the answer "NO" for the query a.

We have the following trivial result:

Proposition 12. *The literals that are facts of a program always get the outcome PP and therefore the answer YES.*

Proof. Since the class of facts is not contradictory, given a fact l, we know by Lemma 7, 1 that $\langle \emptyset, l \rangle$ is an argument and it has no counter-arguments. Therefore, the game for l ends after the *Proponent* chooses $\langle \emptyset, l \rangle$ and since there is no counterargument, she wins. Alternatively, for $\bar{l}$, there can be no arguments so the *Opponent* has no valid moves and the *Proponent* wins again. Therefore, the corresponding element in the lattice is PP, which yields a YES answer.

4   Finite Warrant Games

If we restrict the definition of warrant games in a way that makes them finite, then in each game one of the players has a winning strategy. The reason for this is that finite games are always *determined* [Myc92]. Our set of possible results (or truth-values) for a given query gets reduced to four possibilities: PP, PO, OP and OO.

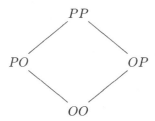

The top and bottom yield, respectively the YES and NO answers to the query. The middle possibilities yield $UNDECIDED$, but with different meanings. If we obtain PO, both the literal l and its negation are warranted and we are facing a (defeasible) contradiction, while if the answer is OP, neither the literal nor its negation can be convincingly supported.

As we noted before, if we just let the defeating relation be $R = \bigcup_{h \in \text{Lit}} R_h$, then all the branches of a warrant game have length 1 (in the case that a literal has an argument for it that is not attacked) or are infinite.

We claim that the following feature of $\leq$ ensures the finiteness of the warrant games:

Definition 13. *The relation $\leq$ is said to be s-acyclic if there is no sequence $\langle \mathcal{A}_0, h_0 \rangle, \langle \mathcal{A}_1, h_1 \rangle, \ldots, \langle \mathcal{A}_{k+1}, h_{k+1} \rangle$ in $\mathrm{Arg}(\mathbb{P})$, such that $\langle \mathcal{A}_j, h_j \rangle \leq \langle \mathcal{A}_{j+1}, h_{j+1} \rangle$, for $j = 0, \ldots, k$, and $\langle \mathcal{A}_{k+1}, h_{k+1} \rangle$ is a subargument of $\langle \mathcal{A}_0, h_0 \rangle$.*

Then:

Proposition 14. *If $\leq$ is s-acyclic, the warrant game for any literal l is finite.*

Proof. Each history h in the warrant game is a sequence $\langle \mathcal{A}_0, h_0 \rangle, \ldots, \langle \mathcal{A}_K, h_K \rangle$, where $\langle \mathcal{A}_j, h_j \rangle \leq \langle \mathcal{A}_{j+1}, h_{j+1} \rangle$, for $j = 0, \ldots, K-1$. Since the set $\mathrm{Arg}(\mathbb{P})$ is finite and $\leq$ is s-acyclic, K must be finite as well.

Instead of imposing conditions on $\leq$, desirable outcomes can be ensured by means of a *protocol* that restricts the admissible actions that may be taken in a warrant game. The main idea here is that one doesn't want to allow repetition of arguments, or subarguments, and each player should maintain internal consistence in the arguments she supports.

Let $\approx$ be the binary relation between arguments defined by $\langle \mathcal{A}_1, h_1 \rangle \approx \langle \mathcal{A}_2, h_2 \rangle$ iff $\langle \mathcal{A}_1, h_1 \rangle \leq \langle \mathcal{A}_2, h_2 \rangle$ and $\langle \mathcal{A}_2, h_2 \rangle \leq \langle \mathcal{A}_1, h_1 \rangle$. DeLP assumes the following protocol:

Definition 15. *In a warrant game for a literal l, a history is $h = \langle \mathcal{A}_0, h_0 \rangle, \ldots, \langle \mathcal{A}_K, h_K \rangle$, where $\langle \mathcal{A}_j, h_j \rangle \leq \langle \mathcal{A}_{j+1}, h_{j+1} \rangle$, for $j = 0, \ldots, K-1$. We say that h is DeLP-admissible if and only if:*

- *Given $\langle \mathcal{A}_j, h_j \rangle, \langle \mathcal{A}_{j+1}, h_{j+1} \rangle$ and $\langle \mathcal{A}_{j+2}, h_{j+2} \rangle$ in h, if $\langle \mathcal{A}_j, h_j \rangle \approx \langle \mathcal{A}_{j+1}, h_{j+1} \rangle$, then it is not the case that $\langle \mathcal{A}_{j+2}, h_{j+2} \rangle \leq \langle \mathcal{A}_{j+1}, h_{j+1} \rangle$. In other words, a blocking defeater can only be followed by a proper defeater.*
- *The sets $\bigcup_{j \geq 0} \mathcal{A}_{2j}$ and $\bigcup_{j \geq 0} \mathcal{A}_{2j+1}$ are not contradictory (concordance).*
- *Every subsequence of any history of the game $h = \langle \mathcal{A}_0, h_0 \rangle, \ldots, \langle \mathcal{A}_K, h_K \rangle$ is an s-acyclic sequence.*

Proposition 16. *A warrant game in which each history is DeLP-admissible is finite.*

Proof. Immediate, since every h is DeLP-admissible, h is a s-acyclic sequence. Given that $\mathrm{Arg}(\mathbb{P})$ is finite, the length of h is finite.

5   The *Defeating* Relation and Truth Values

Now we turn our attention to two extreme cases for the relation $\leq$ and see how it restricts the DeLP-admissibility of histories and therefore the winning conditions of warrant games. That is, how it partitions Lit in classes corresponding each to the four possibilities in finite games.

Proposition 17. *If $\leq = \emptyset$, then a player has a winning strategy for a warrant game for a given literal l if and only if there exists at least one argument for the literal according to the program.*

Proof. Immediate from the fact that the first move for the *Proponent* (in the game for l) or the *Opponent* (in the game for $\bar{l}$) is to state an argument for either l or $\bar{l}$ and there is no argument the other party can use to counterargue, so the game ends. If one of the players cannot find an argument for her literal the other wins.

This means that for each literal l, if there exist arguments for both the literal and its negation, we get the truth value PO, i.e. a defeasible contradiction. Otherwise, if either l or $\bar{l}$ has no supporting arguments in $\mathbb{P}$ while the other has at least one, the truth value of l will be PP or OO.

Proposition 18. *If* $\leq = R = \bigcup_{h \in \text{Lit}} R_h$ *the DeLP-admissibility implies that every terminal history of the warrant games has length at most two.*

Proof. Since $\leq$ coincides with the attacking relation R, all defeaters are *blocking* defeaters, so no history can have length more than two. There can be terminal histories with length one: those consisting of single arguments that have no attackers, as for instance the facts of the program. Finally, we can have trees with a single node with the empty sequence in case the queried literal has no arguments supporting it.

In this case, if the game for a literal l has a terminal history of length one, the *Proponent* can choose it and win both the games for l and $\bar{l}$, i.e. the truth value of l becomes PP. The same is true if the game for $\bar{l}$ has a terminal history of length one: the truth value of l is OO. In the cases in which the lengths of terminal histories are two, each of the players has a winning strategy which yields a win by blocking the first argument in the history. Therefore, OP arises as the truth value of l.

6   Conclusions and Future Work

We have re-casted the basic definitions of DeLP in a simple mathematical language so that we can analyze its underlying assumptions in a new light. We hope that the proposed formal framework can shed some light on how the warrant mechanism works and in particular, on the role of the *defeating* relation. We have borrowed a page from game theory to present the dialectical process in what we believe is a natural way. By initially dropping all restrictions in the dialectical process we uncovered more possibilities for the outcome of a proposed query. We can see these outcomes as truth values, yielding more information on the nature of what a given DeLP program concludes about each literal.

We checked that the facts of the program get a positive answer before turning to the conditions that make our *warrant games* finite. These conditions can come either from the *defeating* relation itself or from an imposed *protocol* on the way the games are constructed. Once we have a way that the games considered are finite, we have fewer truth values. We analyzed the effect of having the extreme cases of an empty *defeating* relation and also the biggest possible one.

Game semantics for DeLP have already been introduced in in [CS04] and [CFS06] The main differences with our approach are that, on one hand, they do not apply standard notions of game theory, while on the other hand they restrict themselves to a single game, that may yield only three truth values. Their notions of strategy and of when a game is won lead to the collapse of several of our truth values into one of theirs. While this is enough for them, since they seek a game semantics independent of the features of the defeating relation, our approach allows to detect fine-grained details of how DeLP works, in particular how it varies according to that relation.

The inference procedure associated to finding winning strategies has a natural "semantical" counterpart. That is, the pair of winners, one for each of the two games can be immediately associated to a truth value as described in the table in section 3. In turn, this means that for each defeating relation we have a partition of the class of literals associated to a DeLP program. As it can be seen from Propositions 17 and 18, the partitions may overlap, even for a pair of defeating relations in which one is a subset of the other.

This framework of analysis can be extended to other argumentative system. In a system as DeLP, where arguments support certain logical formulas[2]. Then, any defeating relation among these arguments may be applied to yield games for a formula and its negation. The properties of the defeating relation determine the actual partition of the class of formulas.

In a more general setting, when arguments are abstract entities there is no "negation" involved. But then, we can still partition the class of arguments in terms of a single game for an argument. If there is a winning strategy for it, it is deemed true.

As a next step we want to study some of the proposed intermediate *defeating* relations like *specificity* and look for desirable properties the defeating relations should have.

References

[AC02]   Amgoud, L., Cayrol, C.: A reasoning model based on the production of acceptable arguments. Annals of Mathematics and Artificial Intelligence 34, 197–215 (2002)

[BH01]   Besnard, P., Hunter, A.: A logic-based theory of deductive arguments. Artificial Intelligence 128, 203–235 (2001)

[Bre01]   Brewka, G.: Dynamic argument systems: a formal model of argumentation processes based on situation calculus. Journal of Logic and Computation 11, 257–282 (2001)

[CDSS03]   Chesñevar, C., Dix, J., Stolzenburg, F., Simari, G.: Relating defeasible and normal logic programming through transformation properties. Theor. Comput. Sci. 290(1), 499–529 (2003)

---

[2] The reason why DeLP is concerned only with literals is the one why LP is: to facilitate the implementation of working interpreters (see [Llo87]).

[CFS06]   Cecchi, L., Fillottrani, P., Simari, G.: On the complexity of delp through game semantics. In: Dix, J., Hunter, A. (eds.) NMR 2006. Proc. 11th Intl. Workshop on Nonmonotonic Reasoning. IfI Technical Report Series, pp. 386–394. Clausthal University, Windermere (2006)

[CML00]   Chesñevar, C., Maguitman, A., Loui, R.: Logical models of argument. ACM Computing Surveys 32(4), 337–383 (2000)

[CS04]   Cecchi, L., Simari, G.: Sobre la relación entre la semántica gs y el razonamiento rebatible. In: X CACiC, pp. 1883–1894. Universidad Nacional de La Matanza (2004)

[Dun95]   Dung, P.: On the acceptability of arguments and its fundamental role in non-monotonic reasoning, logic programming, and n-person games. Artificial Intelligence 77, 321–357 (1995)

[GS04]   García, A., Simari, G.: Defeasible logic programming: An argumentative approach. Theory and Practice of Logic Programming 4(1), 95–138 (2004)

[Ham70]   Hamblin, C.: Fallacies. Methuen, London (1970)

[Ham71]   Hamblin, C.: Mathematical models of dialogue. Theoria 37, 130–155 (1971)

[Hin73]   Hintikka, J.: Language Games and Information. Clarendon Press, London (1973)

[HS97]   Hintikka, J., Sandu, G.: Game-Theoretical Semantics. In: Handbook of Logic and Language, Elsevier, Amsterdam (1997)

[LAP00]   Maudet, N., Amgoud, L., Parsons, S.: Modelling dialogues using argumentation. In: Proceedings of the Fourth International Conference on Multi-Agent Systems. Boston, MA, pp. 31–38 (2000)

[LL78]   Lorenzen, P., Lorenz, K.: Dialogische Logik. Wissenschaftliche Buchgesellschaft, Darmstadt (1978)

[Llo87]   Lloyd, J.: Foundations of Logic Programming, 2nd edn. Springer, Heidelberg (1987)

[Lou98]   Loui, R.: Process and policy: Resource-bounded non-demonstrative reasoning. Computational Intelligence 14, 1–38 (1998)

[MP02]   McBurney, P., Parsons, S.: Games that agents play: A formal framework for dialogues between autonomous agents. Journal of Logic, Language and Information 13, 315–343 (2002)

[Myc92]   Mycielski, J.: Games with perfect information. In: Aumann, R.J., Hart, S. (eds.) Handbook of Game Theory with Economic Applications, ch. 3, vol. 1, pp. 41–70. Elsevier, Amsterdam (1992), http://ideas.repec.org/h/eee/gamchp/1-03.html

[Nut94]   Nute, D.: Defeasible logic. In: Gabbay, D., Hogger, C.J., Robinson, J.A. (eds.) Handbook of Logic in Artificial Intelligence and Logic Programming. Nonmonotonic Reasoning and Uncertain Reasoning, vol. 3, pp. 353–395. Oxford University Press, Oxford (1994)

[OR94]   Osborne, M., Rubinstein, A.: A course in game theory. MIT Press, Cambridge (1994)

[Pol87]   Pollock, J.: Defeasible reasoning. Cognitive Science 11(4), 481–518 (1987)

[Pol95]   Pollock, J.: Cognitive Carpentry: A Blueprint for how to Build a Person. MIT Press, Cambridge (1995)

[Poo85]   Poole, D.: On the comparison of theories: Preferring the most specific explanation. In: IJCAI, pp. 144–147 (1985)

[Pra05]   Prakken, H.: Coherence and flexibility in dialogue games for argumentation. Journal of Logic and Computation 15, 1009–1040 (2005)

[PV02]   Prakken, H., Vreeswijk, G.: Logics for defeasible argumentation. In: Gab-
            bay, D., Guenthner, F. (eds.) Handbook of Philosophical Logic, vol. 4, pp.
            219–318. Kluwer Academic, Dordrecht (2002)
[PWA03]   Parsons, S., Wooldridge, M., Amgoud, L.: Properties and complexity of
            some formal interagent dialogues. Journal of Logic and Computation 13,
            347–376 (2003)
[SGCS03]   Stolzenburg, F., García, A., Chesñevar, C., Simari, G.: Computing gener-
            alized specificity. Journal of Applied Non-Classical Logics 13(1), 87 (2003)
[SL92]   Simari, G., Loui, R.: A mathematical treatment of defeasible reasoning
            and its implementation. Artif. Intell. 53(2–3), 125–157 (1992)
[vB02]   van Benthem, J.: Extensive games as process models. Journal of Logic,
            Language and Information 11, 289–313 (2002)

Appropriate Reasoning about Data Dependencies in Fixed and Undetermined Universes

Joachim Biskup[1] and Sebastian Link[2,*]

[1] Fachbereich Informatik, Universität Dortmund, D-44221 Dortmund, Germany
biskup@ls6.cs.uni-dortmund.de
[2] Information Science Research Centre, Massey University, New Zealand
s.link@massey.ac.nz

Abstract. We study inference systems for the combined class of functional and full hierarchical dependencies in relational databases. Two notions of implication are considered: the original version in which the underlying set of attributes is fixed, and the alternative notion in which this set is left undetermined.

The first main result establishes a finite axiomatisation in fixed universes which clarifies the role of the complementation rule in the combined setting. In fact, we identify inference systems that are appropriate in the following sense: full hierarchical dependencies can be inferred without use of the complementation rule at all or with a single application of the complementation rule at the final step of the inference; and functional dependencies can be inferred without any application of the complementation rule. The second main result establishes a finite axiomatisation for functional and full hierarchical dependencies in undetermined universes.

1 Introduction

Modern database management systems provide commensurate tools to store, manage and process different kinds of data. The core of these systems still relies on the sound technology that is based on the relational model of data [15]. Relations permit the storage of inconsistent data, i.e., data that violate conditions which every legal database instance ought to satisfy. Consequently, additional assertions, called dependencies, are specified by the data administrator in order to restrict the databases to those which are considered meaningful to the application at hand. Most commercial database systems are only capable of enforcing consistency with respect to keys and foreign keys. The reason for this might be that good database design, e.g. Entity-Relationship modeling, will precisely lead to relational database schemata that are in Inclusion Dependency Normal Form [29]. During database normalisation join-related dependencies are explored

---

* This research is supported by the Marsden fund council from Government funding, administered by the Royal Society of New Zealand.

S. Hartmann and G. Kern-Isberner (Eds.): FoIKS 2008, LNCS 4932, pp. 58–77, 2008.

to minimise data redundancy for efficient means of updating. In practice, however, most normalised schemata are subject to denormalisation in order to facilitate the efficient processing of the most common types of queries. Therefore, any remaining join-related dependencies are examined for query optimisation. This describes the trade-off between efficient updates and efficient query processing. Hence, the quality of the target database with respect to these two criteria crucially depends on the ability to reason correctly and appropriately about such dependencies.

Full hierarchical dependencies (FHDs), called full first-order hierarchical decompositions in [16], constitute a large class of relational dependencies. A relation exhibits an FHD precisely when it is the natural join over at least two of its projections that all share the same join attributes. FHDs generalise multivalued dependencies (MVDs) in which the number of such projections is precisely two. The classical notion of an FHD [16] is dependent on the underlying universe R of attributes. For MVDs [18] their dependence on the relation schema R is syntactically reflected by the R-complementation rule which is part of the axiomatisation of MVDs [8]. The R-complementation rule is special in the sense that it is the only inference rule in this axiomatisation which is dependent on R. Further research on this fact has led to an alternative notion of semantic implication in which the underlying universe is left undetermined [12]. In the same paper Biskup shows that this notion can be captured syntactically by a sound and complete set of inference rules, denoted by $\mathfrak{S}$. If $\mathfrak{S}_C$ results from adding the R-complementation rule to $\mathfrak{S}$, then $\mathfrak{S}_C$ is R-sound and R-complete for the R-implication of MVDs for all relation schemata R. In fact, every inference of an MVD by $\mathfrak{S}_C$ can be turned into an inference of the same MVD in which the R-complementation rule is applied at most once, and if it is applied, then in the last step of the inference ($\mathfrak{S}_C$ is said to be R-complementary). This indicates that the R-complementation rule simply reflects a part of the normalisation process, and does not necessarily infer semantically meaningful consequences. This research has been extended recently [23, 24, 26, 30, 31].

Contributions. In this paper we analyse the completeness and appropriateness of inference systems for the *combined* class of functional dependencies (FDs) and full hierarchical dependencies in relational databases.

There are axiomatisations of multivalued and of full hierarchical dependencies for the case where the set of underlying attributes is undetermined [12, 23, 30, 32]. So far, however, no inference system has been proven complete for the combined class of FDs and MVDs, nevermind FDs and FHDs. Among other benefits, the ability to infer all implied FDs and FHDs effectively provides the data administrator with more choices on the final layout of a database schema during the design process. Without this ability the best approximation for efficiently processing the most common types of queries and the most common types of updates may remain unrevealed. Query optimisation in the presence of FDs and FHDs, such as the Chase [17], may benefit from the inference of additional dependencies, too.

Example 1. Suppose the three attributes *Movie*, *Director* and *Actor* represent information about movies titles, the name of their directors and the name of their actors. Moreover, we have specified the multivalued dependency *Movie* ↠ *Director*, stating that the set of *directors* is determined by the *movie* independently of any remaining attributes, and the functional dependency *Actor* → *Director*, enforcing us to store only movie data in which no actor has acted in movies directed by different directors. For every relation schema R that contains (at least) these three attributes every R-relation r that satisfies both the MVD and FD above will also satisfy the FD *Movie* → *Director*. For, if the tuples t_1, t_2 agree on *Movie*, then the MVD above guarantees that there is a tuple t that agrees with t_1 on *Movie* and *Director*, and agrees with t_2 on *Movie* and the rest of the attributes, i.e., at least on *Actor*. However, since t_2 and t agree on *Actor* the FD *Actor* → *Director* guarantees that t_2 and t also agree on *Director*. As t_1 and t also agree on *Director* it follows that t_1 and t_2 agree on *Director*. Hence, a complete set of inference rules for undetermined universes must enable us to infer the FD *Movie* → *Director* that is implied by *Movie* ↠ *Director* and *Actor* → *Director*. □

Moreover, we analyse the appropriateness of existing inference systems for the combined class of FDs and MVDs in fixed universes. In particular, we will clarify the role of the R-complementation rule in the combined setting. It turns out that some systems do not properly reflect the semantics of neither FDs nor MVDs, some properly reflect the semantics of MVDs but not of FDs and some systems properly reflect the semantics of FDs but not of MVDs. In particular, there are systems that require the application of the R-complementation rule in order to infer some implied FDs. Intuitively, an *R-adequate* inference system should be able to avoid such cases since the definition of an FD is independent of the underlying universe. Consequently, if *R-adequate* inference systems do not exist at all, then applications of the R-complementation rule may result in inferences of semantically meaningless functional dependencies.

Example 2. Suppose we fix the relation schema FILM={*Movie,Director,Actor*} together with the MVD and FD from Example 1. We will show that there are axiomatisations in which the FILM-complementation rule must be applied in order to infer the FD *Movie* → *Director*. Therefore, one may argue that the inferred FD is actually meaningless since this line of reasoning is inadequate. □

However, we will identify an inference system that does properly reflect the semantics of both FDs and MVDs in fixed universes, i.e., it is R-complementary and R-adequate for all relation schemata R. Strictly speaking, it is only this fact (i.e. that such an inference system does exist) that confirms our intuition that inferences by previously established axiomatisations always result in meaningful functional and multivalued dependencies. Therefore, there is no need to distrust the semantics of data dependencies that are inferred by complete yet inappropriate inference systems since, by our results, any inappropriate inference can be converted into an appropriate one.

Example 3. Consider the relation schema FILM as well as the set Σ of FDs and MVDs from Example 2 again. The *mixed subset rule* allows us to infer the FD $X \to Y \cap Z$ from the MVD $X \twoheadrightarrow Y$ and the FD $W \to Z$ in case that Y and W are disjoint. For example, the FD *Movie* $\to$ *Director* can be inferred from *Title* $\twoheadrightarrow$ *Director* and *Actor* $\to$ *Director* by means of the mixed subset rule, i.e., without using the FILM-complementation rule.   □

Our results also confirm the intuition that the R-complementation rule is a mere means of database normalisation even in the combined setting of FDs and MVDs. Finally, we extend these results to the combined class of functional and full hierarchical dependencies.

2 Dependencies in Relational Databases

Let $\mathcal{A} = \{A_1, A_2, \ldots\}$ be a (countably) infinite set of symbols, called *attributes*. A *relation schema* is a finite set R of distinct *attributes* from $\mathcal{A}$, which represent column names of a relation. Each attribute A of a relation schema is associated an infinite domain $dom(A)$ which represents the set of possible values that can occur in the column named A. If X and Y are sets of attributes, then we may write XY for $X \cup Y$. If $X = \{A_1, \ldots, A_m\}$, then we may write $A_1 \cdots A_m$ for X. In particular, we may write simply A to represent the singleton $\{A\}$. A *tuple* over the relation schema R (R-tuple or simply tuple, if R is understood) is a function $t : R \to \bigcup_{A \in R} dom(A)$ with $t(A) \in dom(A)$ for all $A \in R$. For $X \subseteq R$ let $t[X]$ denote the restriction of the tuple t over R on X, and $dom(X) = \prod_{A \in X} dom(A)$ the Cartesian product of the domains of attributes in X. A *relation* r over R is a finite set of tuples over R. The relation schema R is also called the domain $Dom(r)$ of the relation r over R. Let $r[X] = \{t[X] \mid t \in r\}$ denote the *projection* of the relation r over R on $X \subseteq R$. For $X, Y \subseteq R$, finite $r_1 \subseteq dom(X)$ and $r_2 \subseteq dom(Y)$ let $r_1 \bowtie r_2 = \{t \in dom(XY) \mid \exists t_1 \in r_1, t_2 \in r_2 \text{ with } t[X] = t_1[X] \text{ and } t[Y] = t_2[Y]\}$ denote the *natural join* of r_1 and r_2. Note that the 0-ary relation $\{()\}$ is the projection $r[\emptyset]$ of any non-empty relation r on $\emptyset$ as well as left and right identity of the natural join operator.

Functional dependencies (FDs) between sets of attributes have played a central role in the study of relational databases [7, 9, 10, 14, 15], and seem to be central for the study of database design in other data models as well [1, 22, 25, 28, 37, 41, 42]. The notion of a functional dependency is well-understood and the semantic interaction between these dependencies has been syntactically captured by Armstrong's well-known axioms [2, 3]. A *functional dependency* (FD) [15] on the relation schema R is an expression $X \to Y$ where $X, Y \subseteq R$. A relation r over R *satisfies* the FD $X \to Y$, denoted by $\models_r X \to Y$, if and only if every pair of tuples in r that agrees on each of the attributes in X also agrees on the attributes in Y. That is, $\models_r X \to Y$ if and only if $t_1[Y] = t_2[Y]$ whenever $t_1[X] = t_2[X]$ holds for any $t_1, t_2 \in r$.

FDs are incapable of modelling many important properties that database users have in mind. Multivalued dependencies (MVDs) provide a more general notion and offer a response to the shortcomings of FDs. A *multivalued dependency* (MVD) [18, 43] on R is an expression $X \twoheadrightarrow Y$ where $X, Y \subseteq R$. A relation r over R *satisfies* the MVD $X \twoheadrightarrow Y$, denoted by $\models_r X \twoheadrightarrow Y$, if and only if for all $t_1, t_2 \in r$ with $t_1[X] = t_2[X]$ there is some $t \in r$ with $t[XY] = t_1[XY]$ and $t[X(R - XY)] = t_2[X(R - XY)]$. Informally, the relation r satisfies $X \twoheadrightarrow Y$ when the value on X determines the set of values on Y independently from the set of values on $R - XY$. This actually suggests that the relation schema R is overloaded in the sense that it carries two independent facts XY and $X(R-XY)$. More precisely, it is shown in [18] that MVDs "provide a necessary and sufficient condition for a relation to be decomposable into two of its projections without loss of information (in the sense that the original relation is guaranteed to be the join of the two projections)". This means that $\models_r X \twoheadrightarrow Y$ if and only if $r = r[XY] \bowtie r[X(R - XY)]$. This characteristic of MVDs is fundamental to relational database design and 4NF [18]. A lot of research has therefore been devoted to studying the behaviour of these dependencies [4, 5, 6, 8, 11, 12, 19, 20, 21, 24, 26, 27, 32, 34, 35, 36, 39, 40]. Full hierarchical dependencies generalise multivalued dependencies [16, 23].

Definition 1. *A full hierarchical dependency (FHD) on a relation schema R is an expression $X : S$ where $X \subseteq R$ and S is a non-empty set of pairwise disjoint subsets of R that are also disjoint from X, i.e., $S \neq \emptyset$, for all $Y \in S$ we have $Y \subseteq R$ and for all $Y, Z \in S \cup \{X\}$ we have $Y \cap Z = \emptyset$. An R-relation $r \subseteq dom(R)$ is said to* satisfy *(or said to be a* model *of) the full hierarchical dependency $X : \{Y_1, \ldots, Y_k\}$ on R, denoted by $\models_r X : \{Y_1, \ldots, Y_k\}$, if and only if for all $t_1, \ldots, t_{k+1} \in r$ the following condition is satisfied: if $t_i[X] = t_j[X]$ for all $1 \leq i, j \leq k + 1$, then there is some $t \in r$ such that $t[XY_i] = t_i[XY_i]$ for $i = 1, \ldots, k$ and $t[X(R - XY_1 \cdots Y_k)] = t_{k+1}[X(R - XY_1 \cdots Y_k)]$.*   □

Notice that Definition 1 reduces to the definition of MVDs in case that $k = 1$. Note that our definition of FHDs is slightly different from what Delobel originally introduced as full first-order hierarchical decompositions [16]. We prefer the form given above for the sake of simplifying the axiomatisation and emphasising the correspondence to the definition of MVDs. The following result is a straightforward generalisation from the MVD case [18].

Theorem 1. *Let $X, Y_1, \ldots, Y_k \subseteq R$ be pairwise disjoint and $k \geq 1$. An R-relation r satisfies the FHD $X : \{Y_1, \ldots, Y_k\}$ on R if and only if $r = r[XY_1] \bowtie \cdots \bowtie r[XY_k] \bowtie r[X(R - XY_1 \cdots Y_k)]$.*   □

Recall that every FHD $X : \{Y_1, \ldots, Y_k\}$ is equivalent to the set $\{X \twoheadrightarrow Y_1, \ldots, X \twoheadrightarrow Y_k\}$ of MVDs [23]. With this in mind functional and full hierarchical dependencies with the same left-hand side permit a *stepwise* decomposition of the underlying relation schema by splitting one current component into two components. Hence, the left-hand side X represents exponentially many decompositions since no order is enforced in which the MVDs $X \twoheadrightarrow Y_i$ are to be applied. This feature distinguishes hierarchical dependencies from general join dependencies which do not have this property [38].

The tree obtained from any such stepwise decomposition provides the data administrator with two kinds of choices on the final layout of the relational database schema. Firstly, the set of relation schemata associated with a full decomposition tree does not permit any data redundancy with respect to the FHDs and, therefore, facilitates efficient updating for the broadest class of updates. However, in order to process queries efficiently several joining operations may be necessarily enforced. Secondly, a truncation of a selected tree at inner nodes of the decomposition tree represents some level of denormalisation that may meet the user's preferences for the level of i) efficient query processing and ii) efficient updating. Complete and appropriate reasoning about the class of functional and full hierarchical dependencies enables the administrator to infer all meaningful dependencies all of which can then be effectively used to determine the final database schema that best approximates the user's preferences for efficiently processing the most common type of queries and updates.

Example 4. Consider the four attributes *Article*, *Manufacturer*, *Location*, and *Costs* representing information about manufacturers that supply articles from their location at a certain cost. Suppose Σ consists of already identified functional dependencies *Article* $\rightarrow$ *Manufacturer* and *Article, Location* $\rightarrow$ *Costs* and the multivalued dependency *Manufacturer* $\twoheadrightarrow$ *Location*. The target database is supposed to process efficiently updates on *Costs* based on the *Article*, e.g., values that result from queries such as

$$\pi_{Costs}(\sigma_{Article=\text{MP3-Player}}(\{Article, Costs\})).$$

Moreover, the target database is most commonly subject to queries about *Article,Location*-information of certain manufacturers such as

$$\pi_{Article,\ Location}(\sigma_{Manufacturer=\text{Sony}}(\{Article, Manufacturer, Location\})).$$

Notice that the FD *Article* $\rightarrow$ *Costs* is implied by Σ. If it is also a reasonable semantic constraint, then a good choice for a target schema would appear to have the two relation schemata $\{Article, Costs\}$ and $\{Article, Manufacturer, Location\}$. The first relation schema enables efficient updates, and the second one offers efficient query processing of our most common types of queries (no joining necessary). The question if the FD *Article* $\rightarrow$ *Costs* does indeed represent an appropriate semantic constraint will be further investigated in Examples 5 and 10. $\square$

For the design of a relational database schema dependencies are normally specified as semantic constraints on the relations which are intended to be instances of the schema. As just explained, the design process requires the data administrator to determine further dependencies which are logically implied by the given ones. In order to emphasise the dependence of implication on the underlying relation schema R we refer to R-implication. Let $lhs(\sigma)$ and $rhs(\sigma)$ denote the attribute sets on the left-hand side and right-hand side, respectively, of a dependency σ, i.e., $lhs(\sigma) = X$ and $rhs(\sigma) = Y_1 \cdots Y_k$ if σ denotes the FHD $X : \{Y_1, \ldots, Y_k\}$, and $lhs(\sigma) = X$ and $rhs(\sigma) = Y$ if σ denotes the FD $X \rightarrow Y$. Let $Attr(\sigma)$ denote the set of attributes affected by σ, i.e., $Attr(\sigma) = lhs(\sigma) \cup rhs(\sigma)$.

Definition 2. *Let $\Sigma \cup \{\varphi\}$ be a set of FDs and FHDs such that $\cup_{\sigma \in \Sigma} Attr(\sigma) \cup Attr(\varphi) \subseteq R$. We say that Σ R-implies φ if and only if each relation r over R that satisfies all $\sigma \in \Sigma$ also satisfies φ.* $\square$

In order to determine the logical consequences of a set of FDs and MVDs with respect to R-implication one can use the inference rules [8, 11, 12] from Table 1. These *inference rules* have the form

$$\frac{\text{premise}}{\text{conclusion}}$$

and inference rules without a premise are called *axioms*. Let $\Sigma \cup \{\sigma\}$ be a set

Table 1. Inference Rules for Functional and Multivalued Dependencies

$$\frac{}{X \to Y} \, Y \subseteq X$$
(reflexivity, $\mathcal{R}_F$)

$$\frac{X \to Y}{X \to XY}$$
(extension, $\mathcal{E}_F$)

$$\frac{X \to Y, Y \to Z}{X \to Z}$$
(transitivity, $\mathcal{T}_F$)

$$\frac{X \twoheadrightarrow Y}{XU \twoheadrightarrow YV} \, V \subseteq U$$
(augmentation, $\mathcal{A}_M$)

$$\frac{X \twoheadrightarrow Y}{X \twoheadrightarrow R - Y}$$
(R-complementation, $\mathcal{C}_M^R$)

$$\frac{X \twoheadrightarrow Y, Y \twoheadrightarrow Z}{X \twoheadrightarrow Z - Y}$$
(pseudo-transitivity, $\mathcal{T}_M$)

$$\frac{X \twoheadrightarrow Y, W \twoheadrightarrow Z}{X \twoheadrightarrow Y \cap Z} \, Y \cap W = \emptyset$$
(subset, $\mathcal{S}_M$)

$$\frac{X \twoheadrightarrow Y, Y \twoheadrightarrow Z}{X \twoheadrightarrow YZ}$$
(additive transitivity, $\mathcal{T}_M^*$)

$$\frac{X \to Y}{X \twoheadrightarrow Y}$$
(implication, $\mathcal{I}_{FM}$)

$$\frac{X \twoheadrightarrow Y, Y \to Z}{X \to Z - Y}$$
(mixed pseudo-transitivity, $\mathcal{T}_{FM}$)

$$\frac{X \twoheadrightarrow Y, W \to Z}{X \to Y \cap Z} \, Y \cap W = \emptyset$$
(mixed subset, $\mathcal{S}_{FM}$)

of FDs and FHDs (FDs and MVDs) on the relation schema R. Furthermore, we use $\mathfrak{S}$ to denote a set of inference rules. In this paper we consider only those sets of inference rules in which the R-complementation rule can be the only inference rule that is dependent on R. In particular, all sets $\mathfrak{S}$ we consider for FDs and MVDs will form a subset of the rule set in Table 1. Let $\Sigma \vdash_{\mathfrak{S}} \sigma$ denote the inference of σ from Σ with respect to $\mathfrak{S}$. Let $\Sigma_{\mathfrak{S}}^+ = \{\sigma \mid \Sigma \vdash_{\mathfrak{S}} \sigma\}$ denote the *syntactic hull* of Σ under inference using only rules from $\mathfrak{S}$. An inference rule is called R-*sound* if the set of dependencies in the premise of the rule R-implies the dependency in the conclusion. The rules of Table 1 are R-sound for all R [8,13]. The set $\mathfrak{S}$ is called R-*sound* for the R-implication of FDs and FHDs if and only if for every set Σ of FDs and FHDs on the relation schema R we have $\Sigma_{\mathfrak{S}}^{+} \subseteq \Sigma_{R}^{*} = \{\sigma \mid \Sigma \ R\text{-implies } \sigma\}$. The set $\mathfrak{S}$ is called R-*complete* for the R-implication of FDs and FHDs if and only if for every set Σ of FDs and FHDs on R we have $\Sigma_{R}^{*} \subseteq \Sigma_{\mathfrak{S}}^{+}$. An inference rule $\mathfrak{R}$ is said to be *independent* of the set $\mathfrak{S}$ if and only if there is some relation schema R and some finite set $\Sigma \cup \{\varphi\}$ of FDs and FHDs on R such that $\varphi \notin \Sigma_{\mathfrak{S}}^{+}$, but $\varphi \in \Sigma_{\mathfrak{S} \cup \{\mathfrak{R}\}}^{+}$. An R-complete set $\mathfrak{S}$ is said to be R-*complementary* if and only if for every set $\Sigma \cup \{\varphi\}$ of FDs and FHDs on R the inference of an FHD φ from Σ using $\mathfrak{S}$ can be turned into an inference of φ from Σ using $\mathfrak{S}$ in which the R-complementation rule $\mathcal{C}_{\mathrm{M}}^{R}$ is applied at most once, and if it is applied, then it is applied in the last step of the inference. In what follows we use $\mathfrak{S}_{C}$ to denote the inference system obtained from the system $\mathfrak{S}$ by adding the R-complementation rule $\mathcal{C}_{\mathrm{M}}^{R}$. The system

$$\mathfrak{F}_{C} = \{\mathcal{R}_{\mathrm{F}}, \mathcal{E}_{\mathrm{F}}, \mathcal{T}_{\mathrm{F}}, \mathcal{A}_{\mathrm{M}}, \mathcal{T}_{\mathrm{M}}, \mathcal{I}_{\mathrm{FM}}, \mathcal{T}_{\mathrm{FM}}, \mathcal{C}_{\mathrm{M}}^{R}\}$$

is known to be both R-sound and R-complete for the R-implication of FDs and MVDs, for all relation schema R [33]. However, $\mathfrak{F}_{C}$ is not R-complementary for all relation schemata R [12]. Biskup conjectured that the set

$$\mathfrak{AC}_{C} = \{\mathcal{R}_{\mathrm{F}}, \mathcal{E}_{\mathrm{F}}, \mathcal{T}_{\mathrm{F}}, \mathcal{A}_{\mathrm{M}}, \mathcal{T}_{\mathrm{M}}, \mathcal{S}_{\mathrm{M}}, \mathcal{T}_{\mathrm{M}}^{*}, \mathcal{I}_{\mathrm{FM}}, \mathcal{S}_{\mathrm{FM}}, \mathcal{C}_{\mathrm{M}}^{R}\}$$

is indeed R-complementary for the R-implication of FDs and MVDs for all relation schemata R [12]. We will formally prove that this conjecture is indeed true. Moreover, we will verify that $\mathfrak{AC}_{C}$ enjoys a further property which makes it appropriate for reasoning about the combined class of FDs and MVDs. A system $\mathfrak{S}_{C}$ of inference rules that is R-complete for the R-implication of FDs and MVDs is said to be R-*adequate* if for every set Σ of FDs and MVDs on R and every FD φ on R such that φ is R-implied by Σ there is an inference of φ from Σ by $\mathfrak{S}$, i.e., an inference in which the R-complementation rule $\mathcal{C}_{\mathrm{M}}^{R}$ is not utilised at all. When denoting inference systems we will use the letter $\mathfrak{A}$ to indicate the adequacy of the system, and the letter $\mathfrak{C}$ to indicate its complementarity.

Example 5. Consider the attributes $\underline{A}rticle$, $\underline{M}anufacturer$, $\underline{L}ocation$, and $\underline{C}osts$ and dependency set $\Sigma = \{A \to M; A, L \to C; M \twoheadrightarrow L\}$ from Example 4 again. A possible inference of the FD $A \to C$ from Σ is the following:

$$
\begin{array}{ll}
& A \to M \\
\hline
\mathcal{I}_{\mathrm{FM}}: & A \twoheadrightarrow M \ \ M \twoheadrightarrow L \\
\hline
\mathcal{T}_{\mathrm{M}}: & A \twoheadrightarrow L \\
\hline
\mathcal{A}_{\mathrm{M}}: & A \twoheadrightarrow A, L \\
\hline
\mathcal{C}_{\mathrm{M}}^{R}: & A \twoheadrightarrow C, M \qquad A, L \to C \\
\hline
\mathcal{S}_{\mathrm{FM}}: & A \to C
\end{array}
$$

One may argue that this line of reasoning is inappropriate since this inference applies the R-complementation rule $\mathcal{C}_{\mathrm{M}}^{R}$ in order to infer the FD $A \to C$.

Consequently, this inference may leave us in doubt whether the FD $A \to C$ is actually meaningful for our application. □

In order to avoid the inference of possibly meaningless MVDs, Biskup [12] introduced the alternative notion of implication in which the underlying set of attributes is left undetermined.

Definition 3. *Let $\Sigma \cup \{\varphi\}$ be a set of FDs and FHDs. We say that Σ implies φ if and only if every relation r satisfies the following condition: if $\cup_{\sigma \in \Sigma} Attr(\sigma) \cup Attr(\varphi) \subseteq Dom(r)$ and r satisfies all $\sigma \in \Sigma$, then r also satisfies φ.* □

The notions of *soundness* and *completeness* are simply adapted to the context of undetermined universes by dropping the reference to the underlying relation schema R from the corresponding notions in the context of fixed universes. While there are axiomatisations for the class of MVDs [12,30] and the class of FHDs [23,38] in undetermined universes, no axiomatisation is known for the combined class of FDs and MVDs nor FDs and FHDs.

Let $\Sigma \cup \{\varphi\}$ be a set of FDs and FHDs, and let R be some relation schema such that $\cup_{\sigma \in \Sigma} Attr(\sigma) \cup Attr(\varphi) \subseteq R$ holds. Based on Definitions 2 and 3 it follows that Σ R-implies φ whenever Σ implies φ. Intuitively, the reverse direction also holds when φ is an FD, but the following Example 6 illustrates that the reverse direction does not hold when φ is an FHD.

Example 6. Let FILM$=\{$Movie,Director,Actor$\}$ and

$$\Sigma = \{\text{Movie} \twoheadrightarrow \text{Director, Actor} \to \text{Director}\}.$$

While Σ FILM-implies Movie $\twoheadrightarrow$ Actor it is relatively simple to give a counterexample for the implication of Movie $\twoheadrightarrow$ Actor by Σ. □

Before concluding this section we present a result on the interaction of FDs and MVDs in undetermined universes. In fixed universes the following fact is well-known. Let R denote a relation schema, and Σ a set of FDs on R. Then Σ R-implies the MVD $X \twoheadrightarrow Y$ if and only if Σ R-implies the FD $X \to Y$ or the FD $X \to R - Y$. This fact takes an even more convincing form in undetermined universes.

Theorem 2. *Let Σ be a set of functional dependencies. Then Σ implies the MVD $X \twoheadrightarrow Y$ if and only if Σ implies the FD $X \to Y$.*

Proof. If Σ implies the FD $X \to Y$, then, by soundness of the implication rule $\mathcal{I}_{\text{FM}}$, Σ also implies the MVD $X \twoheadrightarrow Y$.

Suppose Σ does not imply the FD $X \to Y$. This means that Y is not a subset of the attribute closure $X_{\Sigma}^* = \cup\{B \mid X \to B \in \Sigma^*\}$ of X under Σ, i.e., $Y - X_{\Sigma}^*$ is non-empty. Construct a relation r such that $Dom(r)$ is the disjoint union of $X_{\Sigma}^*, Y - X_{\Sigma}^*$ and a new attribute A, and r consists of exactly two tuples t, t' such that $t[C] = t'[C]$ if and only if $C \in X_{\Sigma}^*$. It is simple to observe that r satisfies Σ and violates $X \twoheadrightarrow Y$ as there is no tuple in r which agrees with t on $X_{\Sigma}^* Y$ and agrees with t' on $X_{\Sigma}^* A$. Consequently, Σ does not imply the MVD $X \twoheadrightarrow Y$. □

3   Inadequate Reasoning in Fixed Universes

Functional dependencies are satisfied by relations independently of the corresponding set of underlying attributes. Since this property should be properly reflected syntactically we may ask that inference systems for FDs and MVDs are R-adequate for all relation schemata R.

We will show in this section that adequacy of an inference system cannot be taken for granted. Let

$$\mathfrak{C} = \{\mathcal{R}_F, \mathcal{E}_F, \mathcal{T}_F, \mathcal{A}_M, \mathcal{T}_M, \mathcal{S}_M, \mathcal{T}_M^*, \mathcal{I}_{FM}, \mathcal{T}_{FM}\}$$

denote the system that is obtained from $\mathfrak{F}$ by adding $\mathcal{S}_M$ and $\mathcal{T}_M^*$.

Lemma 1. *The mixed subset rule $\mathcal{S}_{FM}$ is independent of $\mathfrak{C}$.*

Proof. Let $\Sigma = \{\emptyset \twoheadrightarrow A, B \to A\}$ and $\varphi = \emptyset \to A$. Neglecting all trivial FDs and MVDs with attributes not in AB we represent the closure $\Sigma_{\mathfrak{C}}^+$ of Σ with respect to $\mathfrak{C}$ as two tables. The MVD $X \twoheadrightarrow Y$ (FD $X \to Y$) belongs to $\Sigma_{\mathfrak{C}}^+$ if and only if in the $\twoheadrightarrow$-table ($\to$-table) the entry in row labelled X and column labelled Y is a cross $\times$. The $\to$-table can be obtained as follows. First, we apply $\mathcal{R}_F$ to infer all trivial FDs with attributes in AB. Subsequently, we enter the premise $B \to A$ from Σ. Finally, we apply $\mathcal{E}_F$ to infer $B \to AB$ from $B \to A$. The $\twoheadrightarrow$-table can be obtained as follows. First, we apply $\mathcal{I}_{FM}$ to copy all $\times$ from the $\to$-table into the corresponding entries in the $\twoheadrightarrow$-table. Finally, we enter the premise $\emptyset \twoheadrightarrow A$ from Σ. This set is closed under inference using $\mathfrak{C}$. In particular, φ cannot be inferred from Σ by using $\mathfrak{C}$. In fact, one can observe that both premises in Σ are necessary to infer φ. The only inference rule capable of inferring φ from Σ is $\mathcal{T}_{FM}$, but in order to apply this rule the R-complementation rule $\mathcal{C}_M^R$ must first be applied to $\emptyset \twoheadrightarrow A$. However, $\mathcal{C}_M^R$ is not available in $\mathfrak{C}$.

$\to$	$\emptyset$	A	B	AB
$\emptyset$	$\times$			
A	$\times$	$\times$		
B	$\times$	$\times$	$\times$	$\times$
AB	$\times$	$\times$	$\times$	$\times$

$\twoheadrightarrow$	$\emptyset$	A	B	AB
$\emptyset$	$\times$	$\times$		
A	$\times$	$\times$		
B	$\times$	$\times$	$\times$	$\times$
AB	$\times$	$\times$	$\times$	$\times$

It follows that $\varphi \notin \Sigma_{\mathfrak{C}}^+$ but $\varphi \in \Sigma_{\mathfrak{C} \cup \{\mathcal{S}_{FM}\}}^+$.   □

The next lemma shows that the system $\mathfrak{C}_c$ is inadequate, and thus, also the inadequacy of $\mathfrak{F}_c$.

Lemma 2. *There is a relation schema R, a set Σ of FDs and MVDs on R and an FD φ on R such that $\varphi \in \Sigma_{\mathfrak{C}_c}^+$ but $\varphi \notin \Sigma_{\mathfrak{C}}^+$.*

Proof. Let $R = AB$, and $\Sigma = \{\emptyset \twoheadrightarrow A, B \to A\}$ and $\varphi = \emptyset \to A$. The proof of Lemma 1 has shown that $\varphi \notin \Sigma_{\mathfrak{C}}^+$. It therefore remains to verify that $\varphi \in \Sigma_{\mathfrak{C}_c}^+$. First, we apply $\mathcal{C}_M^R$ to $\emptyset \twoheadrightarrow A$ to infer $\emptyset \twoheadrightarrow B$. Subsequently, we apply $\mathcal{T}_{FM}$ to $\emptyset \twoheadrightarrow B$ and $B \to A$ and infer $\emptyset \to A$.   □

Corollary 1. *The systems $\mathfrak{F}_C$ and $\mathfrak{C}_C$ are not R-adequate for the reasoning about FDs and MVDs for all relation schemata R.*   □

Lemma 2 raises the question whether there is any adequate (or even appropriate) set of inference rules for the R-implication of FDs and MVDs. We will show in Section 4 that $\mathfrak{A}_C = (\mathfrak{F}_C - \{\mathcal{T}_{\text{FM}}\}) \cup \{\mathcal{S}_{\text{FM}}\}$ is adequate and $\mathfrak{A}\mathfrak{C}_C = \mathfrak{A}_C \cup \{\mathcal{S}_{\text{M}}, \mathcal{T}_{\text{M}}^*\}$ is indeed appropriate. Intuitively, the *mixed subset rule* $\mathcal{S}_{\text{FM}}$ allows us to infer those FDs directly that otherwise had to be inferred by using the mixed pseudo-transitivity rule $\mathcal{T}_{\text{FM}}$ and the R-complementation rule $\mathcal{C}_{\text{M}}^R$. This is very much similar to the role of the subset rule $\mathcal{S}_{\text{M}}$ which allows us to infer those MVDs directly that otherwise had to be inferred by using the pseudo-transitivity rule $\mathcal{T}_{\text{M}}$ and the R-complementation rule $\mathcal{C}_{\text{M}}^R$.

Example 7. Let FILM={Movie,Director,Actor} and

$$\Sigma = \{\text{Movie} \twoheadrightarrow \text{Director}, \text{Actor} \rightarrow \text{Director}\}.$$

The FD Movie $\rightarrow$ Actor can be inferred from Σ by first applying the FILM-complementation rule to Movie $\twoheadrightarrow$ Director in order to infer Movie $\twoheadrightarrow$ Actor and then applying the mixed pseudo-transitivity rule to Movie $\twoheadrightarrow$ Actor and Actor $\rightarrow$ Director. However, this line of reasoning is inadequate since the FILM-complementation rule is utilised to infer a functional dependency.   □

4 Appropriate Reasoning in Fixed Universes

Our first main result establishes

$$\mathfrak{A}\mathfrak{C}_C = \{\mathcal{R}_{\text{F}}, \mathcal{E}_{\text{F}}, \mathcal{T}_{\text{F}}, \mathcal{A}_{\text{M}}, \mathcal{T}_{\text{M}}, \mathcal{S}_{\text{M}}, \mathcal{T}_{\text{M}}^*, \mathcal{I}_{\text{FM}}, \mathcal{S}_{\text{FM}}, \mathcal{C}_{\text{M}}^R\}$$

as the first inference system for the R-implication of FDs and MVDs that is indeed appropriate: R-sound, R-complete, R-complementary and R-adequate for all relation schemata R. In particular, this clarifies the role of the R-complementation rule as a mere means of database normalisation in the combined setting of functional and multivalued dependencies.

Lemma 3. *The mixed pseudo-transitivity rule $\mathcal{T}_{FM}$ can be inferred from the following set of inference rules $\{\mathcal{T}_M, \mathcal{I}_{FM}, \mathcal{S}_{FM}\}$.*

Proof

$$\cfrac{X \twoheadrightarrow Y \quad \cfrac{\cfrac{Y \rightarrow Z}{\mathcal{I}_{\text{FM}} : \; Y \twoheadrightarrow Z}}{\mathcal{T}_{\text{M}} : \quad X \twoheadrightarrow Z - Y} \quad Y \rightarrow Z}{\mathcal{S}_{\text{FM}} : \quad X \rightarrow \underbrace{(Z - Y) \cap Z}_{=Z-Y}}$$

This completes the proof.   □

Lemma 3 enables us to obtain another axiomatisation of FDs and MVDs in fixed universes: just replace $\mathcal{T}_{\text{FM}}$ in $\mathfrak{F}_C$ by $\mathcal{S}_{\text{FM}}$.

Corollary 2. *For all relation schemata R, the system*

$$\mathfrak{A}_{\mathcal{C}} = \{\mathcal{R}_F, \mathcal{E}_F, \mathcal{T}_F, \mathcal{A}_M, \mathcal{T}_M, \mathcal{I}_{FM}, \mathcal{S}_{FM}, \mathcal{C}_M^R\}$$

is R-sound and R-complete for the R-implication of FDs and MVDs. $\square$

Theorem 3. *Let R be some relation schema, and let Σ be a set of FDs and MVDs on R. For every inference γ from Σ by the system $\mathfrak{A}_{\mathcal{C}}$ there is an inference ξ from Σ by the system*

$$\mathfrak{A}\mathfrak{C}_{\mathcal{C}} = \mathfrak{A}_{\mathcal{C}} \cup \{\mathcal{S}_M, \mathcal{T}_M^*\}$$

with the following properties:

1. *if γ infers an MVD, then*
   - *γ and ξ infer the same MVD,*
   - *in ξ the R-complementation rule $\mathcal{C}_M^R$ is applied at most once, and*
   - *if $\mathcal{C}_M^R$ is applied in ξ, then $\mathcal{C}_M^R$ is applied as the last rule.*
2. *if γ infers an FD, then*
   - *γ and ξ infer the same FD, and*
   - *in ξ the R-complementation rule $\mathcal{C}_M^R$ is not applied at all.*

Proof (Sketch). The proof is done by induction on the length l of the inference γ. If $l = 1$, then $\xi := \gamma$ has the desired properties. Let $l > 1$, and $\gamma = [\sigma_1, \ldots, \sigma_l]$ be an inference from Σ by $\mathfrak{A}_{\mathcal{C}}$ which has length l. All together, one needs to consider eight cases according to which inference rule in $\mathfrak{A}_{\mathcal{C}}$ was applied to infer σ_l from $[\sigma_1, \ldots, \sigma_{l-1}]$. However, we will only show the most interesting case in which the *mixed subset rule* $\mathcal{S}_{FM}$ eliminates an application of the R-complementation rule $\mathcal{C}_M^R$ during an inference of a functional dependency. Therefore, we assume that σ_l is inferred by applying the *mixed subset rule* $\mathcal{S}_{FM}$ to the premises σ_i and σ_j with $i, j < l$. Let ξ_i (ξ_j) be obtained by using the induction hypothesis for $\gamma_i := [\sigma_1, \ldots, \sigma_i]$ ($\gamma_j := [\sigma_1, \ldots, \sigma_j]$). Consider the inference $\xi := [\xi_i, \xi_j, \sigma_l]$. Then we distinguish between two cases according to the occurrence of the R-complementation rule $\mathcal{C}_M^R$ in ξ_i (assuming that ξ_j infers the FD in the premise). If $\mathcal{C}_M^R$ is not applied in ξ_i, then ξ has the desired properties. It remains to consider the case where $\mathcal{C}_M^R$ is applied in ξ_i (as the last rule), i.e., the last step of ξ_i and the last step of ξ are of the following form:

$$\mathcal{S}_{FM} : \quad \cfrac{\cfrac{X \twoheadrightarrow Y}{\mathcal{C}_M^R : \ X \twoheadrightarrow R - Y} \qquad W \to Z}{X \to \underbrace{(R-Y) \cap Z}_{=Z-Y}} \ (R-Y) \cap W = \emptyset.$$

Since $(R-Y) \cap W = \emptyset$ holds we have $W \subseteq Y$. Hence, these steps can be replaced as follows:

$$\mathcal{T}_{FM} : \quad \cfrac{X \twoheadrightarrow Y \qquad \cfrac{\mathcal{R}_F : Y \to W \overset{W \subseteq Y}{} \qquad W \to Z}{\mathcal{T}_F : \qquad Y \to Z}}{X \to Z - Y}$$

The proof of Lemma 3 shows how the last application of this inference can be replaced by an inference that only uses rules in $\mathfrak{AC}$ (even in $\mathfrak{A}$ already). The result of this replacement is an inference with the desired properties. The cases in which the remaining inference rules are applied can be dealt with similarly.

$\square$

Recall that the system $\mathfrak{C}_{\mathcal{C}} = \mathfrak{F}_{\mathcal{C}} \cup \{\mathcal{S}_{\mathrm{M}}, \mathcal{T}_{\mathrm{M}}^{*}\}$ is R-complementary but not R-adequate for all relation schemata R [12]. On the other hand, the proof of Theorem 3 shows that the system $\mathfrak{A}_{\mathcal{C}}$ is already R-adequate for all relation schemata R, but similar to the case of $\mathfrak{F}_{\mathcal{C}}$ it can be shown that $\mathfrak{A}_{\mathcal{C}}$ is not R-complementary for all relation schemata R [12].

Corollary 3. *The inference system $\mathfrak{C}_{\mathcal{C}}$ satisfies 1. of Theorem 3, but not 2. The inference system $\mathfrak{A}_{\mathcal{C}}$ satisfies 2. of Theorem 3, but not 1.* $\square$

Figure 1 illustrates the connection between the different inference systems and their semantic properties. In summary, one gains complementarity by including the subset rule $\mathcal{S}_{\mathrm{M}}$ and additive transitivity rule $\mathcal{T}_{\mathrm{M}}^{*}$, and adequacy by including the mixed subset rule $\mathcal{S}_{\mathrm{FM}}$.

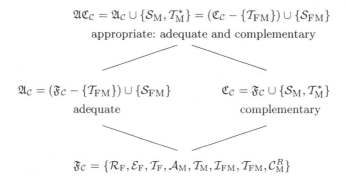

$$\mathfrak{AC}_{\mathcal{C}} = \mathfrak{A}_{\mathcal{C}} \cup \{\mathcal{S}_{\mathrm{M}}, \mathcal{T}_{\mathrm{M}}^{*}\} = (\mathfrak{C}_{\mathcal{C}} - \{\mathcal{T}_{\mathrm{FM}}\}) \cup \{\mathcal{S}_{\mathrm{FM}}\}$$
appropriate: adequate and complementary

$$\mathfrak{A}_{\mathcal{C}} = (\mathfrak{F}_{\mathcal{C}} - \{\mathcal{T}_{\mathrm{FM}}\}) \cup \{\mathcal{S}_{\mathrm{FM}}\} \qquad \mathfrak{C}_{\mathcal{C}} = \mathfrak{F}_{\mathcal{C}} \cup \{\mathcal{S}_{\mathrm{M}}, \mathcal{T}_{\mathrm{M}}^{*}\}$$
adequate   complementary

$$\mathfrak{F}_{\mathcal{C}} = \{\mathcal{R}_{\mathrm{F}}, \mathcal{E}_{\mathrm{F}}, \mathcal{T}_{\mathrm{F}}, \mathcal{A}_{\mathrm{M}}, \mathcal{T}_{\mathrm{M}}, \mathcal{I}_{\mathrm{FM}}, \mathcal{T}_{\mathrm{FM}}, \mathcal{C}_{\mathrm{M}}^{R}\}$$

Fig. 1. Inference Systems and their Properties

Notice that Theorem 3 shows that no R-complete inference system can infer any semantically meaningless FD or MVD since any inappropriate inference of such a dependency can always be converted into an appropriate inference by $\mathfrak{AC}_{\mathcal{C}}$.

5   Nearly Complete Reasoning in Fixed Universes

Among others Theorem 3 shows that $\mathfrak{AC}$ is nearly R-complete for the R-implication of FDs and MVDs on any relation schema R. Indeed, $\mathfrak{AC}$ enables us to infer every R-implied FD. Moreover, for every R-implied MVD $X \twoheadrightarrow Y$ the system $\mathfrak{AC}$ enables us to infer $X \twoheadrightarrow Y$ itself or $X \twoheadrightarrow R - Y$.

Corollary 4. *Let $\Sigma \cup \{\varphi\}$ be a finite set of FDs and MVDs with $\cup_{\sigma \in \Sigma} Attr(\sigma) \cup Attr(\varphi) \subseteq R$. Then*

- *If φ denotes an FD, then: $\varphi \in \Sigma^+_{\mathfrak{AC}_C}$ if and only if $\varphi \in \Sigma^+_{\mathfrak{AC}}$.*
- *If φ denotes the MVD $X \twoheadrightarrow Y$, then: $X \twoheadrightarrow Y \in \Sigma^+_{\mathfrak{AC}_C}$ if and only if $X \twoheadrightarrow Y \in \Sigma^+_{\mathfrak{AC}}$ or $X \twoheadrightarrow (R - Y) \in \Sigma^+_{\mathfrak{AC}}$.*   □

Another interpretation of Corollary 4 is the following: if $\mathfrak{AC}$ is utilised to infer FDs, then the underlying universe does not need to be fixed at all; and if $\mathfrak{AC}$ is utilised to infer MVDs, then the fixing of a universe can be deferred until the very last step of the inference.

Example 8. Let $\textsc{Film} = \{\text{Movie,Director,Actor}\}$ and

$$\Sigma = \{\text{Movie} \twoheadrightarrow \text{Director}, \text{Actor} \to \text{Director}\}.$$

It follows that $\text{Movie} \to \text{Director} \in \Sigma^+_{\mathfrak{AC}}$, and $\text{Movie} \twoheadrightarrow \text{Actor} \notin \Sigma^+_{\mathfrak{AC}}$ (cf. Lemma 4) but $\text{Movie} \twoheadrightarrow \text{Actor} \in \Sigma^+_{\mathfrak{AC}_C}$. In the last inference we eventually commit ourselves to the relation schema $\textsc{Film}$ by applying the $\textsc{Film}$-complementation rule in the final (only) step of the inference.   □

6   Complete Reasoning in Undetermined Universes

In this section we establish the first axiomatisation for the combined class of FDs and MVDs in undetermined universes. While we have seen in the previous section that $\mathfrak{AC}$ is nearly R-complete for the R-implication on all relation schemata R it turns out that $\mathfrak{AC}$ is indeed complete for the implication of FDs and MVDs in undetermined universes.

Before sketching the proof we shall mention a lemma. Its correctness can easily be observed by inspecting the syntactic definitions of the inference rules in $\mathfrak{AC}$. For each of the rules, the right-hand side of the conclusion does not contain any attribute that did not already occur in the left-hand side of the conclusion or in the right-hand side of at least one of the premises.

Lemma 4. *Let $\Sigma \cup \{\varphi\}$ be a finite set of FDs and MVDs. If $\varphi \in \Sigma^+_{\mathfrak{AC}}$, then $rhs(\varphi) \subseteq \cup_{\sigma \in \Sigma} rhs(\sigma) \cup lhs(\varphi)$.*   □

We are now prepared to establish the first axiomatisation of FDs and MVDs in undetermined universes.

Theorem 4. *The set $\mathfrak{AC} = \{\mathcal{R}_F, \mathcal{E}_F, \mathcal{T}_F, \mathcal{A}_M, \mathcal{T}_M, \mathcal{S}_M, \mathcal{T}^*_M, \mathcal{I}_{FM}, \mathcal{S}_{FM}\}$ is sound and complete for the implication of FDs and MVDs in undetermined universes.*

Proof (Sketch). The soundness of the inference rules in $\mathfrak{AC}$ has been established in previous work [12, 13]. For the soundness of $\mathfrak{AC}$ one needs to show that every $\varphi \in \Sigma^+_{\mathfrak{AC}}$ is implied by Σ. That is, every relation r that satisfies $T := \cup_{\sigma \in \Sigma} Attr(\sigma) \cup Attr(\varphi) \subseteq Dom(r)$ and $\models_r \sigma$ for all $\sigma \in \Sigma$ also satisfies

$\models_r \varphi$. One can show that there is an inference γ of φ from Σ by $\mathfrak{AC}$ such that $Attr(\psi) \subseteq T \subseteq Dom(r)$ holds for every ψ occurring in γ. Since each rule of $\mathfrak{AC}$ is sound we can therefore conclude by induction that each ψ occurring in γ is satisfied by r. In particular, r also satisfies φ.

For the completeness of $\mathfrak{AC}$ we assume that $\varphi \notin \Sigma_{\mathfrak{AC}}^+$. Let $R \subseteq \mathcal{A}$ be a finite set of attributes such that T is a proper subset of R, i.e., $T \subset R$. In particular, it follows that $R - Y$ is not a subset of T.

If φ denotes a functional dependency, then Corollary 4 shows that $\varphi \notin \Sigma_{\mathfrak{AC}_C}^+$. However, $\mathfrak{AC}_C$ is R-complete for the R-implication of FDs and MVDs. Hence, it follows that Σ does not R-imply φ. Consequently, Σ does not imply φ.

If φ denotes the multivalued dependency $X \twoheadrightarrow Y$, then Lemma 4 shows that $X \twoheadrightarrow R - Y \notin \Sigma_{\mathfrak{AC}}^+$ since $R - Y$ is not a subset of T. From $X \twoheadrightarrow Y \notin \Sigma_{\mathfrak{AC}}^+$ and $X \twoheadrightarrow R - Y \notin \Sigma_{\mathfrak{AC}}^+$ we conclude $X \twoheadrightarrow Y \notin \Sigma_{\mathfrak{AC}_C}^+$ by Corollary 4. However, $\mathfrak{AC}_C$ is R-complete for the R-implication of FDs and MVDs. Hence, it follows that Σ does not R-imply φ. Consequently, Σ does not imply φ. $\qquad\square$

Theorem 4 proves the conjecture of Biskup in [12] for the combined class of FDs and MVDs.

Example 9. Let $\Sigma = \{\text{Movie} \twoheadrightarrow \text{Director}, \text{Actor} \rightarrow \text{Director}\}$. The multivalued dependency Movie $\twoheadrightarrow$ Actor is not implied by Σ and, thus, not derivable by using the inference rules in $\mathfrak{AC}$. Moreover, the FD Movie $\rightarrow$ Director is indeed implied by Σ and, consequently, also derivable from Σ by using $\mathfrak{AC}$. $\qquad\square$

The inference system $\mathfrak{AC}$ does not permit the application of the R-complementation rule, and does therefore not result in the inference of FDs or MVDs that are possibly semantically meaningless.

Example 10. Consider the attributes $\underline{A}rticle, \underline{M}anufacturer, \underline{L}ocation,$ and $\underline{C}osts$ and dependency set $\Sigma = \{A \rightarrow M; A, L \rightarrow C; M \twoheadrightarrow L\}$ from Example 4 again. Recall, that the inference from Example 5 has left us in doubt about the meaningfulness of the FD $A \rightarrow C$. The inference

$$
\begin{array}{c}
A \rightarrow M \\
\hline
\mathcal{I}_{\text{FM}}:\ A \twoheadrightarrow M \quad M \twoheadrightarrow L \\
\hline
\mathcal{T}_{\text{M}}:\qquad A \twoheadrightarrow L \qquad\qquad A, L \rightarrow C \\
\hline
\mathcal{A}_{\text{M}}:\ A \twoheadrightarrow A, L \qquad \mathcal{I}_{\text{FM}}:\ A, L \twoheadrightarrow C \\
\hline
\mathcal{T}_{\text{M}}:\qquad\qquad A \twoheadrightarrow C \qquad\qquad A, L \rightarrow C \\
\hline
\mathcal{S}_{\text{FM}}:\qquad\qquad A \rightarrow C
\end{array}
$$

exhibits an adequate line of reasoning, the FD $A \rightarrow C$ is an appropriate consequence of Σ, and the target schema suggested in Example 4 represents therefore an excellent choice. $\qquad\square$

7   Extension to Full Hierarchical Dependencies

Complete reasoning techniques for the class of full hierarchical dependencies have only been established independently of any other class of dependencies [23, 38].

In this section we extend the results from previous sections to the combined class of FDs and FHDs. The main difference between MVDs and FHDs is that the latter permit several (disjoint) attribute sets on the right-hand side while MVDs only permit a single attribute set. One can observe that the inference rules for FDs and FHDs are similar to those for FDs and MVDs. What is required additionally, are inference rules that deal with the interactions of the several attribute sets on the right-hand side of an FHD. In fact, the merging of two attribute sets (by $\mathcal{M}_H$) and the removal of an arbitrary attribute set (by $\mathcal{O}_H$) capture these interactions completely. Further differences between the inference rules mainly result from the disjointness that we require (by definition) for attribute sets of an FHD while we do not require disjoint left- and right-hand sides in an MVD.

Table 2. Inference Rules for Functional and Full Hierarchical Dependencies

$$\frac{X : \{Y_1, \ldots, Y_k\}}{XZ : \{Y_1 - Z, \ldots, Y_k - Z\}}$$
(augmentation, $\mathcal{A}_H$)

$$\frac{XY : \{Y_1, \ldots, Y_k\}, X : \{Y\}}{X : \{Y_1, \ldots, Y_k, Y\}}$$
(transitivity, $\mathcal{T}_H$)

$$\frac{X : \{Y_1, \ldots, Y_k, Y\}}{X : \{Y_1, \ldots, Y_k\}}$$
(omission, $\mathcal{O}_H$)

$$\frac{X : \{Y_1, \ldots, Y_k, Y_{k+1}\}}{X : \{Y_1, \ldots, Y_k Y_{k+1}\}}$$
(merging, $\mathcal{M}_H$)

$$\frac{X : \{Y_1, \ldots, Y_k\}}{X : \{Y_1, \ldots, Y_{k-1}, R - XY_1 \cdots Y_k\}}$$
(R-complementation, $\mathcal{C}_H^R$)

$$\frac{X \rightarrow Y}{X : \{Y - X\}}$$
(implication, $\mathcal{I}_{FH}$)

$$\frac{X : \{Y\}, W \rightarrow Z}{X \rightarrow Y \cap Z} Y \cap W = \emptyset$$
(mixed subset, $\mathcal{S}_{FH}$)

$$\frac{X : \{Y\}, XY \rightarrow Z}{X \rightarrow Z - Y}$$
(mixed pseudo-transitivity, $\mathcal{T}_{FH}$)

$$\frac{X : \{Y\}, W : \{Y_1, \ldots, Y_k\}}{X : \{Y \cap Y_1, \ldots, Y \cap Y_k, Y - Y_1 \cdots Y_k\}} Y \cap W = \emptyset$$
(subset, $\mathcal{S}_H$)

To the authors' best knowledge Theorem 5 establishes the first axiomatisation for the combined class of FDs and FHDs extending a result for FHDs only [23].

Theorem 5. *For all relation schemata R, the inference systems*

- $\mathfrak{H}\mathfrak{F}_C = \{\mathcal{R}_F, \mathcal{E}_F, \mathcal{T}_F, \mathcal{A}_H, \mathcal{T}_H, \mathcal{O}_H, \mathcal{I}_{FH}, \mathcal{T}_{FH}, \mathcal{C}_H^R\}$,
- $\mathfrak{H}\mathfrak{C}_C = \mathfrak{H}\mathfrak{F}_C \cup \{\mathcal{S}_H, \mathcal{M}_H\}$,
- $\mathfrak{H}\mathfrak{A}_C = (\mathfrak{H}\mathfrak{F}_C - \{\mathcal{T}_{FH}\}) \cup \{\mathcal{S}_{FH}\}$, *and*
- $\mathfrak{H}\mathfrak{A}\mathfrak{C}_C = \mathfrak{H}\mathfrak{A}_C \cup \{\mathcal{S}_H, \mathcal{M}_H\} = (\mathfrak{H}\mathfrak{C}_C - \{\mathcal{T}_{FH}\}) \cup \{\mathcal{S}_{FH}\}$.

are R-sound and R-complete for the R-implication of FDs and FHDs. $\square$

The next result shows that the systems for FDs and FHDs have the same properties as their counterparts for FDs and MVDs (cf. Figure 1). The definitions of R-adequacy and R-complementarity are easily extended to FDs and FHDs.

Theorem 6. *For all relation schemata R, the inference system*

- $\mathfrak{H}\mathfrak{C}_C$ *is R-complementary,*
- $\mathfrak{H}\mathfrak{A}_C$ *is R-adequate, and*
- $\mathfrak{H}\mathfrak{A}\mathfrak{C}_C$ *is R-complementary and R-adequate*

for the R-implication of FDs and FHDs. $\square$

Example 11. We illustrate how an application of the subset rule can shift applications of the R-complementation rule. Suppose we have the following inference

$$\frac{\dfrac{X:\{Y\}}{\mathcal{C}_H^R:\quad X:\{R-XY\}}\quad X(R-XY):\{Y_1,\dots,Y_k\}}{\mathcal{T}_H:\qquad X:\{Y_1,\dots,Y_k,R-XY\}}$$

Applying the subset rule $\mathcal{S}_H$ instead one may infer the same FHD as follows:

$$\frac{\dfrac{X:\{Y\}\qquad X(R-XY):\{Y_1,\dots,Y_k\}}{\mathcal{S}_H:\quad X:\{Y_1\cap Y,\dots,Y_k\cap Y,Y-Y_1\cdots Y_k\}}}{\mathcal{C}_H^R:\quad X:\{Y_1,\dots,Y_k,\underbrace{R-XY_1\cdots Y_k(Y-Y_1\cdots Y_k)}_{=R-XYY_1\cdots Y_k=R-XY}\}}$$

Note that Y and $X(R-XY)$ are disjoint, and $Y_1\cdots Y_k$ and $X(R-XY)$ are disjoint, too. Hence, $Y_1\cdots Y_k\subseteq Y$, and thus $Y_i\cap Y=Y_i$ for all $i=1,\dots,k$. $\square$

Example 12. We illustrate how an application of the mixed subset rule can eliminate applications of the R-complementation rule. Consider the inference:

$$\frac{\dfrac{X:\{Y\}}{\mathcal{C}_H^R:\quad X:\{R-XY\}}\quad X(R-XY)\to Z}{\mathcal{T}_{FH}:\qquad X\to\underbrace{Z-X(R-XY)}_{=Y\cap Z}}$$

For this inference notice that X and Y are disjoint. Hence, $X(R-XY)=R-Y$ and, consequently, $Z-(R-Y)=Y\cap Z$. Applying the mixed subset rule $\mathcal{S}_{FH}$ instead one may infer the same FHD by the following inference step:

$$\frac{X:\{Y\}\qquad X(R-XY)\to Z}{\mathcal{S}_{FH}:\qquad X\to Y\cap Z}$$

This eliminates the application of the R-complementation rule. $\square$

In undetermined universes the following result extends Theorem 4.

Theorem 7. *The system $\mathfrak{H}\mathfrak{A}\mathfrak{C}$ of inference rules is sound and complete for the implication of FDs and FHDs in undetermined universes.* $\square$

8   Conclusion

We have extended previous research on the appropriateness of inference systems for MVDs to the combined class of FDs and FHDs. In particular, we have established the first appropriate axiomatisation of FDs and FHDs in fixed universes, and the first ever axiomatisation of FDs and FHDs in undetermined universes. Our results demonstrate that the complementation rule is a mere means for achieving database normalisation: to infer an FHD at most one application of the complementation rule is necessary in the very last step of the inference; and to infer an FD the complementation rule does not need to be applied at all. Most importantly, we have formally demonstrated that previous axiomatisations for the class of FDs and FHDs (FDs and MVDs) cannot infer any semantically meaningless data dependencies since any inappropriate inference can be converted into an appropriate inference.

A very interesting treatment of MVDs and FHDs in the context of Entity-Relationship modeling can be found in [39]. There, the R-complete inference rules do not directly apply an R-complementation rule but make use of R's partitions into components and attributes where R denotes some relationship type. This is another way of indicating the dependence of implication on the underlying universe R. In this context it would therefore be very interesting to investigate the notion of implication in undetermined universes.

References

1. Arenas, M., Libkin, L.: A normal form for XML documents. ACM Trans. Database Syst. 29(1), 195–232 (2004)
2. Armstrong, W.W.: Dependency structures of database relationships. Information Processing 74, 580–583 (1974)
3. Armstrong, W.W., Nakamura, Y., Rudnicki, P.: Armstrong's axioms. Journal of formalized Mathematics 14 (2002)
4. Balcázar, J., Baixeries, J.: Characterizations of multivalued dependencies and related expressions. In: Suzuki, E., Arikawa, S. (eds.) DS 2004. LNCS (LNAI), vol. 3245, pp. 306–313. Springer, Heidelberg (2004)
5. Balcázar, J., Baixeries, J.: Characterization and armstrong relations for degenerate multivalued dependencies using formal concept analysis. In: Ganter, B., Godin, R. (eds.) ICFCA 2005. LNCS (LNAI), vol. 3403, pp. 162–175. Springer, Heidelberg (2005)
6. Beeri, C.: On the membership problem for functional and multivalued dependencies in relational databases. ACM Trans. Database Syst. 5(3), 241–259 (1980)
7. Beeri, C., Bernstein, P.A.: Computational problems related to the design of normal form relational schemata. ACM Trans. Database Syst. 4(1), 30–59 (1979)
8. Beeri, C., Fagin, R., Howard, J.H.: A complete axiomatization for functional and multivalued dependencies in database relations. In: SIGMOD Conference, pp. 47–61. ACM, New York (1977)
9. Bernstein, P.: Synthesizing third normal form relations from functional dependencies. ACM Trans. Database Syst. 1(4), 277–298 (1976)
10. Bernstein, P.A., Goodman, N.: What does Boyce-Codd normal form do? In: VLDB Conference, pp. 245–259. IEEE Computer Society, Los Alamitos (1980)

11. Biskup, J.: On the complementation rule for multivalued dependencies in database relations. Acta Inf. 10(3), 297–305 (1978)
12. Biskup, J.: Inferences of multivalued dependencies in fixed and undetermined universes. Theor. Comput. Sci. 10(1), 93–106 (1980)
13. Biskup, J.: Grundlagen von Informationssystemen. Vieweg (1995)
14. Biskup, J., Dayal, U., Bernstein, P.: Synthesizing independent database schemas. In: SIGMOD Conference, pp. 143–151 (1979)
15. Codd, E.F.: A relational model of data for large shared data banks. Commun. ACM 13(6), 377–387 (1970)
16. Delobel, C.: Normalisation and hierarchical dependencies in the relational data model. ACM Trans. Database Syst. 3(3), 201–222 (1978)
17. Deutsch, A., Popa, L., Tannen, V.: Query reformulation with constraints. SIGMOD Record 35(1), 65–73 (2006)
18. Fagin, R.: Multivalued dependencies and a new normal form for relational databases. ACM Trans. Database Syst. 2(3), 262–278 (1977)
19. Fischer, P.C., Saxton, L.V., Thomas, S.J., Van Gucht, D.: Interactions between dependencies and nested relational structures. J. Comput. Syst. Sci. 31(3), 343–354 (1985)
20. Galil, Z.: An almost linear-time algorithm for computing a dependency basis in a relational database. J. ACM 29(1), 96–102 (1982)
21. Hagihara, K., Ito, M., Taniguchi, K., Kasami, T.: Decision problems for multivalued dependencies in relational databases. SIAM J. Comput. 8(2), 247–264 (1979)
22. Hara, C., Davidson, S.: Reasoning about nested functional dependencies. In: PODS Conference, pp. 91–100. ACM, New York (1999)
23. Hartmann, S., Köhler, H., Link, S.: Full hierarchical dependencies in fixed and undetermined universes. Ann. Math. Artif. Intell. 50(1–2), 195–226 (2007)
24. Hartmann, S., Link, S.: On a problem of Fagin concerning multivalued dependencies in relational databases. Theor. Comput. Sci. 353(1–3), 53–62 (2006)
25. Hartmann, S., Link, S., Schewe, K.-D.: Axiomatisations of functional dependencies in the presence of records, lists, sets and multisets. Theor. Comput. Sci. 355(2), 167–196 (2006)
26. Hartmann, S., Link, S., Schewe, K.-D.: Functional and multivalued dependencies in nested databases generated by record and list constructor. Ann. Math. Artif. Intell. 46(1–2), 114–164 (2006)
27. Lakshmanan, V., VeniMadhavan, C.: An algebraic theory of functional and multivalued dependencies in relational databases. Theor. Comput. Sci. 54, 103–128 (1987)
28. Levene, M., Loizou, G.: Axiomatisation of functional dependencies in incomplete relations. Theor. Comput. Sci. 206(1–2), 283–300 (1998)
29. Levene, M., Vincent, M.: Justification for inclusion dependency normal form. IEEE Trans. Knowl. Data Eng. 12(2), 281–291 (2000)
30. Link, S.: On multivalued dependencies in fixed and undetermined universes. In: Dix, J., Hegner, S.J. (eds.) FoIKS 2006. LNCS, vol. 3861, pp. 257–276. Springer, Heidelberg (2006)
31. Link, S.: On the logical implication of multivalued dependencies with null values. In: Twelfth Computing: The Australasian Theory Symposium. Conferences in Research and Practice in Information Technology, vol. 51, pp. 113–122 (2006)
32. Mendelzon, A.: On axiomatising multivalued dependencies in relational databases. J. ACM 26(1), 37–44 (1979)
33. Paredaens, J., De Bra, P., Gyssens, M., Van Gucht, D.: The Structure of the Relational Database Model. Springer, Heidelberg (1989)

34. Sagiv, Y.: An algorithm for inferring multivalued dependencies with an application to propositional logic. J. ACM 27(2), 250–262 (1980)
35. Sagiv, Y., Delobel, C., Parker Jr., D.S., Fagin, R.: An equivalence between relational database dependencies and a fragment of propositional logic. J. ACM 28(3), 435–453 (1981)
36. Sözat, M., Yazici, A.: A complete axiomatization for fuzzy functional and multivalued dependencies in fuzzy database relations. ACM Fuzzy Sets and Systems 117(2), 161–181 (2001)
37. Tari, Z., Stokes, J., Spaccapietra, S.: Object normal forms and dependency constraints for object-oriented schemata. ACM Trans. Database Syst. 22, 513–569 (1997)
38. Thalheim, B.: Dependencies in Relational Databases. Teubner-Verlag (1991)
39. Thalheim, B.: Conceptual treatment of multivalued dependencies. In: Song, I.-Y., Liddle, S.W., Ling, T.-W., Scheuermann, P. (eds.) ER 2003. LNCS, vol. 2813, pp. 363–375. Springer, Heidelberg (2003)
40. Vincent, M., Liu, J., Liu, C.: A redundancy free 4NF for XML. In: Bellahsène, Z., Chaudhri, A.B., Rahm, E., Rys, M., Unland, R. (eds.) Database and XML Technologies. LNCS, vol. 2824, pp. 254–266. Springer, Heidelberg (2003)
41. Weddell, G.: Reasoning about functional dependencies generalized for semantic data models. ACM Trans. Database Syst. 17(1), 32–64 (1992)
42. Wijsen, J.: Temporal FDs on complex objects. ACM Trans. Database Syst. 24(1), 127–176 (1999)
43. Zaniolo, C., Melkanoff, M.: On the design of relational database schemata. ACM Trans. Database Syst. 6(1), 1–47 (1981)

Autonomous Sets –
A Method for Hypergraph Decomposition with Applications in Database Theory

Henning Koehler

Massey University, Palmerston North, New Zealand
h.koehler@massey.ac.nz

Abstract. We present a method for decomposing a hypergraph with certain regularities into smaller hypergraphs. By applying this to the set of all canonical covers of a given set of functional dependencies, we obtain more efficient methods for solving several optimization problems in database design. These include finding one or all "optimal" covers w.r.t. different criteria, which can help to synthesize better decompositions, and to reduce the cost of constraint checking.

1 Introduction

Many data structures can be modeled as hypergraphs, and such hypergraphs often display certain regularities. To make such regularities explicit, we introduce the notion of autonomous vertex sets. This allows us to decompose a given hypergraph into smaller hypergraphs, which can be stored and manipulated more efficiently.

In particular, the set of all solutions to a given problem often forms a hypergraph with the type of regularities we are interested in. While in this case the hypergraph is not given explicitly, determining the autonomous sets can help to split the problem into smaller sub-problems, which can be solved independently.

One such hypergraph, which arises in database theory, is formed by the set of all canonical covers of a given set of functional dependencies. Covers, in particular canonical covers, play an important role in database design. They can determine the decomposition when following the synthesis approach [4], and are needed for constraint checking. To find good decompositions, or speed up constraint checking, it thus is vital to use the right cover, which is often in canonical form [8,15] or can immediately be derived from a canonical cover [12].

We will investigate this hypergraph and show how autonomous sets can be found efficiently. Based on these we indicate how certain covers can be found which are "optimal" in some sense. As a result of using optimal covers, we can improve the speed of database updates.

The rest of the paper is organized as follows. In Section 2 we first introduce some basic terminology related to hypergraphs. We then define autonomous sets in Section 2.1 and provide a characterization. An algorithm for computing the

S. Hartmann and G. Kern-Isberner (Eds.): FoIKS 2008, LNCS 4932, pp. 78–95, 2008.

minimal autonomous sets of a given hypergraph is given in Section 2.2. In Section 3 we then apply our hypergraph theory to canonical covers and characterize certain autonomous sets. These autonomous sets need not be minimal though, and we will show that identifying the minimal autonomous sets is co-NP-hard in Section 4. Finally we present some problems vital to database design in Section 5, which can be simplified using the autonomous sets found. Related work is mentioned in Section 6.

2   Hypergraph Decomposition

We introduce some basic terminology and well known results about hypergraphs.

Definition 1. *A hypergraph H on a vertex set V is a set of subsets of V, i.e., $H \subseteq \mathcal{P}(V)$. The elements of H are called* edges. *A hypergraph is called* simple *if none of its edges is included in another.*

Definition 2. *The set ϑ_H of vertices actually appearing in edges of a hypergraph H is called the* support *of H:*

$$\vartheta_H := \bigcup_{e \in H} e$$

Note that we *do* allow the empty edge in a hypergraph, and that we do *not* require that $V = \vartheta_H$. This simplifies some arguments, but has no significant impact on the results.

Definition 3. *Let H, G be hypergraphs. We define the* cross-union $H \vee G$ *of H and G as*

$$H \vee G := \{e_H \cup e_G \mid e_H \in H, e_G \in G\}$$

If V_H, V_G are the vertex sets of H and G then $H \vee G$ is a hypergraph on $V_H \cup V_G$.

Definition 4. *Let H be a hypergraph on vertex set V. The* projection $H[S]$ *of H onto S is*

$$H[S] := \{e \cap S \mid e \in H\},$$

which is a hypergraph on $V \cap S$.

Definition 5. *Let H be a hypergraph on V. A set $t \subseteq V$ is a* transversal *of H if t intersects with every edge of H. We denote the set of all minimal transversals (w.r.t. inclusion) by $Tr(H)$, and call $Tr(H)$ the* transversal hypergraph *of H.*

Clearly $Tr(H)$ is a simple hypergraph on V, even if H is not simple.

Lemma 1. *[3] Let H be a simple hypergraph. Then $Tr(Tr(H)) = H$.*

2.1 Autonomous Sets

We shall introduce the concept of an autonomous vertex set. Note that our definition is not meant to extend any use of the term "autonomous set" in the context of graphs, where it is better known as "module", and characterizes vertex sets M in which each vertex $v \in M$ has the same neighbors outside M [7]. Using our terminology, autonomous sets are only interesting for hypergraphs but not for graphs. Essentially the only graphs with non-trivial autonomous sets are complete bipartite graphs.

Definition 6. *Let H be a hypergraph on the vertex set V. We call a vertex subset $S \subseteq V$ autonomous if $H = H[S] \vee H[\overline{S}]$ where $\overline{S} := V \setminus S$ denotes the complement of S.*

Clearly the complement of an autonomous set is itself autonomous, and

$$H \subseteq H[S] \vee H[\overline{S}]$$

for any $S \subseteq V$. The sets $\emptyset, V$ are autonomous for any hypergraph H on V, as are all subsets of $V \setminus \vartheta_H$ and their complements.

Example 1. Consider the vertex set $V = ABCDE$, and on it the hypergraph

$$H = \{AC, AD, BC, BD\}$$

H is simple, and its support is $\vartheta_H = ABCD$. The set $S = AB$ is autonomous for H, as is its complement $\overline{S} = CDE$, since

$$H[AB] \vee H[CDE] = \{A, B\} \vee \{C, D\} = \{AC, AD, BC, BD\} = H$$

Lemma 2. *Let $S, T \subseteq V$ be autonomous. Then $S \cap T$ is autonomous as well.*

Proof. We need to show that for every pair of edges $e_1, e_2 \in H$ the edge $e_1' \cup e_2'$ with $e_1' := e_1 \cap (S \cap T)$ and $e_2' := e_2 \cap \overline{S \cap T}$ lies in H as well. Since S is autonomous, the edge $e' := (e_1 \cap S) \cup (e_2 \cap \overline{S})$ lies in H. Thus, since T is autonomous, the edge $e'' := (e' \cap T) \cup (e_2 \cap \overline{T})$ lies in H. Clearly

$$e'' \cap (S \cap T) = e' \cap (S \cap T) = e_1 \cap (S \cap T) = e_1'$$

and similarly

$$
\begin{aligned}
e'' \cap \overline{S \cap T} &= (e' \cap (T \setminus S)) \cup (e_2 \cap \overline{T}) \\
&= (e_2 \cap (T \setminus S)) \cup (e_2 \cap \overline{T}) \\
&= e_2'
\end{aligned}
$$

which shows $e_1' \cup e_2' = e'' \in H$.

Corollary 1. *Let $S, T \subseteq V$ be autonomous. Then $S \cup T$ is autonomous.*

Proof. The complements and intersections of autonomous sets are autonomous by definition and by Lemma 2, respectively, and we have

$$S \cup T = \overline{\overline{S} \cap \overline{T}}$$

Proposition 1. *Let H, G be hypergraphs and S_1, S_2 vertex sets. Then we have*

(i) $H[S_1][S_2] = H[S_1 \cap S_2]$
(ii) $(H \vee G)[S_1] = H[S_1] \vee G[S_1]$

Lemma 3. *Let H be a hypergraph on V and $S \subseteq V$ autonomous for H. Then for any $T \subseteq V$ the set $S \cap T$ is autonomous for $H[T]$.*

Proof. Since S is autonomous for H we have $H = H[S] \vee H[\overline{S}]$. Thus

$$\begin{aligned}
H[T] &= (H[S] \vee H[\overline{S}])[T] \\
&= H[S][T] \vee H[\overline{S}][T] \\
&= H[T][S \cap T] \vee H[T][\overline{S \cap T}]
\end{aligned}$$

Theorem 1. *Let H be a hypergraph on V and $\{S_1, \ldots, S_n\}$ a partition of V into autonomous sets. Then*

$$H = H[S_1] \vee \ldots \vee H[S_n]$$

Proof. By induction on n. The equation hold trivially for $n = 1$. Assume now the theorem holds for a fixed value of n. To show the theorem for $n+1$ we use that $S_n \cup S_{n+1}$ is autonomous by Corollary 1, so that by assumption we have

$$H = H[S_1] \vee \ldots \vee H[S_{n-1}] \vee H[S_n \cup S_{n+1}]$$

By Lemma 3 S_n is autonomous for $H[S_n \cup S_{n+1}]$, i.e., we have

$$H[S_n \cup S_{n+1}] = H[S_n] \vee H[S_{n+1}]$$

which shows the theorem for $n+1$.

When talking about *minimal autonomous sets*, we will always mean minimal w.r.t. inclusion among all non-empty autonomous sets, even though the empty set is always autonomous by definition. While it would be more precise to call them minimal *non-empty* autonomous sets, this quickly becomes tedious.

Theorem 2. *Every hypergraph H has a finest partition $\{S_1, \ldots, S_n\}$ into minimal autonomous sets. The autonomous sets of H are just the unions of these sets.*

Proof. Let $S_1, \ldots, S_n$ be the minimal autonomous sets of H. By Lemma 2 they are pairwise disjoint. The union of autonomous sets is itself autonomous by Corollary 1, in particular

$$S := \bigcup_{i=1}^{n} S_i$$

Furthermore the complement of S is autonomous, and since $\overline{S}$ does not include any minimal autonomous set it is empty, i.e., $S = V$. Thus the sets $S_1, \ldots, S_n$ form a partition of V.

Whenever an autonomous set T intersects with some S_i it must include it completely, since otherwise $T \cap S_i$ would be a smaller non-empty autonomous set by Lemma 2. Thus each autonomous set is the union of those S_i it intersects with.

Example 2. Consider again $H = \{AC, AD, BC, BD\}$ on $V = ABCDE$. Then

$$H = H[AB] \vee H[CD] \vee H[E] = \{A, B\} \vee \{C, D\} \vee \{\emptyset\}$$

so the minimal autonomous sets of H are AB, CD, E. Thus H has a total of 2^3 autonomous sets:

$$\emptyset, AB, CD, E, ABCD, ABE, CDE, ABCDE$$

We now consider another type of decomposition, which will help us in characterizing the autonomous sets of simple hypergraphs.

Definition 7. *Let H be a hypergraph on vertex set V. The* subhypergraph $H \langle S \rangle$ *of H* induced *by $S \subseteq V$ is*

$$H \langle S \rangle := \{e \in H \mid e \subseteq S\}$$

Definition 8. *Let H be a hypergraph on the vertex set V. We call a vertex subset $S \subseteq V$* isolated *if $H = H \langle S \rangle \cup H \langle \overline{S} \rangle$.*

Clearly S is isolated if and only if every edge it intersects with lies completely in S. As with minimal autonomous sets, we will mean by *minimal isolated sets* the minimal sets (w.r.t. inclusion) among all non-empty isolated sets.

Definition 9. *As with graphs, we say that two vertices v_1, v_n in a hypergraph H are* connected *if there exists a sequence $v_1, v_2, \ldots, v_n$ such v_i, v_{i+1} always lie in some common hyperedge of H. H is* connected *if all its vertices are connected. The* connected components *of H are its connected subhypergraphs.*

It follows immediately that the minimal isolated sets of H are the vertex sets of its maximal connected components, and that the isolated sets of H are the unions of them.

Recall that $Tr(H)$ denotes the transversal hypergraph of H.

Theorem 3. *Let H, G be hypergraphs on disjoint vertex sets V_H and V_G. Then*

$$Tr(H \vee G) = Tr(H) \cup Tr(G)$$
$$Tr(H \cup G) = Tr(H) \vee Tr(G)$$

Proof. (1) We first show that a set $t \subseteq V := V_H \cup V_G$ is a transversal of $H \vee G$ iff it intersects with every edge of H or with every edge of G. For the "if" part, assume w.l.o.g. that t intersects with every edge of H. Since every edge $e \in H \vee G$ is of the form $e = e_H \cup e_G$ with $e_H \in H, e_G \in G$, t intersects with e because it intersects with e_H. We show the "only if" part by contraposition and assume that there be edges $e_H \in H, e_G \in G$ such that t intersects with neither of them. But then t does not intersect $e_H \cup e_G \in H \vee G$ either, i.e., t is not a transversal of $H \vee G$.

We thus have that the transversals of $H \vee G$ are the transversals of H plus the transversals of G. Thus the minimal transversals of $H \vee G$ are the minimal elements of $Tr(H) \cup Tr(G)$. Since V_H and V_G are disjoint, all elements of $Tr(H) \cup Tr(G)$ are minimal. Thus

$$Tr(H \vee G) = Tr(H) \cup Tr(G)$$

(2) By definition a set $t \subseteq V$ is a transversal of $H \cup G$ iff it is a transversal of both H and G. Thus the transversals of $H \cup G$ are the unions of transversals of H with transversals of G. The minimal transversals of $H \cup G$ are therefore the minimal elements of $Tr(H) \vee Tr(G)$. Since V_H and V_G are disjoint, all elements of $Tr(H) \vee Tr(G)$ are minimal. Thus

$$Tr(H \cup G) = Tr(H) \vee Tr(G)$$

We are now able to characterize the autonomous sets of a simple hypergraph.

Theorem 4. *Let H be a simple hypergraph. Then the autonomous sets of H are the isolated sets of its transversal hypergraph $Tr(H)$.*

Proof. Let $S \subseteq V$ be autonomous for H, i.e., $H = H[S] \vee H[\overline{S}]$. Then

$$Tr(H) = Tr(H[S]) \cup Tr(H[\overline{S}])$$

by Theorem 3, so S is isolated for $Tr(H)$.

Conversely let S be any isolated set of $Tr(H)$. Then

$$Tr(H) = Tr(H) \langle S \rangle \cup Tr(H) \langle \overline{S} \rangle$$

and by Theorem 3 we have

$$Tr(Tr(H)) = Tr(Tr(H) \langle S \rangle) \vee Tr(Tr(H) \langle \overline{S} \rangle)$$

Thus S is autonomous for $Tr(Tr(H)) = H$.

Example 3. The requirement that H be simple in Theorem 4 is necessary: Consider the hypergraphs

$$H = \{AC, AD, BC, BD\}$$
$$\text{and}$$
$$H' = H \cup \{ABC\}$$

Both H and H' have the same minimal transversals

$$Tr(H) = Tr(H') = \{AB, CD\}$$

Clearly AB, CD are isolated sets of $Tr(H')$, but AB and CD are not autonomous for H':

$$\begin{aligned} H'[AB] \vee H'[CD] &= \{A, B, AB\} \vee \{C, D\} \\ &= \{AC, AD, BC, BD, ABC, ABD\} \\ &= H' \cup \{ABD\} \neq H' \end{aligned}$$

Since graphs are just special hypergraphs, our theory of autonomous sets applies to them as well. Clearly all complete bipartite graphs have a non-trivial partition into two autonomous sets, but one may wonder whether there are others.

Lemma 4. *A simple graph G without isolated vertices has a non-trivial partition into autonomous sets iff it is complete bipartite.*

Proof. Let $S \notin \{\emptyset, \vartheta_G\}$ be autonomous. Since G contains no isolated vertices, $G[S]$ and $G[\overline{S}]$ contain non-empty edges. As all edges in a simple graph contain exactly two vertices, the edges of $G[S]$ and $G[\overline{S}]$ contain exactly one vertex each. Thus $G = G[S] \vee G[\overline{S}]$ is complete bipartite.

We note that non-simple graphs with non-trivial autonomous partition may also have loops on all vertices of one side of the bipartition, as well as isolated vertices.

2.2 Computing Autonomous Sets

To complete this section, we now address the question of computing the minimal autonomous sets of H. While Theorem 4 suggests an approach (at least for simple hypergraphs), computing the transversal hypergraph can lead to exponential runtime. Instead, we shall utilize the following observation.

Lemma 5. *Let H be a hypergraph on V and $P = \{S_1, \ldots, S_n\}$ the partition of V into minimal autonomous sets. Let further $H_1' \subseteq H[S_1]$ be non-empty, and $H' \subseteq H$ be the hypergraph*

$$H' := H_1' \vee H[S_2] \vee \ldots \vee H[S_n].$$

Then $S_2, \ldots, S_n$ are minimal autonomous sets of H'.

Proof. By definition $S_2, \ldots, S_n$ are autonomous for H'. If one of those S_i were not minimal for H', i.e., could be partitioned into smaller autonomous sets $T_1, \ldots, T_k$, then

$$H[S_i] = H[T_1] \vee \ldots \vee H[T_k]$$

and thus the sets T_i would be autonomous for H as well, contradicting the minimality of S_i.

We use this to compute the partition of V into minimal autonomous sets as follows. We pick some vertex $v \in V$ and split H into two hypergraphs, one containing all the edges which contain v, the other one containing all those edges which do not contain v. We will need only one of them, so let H_v be the smaller one of the two, i.e., the one with fewer edges (if both contain exactly the same number of edges we may choose either one):

$$H_v := \text{smaller of } \begin{cases} \{e \in H \mid v \in e\} \\ \{e \in H \mid v \notin e\} \end{cases}$$

If H_v is empty, then v lies in all or no edges of H, and in both cases the set $\{v\}$ is autonomous for H. This reduces the problem of finding the minimal autonomous sets of H to finding the minimal autonomous sets of $H[\overline{v}]$, where $\overline{v} := \vartheta_H \setminus \{v\}$, as they are also minimal autonomous sets of H.

Consider now the case where H_v is not empty. Let S_1 be the minimal autonomous set of H containing v. Then H_v has the same form as H' in Lemma 5, where H_1' contains either the edges of $H[S_1]$ which do or those which do not contain v. We now compute the minimal autonomous sets of H_v, and check for each set whether it is autonomous for H. By Lemma 5 the sets autonomous for H are exactly the $S_2, \ldots, S_n$, while the sets not autonomous for H partition S_1. Taking the union of those non-autonomous sets and keeping the autonomous ones thus gives us the minimal autonomous sets of H. Note that the set $\{v\}$ is always autonomous for H_v, as v is contained in either all or no edges of H_v. Thus it suffices to compute the minimal autonomous sets of $H_v[\overline{v}]$.

In either case we have reduced the problem of finding the minimal autonomous set of H to that of finding the minimal autonomous sets of a hypergraph with fewer vertices. This gives us the following recursive algorithm.

Algorithm "Recursive Autonomous Partitioning"
    INPUT: hypergraph H
    OUTPUT: partition of ϑ_H into minimal autonomous sets

    function RAP(H)
    select vertex $v \in \vartheta_H$
    $H_v := \text{smaller of } \begin{cases} \{e \in H \mid v \in e\} \\ \{e \in H \mid v \notin e\} \end{cases}$
    if $H_v = \emptyset$ then
      return $\{\{v\}\} \cup RAP(H[\overline{v}])$
    else
      $Aut := \emptyset, S_1 := \{v\}$
      $Aut_v := RAP(H_v[\overline{v}])$
      for all $S \in Aut_v$ do
        if S autonomous for H then
          $Aut := Aut \cup \{S\}$
        else
          $S_1 := S_1 \cup S$
    end
    return $\{S_1\} \cup Aut$

While the test whether a set S is autonomous for H can be performed by computing $H' := H[S] \vee H[\overline{S}]$ and comparing it to H, the resulting set can easily contain up to $|H|^2$ edges if S is not autonomous. We observe that always $H \subseteq H'$ and thus $H = H'$ iff $|H| = |H'|$. Since $|H'| = |H[S]| \cdot |H[\overline{S}]|$, the later condition can be checked faster without actually computing H'.

Theorem 5. *Let H be a hypergraph with k vertices and n edges. Then the "Recursive Autonomous Partitioning" algorithm computes the partition of ϑ_H into minimal autonomous sets of H in time $O(nk^2)$.*

Proof. We have already argued that the algorithm computes the minimal autonomous sets of H correctly, so we only need to show the time bound.

The depth of recursion is at most k. In each call we compute H_v, which can be done in $O(n)$. If $H_v = \emptyset$ we only need to compute $H[\overline{v}]$, which is possible in $O(nk)$. Thus this part of the algorithm can be performed in $O(nk^2)$.

If $H_v \neq \emptyset$ we need to test each set found to be autonomous for H_v whether it is autonomous for H. The number of such tests is at most k, and each test can be performed in $O(nk)$, by computing $H[S]$ and $H[\overline{S}]$ and testing whether $|H| = |H[S]| \cdot |H[\overline{S}]|$. This leads to a complexity of $O(nk^2)$. Since the number of edges of H_v is at most half of the number of edges of H, the number of steps required for performing the tests on H_v (or the next subgraph in the recursion for which tests are required) is at most half as many. This leads to a total complexity of

$$O((n + \frac{n}{2} + \frac{n}{4} + \ldots)k^2) = O(nk^2)$$

2.3 Superedges and Partial Superedges

While canonical covers will form the edges in our hypergraph, we will have to argue about sets of FDs which form a cover, but may contain more FDs than needed. We call such supersets of edges "superedges".

Definition 10. *Let H be a hypergraph on V. A set $E \subseteq V$ is called a superedge of H if it includes some edge $e \in H$, i.e. $e \subseteq E$. We call $E \subseteq S \subseteq V$ a partial (super)edge on S if E is a (super)edge of $H[S]$.*

Lemma 6. *Let H be a hypergraph on V and $S \subseteq V$. A set $S' \subseteq S$ is a partial superedge on S iff $S' \cup \overline{S}$ is a superedge.*

Proof. By definition S' is a partial superedge on S iff it includes a partial edge $e_S \in H[S]$, i.e. iff there exists an edge $e \in H$ with $e \cap S = e_S \subseteq S'$. Since

$$e = e_S \cup (e \cap \overline{S}) \subseteq S' \cup \overline{S}$$

this implies that $S' \cup \overline{S}$ is a superedge. Conversely, if $S' \cup \overline{S}$ is a superedge, it includes an edge $e \in H$, which gives us

$$e \cap S \subseteq (S' \cup \overline{S}) \cap S = S'$$

Lemma 7. *Let H be a hypergraph on V and $P = \{S_1, \ldots, S_n\}$ a partition of V into autonomous sets. A set $E \subseteq V$ is a superedge iff $E_i := E \cap S_i$ is a partial superedge on S_i for $i = 1, \ldots, n$.*

Proof. If E is a superedge then it includes an edge $e \in H$. Thus E_i includes $e \cap S_i \in H[S_i]$, which makes E_i a partial superedge on S_i.

Now for each $i = 1, \ldots, n$ let E_i be a partial superedge on S_i, including the partial edge $e_i \in H[S_i]$. Thus E includes

$$e := e_1 \cup \ldots \cup e_n \in H[S_1] \vee \ldots \vee H[S_n] \overset{(\text{thm } 1)}{=} H$$

which makes E a superedge of H.

We can therefore strengthen Lemma 6 when S is autonomous:

Lemma 8. *Let H be a hypergraph on V, $E \subseteq V$ a superedge and $S \subseteq V$ autonomous. A set $S' \subseteq S$ is a partial superedge on S iff $S' \cup (E \setminus S)$ is a superedge.*

Proof. By Lemma 7, the set $S' \cup (E \setminus S)$ is a superedge iff S' is a partial superedge on S and $E \setminus S$ is a partial superedge on $\overline{S}$. Since E is a superedge, Lemma 7 assures that $E \setminus S = E \cap \overline{S}$ is a partial superedge on $\overline{S}$.

3   Canonical Covers

We shall now apply our theory of hypergraph decomposition to the set of all canonical covers. For this, we begin by introducing basic terminology.

A functional dependency (FD) on a set of attributes R is an expression of the form $X \to Y$ (read "X *determines* Y") where X and Y are subsets of R. Functional dependencies are frequently used in database systems, where they restrict which data tables can be stored [10,13,14]. For attribute sets X, Y and attribute A we will write XY short for $X \cup Y$ and A short for $\{A\}$.

A set $X \subseteq R$ is a *key* of R w.r.t. a set Σ of integrity constraints on R, if Σ implies $X \to R$. Note that some authors use the term 'key' only for minimal keys, and call keys which may not be minimal 'superkeys'.

A set Σ of FDs can imply other FDs. Implication of FDs can be characterized by the following derivation rules, known as the Armstrong Axioms [1]:

$$\frac{}{X \to Y} Y \subseteq X, \quad \frac{X \to Y}{XW \to YW}, \quad \frac{X \to Y \quad Y \to Z}{X \to Z} \tag{1}$$

Here the FDs on the top imply the FD at the bottom, and $Y \subseteq X$ is a side condition which needs to hold for the first rule to be applicable. We write Σ^* for the set of all FDs on R implied by Σ.

Two sets of FDs Σ, Σ' are called *equivalent* or *covers* of each other if they imply each other. This can be written as $\Sigma \vDash \Sigma'$ and $\Sigma' \vDash \Sigma$, or as $\Sigma^* = \Sigma'^*$.

We will use letters at the end of the alphabet $(\ldots, X, Y, Z)$ to denote subsets of R, while letters at the beginning $(A, B, C, \ldots)$ denote single attributes.

Definition 11. *We use the following terminology:*

(i) A FD $X \rightarrow A$ is called singular.
(ii) A non-trivial singular FD $X \rightarrow A \in \Sigma^*$ is called atomic, *if and only if for all $Y \subsetneq X$ we have $Y \rightarrow A \notin \Sigma^*$.*
(iii) The atomic closure Σ^{*a} *of Σ is the set of all atomic FDs in Σ^**
(iv) A set $G \subseteq \Sigma^{*a}$ of atomic FDs is called canonical cover *if it is a cover of Σ which is minimal w.r.t. set inclusion, i.e., for all $H \subsetneq G$ the set H is not a cover of Σ.*

Note that atomic FDs have also been called "elemental" FDs [17].

Definition 12. *Let Σ be a set of FDs. We denote the set of all canonical covers of Σ by*

$$CC(\Sigma) := \{G \subseteq \Sigma^{*a} \mid G \text{ is a canonical cover of } \Sigma\}$$

When given a set Σ of functional dependencies for schema decomposition, instance validation or similar tasks, we may choose to use a cover Σ' of functional dependencies equivalent to Σ instead. The choice of Σ' is important, as it determines the result of the decomposition, the speed of updates, and generally can have a huge impact on database performance. Optimal results are usually achieved by covers which are in some standard form, often canonical, but finding these optimal covers is often NP-hard.

To simplify these problems, we now wish to find autonomous sets of $CC(\Sigma)$. While algorithm "Recursive Autonomous Partitioning" allows us to compute the minimal autonomous sets of a given hypergraph, $CC(\Sigma)$ is usually not given directly. Instead we will assume that only Σ is given, and develop alternative means for finding autonomous sets of $CC(\Sigma)$ without computing the entire hypergraph first.

3.1   Partial Covers

The set of all canonical covers of Σ forms a simple hypergraph on the FDs in Σ^{*a}. We may thus use the terms defined for hypergraphs for canonical covers as well. In particular, we shall talk about autonomous sets of FDs, and (partial) superedges. Note that in this context the superedges are the atomic covers, while the edges are the canonical covers.

Definition 13. *We call a set of FDs in Σ^{*a}* autonomous *if it is autonomous for the hypergraph $CC(\Sigma)$. When talking about transversals, we always mean transversals of $CC(\Sigma)$.*

Lemma 9. *A set $G \subseteq \Sigma^{*a}$ is a cover of Σ iff it intersects with all minimal transversals of $CC(\Sigma)$.*

Proof. G is a cover iff it is a superedge of $CC(\Sigma)$. Furthermore, $CC(\Sigma)$ is simple, and by Lemma 1 the edges of a simple hypergraph are the minimal sets which intersect with all minimal transversals. Thus superedges are simply sets (not necessarily minimal) which intersect with all minimal transversals.

As superedges become (atomic) covers for the hypergraph $CC(\Sigma)$, partial superedges become partial covers.

Definition 14. *Let Σ be a set of FDs and $G \subseteq S \subseteq \Sigma^{*a}$. We call G a partial cover of Σ on S if G is a partial superedge of $CC(\Sigma)$ on S.*

When S is autonomous, testing whether a set of FDs is a partial cover on S is easy:

Lemma 10. *Let $S \subseteq \Sigma^{*a}$ be autonomous, and let $\Sigma' \subseteq \Sigma^{*a}$ be an atomic cover of Σ. Then a set $G \subseteq S$ is a partial cover on S iff $G \cup (\Sigma' \setminus S)$ is a cover of Σ.*

Proof. Follows directly from Lemma 8.

Clearly $G \cup (\Sigma' \setminus S)$ is a cover of Σ iff $G \cup (\Sigma' \setminus S) \vDash \Sigma' \cap S$, which allows us to perform this test quickly.

We will identify some autonomous (but not necessarily minimal) sets of $CC(\Sigma)$. Theorem 4 relates autonomous sets to the minimal transversals of $CC(\Sigma)$. The following lemmas establish some results about the form of these minimal transversals.

Lemma 11. *Let $S \subseteq \Sigma^{*a}$ be a minimal transversal of $CC(\Sigma)$ and $X \to A \in S$. Then $\overline{S} = \Sigma^{*a} \setminus S$ is not a cover of Σ, but $\overline{S} \cup \{X \to A\}$ is.*

Proof. By Lemma 9, $\overline{S}$ is not a cover of Σ since it does not intersect with S. If $\overline{S} \cup \{X \to A\}$ were not a cover, then every cover would contain a FD in

$$\overline{\overline{S} \cup \{X \to A\}} = S \setminus \{X \to A\}$$

Thus $S \setminus \{X \to A\}$ would be a transversal, which contradicts the minimality of S.

Definition 15. *The sets of attributes X and Y are equivalent under a set of FDs Σ, written $X \leftrightarrow Y$, if $X \to Y$ and $Y \to X$ lie in Σ^*.*

Lemma 12. *Let $X \to A, Y \to B$ be contained in a common minimal transversal $S \subseteq \Sigma^{*a}$ of $CC(\Sigma)$. Then X and Y are equivalent under $\overline{S} = \Sigma^{*a} \setminus S$.*

Proof. By Lemma 11 we have

$$\begin{aligned} \overline{S} &\nvDash Y \to B \\ \overline{S} \cup \{X \to A\} &\vDash Y \to B \end{aligned} \tag{2}$$

Let us denote the closure of Y under $\overline{S}$ by $Y^{*\overline{S}}$. If $X \not\subseteq Y^{*\overline{S}}$ then

$$Y^{*\overline{S}} = Y^{*\overline{S} \cup \{X \to A\}}$$

which contradicts (2). Thus $\overline{S} \vDash Y \to X$, and by symmetry $\overline{S} \vDash X \to Y$.

Definition 16. *Let Σ be a set of FDs on R. We denote the set of FDs in Σ^{*a} with LHS equivalent to $X \subseteq R$ as*

$$EQ_X := \{Y \to Z \in \Sigma^{*a} \mid Y \leftrightarrow X\}$$

The partition of Σ^{*a} into non-empty equivalence sets is denoted as

$$EQ := \{EQ_X \mid \exists Y.X \to Y \in \Sigma^{*a}\}$$

Theorem 6. *Let Σ be a set of FDs on R. Then every set $EQ_X \in EQ$ is autonomous.*

Proof. By Lemma 12 all FDs in a (maximal) connected component of $Tr(CC(\Sigma))$ have equivalent LHSs under Σ. Thus EQ_X is the union of vertex sets of maximal connected components of $Tr(CC(\Sigma))$, and therefore an isolated set of $Tr(CC(\Sigma))$. By Theorem 4 isolated sets of $Tr(CC(\Sigma))$ are autonomous for $CC(\Sigma)$.

From this we can quickly obtain the following:

Theorem 7. *Let Σ be a set of FDs on R. A set $G \subseteq \Sigma^{*a}$ is a cover of Σ iff $G \cap EQ_X$ is a partial cover of Σ on EQ_X for every $EQ_X \in EQ$.*

Proof. By Theorem 6 the sets EQ_X form a partition of Σ^{*a} into autonomous sets, so the theorem is a special case of Lemma 7.

4   An NP-Hardness Result

While we identified the equivalence classes EQ_X as autonomous sets of $CC(\Sigma)$, they need not be minimal. In the following we will show that finding the minimal autonomous sets of $CC(\Sigma)$ is difficult, given Σ.

Definition 17. *We call an atomic FD $X \to A \in \Sigma^{*a}$ essential iff it appears in some canonical cover of Σ. Otherwise, we call it inessential.*

We will first show that testing essentiality is NP-hard. We do so by reducing the following problem to it, which is known to be NP-complete [11]:

Problem "prime attribute"
    Given a set Σ of FDs on schema R and an attribute $A \in R$, is A a prime attribute, i.e., does A lie in a minimal key of R?

Theorem 8. *Given a set Σ of FDs, the problem of deciding whether a FD is essential is NP-complete.*

Proof. To verify that a FD is essential, we only need to guess the canonical cover containing it. By [5] the size of any canonical cover is polynomial in Σ, so the problem lies in NP.

We prove completeness by reducing the "prime attribute" problem to it. Let G be a set of FDs on R, and $A \notin R$ an additional attribute. We construct Σ as

$$\Sigma := G \cup \{A \to A_i \mid A_i \in R\}$$

We claim that A_i is a prime attribute w.r.t. G iff $A \to A_i$ is essential w.r.t. Σ. Since A does not appear in any FD in G we have

$$\Sigma^{*a} = G^{*a} \cup \{A \to A_i \mid A_i \in R\}$$

Now let Σ' be any canonical cover of Σ, and let $\Sigma'_A := \{A \to A_i \in \Sigma'\}$. Clearly the FD $A \to R \in \Sigma^*$ is implied by Σ' iff $A \to K$ is implied by Σ'_A for some minimal key K of R (w.r.t. G). This is the case iff Σ'_A consists of exactly (since Σ' is non-redundant) those FDs $A \to A_i$ for which $A_i \in K$. Thus $A \to A_i$ is essential iff A_i is prime.

From this, we can deduce that identifying the (minimal) autonomous sets of $CC(\Sigma)$ is difficult as well.

Theorem 9. *Given a set Σ of FDs and an atomic FD $X \to A \in \Sigma^{*a}$, the problem of deciding whether the set $\{X \to A\}$ is autonomous for $CC(\Sigma)$ is co-NP-complete.*

Proof. $\{X \to A\}$ is autonomous iff $X \to A$ appears in all or no canonical covers of Σ. Thus, if $\{X \to A\}$ is not autonomous, we only need to guess one canonical cover which contains $X \to A$, and one which does not. By [5] the size of these canonical covers is polynomial in Σ. This shows that the problem lies in co-NP.

To show co-NP-hardness, we use that it is NP-hard to decide whether a FD $X \to A$ is essential. Let Σ' be any canonical cover of Σ. Such a canonical cover Σ' can be computed in polynomial time [13]. By definition, $X \to A$ is essential iff it appears in some, but not necessarily all canonical covers of Σ. We distinguish two cases.

(1) If $X \to A \in \Sigma'$ then $X \to A$ is essential.
(2) If $X \to A \notin \Sigma'$ then $X \to A$ is essential iff it appears in some but not all canonical covers of Σ, i.e., iff $\{X \to A\}$ is not autonomous.

We thus reduced the NP-hard problem of deciding whether $X \to A$ is essential to the problem of deciding whether $\{X \to A\}$ is not autonomous.

5   LR-Reduced Covers and Applications

The general purpose of finding all canonical covers is to find covers which are best in some sense. This approach only works if we can be sure that the best cover (or at least one of the best if there are multiple optimal solutions) for a given problem is canonical. This is often the case, but not always. Another common type of cover are *LR-reduced* covers.

Definition 18. *A set Σ of FDs is called* LR-reduced *if no attribute can be removed from any FD in Σ while maintaining the property of being a cover.*

While we only considered the set $CC(\Sigma)$ of all canonical covers, the set of all LR-reduced covers can be constructed from $CC(\Sigma)$ easily. By "splitting" a FD $X \to Y$ into singular FDs, we mean to replace it by $\{X \to A \mid A \in Y\}$.

Lemma 13. *[12] Splitting FDs into singular FDs turns an LR-reduced cover into a canonical cover. Combining FDs with identical LHSs turns a canonical cover into an LR-reduced cover.*

As an example for the usefulness of LR-reduced covers, one may consider the problem of finding a cover with minimal number of attributes, which is known to be NP-hard [12].

Definition 19. *The* area *of a FD $X \to Y$ is $|X| + |Y|$, i.e., the number of attributes appearing in $X \to Y$. The* area *of a set of FDs Σ is the sum of the areas of all FDs in Σ.*

Definition 20. *A set Σ of FDs is called* area optimal *if there exists no cover G of Σ with smaller area.*

Area optimal covers can help in reducing the workload for checking whether dependencies hold on a relation, as well as speed up various algorithms [12].

   Area optimal covers are rarely canonical, since combining FDs with equal LHS reduced the area. It is easy to see though that they are always LR-reduced.

Lemma 14. *[12] An area optimal set of FDs is LR-reduced.*

Thus, we may combine FDs with identical LHSs to obtain all LR-reduced covers in which no FDs have identical LHSs. Clearly those include all area optimal covers. Our approach helps here, since we may determine area optimal partial covers for each EQ_X separately to get an area optimal covers.

   In [2] Ausiello, D'Atri and Sacca introduce several similar minimality criteria for covers, and show that some of the corresponding decision problems are NP-hard. While these minimal covers are not always canonical or LR-reduced, it is easy to see that some of them always are. Thus, while we may not find all minimal covers (w.r.t. the minimality criteria in [2]), we can always find some.

   If nothing else, representing the set of all canonical covers efficiently in de-composed form can help a designer to compare the choices he or she has when selecting a cover:

Example 4. Let Σ consist of the following FDs:

$$\Sigma = \{AB \to C, C \to A, A \to D, D \to E, E \to A\}$$

Then Σ has the following canonical covers:

$$CC(\Sigma) = \left\{ \begin{array}{l}
\{AB \to C, C \to A, A \to D, D \to E, E \to A\}, \\
\{AB \to C, C \to A, A \to E, E \to D, D \to A\}, \\
\{AB \to C, C \to A, A \to D, D \to A, A \to E, E \to A\}, \\
\{AB \to C, C \to A, A \to D, D \to A, D \to E, E \to D\}, \\
\{AB \to C, C \to A, A \to E, E \to A, D \to E, E \to D\}, \\
\{DB \to C, C \to A, A \to D, D \to E, E \to A\}, \\
\{DB \to C, C \to A, A \to E, E \to D, D \to A\}, \\
\{DB \to C, C \to A, A \to D, D \to A, A \to E, E \to A\}, \\
\{DB \to C, C \to A, A \to D, D \to A, D \to E, E \to D\}, \\
\{DB \to C, C \to A, A \to E, E \to A, D \to E, E \to D\}, \\
\{EB \to C, C \to A, A \to D, D \to E, E \to A\}, \\
\{EB \to C, C \to A, A \to E, E \to D, D \to A\}, \\
\{EB \to C, C \to A, A \to D, D \to A, A \to E, E \to A\}, \\
\{EB \to C, C \to A, A \to D, D \to A, D \to E, E \to D\}, \\
\{EB \to C, C \to A, A \to E, E \to A, D \to E, E \to D\}, \\
\{AB \to C, C \to D, A \to D, D \to E, E \to A\}, \\
\{AB \to C, C \to D, A \to E, E \to D, D \to A\}, \\
\{AB \to C, C \to D, A \to D, D \to A, A \to E, E \to A\}, \\
\{AB \to C, C \to D, A \to D, D \to A, D \to E, E \to D\}, \\
\{AB \to C, C \to D, A \to E, E \to A, D \to E, E \to D\}, \\
\{DB \to C, C \to D, A \to D, D \to E, E \to A\}, \\
\{DB \to C, C \to D, A \to E, E \to D, D \to A\}, \\
\{DB \to C, C \to D, A \to D, D \to A, A \to E, E \to A\}, \\
\{DB \to C, C \to D, A \to D, D \to A, D \to E, E \to D\}, \\
\{DB \to C, C \to D, A \to E, E \to A, D \to E, E \to D\}, \\
\{EB \to C, C \to D, A \to D, D \to E, E \to A\}, \\
\{EB \to C, C \to D, A \to E, E \to D, D \to A\}, \\
\{EB \to C, C \to D, A \to D, D \to A, A \to E, E \to A\}, \\
\{EB \to C, C \to D, A \to D, D \to A, D \to E, E \to D\}, \\
\{EB \to C, C \to D, A \to E, E \to A, D \to E, E \to D\}, \\
\{AB \to C, C \to E, A \to D, D \to E, E \to A\}, \\
\{AB \to C, C \to E, A \to E, E \to D, D \to A\}, \\
\{AB \to C, C \to E, A \to D, D \to A, A \to E, E \to A\}, \\
\{AB \to C, C \to E, A \to D, D \to A, D \to E, E \to D\}, \\
\{AB \to C, C \to E, A \to E, E \to A, D \to E, E \to D\}, \\
\{DB \to C, C \to E, A \to D, D \to E, E \to A\}, \\
\{DB \to C, C \to E, A \to E, E \to D, D \to A\}, \\
\{DB \to C, C \to E, A \to D, D \to A, A \to E, E \to A\}, \\
\{DB \to C, C \to E, A \to D, D \to A, D \to E, E \to D\}, \\
\{DB \to C, C \to E, A \to E, E \to A, D \to E, E \to D\}, \\
\{EB \to C, C \to E, A \to D, D \to E, E \to A\}, \\
\{EB \to C, C \to E, A \to E, E \to D, D \to A\}, \\
\{EB \to C, C \to E, A \to D, D \to A, A \to E, E \to A\}, \\
\{EB \to C, C \to E, A \to D, D \to A, D \to E, E \to D\}, \\
\{EB \to C, C \to E, A \to E, E \to A, D \to E, E \to D\}
\end{array} \right\}$$

Instead of using this bulky "direct" representation, we decompose $CC(\Sigma)$ into sets of partial covers:

$$\left\{ \begin{array}{l} \{AB \to C\}, \\ \{DB \to C\}, \\ \{EB \to C\} \end{array} \right\} \vee \left\{ \begin{array}{l} \{C \to A\}, \\ \{C \to D\}, \\ \{C \to E\} \end{array} \right\} \vee \left\{ \begin{array}{l} \{A \to D, D \to E, E \to A\}, \\ \{A \to E, E \to D, D \to A\}, \\ \{A \to D, D \to A, A \to E, E \to A\}, \\ \{A \to D, D \to A, D \to E, E \to D\}, \\ \{A \to E, E \to A, D \to E, E \to D\} \end{array} \right\}$$

Using this decomposed representation, dealing with $CC(\Sigma)$ becomes much easier.

Finally, autonomous sets can be useful in many other settings. It has been suggested that the approach may be applicable in studying association rules as they emerge in frequent items mining tasks, and we are currently investigating how the theory can be applied to several graph problems, such as that of finding minimal feedback sets.

6   Related Work

The idea of using autonomous sets to simplify problems has been used before, though not in the context of a general frame work, and not always for the hypergraph of canonical covers.

Saiedian and Spencer describe an approach in [16], where they determine autonomous sets for the hypergraph consisting of all minimal keys. The same autonomous sets are used again by Gottlob, Pichler and Wei in [6].

A unique representation for a set of unitary FDs (i.e., FDs with only a single attribute in their left hand side) is given in [9]. This representation is obtained by factoring the attribute determination graph via the equivalence relation on attributes induced by Σ. From this, partial canonical covers for the equivalence classes could be constructed as minimal strongly connected directed graphs. Our work generalizes this to arbitrary functional dependencies.

Maier already noted in [12] that there is a correspondence between the equivalence classes of non-redundant covers. Our work generalizes these results by investigating the projections of (canonical) covers onto arbitrary autonomous sets, and placing them into a more general theoretic framework.

References

1. Armstrong, W.W.: Dependency structures of data base relationships. In: IFIP Congress, pp. 580–583 (1974)
2. Ausiello, G., D'Atri, A., Saccà, D.: Minimal representation of directed hypergraphs. SIAM J. Comput. 15(2), 418–431 (1986)
3. Berge, C.: Hypergraphs: Combinatorics of Finite Sets. Elsevier Science Pub. Co., Amsterdam (1989)

4. Biskup, J., Dayal, U., Bernstein, P.A.: Synthesizing independent database schemas. In: SIGMOD Conference, pp. 143–151 (1979)
5. Gottlob, G.: On the size of nonredundant FD-covers. Inf. Process. Lett. 24(6), 355–360 (1987)
6. Gottlob, G., Pichler, R., Wei, F.: Tractable database design through bounded treewidth. In: PODS, pp. 124–133 (2006)
7. Habib, M., de Montgolfier, F., Paul, C.: A simple linear-time modular decomposition algorithm for graphs, using order extension. In: Hagerup, T., Katajainen, J. (eds.) SWAT 2004. LNCS, vol. 3111, pp. 187–198. Springer, Heidelberg (2004)
8. Koehler, H.: Finding faithful Boyce-Codd normal form decompositions. In: Cheng, S.-W., Poon, C.K. (eds.) AAIM 2006. LNCS, vol. 4041, pp. 102–113. Springer, Heidelberg (2006)
9. Lechtenbörger, J.: Computing unique canonical covers for simple FDs via transitive reduction. Inf. Process. Lett. 92(4), 169–174 (2004)
10. Levene, M., Loizou, G.: A Guided Tour of Relational Databases and Beyond. Springer, Heidelberg (1999)
11. Lucchesi, C.L., Osborn, S.L.: Candidate keys for relations. Journal of Computer and System Sciences 17(2), 270–279 (1978)
12. Maier, D.: Minimum covers in the relational database model. Journal of the ACM 27(4), 664–674 (1980)
13. Maier, D.: The Theory of Relational Databases. Computer Science Press (1983)
14. Mannila, H., Räihä, K.-J.: The Design of Relational Databases. Addison-Wesley, Reading (1987)
15. Osborn, S.L.: Testing for existence of a covering Boyce-Codd normal form. Information Processing Letters 8(1), 11–14 (1979)
16. Saiedian, H., Spencer, T.: An efficient algorithm to compute the candidate keys of a relational database schema. Comput. J. 39(2), 124–132 (1996)
17. Zaniolo, C.: A new normal form for the design of relational database schemata. ACM Trans. Database Syst. 7(3), 489–499 (1982)

Cost-Minimising Strategies for Data Labelling: Optimal Stopping and Active Learning

Christos Dimitrakakis and Christian Savu-Krohn

Chair of Information Technology, University of Leoben
Leoben A-8700, Austria
{christos.dimitrakakis,christian.savu-krohn}@mu-leoben.at

Abstract. Supervised learning deals with the inference of a distribution over an output or label space $\mathcal{Y}$ conditioned on points in an observation space $\mathcal{X}$, given a training dataset D of pairs in $\mathcal{X} \times \mathcal{Y}$. However, in a lot of applications of interest, acquisition of large amounts of observations is easy, while the process of generating labels is time-consuming or costly. One way to deal with this problem is *active* learning, where points to be labelled are selected with the aim of creating a model with better performance than that of an model trained on an equal number of randomly sampled points. In this paper, we instead propose to deal with the labelling cost directly: The learning goal is defined as the minimisation of a cost which is a function of the expected model performance and the total cost of the labels used. This allows the development of general strategies and specific algorithms for (a) optimal stopping, where the expected cost dictates whether label acquisition should continue (b) empirical evaluation, where the cost is used as a performance metric for a given combination of inference, stopping and sampling methods. Though the main focus of the paper is optimal stopping, we also aim to provide the background for further developments and discussion in the related field of active learning.

1 Introduction

Much of classical machine learning deals with the case where we wish to learn a target concept in the form of a function $f : \mathcal{X} \to \mathcal{Y}$, when all we have is a finite set of examples $D = \{(x_i, y_i)\}_{i=1}^n$. However, in many practical settings, it turns out that for each example i in the set only the observations x_i are available, while the availability of observations y_i is restricted in the sense that either (a) they are only observable for a subset of the examples (b) further observations may only be acquired at a cost. In this paper we deal with the second case, where we can actually obtain labels for any $i \in D$, but doing so incurs a cost. Active learning algorithms (i.e. [1, 2]) deal indirectly with this by selecting examples which are expected to increase accuracy the most. However, the basic question of whether new examples should be queried at all is seldom addressed.

This paper deals with the labelling cost explicitly. We introduce a cost function that represents the trade-off between final performance (in terms of generalisation error) and querying costs (in terms of the number of labels queried). This

S. Hartmann and G. Kern-Isberner (Eds.): FoIKS 2008, LNCS 4932, pp. 96–111, 2008.

is used in two ways. Firstly, as the basis for creating cost-dependent stopping rules. Secondly, as the basis of a comparison metric for learning algorithms and associated stopping algorithms.

To expound further, we decide when to stop by estimating the expected performance gain from querying additional examples and comparing it with the cost of acquiring more labels. One of the main contributions is the development of methods for achieving this in a Bayesian framework. While due to the nature of the problem there is potential for misspecification, we nevertheless show experimentally that the stopping times we obtain are close to the optimal stopping times.

We also use the trade-off in order to address the lack of a principled method for comparing different active learning algorithms under conditions similar to real-world usage. For such a comparison a method for choosing stopping times independently of the test set is needed. Combining stopping rules with active learning algorithms allows us to objectively compare active learning algorithms for a range of different labelling costs.

The paper is organised as follows. Section 1.1 introduces the proposed cost function for when labels are costly, while Section 1.2 discusses related work. Section 2 derives a Bayesian stopping method that utilises the proposed cost function. Some experimental results illustrating the proposed evaluation methodology and demonstrating the use of the introduced stopping method are presented in Section 3. The proposed methods are not flawless, however. For example, the algorithm-independent stopping rule requires the use of i.i.d. examples, which may interfere with its coupling to an active learning algorithm. We conclude with a discussion on the applicability, merits and deficiencies of the proposed approach to optimal stopping and of principled testing for active learning.

1.1   Combining Classification Error and Labelling Cost

There are many applications where raw data is plentiful, but labelling is time consuming or expensive. Classic examples are speech and image recognition, where it is easy to acquire hours of recordings, but for which transcription and labelling are laborious and costly. For this reason, we are interested in querying labels from a given dataset such that we find the optimal balance between the cost of labelling and the classification error of the hypothesis inferred from the labelled examples. This arises naturally from the following cost function.

Let some algorithm F which queries labels for data from some unlabelled dataset D, incurring a cost $\gamma \in [0, \infty)$ for each query. If the algorithm stops after querying labels of examples $d_1, d_2, \ldots, d_t$, with $d_i \in [1, |D|]$.it will suffer a total cost of γt, plus a cost depending on the generalisation error. Let $f(t)$ be the hypothesis obtained after having observed t examples and corresponding to the generalisation error $\mathbf{E}[R|f(t)]$ be the generalisation error of the hypothesis. Then, we define the total cost for this specific hypothesis as

$$\mathbf{E}[C_\gamma|f(t)] = \mathbf{E}[R|f(t)] + \gamma t. \tag{1}$$

We may use this cost as a way to compare learning and stopping algorithms, by calculating the expectation of C_γ conditioned on different algorithm combinations, rather than on a specific hypothesis.

In addition, this cost function can serve as a formal framework for active learning. Given a particular dataset D, the optimal subset of examples to be used for training will be $D^* = \arg\min_i \mathbf{E}(R|F, D_i) + \gamma|D_i|$. The ideal, but unrealisable, active learner in this framework would just use labels of the subset D^* for training.

Thus, these notions of optimality can in principle be used both for deriving stopping and sampling algorithms and for comparing them. Suitable metrics of expected real-world performance will be discussed in the next section. Stopping methods will be described in Section 2.

1.2   Related Work

In the active learning literature, the notion of an objective function for trading off classification error and labelling cost has not yet been adopted. However, a number of both qualitative and quantitative metrics were proposed in order to compare active learning algorithms. Some of the latter are defined as summary statistics over some subset $\mathcal{T}$ of the possible stopping times. This is problematic as it could easily be the case that there exists $\mathcal{T}_1, \mathcal{T}_2$ with $\mathcal{T}_1 \subset \mathcal{T}_2$, such that when comparing algorithms over $\mathcal{T}_1$ we get a different result than when we are comparing them over a larger set $\mathcal{T}_2$. Thus, such measures are not easy to interpret since the choice of $\mathcal{T}$ remains essentially arbitrary. Two examples are (a) the *percentage reduction in error*, where the percentage reduction in error of one algorithm over another is averaged over the whole learning curve [3, 4] and (b) the average number of times one algorithm is significantly better than the other during an arbitrary initial number of queries, which was used in [5]. Another metric is the *data utilisation ratio* used in [5, 4, 6], which is the amount of data required to reach some specific error rate. Note that the selection of the appropriate error rate is essentially arbitrary; in both cases the concept of the *target error rate* is utilised, which is the average test error when almost all the training set has been used.

Our setting is more straightforward, since we can use (1) as the basis for a performance measure. Note that we are not strictly interested in comparing hypotheses f, but algorithms F. In particular, we can calculate the expected cost given a learning algorithm F and an associated stopping algorithm $Q_F(\gamma)$, which is used to select the *stopping time* T. From this follows that the expected cost of F when coupled with $Q_F(\gamma)$ is

$$v_e(\gamma, F, Q_F) \equiv \mathbf{E}[C_\gamma|F, Q_F(\gamma)] = \sum_t \left(\mathbf{E}[R|f(t)] + \gamma t\right) \mathbf{P}[T = t \mid F, Q_F(\gamma)] \quad (2)$$

By keeping one of the algorithms fixed, we can vary the other in order to obtain objective estimates of their performance difference. In addition, we may want to calculate the expected performance of algorithms for a range of values of γ, rather than a single value, in a manner similar to what [7] proposed as an alternative to ROC curves. This will require a stopping method $Q_F(\gamma)$ which will ideally stop querying at a point that minimises $\mathbf{E}(C_\gamma)$.

The stopping problem is not usually mentioned in the active learning literature and there are only a few cases where it is explicitly considered. One such case is [2], where it is suggested to stop querying when no example lies within the SVM margin. The method is used indirectly in [8], where if this event occurs the algorithm tests the current hypothesis[1], queries labels for a new set of unlabelled examples[2] and finally stops if the error measured there is below a given threshold; similarly, [9] introduced a bounds-based stopping criterion that relies on an allowed error rate. These are reasonable methods, but there exists no formal way of incorporating the cost function considered here within them. For our purpose we need to calculate the expected reduction in classification error when querying new examples and compare it with the labelling cost. This fits nicely within the statistical framework of optimal stopping problems.

2   Stopping Algorithms

An optimal stopping problem under uncertainty is generally formulated as follows. At each point in time t, the experimenter needs to make a decision $a \in A$, for which there is a *loss function* $\mathcal{L}(a|w)$ defined for all $w \in \Omega$, where Ω is the set of all possible universes. The experimenter's uncertainty about which $w \in \Omega$ is true is expressed via the distribution $\mathbf{P}(w|\xi_t)$, where ξ_t represents his belief at time t. The *Bayes risk* of taking an action at time t can then be written as $\rho_0(\xi_t) = \min_a \sum_w \mathcal{L}(a, w) \mathbf{P}(w|\xi_t)$. Now, consider that instead of making an immediate decision, he has the opportunity to take k more observations D_k from a sample space S^k, at a cost of γ per observation, thus allowing him to update his belief to $\mathbf{P}(w|\xi_{t+k}) \equiv \mathbf{P}(w|D_k, \xi_t)$. What the experimenter must do in order to choose between immediately making a decision a and continuing sampling, is to compare the risk of making a decision now with the cost of making k observations plus the risk of making a decision after k timesteps, when the extra data would enable a more informed choice. In other words, one should stop and make an immediate decision if the following holds for all k:

$$\rho_0(\xi_t) \leq \gamma k + \int_{S^k} p(D_k = s|\xi_t) \min_a \left[\sum_w \mathcal{L}(a, w) \mathbf{P}(w|D_k = s, \xi_t) \right] ds. \quad (3)$$

We can use the same formalism in our setting. In one respect, the problem is simpler, as the only decision to be made is when to stop and then we just use the currently obtained hypothesis. The difficulty lies in estimating the expected error. Unfortunately, the metrics used in active learning methods for selecting

---

[1] i.e. a classifier for a classification task.

[2] Though this is not really an i.i.d. sample from the original distribution except when $|D| - t$ is large.

new examples (see [5] for a review) do not generally include calculations of the expected performance gain due to querying additional examples.

There are two possibilities for estimating this performance gain. The first is an algorithm-independent method, described in detail in Sec. 2.1, which uses a set of convergence curves, arising from theoretical convergence properties. We employ a Bayesian framework to infer the probability of each convergence curve through observations of the error on the next randomly chosen example to be labelled. The second method, outlined in Sec. 4, relies upon a classifier with a probabilistic expression of its uncertainty about the class of unlabelled examples, but is much more computationally expensive.

2.1   When No Model Is Perfect: Bayesian Model Selection

The presented Bayesian formalism for optimal sequential decisions follows [10]. We require maintaining a belief ξ_t in the form of a probability distribution over the set of possible universes $w \in \Omega$. Furthermore, we require the existence of a well-defined cost for each w. Then we can write the Bayes risk as in the left side of (3), but ignoring the minimisation over A as there is only one possible decision to be made after stopping,

$$\rho_0(\xi_t) = \mathbf{E}(R_t \mid \xi_t) = \sum_{w \in \Omega} \mathbf{E}(R_t \mid w)\, \mathbf{P}(w \mid \xi_t), \tag{4}$$

which can be extended to continuous measures without difficulty. We will write the expected risk according to our belief at time t for the optimal procedure taking at most k more samples as

$$\rho_{k+1}(\xi_t) = \min\left\{\rho_0(\xi_t), \mathbf{E}[\rho_k(\xi_{t+1}) \mid \xi_t] + \gamma\right\}. \tag{5}$$

This implies that at any point in time t, we should ignore the cost for the t samples we have paid for and are only interested in whether we should take additional samples. The general form of the stopping algorithm is defined in Alg. 1. Note that the horizon K is a necessary restriction for computability. A larger value of K leads to potentially better decisions, as when $K \to \infty$, the bounded horizon optimal decision approaches that of the optimal decision in the unbounded horizon setting, as shown for example in Chapter 12 of [10]. Even with finite $K > 1$, however, the computational complexity is considerable, since we will have to additionally keep track of how our future beliefs $\mathbf{P}(w \mid \xi_{t+k})$ will evolve for all $k \leq K$.

2.2   The OBSV Algorithm

In this paper we consider a specific one-step bounded stopping algorithm that uses independent validation examples for observing the empirical error estimate r_t, which we dub OBSV and is shown in detail in Alg. 2. The algorithm considers hypotheses $w \in \Omega$ which model how the generalisation error r_t of the learning algorithm changes with time. We assume that the initial error is r_0 and that the

---

Algorithm 1. A general bounded stopping algorithm using Bayesian inference.

Given a dataset D and any learning algorithm F, an initial belief $\mathbf{P}(w \mid \xi_0)$ and a method for updating it, and additionally a known query cost γ, and a horizon K,

1: **for** $t = 1, 2, \ldots$ **do**
2:     Use F to query a new example $i \in D$ and obtain $f(t)$.
3:     Observe the empirical error estimate v_t for $f(t)$.
4:     Calculate $\mathbf{P}(w \mid \xi_t) = \mathbf{P}(w \mid v_t, \xi_{t-1})$
5:     **if** $\nexists\, k \in [1, K] : \rho_k(\xi_t) < \rho_0(\xi_t)$ **then**
6:         Exit.
7:     **end if**
8: **end for**

---

algorithm always converges to some unknown $r_\infty \equiv \lim_{t \to \infty} r_t$. Furthermore, we need some observations v_t that will allow us to update our beliefs over Ω. The remainder of this section discusses the algorithm in more detail.

Steps 1-5, 11-12. Initialisation and Observations. We begin by splitting the training set D in two parts: D_A, which will be sampled without replacement by the *active learning algorithm* (if there is one) and D_R, which will be *uniformly* sampled without replacement. This condition is necessary in order to obtain i.i.d. samples for the inference procedure outlined in the next section. However, if we only sample randomly, and we are not using an active learning algorithm then we do not need to split the data and we can set $D_A = \emptyset$.

At each timestep t, we will use a sample from D_R to update $p(w)$. If we then expect to reduce our future error sufficiently, we will query an example from D_A using F and subsequently update the classifier f with both examples. Thus, not only are the observations used for inference independent and identically distributed, but we are also able to use them to update the classifier f.

Step 6. Updating the Belief. We model the learning algorithm as a process which asymptotically converges from an initial error r_0 to the unknown final error r_∞. Each model w will be a *convergence estimate*, a model of how the error converges from the initial to the final error rate. More precisely, each w corresponds to a function $h_w : \mathbb{N} \to [0, 1]$ that models how close we are to convergence at time t. The predicted error at time t according to w, given the initial error r_0 and the final error r_∞, will be

$$g_w(t \mid r_0, r_\infty) = r_0 h_w(t) + r_\infty[1 - h_w(t)]. \tag{6}$$

We find it reasonable to assume that $p(w, r_0, r_\infty) = p(w)p(r_0)p(r_\infty)$, i.e. that the convergence rates do not depend upon the initial and final errors.

We may now use these predictions together with some observations to update $p(w, r_\infty \mid \xi)$. More specifically, if $\mathbf{P}[r_t = g_w(t \mid r_0, r_\infty) \mid r_0, r_\infty, w] = 1$ and we take m_t independent observations $\mathbf{z}_t = (z_t(1), z_t(2), \ldots, z_t(m_t))$ of the error with mean v_t, the likelihood will be given by the Bernoulli density

$$p(\mathbf{z}_t \mid w, r_0, r_\infty) = \left(g_w(t \mid r_0, r_\infty)^{v_t} [1 - g_w(t \mid r_0, r_\infty)]^{1-v_t}\right)^{m_t}. \qquad (7)$$

Then it is simple to obtain a posterior density for both w and r_∞,

$$p(w \mid \mathbf{z}_t) = \frac{p(w)}{p(\mathbf{z}_t)} \int_0^1 p(\mathbf{z}_t \mid w, r_0, r_\infty = u)\, p(r_\infty = u \mid w)\, du \qquad (8a)$$

$$p(r_\infty \mid \mathbf{z}_t) = \frac{p(r_\infty)}{p(\mathbf{z}_t)} \int_\Omega p(\mathbf{z}_t \mid w, r_0, r_\infty)\, p(w \mid r_\infty)\, dw. \qquad (8b)$$

Starting with a prior distribution $p(w \mid \xi_0)$ and $p(r_\infty \mid \xi_0)$, we may sequentially update our belief using (8) as follows:

$$p(w \mid \xi_{t+1}) \equiv p(w \mid \mathbf{z}_t, \xi_t) \qquad (9a)$$

$$p(r_\infty \mid \xi_{t+1}) \equiv p(r_\infty \mid \mathbf{z}_t, \xi_t). \qquad (9b)$$

The realised convergence for a particular training data set may differ substantially from the expected convergence: the average convergence curve will be smooth, while any specific instantiation of it will not be. More formally, the *realised error* given a specific training dataset is $q_t \equiv \mathbf{E}[R_t \mid D^t]$, where $D^t \sim \mathcal{D}^t$, while the *expected error* given the data distribution is $r_t \equiv \mathbf{E}[R_t] = \int_{S^t} \mathbf{E}[R_t \mid D^t]\, \mathbf{P}(D_t)\, dD_t$. The smooth convergence curves that we model would then correspond to models for r_t.

Fortunately, in our case we can estimate a distribution over r_t without having to also estimate a distribution for q_t, as this is integrated out for observations $z \in \{0, 1\}$

$$p(z \mid q_t) = q_t^z\, (1 - q_t)^{1-z} \qquad (10a)$$

$$p(z \mid r_t) = \int_0^1 p(z \mid q_t) p(q_t = u \mid r_t)\, du = r_t^z\, (1 - r_t)^{1-z}. \qquad (10b)$$

Step 5. Deciding whether to Stop. We may now use the distribution over the models to predict the error should we choose to add k more examples. This is simply

$$\mathbf{E}[R_{t+k} \mid \xi_t] = \int_0^1 \int_\Omega g_w(t + k \mid r_0, r_\infty) p(w \mid \xi_t) p(r_\infty \mid \xi_t)\, dw\, dr_\infty.$$

The calculation required for step 8 of OBSV follows trivially.

Specifics of the Model. What remains unspecified is the set of convergence curves that will be employed. We shall make use of curves related to common theoretical convergence results. It is worthwhile to keep in mind that we simply aim to find the combination of the available estimates that gives the best predictions. While none of the estimates might be particularly accurate, we expect to obtain reasonable stopping times when they are optimally combined in the manner described in the previous section. Ultimately, we expect to end up with a fairly narrow distribution over the possible convergence curves.

Algorithm 2. OBSV, a specific instantiation of the bounded stopping algorithm.

Given a dataset D with examples in N_c classes and any learning algorithm F, initial beliefs $\mathbf{P}(w \mid \xi_0)$ and $\mathbf{P}(r_\infty \mid \xi_0)$ and a method for updating them, and additionally a known query cost γ for discovering the class label $y_i \in [1, \ldots, n]$ of example $i \in D$,

1: Split D into D_A, D_R.
2: $r_0 = 1 - 1/N_c$.
3: Initialise the classifier f.
4: **for** $t = 1, 2, \ldots$ **do**
5:     Sample $i \in D_R$ without replacement and observe $f(x_i), y_i$ to calculate v_t.
6:     Calculate $\mathbf{P}(w, r_\infty \mid \xi_t) \equiv \mathbf{P}(w, r_\infty \mid v_t, \xi_{t-1})$.
7:     If $D_A \neq \emptyset$, set $k = 2$, otherwise $k = 1$.
8:     **if** $\mathbf{E}[R_{t+k} \mid \xi_t] + k\gamma < \mathbf{E}[R_t \mid \xi_t]$ **then**
9:         Exit.
10:   **end if**
11:   If $D_A \neq \emptyset$, use F to query a new example $j \in D_A$ without replacement, $D_T \leftarrow D_T \cup j$.
12:     $D_T \leftarrow D_T \cup i$, $f \leftarrow F(D_T)$.
13: **end for**

One of the weakest convergence results [11] is for sample complexity of order $\mathcal{O}(1/\epsilon_t^2)$, which corresponds to the convergence curve

$$h_q(t) = \sqrt{\frac{\kappa}{t + \kappa}}, \quad \kappa \geq 1 \tag{11}$$

Another common type is for sample complexity of order $\mathcal{O}(1/\epsilon_t)$, which corresponds to the curve

$$h_g(t) = \frac{\lambda}{t + \lambda}, \quad \lambda \geq 1 \tag{12}$$

A final possibility is that the error decreases exponentially fast. This is theoretically possible in some cases, as was proven in [9]. The resulting sample complexity of order $\mathcal{O}(\log(1/\epsilon_t))$ corresponds to the convergence curve

$$h_{exp}(t) = \beta^t, \quad \beta \in (0, 1). \tag{13}$$

Since we do not know what appropriate values of the constants β, λ and κ, are, we will model this uncertainty as an additional distribution over them, i.e. $p(\beta \mid \xi_t)$. This would be updated together with the rest of our belief distribution and could be done in some cases analytically. In this paper however we consider approximating the continuous densities by a sufficiently large set of models, one for each possible value of the unknown constants.

As a simple illustration, we examined the performance of the estimation and the stopping criterion in a simple classification problem with data of 10 classes, each with an equivariant Gaussian distribution in an 8-dimensional space. Each unknown point was simply classified as having the label closest to the empirical mean of the observations for each class. Examples were always chosen randomly.

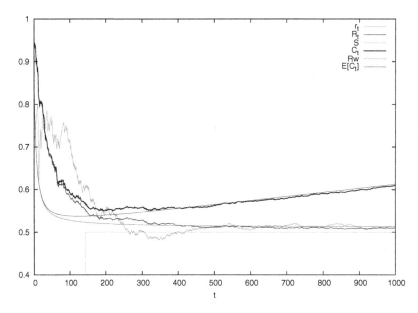

Fig. 1. Illustration of the estimated error on a 10-class problem with a cost per label of $\gamma = 0.001$. On the vertical axis, r_t is the **history** of the predicted generalisation error, i.e $\mathbf{E}[r_t \mid \xi_{t-1}]$, while R_t is the **generalisation error** measured on a test-set of size 10,000 and C_t is the corresponding actual **cost**. Finally, R_w and $E[C_t]$ are the final **estimated** convergence and cost *curves* given all the observations. The stopping time is indicated by S, which equals 0.5 whenever Alg. 2 decides to stop and t is the number of iterations.

As can be seen in Fig. 1, at the initial stages the estimates are inaccurate. This is because of two reasons: (a) The distribution over convergence rates is initially dominated by the prior. As more data is accumulated, there is better evidence for what the final error will be. (b) As we mentioned in the discussion of step 6, the realised convergence curve is much more random than the expected convergence curve which is actually modelled. However, as the number of examples approaches infinity, the expected and realised errors converge. The stopping time for Alg. 2 (indicated by S) is nevertheless relatively close to the optimal stopping time, as C_t appears to be minimised near 200. The following section presents a more extensive evaluation of this stopping algorithm.

3   Experimental Evaluation

The main purpose of this section is to evaluate the performance of the OBSV stopping algorithm. This is done by examining its cost and stopping time when compared to the optimal stopping time. Another aim of the experimental evaluation was to see whether mixed sampling strategies have an advantage compared to random sampling strategies with respect to the cost, when the stopping time is decided using a stopping algorithm that takes into account the labelling cost.

Following [7], we plot performance curves for a range of values of γ, utilising multiple runs of cross-validation in order to assess the sensitivity of the results to the data. For each run, we split the data into a training set D and test set D_E, the training set itself being split into random and mixed sampling sets whenever appropriate.

More specifically, we compare the OBSV algorithm with the **oracle** stopping time. The latter is defined simply as the stopping time minimising the cost as this is measured on the independent test set for that particular run. We also compare **random** sampling with **mixed** sampling. In random sampling, we simply query unlabelled examples without replacement. For the mixed sampling procedure, we actively query an additional label for the example from D_A closest to the decision boundary of the current classifier, also without replacement. This strategy relies on the assumption that those labels are most informative [6], [4], [5] and thus convergence will be faster. Stopping times and cost ratio curves are shown for a set of γ values, for costs as defined in (2). These values of γ are also used as input to the stopping algorithm. The ratios are used both to compare stopping algorithms (OBSV versus the oracle) and sampling strategies (random sampling, where $D_A = \emptyset$, and mixed sampling, with $|D_A| = |D_R|$). Average test error curves are also plotted for reference.

For the experiments we used two data sets from the UCI repository[3]: the Wisconsin breast cancer data set (`wdbc`) with 569 examples and the spambase database (`spam`) with 4601 examples. We evaluated `wdbc` and `spam` using 5 and 3 randomised runs of 3-fold stratified cross-validation respectively. The classifier used was AdaBoost [12] with 100 decision stumps as base hypotheses. Hence we obtain a total of 15 runs for `wdbc` and 9 for `spam`. We ran experiments for values of $\gamma \in \{9 \cdot 10^{-k}, 8 \cdot 10^{-k}, \ldots, 1 \cdot 10^{-k}\}$, with $k = 1, \ldots, 7$, and $\gamma = 0$. For every algorithm and each value of γ we obtain a different stopping time t_γ for each run. We then calculate $v_e(\gamma, F, t_\gamma)$ as given in (2) on the corresponding test set of the run. By examining the averages and extreme values over all runs we are able to estimate the sensitivity of the results to the data.

The results comparing the oracle with OBSV for the **random** sampling strategy[4] are shown in Fig. 2. In Fig. 2(a), 2(b) it can be seen that the stopping times of OBSV and the oracle increase at a similar rate. However, although OBSV is reasonably close, on average it regularly stops earlier. This may be due to a number of reasons. For example, due to the prior, OBSV stops immediately when $\gamma > 3 \cdot 10^{-2}$. At the other extreme, when $\gamma \to 0$ the cost becomes the test error and therefore the oracle always stops at latest at the minimum test error[5]. This is due to the stochastic nature of the realised error curve, which cannot be modelled; there, the perfect information that the oracle enjoys accounts for most of the performance difference. As shown in Fig. 2(c), 2(d), the extra cost induced by using OBSV instead of the oracle is bounded from above for most of the runs by factors of 2 to 5 for `wdbc` and around 0.5 for `spam`. The rather higher

---

[3] http://mlearn.ics.uci.edu/MLRepository.html
[4] The corresponding average test errors can be seen in Fig. 4(a), 4(b).
[5] This is obtained after about 260 labels on `wdbc` and 2400 labels on `spam`.

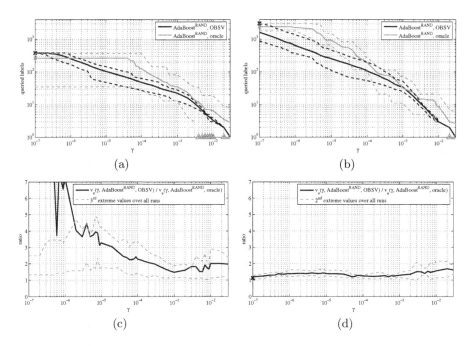

Fig. 2. Results for **random sampling** on the wdbc (left column) and the spam data (right column) as obtained from the 15 (wdbc) and 9 (spam) runs of AdaBoost with 100 decision stumps. The first row (a), (b), plots the average stopping times from OBSV and the oracle as a function of the labelling cost γ. For each γ the extreme values from all runs are denoted by the dashed lines. The second row, (c), (d), shows the corresponding average ratio in v_e over all runs between OBSV and the oracle, where for each γ the 3rd (wdbc) / 2nd (spam) extreme values from all runs are denoted by the dashed lines. Note a zero value on a logarithmic scale is denoted by a cross or by a triangle. Note for wdbc and smaller values of γ the average ratio in v_e sometimes exceeds the denoted extreme values due to a zero test error occurred in one run.

difference on wdbc is partially a result of the small dataset. Since we can only measure an error in quanta of $1/|D_E|$, any actual performance gain lower than this will be unobservable. This explains why the number of examples queried by the oracle becomes constant for a value of γ smaller than this threshold. Finally, this fact also partially explains the greater variation of the oracle's stopping time in the smaller dataset. We expect that with larger test sets, the oracle's behaviour would be smoother.

The corresponding comparison for the **mixed** sampling strategies is shown in Fig. 4(a), 4(b). We again observe the stopping times to increase at a similar rate, and OBSV to stop earlier on average than the oracle for most values of γ (Fig. 3(a), 3(b)). Note that the oracle selects the minimum test error at around 180 labels from wdbc and 1300 labels from spam, which for both data sets is only about a half of the number of labels the random strategy needs. OBSV tracks these stopping times closely. Over all, the fact that in both mixed and random

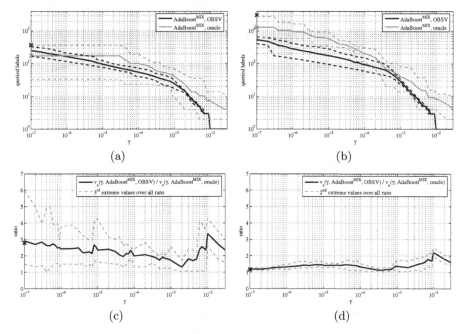

Fig. 3. Results for **mixed sampling** on the wdbc (left column) and the spam data (right column) as obtained from the 15 (wdbc) and 9 (spam) runs of AdaBoost with 100 decision stumps. The first row (a), (b), plots the average stopping times from OBSV and the oracle as a function of the labelling cost γ. For each γ the extreme values from all runs are denoted by the dashed lines. The second row, (c), (d), shows the corresponding average ratio in v_e over all runs between OBSV and the oracle, where for each γ the 3rd (wdbc) / 2nd (spam) extreme values from all runs are denoted by the dashed lines. Note a zero value on a logarithmic scale is denoted by a cross.

sampling, the stopping times of OBSV and the oracle are usually well within the extreme value ranges, indicates a satisfactory performance.

Finally we compare the two sampling strategies directly as shown in Fig. 4, using the practical OBSV algorithm. As one might expect from the fact that the mixed strategy converges faster to a low error level, OBSV stops earlier or around the same time using the mixed strategy than it does for the random (Fig. 4(c), 4(d)). Those two facts together indicate that OBSV works as intended, since it stops earlier when convergence is faster. The results also show that when using OBSV as a stopping criterion mixed sampling is equal to or better than random sampling [Fig. 4(e), 4(f)]. However the differences are mostly not very significant.

4   Discussion

This paper discussed the interplay between a well-defined cost function, stopping algorithms and objective evaluation criteria and their relation to active learning.

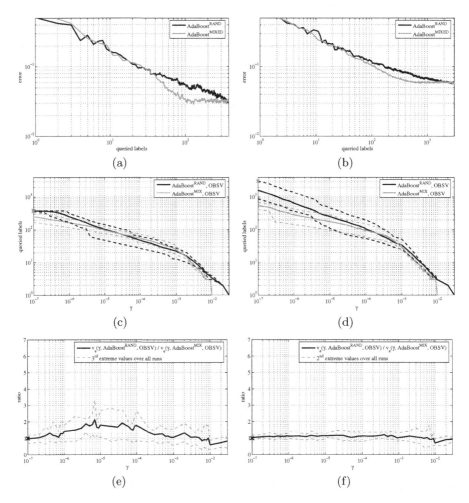

Fig. 4. Results comparing random (**RAND**) and mixed (**MIX**) sampling on the wdbc (left column) and the spam data (right column) as obtained from the 15 (wdbc) and 9 (spam) runs of AdaBoost with 100 decision stumps. The first row (a), (b), shows the test error of each sampling strategy averaged over all runs. The second row (a), (b), plots the average stopping times from OBSV and the oracle as a function of the labelling cost γ. For each γ the extreme values from all runs are denoted by the dashed lines. The third row, (c), (d), shows the corresponding average ratio in v_e over all runs between OBSV and the oracle, where for each γ the 3rd (wdbc) / 2nd (spam) extreme values from all runs are denoted by the dashed lines. Note a zero value on a logarithmic scale is denoted by a cross.

Specifically, we have argued that (a) learning when labels are costly is essentially a stopping problem (b) it is possible to use optimal stopping procedures based on a suitable cost function (c) the goal of active learning algorithms could also be represented by this cost function, (d) metrics on this cost function should be used to evaluate performance and finally that, (e) the stopping problem cannot be separately considered from either the cost function or the evaluation. To our current knowledge, these issues have not yet been sufficiently addressed.

For this reason, we have proposed a suitable cost function and presented a practical stopping algorithm which aims to be optimal with respect to this cost. Experiments with this algorithm for a specific prior show that it suffers only small loss compared to the optimal stopping time and is certainly a step forward from ad-hoc stopping rules.

On the other hand, while the presented stopping algorithm is an adequate first step, its combination with active learning is not perfectly straightforward since the balance between active and uniform sampling is a hyperparameter which is not obvious how to set.[6] An alternative is to use model-specific stopping methods. This could be done if we restrict ourselves to probabilistic classifiers, as for example in [1]; in this way we may be able to simultaneously perform optimal example selection and stopping. If such a classifier is not available for the problem at hand, then judicious use of frequentist techniques such as bootstrapping [13] may provide a sufficiently good alternative for estimating probabilities. Such an approach was advocated by [14] in order to optimally select examples; however in our case we could extend this to optimal stopping. Briefly, this can be done as follows. Let our belief at time t be ξ_t, such that for any point $x \in \mathcal{X}$, we have a distribution over $\mathcal{Y}$, $\mathbf{P}(y \mid x, \xi_t)$. We may now calculate this over the whole dataset to estimate the realised generalisation error as the *expected error given the empirical data distribution and our classifier*

$$\mathbf{E}_D(v_t \mid \xi_t) = \frac{1}{|D|} \sum_{i \in D} [1 - \arg\max_y \mathbf{P}(y_i = y \mid x_i, \xi_t)]. \tag{14}$$

We can now calculate (14) for each one of the different possible labels. So we calculate the *expected error on the empirical data distribution if we create a new classifier from ξ_t by adding example i* as

$$\mathbf{E}_D(v_t \mid x_i, \xi_t) = \sum_{y \in \mathcal{Y}} \mathbf{P}(y_i = y \mid x_i, \xi_t) \, \mathbf{E}_D(v_t \mid x_i, y_i = y, \xi_t) \tag{15}$$

Note that $\mathbf{P}(y_i = y \mid x_i, \xi_t)$ is just the probability of example i having label y according to our current belief, ξ_t. Furthermore, $\mathbf{E}_D(v_t \mid x_i, y_i = y, \xi_t)$ results from calculating (14) using the classifier resulting from ξ_t and the added example i with label y. Then $\mathbf{E}_D(v_t, \xi_t) - \mathbf{E}_D(v_t \mid x_i, \xi_t)$ will be the expected gain from using i to train. The (subjectively) optimal 1-step stopping algorithm is as follows: Let $i^* = \arg\min_i \mathbf{E}_D(v_t \mid x_i, \xi_t)$. Stop if $\mathbf{E}_D(v_t \mid \xi_t) - \mathbf{E}_D(v_t \mid x_{i^*}, \xi_t) < \gamma$.

A particular difficulty in the presented framework, and to some extent also in the field of active learning, is the choice of hyperparameters for the classifiers themselves. For Bayesian models it is possible to select those that maximise the marginal likelihood.[7] One could alternatively maintain a set of models with

---

[6] In this paper, the active and uniform sampling rates were equal.

[7] Other approaches require the use of techniques such as cross-validation, which creates further complications.

different hyperparameter choices and separate convergence estimates. In that case, training would stop when there were no models for which the expected gain was larger than the cost of acquiring another label. Even this strategy, however, is problematic in the active learning framework, where each model may choose to query a different example's label. Thus, the question of hyperparameter selection remains open and we hope to address it in future work.

On another note, we hope that the presented exposition will at the very least increase awareness of optimal stopping and evaluation issues in the active learning community, lead to commonly agreed standards for the evaluation of active learning algorithms, or even encourage the development of example selection methods incorporating the notions of optimality suggested in this paper. Perhaps the most interesting result for active learning practitioners is the very narrow advantage of mixed sampling when a realistic algorithm is used for the stopping times. While this might only have been an artifact of the particular combinations of stopping and sampling algorithm and the datasets used, we believe that it is a matter which should be given some further attention.

Acknowledgements. We would like to thank Peter Auer for helpful discussions, suggestions and corrections. This work was supported by the FSP/JRP Cognitive Vision of the Austrian Science Funds (FWF, Project number S9104-N13). This work was also supported in part by the IST Programme of the European Community, under the PASCAL Network of Excellence, IST-2002-506778. This publication only reflects the authors' views.

References

[1] Cohn, D.A., Ghahramani, Z., Jordan, M.I.: Active learning with statistical models. In: Tesauro, G., Touretzky, D., Leen, T. (eds.) Advances in Neural Information Processing Systems, vol. 7, pp. 705–712. The MIT Press, Cambridge (1995)

[2] Schohn, G., Cohn, D.: Less is more: Active learning with support vector machines. In: ICML 2000. Proceedings of the 17th International Conference on Machine Learning, pp. 839–846 (2000)

[3] Saar-Tsechansky, M., Provost, F.: Active learning for class probability estimation and ranking. In: IJCAI 2001. Proceedings of the 17th international joint conference on articial intelligence, pp. 911–920 (2001)

[4] Melville, P., Mooney, R.J.: Diverse ensembles for active learning. In: ICML 2004. Proceedings of the 21st International Conference on Machine Learning, pp. 584–591 (2004)

[5] Körner, C., Wrobel, S.: Multi-class ensemble-based active learning. In: Fürnkranz, J., Scheffer, T., Spiliopoulou, M. (eds.) ECML 2006. LNCS (LNAI), vol. 4212, pp. 687–694. Springer, Heidelberg (2006)

[6] Abe, N., Mamitsuka, H.: Query learning strategies using boosting and bagging. In: ICML 1998. Proceedings of the Fifteenth International Conference on Machine Learning, pp. 1–9. Morgan Kaufmann Publishers Inc., San Francisco, CA, USA (1998)

[7] Bengio, S., Mariéthoz, J., Keller, M.: The expected performance curve. In: ICML 2007. International Conference on Machine Learning (2005)

[8] Campbell, C., Cristianini, N., Smola, A.: Query learning with large margin classi-
     fiers. In: ICML 2000. Proceedings of the 17th International Conference on Machine
     Learning, pp. 111–118 (2000)

[9] Balcan, M.F., Beygelzimer, A., Langford, J.: Agnostic active learning. In: Cohen,
     W.W., Moore, A. (eds.) 23rd International Conference on Machine Learning, pp.
     65–72. ACM, New York (2006)

[10] De Groot, M.H.: Optimal Statistical Decisions. John Wiley & Sons, Chichester
     (1970) (republished in 2004)

[11] Kääriäinen, M.: Active learning in the non-realizable case. In: Proceedings of the
     17th International Conference on Algorithmic Learning Theory (2006)

[12] Freund, Y., Schapire, R.E.: A decision-theoretic generalization of on-line learning
     and an application to boosting. Journal of Computer and System Sciences 55(1),
     119–139 (1997)

[13] Efron, B., Tibshirani, R.J.: An Introduction to the Bootstrap. In: Monographs on
     Statistics & Applied Probability, vol. 57. Chapmann & Hall, Sydney (1993)

[14] Roy, N., McCallum, A.: Toward optimal active learning through sampling estima-
     tion of error reduction. In: Proc. 18th International Conf. on Machine Learning,
     pp. 441–448. Morgan Kaufmann, San Francisco, CA (2001)

Information-Optimal Reflections
of View Updates
on Relational Database Schemata

Stephen J. Hegner

Umeå University, Department of Computing Science
SE-901 87 Umeå, Sweden
hegner@cs.umu.se
http://www.cs.umu.se/~hegner

Abstract. For the problem of reflecting an update on a database view to the main schema, the constant-complement strategies are precisely those which avoid all update anomalies, and so define the gold standard for well-behaved solutions to the problem. However, the families of view updates which are supported under such strategies are limited, so it is sometimes necessary to go beyond them, albeit in a systematic fashion. In this work, an investigation of such extended strategies is initiated for relational schemata. The approach is to characterize the information content of a database instance, and then require that the optimal reflection of a view update to the main schema embody the least possible change of information. To illustrate the utility of the idea, sufficient conditions for the existence of optimal insertions in the context of families of extended embedded implicational dependencies (XEIDs) are established. It is furthermore established that all such optimal insertions are equivalent up to a renaming of the new constant symbols which were introduced in support of the insertion.

Keywords: update, view.

1   Introduction

The problem of reflecting view updates to the main schema of a database system is a difficult one whose solution invariably involves compromise. The constant-complement approach [1] is exactly the one which avoids all so-called update anomalies [2], and so is the gold standard for well-behaved strategies. On the other hand, it is also quite conservative regarding the updates which it admits.

Substantial research has been conducted on allowing more general view updates in a systematic fashion. In the classical relational context, much of this work, such as [3], [4], [5], [6], and [7], focuses upon translations via the relational algebra. In this work, a quite different, logic-based approach is undertaken. The fundamental point of departure is that an optimal reflection of a view update is one which minimizes the information change in the main schema, with the information content of a database measured by the set of sentences in a certain family which it satisfies. An example will help clarify the main ideas.

S. Hartmann and G. Kern-Isberner (Eds.): FoIKS 2008, LNCS 4932, pp. 112–131, 2008.

Example 1.1. Let $\mathbf{E_0}$ be the relational schema with relations $R[ABC]$ and $S[CD]$, constrained by the inclusion dependency $R[C] \subseteq S[C]$. Regard a database as a set of ground atoms over the associated logic. For example, $M_{00} = \{R(a_0, b_0, c_0), R(a_1, b_1, c_1), S(c_0, d_0), S(c_1, d_1)\}$ is such a database. Now, let K be a set of constants in the underlying logical language, regarded as domain elements for this schema. The *information content* of a database M relative to K is the set of all positive (i.e., no negation, explicit or implicit), existential, and conjunctive sentences which are implied by M. Using the notation to be introduced in Definition 3.3, this information content is denoted $\mathsf{Info}\langle M, \mathsf{WFF}(\mathbf{E_0}, K, \exists\wedge+)\rangle$. A *basis* for this information content is a subset $\Phi \subseteq \mathsf{Info}\langle M, \mathsf{WFF}(\mathbf{E_0}, K, \exists\wedge+)\rangle$ such that Φ and $\mathsf{Info}\langle M, \mathsf{WFF}(\mathbf{E_0}, K, \exists\wedge+)\rangle$ are logically equivalent. For $K_{00} = \{a_0, a_1, b_0, b_1, c_0, c_1, d_0, d_1\}$, the set of all constant symbols of M_{00}, the set M_{00} itself is clearly a basis for $\mathsf{Info}\langle M_{00}, \mathsf{WFF}(\mathbf{E_0}, K_{00}, \exists\wedge+)\rangle$. On the other hand, with $K'_{00} = \{a_0, a_1, b_0, b_1, c_0, d_0\}$, a basis for $\mathsf{Info}\langle M_{00}, \mathsf{WFF}(\mathbf{E_0}, K'_{00}, \exists\wedge+)\rangle$ is $\{R(a_0, b_0, c_0), S(c_0, d_0), (\exists y)(\exists x)(R(a_1, b_1, x)\wedge S(x, y))\}$. Note that the constants in $K_{00} \setminus K'_{00}$ have been replaced by existentially quantified variables.

To see how this idea is useful in the context of view updates, let $\Pi_{R[AB]} = (R[AB], \pi_{R[AB]})$ be the view of $\mathbf{E_0}$ which projects $R[ABC]$ onto $R[AB]$ and which drops the relation S entirely. Consider M_{00} to be the initial state of schema $\mathbf{E_0}$; its image state in the view is then $N_{00} = \{R(a_0, b_0), R(a_1, b_1)\}$. Now, suppose that the view update $\mathsf{Insert}\langle R(a_2, b_2)\rangle$ is requested, so that $N_{01} = N_{00} \cup \{R(a_2, b_2)\}$ is the desired new view state, and consider $M_{01} = M_{00} \cup \{R(a_2, b_2, c_2), S(c_2, d_2)\}$ as a proposed reflection to the main schema $\mathbf{E_0}$. Relative to its entire set $K_{01} = \{a_0, a_1, a_2, b_0, b_1, b_2, c_0, c_1, c_2, d_0, d_1, d_2\}$ of constant symbols, a basis for $\mathsf{Info}\langle M_{01}, \mathsf{WFF}(\mathbf{E_0}, K_{01}, \exists\wedge+)\rangle$ is just M_{01} itself. Similarly, for $M_{02} = M_{00} \cup \{R(a_2, b_2, c_3), S(c_3, d_3)\}$ with $K_{02} = \{a_0, a_1, a_2, b_0, b_1, b_2, c_0, c_1, c_3, d_0, d_1, d_3\}$ a basis for $\mathsf{Info}\langle M_{02}, \mathsf{WFF}(\mathbf{E_0}, K_{02}, \exists\wedge+)\rangle$ is just M_{02} itself. Observe that the proposed updates M_{01} and M_{02} are identical up to a renaming of the new constants. The utility of information measure is that it provides a means to recapture this idea formally; the information content of each, relative to the set K_{00} of constant symbols of the original state M_{00}, is the same. More precisely, $\mathsf{Info}\langle M_{01}, \mathsf{WFF}(\mathbf{E_0}, K_{00}, \exists\wedge+)\rangle = \mathsf{Info}\langle M_{02}, \mathsf{WFF}(\mathbf{E_0}, K_{00}, \exists\wedge+)\rangle$. A basis for each of these is $I_1 = M_{00} \cup \{(\exists x)(\exists y)(R(a_2, b_2, x)\wedge S(x, y))\}$. In effect, this measure is indifferent to whether c_2 and d_2 or c_3 and d_3 are used.

Now, consider the alternative solution $M_{03} = M_{00} \cup \{R(a_2, b_2, c_3), S(c_3, d_1)\}$ to this view-update problem. A basis for $\mathsf{Info}\langle M_{03}, \mathsf{WFF}(\mathbf{E_0}, K_{00}, \exists\wedge+)\rangle$ is $I_3 = M_{00} \cup \{(\exists x)(R(a_2, b_2, x)\wedge S(x, d_1))\}$, which is strictly stronger than I_1, since $(\exists x)(R(a_2, b_2, x)\wedge S(x, d_1)) \models (\exists x)(\exists y)(R(a_2, b_2, x)\wedge S(x, y))$, but not conversely. Thus, relative to the information measure defined by K_{00}, M_{03} adds more information to M_{00} than does M_{01} or M_{02}. Similarly, $M_{04} = M_{00} \cup \{R(a_2, b_2, c_0))\}$ adds more information than does M_{01} or M_{02}, since a basis for its information content is just M_{04} itself, which is stronger than I_1, since $R(a_2, b_2, c_0)\wedge S(c_0, d_0) \models (\exists x)(\exists y)(R(a_2, b_2, x)\wedge S(x, y)))$, but not conversely.

The first and primary measure of quality of a reflected update is the change of information content which is induces. Under this measure, M_{01} and M_{02} are equivalent, and both are superior to either of M_{03} or M_{04}. However, this is by itself not quite adequate. Rather, there is an additional measure of quality which must be taken into account. To illustrate, consider the proposed solution $M_{05} = M_{01} \cup M_{02} = M_{00} \cup \{R(a_2, b_2, c_2), R(a_2, b_2, c_3), S(c_2, d_2), S(c_3, d_3)\}$ to this update problem. It has the same information content, I_1, relative to K_{00}, as do M_{01} and M_{02}. The information measure cannot distinguish the insertion of two new tuples with completely new constants from the insertion of just one. However, it is clear that M_{05} should be considered inferior to both M_{01} and M_{02} as a solution to the given update problem, since it is a proper superset of each. Therefore, a second criterion of quality is invoked; namely that no solution whose set of changes is a proper superset of those of another can be considered to be superior. It is important to emphasize that it is a inclusion relationship which applies here, and not simply a counting argument. For example, consider again the proposed solution M_{04}. From a strict counting point of view, M_{04} involves fewer changes than do M_{01} or M_{02}. However, neither M_{01} nor M_{02} is a superset of M_{04}. Thus, the superiority of M_{01} and M_{02} is not contradicted. In other words, only solutions which are *tuple minimal*, in the sense that no proper subset of the changes is also an admissible solution, are permitted.

The main modelling premise of this paper is that the quality of a view update can be measured by the amount of change in information content which it induces, and so an optimal reflection of a view update request is one which is both tuple minimal and which induces the least amount of change of information content. Under this premise, both M_{01} and M_{02} are superior to either of M_{03} or M_{04}. Furthermore, since M_{01} and M_{02} induce the same change in information content, they are equivalent. In Section 3, it is established that, under suitable conditions, all such optimal solutions are equivalent, up to a renaming of the constant symbols. In Section 4, it is established, again under suitable conditions, that for insertions, a minimal solution (in terms of change of information content) must be optimal. These conditions include in particular schemata constrained by XEIDs — the extended embedded implicational dependencies of Fagin [8], which include virtually all other classes of classical database dependencies.

In summary, there are two conditions which must be met for optimality of a proposed update reflection u. First, it must be *tuple minimal*, in that there can be no other solution whose set of changes is a proper subset of those of u. Second, it must be *information least* in terms of a specific set of sentences. This approach applies also to deletions and updates which involve both insertion and deletion, and this generality is incorporated into the formalism which is presented. However, for deletions the two measures will coincide, since no new constants are involved in a deletion.

2   The Relational Model

Discussion 2.1 (Two representations of the traditional relational model). In the traditional approach to the relational model [9] [10], the starting point is a set $\mathcal{A}$ of attributes, a finite nonempty set Rels of of relation symbols, and a function Ar : Rels $\rightarrow$ $\mathcal{A}$ which assigns to each $R \in$ Rels a set Ar$(R) \subseteq \mathcal{A}$, the *attributes* of R. Furthermore, to each $A \in \mathcal{A}$ is associated a (usually countably infinite) *domain* dom$(\mathcal{A})$. An *R-tuple* is then a function $t :$ Ar$(R) \rightarrow$ dom$(\mathcal{A})$, and a *relational database* over $(\mathcal{A}, \text{dom})$ is a collection of R-tuples for each $R \in \mathcal{A}$.

From a logical point of view, there are two common interpretations of the domain elements. In logic programming, they are usually taken to be constant symbols of the underlying logic. Tuples then become ground atoms, with (extensional) databases finite sets of such atoms [11]. Furthermore, in that context, the set of all constant symbols is usually taken to be finite, in order to allow first-order axiomatization of domain closure. On the other hand, for model-theoretic constructions, such as those of [8], it is necessary to interpret the relational domain elements as members of the underlying set of a structure [12, Def. 11.1], and to allow these sets to be countably infinite is essential. Both representations of tuples are crucial to this paper, so it is necessary to establish a bijective correspondence between them. To accomplish this, it is first necessary to establish a bijective correspondence between the elements of the structure and the constant symbols. This requires some care, since this condition cannot be stated using finite sentences in first-order logic. The solution employed in this paper is to use the same countable underlying set for all structures, and then to fix a bijection between the constant symbols and the structure elements. This bijection is invariant across all the main schemata and the view to be updated. Once such a bijection of elements and constants is established, a corresponding bijection of tuples and ground atoms, and consequently of databases represented in these two distinct formats, follows directly.

Definition 2.2 (Relational contexts and constant interpretations). A relational context contains the logical information which is shared amongst the distinct schemata and the corresponding database mappings: the attribute names, the variables, and constant symbols. Formally, a relational context $\mathcal{D}$ consists of attribute names $\mathcal{A}_\mathcal{D}$, variables Vars$(\mathcal{D})$, and for each $A \in \mathcal{A}_\mathcal{D}$, a set Const$_\mathcal{D}(A)$ of constant symbols. The variables Vars$(\mathcal{D})$ are further partitioned into two disjoint sets; a countable set GenVars$(\mathcal{D}) = \{x_0, x_1, x_2, \ldots\}$ of *general variables*, and special $\mathcal{A}_\mathcal{D}$-indexed set AttrVars$(\mathcal{D}) = \{x_A \mid A \in \mathcal{A}_\mathcal{D}\}$ of *attribute variables*. The latter are used in the definition of interpretation mappings; see Definition 2.5 for details.

A constant interpretation provides a model-theoretic interpretation for the constant symbols, in the sense of [12, Def. 11.1]. It is also fixed over all databases of all schemata. Formally, a *constant interpretation* for the relational context $\mathcal{D}$ is a pair $\mathcal{I} = (\text{Dom}_\mathcal{I}, \text{IntFn}_\mathcal{I})$ in which Dom$_\mathcal{I}$ is a countably infinite set, called the *domain* of $\mathcal{I}$, and IntFn$_\mathcal{I} :$ Const$(\mathcal{D}) \rightarrow$ Dom$_\mathcal{I}$ is a bijective function, called the *interpretation function* of $\mathcal{I}$. Note that the latter effectively stipulates the following two well-known conditions [13, p. 120]:

Domain closure: $(\forall x)(\bigvee_{a \in \text{Const}(\mathcal{D})} x = a)$ $\hfill$ (DCA$(\mathcal{D})$)

Unique naming: $(\neg(a = b))$ for distinct $a, b \in$ Const$(\mathcal{D})$ $\hfill$ (UNA$(\mathcal{D})$)

Since there are countably many constant symbols, the domain closure axiom is not a finite disjunction. This is not a problem however, since it is never used in an otherwise first-order set of constraints. Except for the extended tuple databases of Definition 4.4, in which this constraint is relaxed, the assignment of domain values to constants is fixed, and so it is not necessary to verify that it holds.

For $A \in \mathcal{A}_\mathcal{D}$, define $\mathsf{Dom}_\mathcal{I}(A) = \{z \in \mathsf{Dom}_\mathcal{I} \mid \mathsf{IntFn}_I(z) \in \mathsf{Const}_\mathcal{D}(A)\}$. Thus, $\mathsf{Dom}_\mathcal{I}(A)$ is the set of all domain values which are associated with attribute A.

As a notational convention, from this point on, unless stated otherwise, fix relational context $\mathcal{D}$ and a constant interpretation $\mathcal{I} = (\mathsf{Dom}_\mathcal{I}, \mathsf{IntFn}_\mathcal{I})$ for it.

Definition 2.3 (Tuples and databases). An *unconstrained relational schema* over $(\mathcal{D}, \mathcal{I})$ is a pair $\mathbf{D} = (\mathsf{Rels}(\mathbf{D}), \mathsf{Ar}_\mathbf{D})$ in which $\mathsf{Rels}(\mathbf{D})$ is finite set of relational symbols and $\mathsf{Ar}_\mathbf{D} : \mathsf{Rels}(\mathbf{D}) \to 2^{\mathcal{A}_\mathcal{D}}$ a function which assigns an *arity*, a set of distinct attributes from $\mathcal{A}_\mathcal{D}$, to each $R \in \mathsf{Rels}(\mathbf{D})$.

It is now possible to address the problem of modelling databases in the two distinct ways identified in Discussion 2.1 above. For $R \in \mathsf{Rels}(\mathbf{D})$, an *R-tuple* is a function t on $\mathsf{Ar}_\mathbf{D}(R)$ with the property that $t[A] \in \mathsf{Dom}_I(A)$ for every $A \in \mathsf{Ar}_\mathbf{D}$. Similarly, an *R-atom* is such a function with the property that $t[A] \in \mathsf{Const}_\mathcal{D}(A) \cup \mathsf{GenVars}(\mathcal{D}) \cup \{x_A\}$. A *ground R-atom* contains no variables, so $t[A] \in \mathsf{Const}_\mathcal{D}(A)$. The set of all R-tuples (resp. R-atoms, resp. ground R-atoms) is denoted $\mathsf{Tuples}(\mathbf{D})$, (resp. $\mathsf{Atoms}(\mathbf{D})$, resp. $\mathsf{GrAtoms}(\mathbf{D})$). In view of Definition 2.2 above, it is easy to see that there is a bijective correspondence between $\mathsf{GrAtoms}(\mathbf{D})$ and $\mathsf{Tuples}(\mathbf{D})$ given by $t(a_1, a_2, \ldots, a_n) \mapsto t(\mathsf{IntFn}_I(a_1), \mathsf{IntFn}_I(a_2), \ldots \mathsf{IntFn}_I(a_n))$.

It will be necessary to work with sets of R-tuples and sets of R-atoms, with R ranging over distinct relation symbols. A $\mathbf{D}$-*tuple* is an R-tuple for some $R \in \mathsf{Rels}(\mathbf{D})$, with the set of all $\mathbf{D}$-tuples denoted $\mathsf{Tuples}(\mathbf{D})$. A *tuple database for* $\mathbf{D}$ is a finite subset of $\mathsf{Tuples}(\mathbf{D})$, with the set of all tuple databases for $\mathbf{D}$ denoted $\mathsf{TDB}(\mathbf{D})$. The $\mathbf{D}$-atoms and ground $\mathbf{D}$-atoms are defined analogously, with the corresponding sets denoted $\mathsf{Atoms}(\mathbf{D})$ and $\mathsf{GrAtoms}(\mathbf{D})$, respectively. An *atom database for* $\mathbf{D}$ is a finite subset of $\mathsf{GrAtoms}(\mathbf{D})$; the set of all atom databases for $\mathbf{D}$ is denoted $\mathsf{DB}(\mathbf{D})$.

In the above definitions, it is necessary to be able to recover the associated relation from a tuple, and so *tagging* is employed, in which tuples are marked with the associated relation. Formally, this is accomplished by introducing a new attribute $\mathsf{RName} \notin \mathcal{A}_\mathcal{D}$, and then regarding an R-tuple not as a function t just on $\mathsf{Ar}_\mathbf{D}(R)$ but rather as one on on $\{\mathsf{RName}\} \cup \mathsf{Ar}_\mathbf{D}(R)$ with the property that $t[\mathsf{RName}] = R$. Tagging of R-atoms is defined analogously; both will be used from this point on throughout the paper. Nevertheless, in writing tuples, the more conventional notation $R(\tau_1, \tau_2, \ldots, \tau_n)$ will be used in lieu of the technically more correct $(R, \tau_1, \tau_2, \ldots, \tau_n)$, although tags will be used in formal constructions.

For the product construction of Definition 4.5, it is necessary to restrict attention to nonempty databases. To this end, call $M \in \mathsf{TDB}(\mathbf{D})$ (resp. $M \in \mathsf{DB}(\mathbf{D})$) *relationwise nonempty* if for each $R \in \mathsf{Rels}(\mathbf{D})$, there is at least one R-tuple (resp. R-atom) in M, and define $\mathsf{RNeTDB}(\mathbf{D})$ (resp. $\mathsf{RNeDB}(\mathbf{D})$) to be the set of all relationwise nonempty members of $\mathsf{TDB}(\mathbf{D})$ (resp. $\mathsf{DB}(\mathbf{D})$).

The first-order language associated with the relational schema $\mathbf{D}$ is defined in the natural way; however, it is useful to introduce some notation which identifies particular sets of formulas. Define $\mathsf{WFF}(\mathbf{D})$ to be the set of all well-formed first-order formulas with equality, in the language whose set of relational symbols is $\mathsf{Rels}(\mathbf{D})$, whose set of constant symbols is $\mathsf{Const}(\mathcal{D})$, and which contains no non-nullary function symbols. The variables are those of $\mathcal{D}$. Additional arguments may be given to restrict this set. If $S \subseteq \mathsf{Const}(\mathcal{D})$, then $\mathsf{WFF}(\mathbf{D}, S)$ denotes the formulas in $\mathsf{WFF}(\mathbf{D})$ which involve only constant symbols from S. In particular, $\mathsf{WFF}(\mathbf{D}, \emptyset)$ denotes the set of formulas which do not contain constant symbols. Arguments are also used to limit the logical connectives. $\mathsf{WFF}(\mathbf{D}, \exists+)$ identifies those formulas which are built up from the connectives $\wedge$ and $\vee$, using at most existential quantifiers. $\mathsf{WFF}(\mathbf{D}, \exists\wedge+)$ enforces the further restriction that disjunction is not allowed. It will be furthermore assumed, in $\mathsf{WFF}(\mathbf{D}, \exists\wedge+)$, that equality atoms (i.e., atoms of the form $x_i = x_j$ and $x_i = a$) are not allowed. This is not an essential limitation; such equality can always be represented by setting the terms to be equal in the atoms in which they are used. These notations may be combined, with the obvious semantics. For example, $\mathsf{WFF}(\mathbf{D}, \emptyset, \exists\wedge+)$ denotes the members of $\mathsf{WFF}(\mathbf{D}, \exists\wedge+)$ which do not involve constant symbols.

$\mathsf{WFS}(\mathbf{D})$ denotes the subset of $\mathsf{WFF}(\mathbf{D})$ consisting of sentences; that is, formulas with no free variables. The conventions regarding additional arguments applies to sets of sentences as well. For example, $\mathsf{WFS}(\mathbf{D}, \emptyset, \exists\wedge+)$ is the subset of $\mathsf{WFF}(\mathbf{D}, \emptyset, \exists\wedge+)$ consisting of sentences.

$M \in \mathsf{TDB}(\mathbf{D})$ is an $\mathcal{I}$-model of $\varphi \in \mathsf{WFS}(\mathbf{D})$ if it is a model of φ in the ordinary sense which furthermore interprets the constant symbols according to $\mathcal{I}$. The set of all $\mathcal{I}$-models of φ is denoted $\mathsf{Mod}_{\mathcal{I}}(\varphi)$. In view of the bijection $\mathsf{GrAtoms}(\mathbf{D}) \rightarrow \mathsf{Tuples}(\mathbf{D})$ identified in Definition 2.3 above, it is possible to identify $\mathcal{I}$-models with finite sets of ground atoms. More precisely, define the *atomic $\mathcal{I}$-models* of φ to be $\mathsf{AtMod}_{\mathcal{I}}(\varphi) = \{M \in \mathsf{DB}(\mathbf{D}) \mid M \cup \{\varphi\} \cup \mathsf{UNA}(\mathcal{D})$ is consistent$\}$. Clearly, $\mathsf{IntFn}_{\mathcal{I}}(M) \in \mathsf{Mod}_{\mathcal{I}}(\varphi)$ iff $M \in \mathsf{AtMod}_{\mathcal{I}}(\varphi)$. The relationally nonempty versions $\mathsf{RNeMod}_{\mathcal{I}}(\varphi)$ and $\mathsf{RNeAtMod}_{\mathcal{I}}(\varphi)$ are defined analogously. Furthermore, all of these definitions of model extend to sets Φ of sentences in the obvious way; notation such as $\mathsf{Mod}_{\mathcal{I}}(\Phi)$, $\mathsf{RNeMod}_{\mathcal{I}}(\Phi)$, $\mathsf{AtMod}_{\mathcal{I}}(\Phi)$, and $\mathsf{RNeAtMod}_{\mathcal{I}}(\Phi)$ will be used throughout.

Definition 2.4 (Schemata with constraints and constrained databases). A *relational schema* over $(\mathcal{D}, \mathcal{I})$ is a triple $\mathbf{D} = (\mathsf{Rels}(\mathbf{D}), \mathsf{Ar}_{\mathbf{D}}, \mathsf{Constr}(\mathbf{D}))$ in which $(\mathsf{Rels}(\mathbf{D}), \mathsf{Ar}_{\mathbf{D}})$ is an unconstrained relational schema over $(\mathcal{D}, \mathcal{I})$ and $\mathsf{Constr}(\mathbf{D}) \subseteq \mathsf{WFS}(\mathbf{D}, \emptyset)$ is the set of *dependencies* or *constraints* of $\mathbf{D}$. Note that constant symbols are not allowed in the constraints.

In representing a database as a set of $\mathbf{D}$-atoms, the closed-world assumption is implicit. On the other hand, to express what it means for such a representation to satisfy a set of constraints, it is necessary to state explicitly which atoms are not true as well. Formally, for $M \in \mathsf{DB}(\mathbf{D})$, define the *diagram* of M to be $\mathsf{Diagram}_{\mathbf{D}}(M) = M \cup \{\neg t \mid t \in \mathsf{GrAtoms}(\mathbf{D}) \setminus M\}$. Define the *legal* (or *constrained*) *databases* of $\mathbf{D}$ to be $\mathsf{LDB}(\mathbf{D}) = \{M \in \mathsf{DB}(\mathbf{D}) \mid \mathsf{Diagram}_{\mathbf{D}}(M) \cup$ $\mathsf{Constr}(\mathbf{D})$ has an $\mathcal{I}$-model$\}$ and the *nonempty legal databases* to be $\mathsf{RNeLDB}(\mathbf{D}) = \mathsf{LDB}(\mathbf{D}) \cap \mathsf{RNeDB}(\mathbf{D})$.

Definition 2.5 (Database morphisms and views). Database morphisms are defined using expressions in the relational calculus; more formally, they are interpretations of the theory of the view into the theory of the main schema [14]. Let $\mathbf{D}_1$ and $\mathbf{D}_2$ be relational schemata over $(\mathcal{D}, \mathcal{I})$. Given $R \in \mathsf{Rels}(\mathbf{D}_2)$, an *interpretation* for R into $\mathbf{D}_1$ is a $\varphi \in \mathsf{WFF}(\mathbf{D})$ in which precisely the variables $\{x_A \mid A \in \mathsf{Ar}_{\mathbf{D}}(R)\}$ are free, with x_A is used to mark the position in the formula which is bound to attribute A. The set of all interpretations of R into $\mathbf{D}_1$ is denoted $\mathsf{Interp}(R, \mathbf{D}_1)$. A *syntactic morphism* $f : \mathbf{D}_1 \rightarrow \mathbf{D}_2$ is a family $f = \{f^R \mid R \in \mathsf{Rels}(\mathbf{D}_2) \text{ and } f^R \in \mathsf{Interp}(R, \mathbf{D}_1)\}$.

Let $t \in \mathsf{GrAtoms}(R, \mathbf{D}_2)$. The *substitution* of t into f, denoted $\mathsf{Subst}\langle f, t \rangle$, is the sentence in $\mathsf{WFS}(\mathbf{D}_1)$ obtained by substituting $t[A]$ for x_A, for each $A \in \mathsf{Ar}_{\mathbf{D}}(R)$. For $M \in \mathsf{DB}(\mathbf{D}_1)$, define $f(M) = \{t \in \mathsf{Atoms}(\mathbf{D}_2) \mid \mathsf{Subst}\langle f, t \rangle \cup \mathsf{Diagram}_{\mathbf{D}_1}(M)$ has an $\mathcal{I}$-model$\}$. f is called a *semantic morphism* if it maps legal databases to legal databases; formally, $f(M) \in \mathsf{LDB}(\mathbf{D}_2)$ for each $M \in \mathsf{LDB}(\mathbf{D}_1)$.

Say that f is of class $\exists+$ (resp. $\exists\wedge+$) if each $f^R \in \mathsf{WFF}(\mathbf{D}_1, \exists+)$ (resp. $f^R \in \mathsf{WFF}(\mathbf{D}_1, \exists\wedge+)$). It is easy to see that if f is of class $\exists+$ (resp. $\exists\wedge+$), then for each $t \in \mathsf{Atoms}(\mathbf{D}_2)$, $\mathsf{Subst}\langle f, t \rangle \in \mathsf{WFS}(\mathbf{D}_1, \exists+)$ (resp. $\mathsf{Subst}\langle f, t \rangle \in \mathsf{WFS}(\mathbf{D}_1, \exists\wedge+)$).

Let $\mathbf{D}$ be a relational schema over $(\mathcal{D}, \mathcal{I})$. A *(relational) view* of $\mathbf{D}$ is a pair $\Gamma = (\mathbf{V}, \gamma)$ in which $\mathbf{V}$ is a relational schema over $(\mathcal{D}, \mathcal{I})$ and $\gamma : \mathbf{D} \rightarrow \mathbf{V}$ is a semantic morphism which is furthermore *semantically surjective* in the sense that for every $N \in \mathsf{LDB}(\mathbf{V})$, there is an $M \in \mathsf{LDB}(\mathbf{D})$ with $f(M) = N$. Γ is of class $\exists+$ (resp. class $\exists\wedge+$) precisely in the case that γ is of that same class.

3   The General Theory of Updates

In this section, the general ideas concerning the information content of a database state, and the ideas of optimizing an update relative to such content, are developed. It is furthermore established that for a wide class of schemata and views, all optimal updates are isomorphic in a natural way.

Notation 3.1. Throughout the rest of this paper, unless stated specifically to the contrary, take $\mathbf{D}$ to be a relational schema over $(\mathcal{D}, \mathcal{I})$ and $\Gamma = (\mathbf{V}, \gamma)$ to be a (relational) view of $\mathbf{D}$.

For X an entity (for example, an atom, a formula, a database, etc.), $\mathsf{ConstSym}(X)$ denotes the set of all $a \in \mathsf{Const}(\mathcal{D})$ which occur in X. Similarly, $\mathsf{Vars}(X)$ denotes the set of all variables which occur in X. This will not be formalized further, but the meaning should always be unambiguous.

Definition 3.2 (Updates and reflections). An *update* on $\mathbf{D}$ is a pair $(M_1, M_2) \in \mathsf{LDB}(\mathbf{D}) \times \mathsf{LDB}(\mathbf{D})$. M_1 is the current state, and M_2 the new state. It is an *insertion* if $M_1 \subseteq M_2$, and a *deletion* if $M_2 \subseteq M_1$.

To describe the situation surrounding an update request on Γ, it is sufficient to specify the current state M_1 of the main schema and the desired new state N_2

of the view schema $\mathbf{V}$. The current state of the view can be computed as $\gamma(M_1)$; it is only the new state M_2 of the main schema (subject to $N_2 = \gamma(M_2)$) which must be obtained from an update strategy. Formally, an *update request* from Γ to $\mathbf{D}$ is a pair (M_1, N_2) in which $M_1 \in \mathsf{LDB}(\mathbf{D})$ (the old state of the main schema) and $N_2 \in \mathsf{LDB}(\mathbf{V})$ (the new state of the view schema). If $\gamma(M_1) \subseteq N_2$, it is called an *insertion request*, and if $N_2 \subseteq \gamma(M_1)$, it is called a *deletion request*. Collectively, insertion requests and deletion requests are termed *unidirectional update requests*. A *realization* of (M_1, N_2) along Γ is an update (M_1, M_2) on $\mathbf{D}$ with the property that $\gamma(M_2) = N_2$. The update (M_1, M_2) is called a *reflection* (or *translation*) of the view update $(\gamma(M_1), N_2)$. The set of all realizations of (M_1, N_2) along Γ is denoted $\mathsf{UpdRealiz}\langle M_1, N_2, \Gamma \rangle$. The subset of $\mathsf{UpdRealiz}\langle M_1, N_2, \Gamma \rangle$ consisting of insertions (resp. deletions) is denoted $\mathsf{InsRealiz}\langle M_1, N_2, \Gamma \rangle$ (resp. $\mathsf{DelRealiz}\langle M_1, N_2, \Gamma \rangle$).

Definition 3.3 (Information content and Φ-equivalence). Let $\Phi \subseteq \mathsf{WFS}(\mathbf{D})$ and let $M \in \mathsf{DB}(\mathbf{D})$. The *information content* of M relative to Φ is the set of all sentences in Φ which are true for M. More precisely, $\mathsf{Info}\langle M, \Phi \rangle = \{\varphi \in \Phi \mid M \in \mathsf{AtMod}_\mathcal{I}(\varphi)\}$. For $\varphi \in \mathsf{WFS}(\mathbf{D})$, $\mathsf{Info}\langle M, \varphi \rangle$ denotes $\mathsf{Info}\langle M, \{\varphi\} \rangle$. M_1 and M_2 are *Φ-equivalent* if they have the same information content relative to Φ; i.e., $\mathsf{Info}\langle M_1, \Phi \rangle = \mathsf{Info}\langle M_2, \Phi \rangle$.

Definition 3.4 (Update difference and optimal reflections). The update difference of an update (M_1, M_2) on $\mathbf{D}$ with respect to a set $\Phi \subseteq \mathsf{WFS}(\mathbf{D})$ is a measure of how much M_2 differs from M_1 in terms of satisfaction of the sentences of Φ. Formally, the *positive (Δ^+), negative (Δ^-), and total (Δ) update differences* of (M_1, M_2) with respect to Φ are defined as follows:

$$\Delta^+\langle (M_1, M_2), \Phi \rangle = \mathsf{Info}\langle M_2, \Phi \rangle \setminus \mathsf{Info}\langle M_1, \Phi \rangle$$
$$\Delta^-\langle (M_1, M_2), \Phi \rangle = \mathsf{Info}\langle M_1, \Phi \rangle \setminus \mathsf{Info}\langle M_2, \Phi \rangle$$
$$\Delta\langle (M_1, M_2), \Phi \rangle = \Delta^+\langle (M_1, M_2), \Phi \rangle \cup \Delta^-\langle (M_1, M_2), \Phi \rangle$$

Note that, given $\varphi \in \Delta\langle (M_1, M_2), \Phi \rangle$, it is always possible to determine whether $\varphi \in \Delta^+\langle (M_1, M_2), \Phi \rangle$ or $\varphi \in \Delta^-\langle (M_1, M_2), \Phi \rangle$ by checking whether or not $M_1 \in \mathsf{AtMod}_\mathcal{I}(\varphi)$. Given an update request (M_1, N_2), the quality of a realization (M_1, M_2) is measured by its update difference. Formally, let $\Phi \subseteq \mathsf{WFS}(\mathbf{D})$, let (M_1, N_2) be an update request from Γ to $\mathbf{D}$, let $T \subseteq \mathsf{UpdRealiz}\langle M_1, N_2, \Gamma \rangle$, and let $(M_1, M_2) \in T$.

(a) (M_1, M_2) is *minimal* in T with respect to Φ if for any $(M_1, M_2') \in T$, if $\Delta\langle (M_1, M_2'), \Phi \rangle \subseteq \Delta\langle (M_1, M_2), \Phi \rangle$, then $\Delta\langle (M_1, M_2'), \Phi \rangle = \Delta\langle (M_1, M_2), \Phi \rangle$.

(b) (M_1, M_2) is *least* in T with respect to Φ if for all $(M_1, M_2') \in T$, $\Delta\langle (M_1, M_2), \Phi \rangle \subseteq \Delta\langle (M_1, M_2'), \Phi \rangle$.

Definition 3.5 (Information-monotone   sentences   and   update classifiers). For the above definitions of minimal and least to be useful, it is necessary to place certain restrictions on the nature of Φ. As a concrete example of the problems, define $\mathsf{GrAtoms}^{\neg}(\mathbf{D}) = \{\neg t \mid t \in \mathsf{GrAtoms}(\mathbf{D})\}$, with $\mathsf{GrAtoms}^{\pm}(\mathbf{D}) = \mathsf{GrAtoms}(\mathbf{D}) \cup \mathsf{GrAtoms}^{\neg}(\mathbf{D})$. In the context of Definition 3.4 above, it is easy to see that every reflection (M_1, M_2) is minimal with respect to $\mathsf{GrAtoms}^{\pm}(\mathbf{D})$, while only identity updates (with $M_1 = M_2$) are least. Any $\Phi \subseteq \mathsf{WFS}(\mathbf{D})$ with the property that $\mathsf{GrAtoms}^{\pm}(\mathbf{D}) \subseteq \Phi$ will have this same property. The problem is that the sentences in $\mathsf{GrAtoms}^{\neg}(\mathbf{D})$ are not information-monotone; adding new tuples reduces the information content. The sentence $\varphi \in \mathsf{WFS}(\mathbf{D})$ is *information monotone* if for any $M_1, M_2 \in \mathsf{DB}(\mathbf{D})$ if $M_1 \subseteq M_2$, then $\mathsf{Info}\langle M_1, \varphi \rangle \subseteq \mathsf{Info}\langle M_2, \varphi \rangle$. The set $\Phi \subseteq \mathsf{WFS}(\mathbf{D})$ is *information monotone* if each $\varphi \in \Phi$ has this property. Any $\varphi \in \mathsf{WFS}(\mathbf{D})$ which does not involve negation, either explicitly or implicitly (via implication, for example), is information monotone. Thus, in particular, for any $S \subseteq \mathsf{Const}(\mathcal{D})$, $\mathsf{WFS}(\mathbf{D}, S, \exists +)$, $\mathsf{WFS}(\mathbf{D}, S, \exists \wedge +)$, and $\mathsf{GrAtoms}(\mathbf{D})$ all consist of information-monotone sentences. The total absence of negation is not necessary, however. Sentences which allow negation of equality terms (e.g., $\neg(x_i = x_j)$) but only existential quantification are also information monotone.

An *update classifier* for $\mathbf{D}$ is simply a set Σ of information-monotone sentences. The idea is simple: updates which involve less change of information are to be preferred to those which involve more. However, as illustrated in the example of Example 1.1, there are two distinct measures of optimality. On the one hand, an optimal realization (M_1, M_2) of an update request (M_1, N_2) must be least with respect to the update classifier, which in that example is $\mathsf{WFS}(\mathbf{D}, \mathsf{ConstSym}(M_{00}), \exists \wedge +)$. Unfortunately, this measure cannot always eliminate solutions which contain two "isomorphic" copies of the same update, such as M_{05} of that example. To remedy this, the update must also be minimal with respect to $\mathsf{Atoms}(\mathbf{D})$; or, equivalently, with respect to the symmetric difference $M_1 \triangle M_2 = (M_1 \setminus M_2) \cup (M_2 \setminus M_1)$. Formally, let (M_1, N_2) be an update request from Γ to $\mathbf{D}$, let $T \subseteq \mathsf{UpdRealiz}\langle M_1, N_2, \Gamma \rangle$, and let $(M_1, M_2) \in T$.

(a) (M_1, M_2) is $\langle \Sigma, T \rangle$-*admissible* if it is minimal in T with respect to both Σ and $\mathsf{Atoms}(\mathbf{D})$.

(b) (M_1, M_2) is $\langle \Sigma, T \rangle$-*optimal* if it is $\langle \Sigma, T \rangle$-admissible and least in T with respect to Σ.

Roughly, (M_1, M_2) is admissible if no other realization is better, and it is optimal if it is better than all others, up to the equivalence defined by Σ. Observe that if some update request is $\langle \Sigma, T \rangle$-optimal, then all $\langle \Sigma, T \rangle$-admissible update requests are $\langle \Sigma, T \rangle$-optimal.

As a notational shorthand, if $T = \mathsf{InsRealiz}\langle M_1, N_2, \Gamma \rangle$ (resp. $T = \mathsf{DelRealiz}\langle M_1, N_2, \Gamma \rangle$), that is, if T is the set of all possible insertions (resp. deletions) which realize (M_1, N_2), then $\langle \Sigma, T \rangle$-admissible and $\langle \Sigma, T \rangle$-optimal will be abbreviated to $\langle \Sigma, \uparrow \rangle$-admissible and $\langle \Sigma, \uparrow \rangle$-optimal (resp. $\langle \Sigma, \downarrow \rangle$-admissible and $\langle \Sigma, \downarrow \rangle$-optimal).

Example 3.6 (Update classifiers). For $M_1 \in \mathsf{LDB}(\mathbf{D})$, the *standard M_1-based update classifier* is $\mathsf{StdUCP}(\mathbf{D}, M_1) = \mathsf{WFS}(\mathbf{D}, \mathsf{ConstSym}(M_1), \exists \wedge +)$. As illustrated in Example 1.1, this classifier is appropriate for characterizing optimal insertions. Because it "hides" new constants, optimal solutions which are unique up to constant renaming are easily recaptured.

A much simpler example is GrAtoms($\mathbf{D}$). It yields optimal solutions only in the case that such solutions are truly unique. For deletions, this equivalence is adequate. In fact, for deletions, StdUCP($\mathbf{D}, M_1$) and GrAtoms($\mathbf{D}$) always identify the same optimal solutions.

There are other possibilities which provide different notions of optimality. Let $\mathbf{E}_1$ be the schema which is identical to $\mathbf{E}_0$ of Example 1.1, save that it includes an additional relation symbol $S'[CD]$, and the inclusion dependency $R[C] \subseteq S[C]$ is replaced with $R[C] \subseteq S[C] \cup S'[C]$. Let M'_{00} be the state of $\mathbf{E}_1$ which is the extension of M_{00} in which the relation of S' is empty. The view $\Pi_{R[AB]} = (R[AB], \pi_{R[AB]})$ is unchanged. Under the update classifier WFS($\mathbf{E}_1$, ConstSym(M'_{00}), $\exists\wedge+$), the update request (M'_{00}, N_{01}) (using N_{01} from Example 1.1) no longer has an optimal solution, since a minimal solution involves adding a tuple either to S or to S' but not to both. However, optimality can be recovered formally via an alternative update classifier. Let $\varXi_1$ denote the subset of WFS($\mathbf{E}_1, \exists+$) obtained from WFS($\mathbf{E}_0, \exists\wedge+$) by replacing each occurrence of the form $S(\tau_1, \tau_2)$ by $(S(\tau_1, \tau_2) \vee S'(\tau_1, \tau_2))$. Here τ_1 and τ_2 are arbitrary terms (i.e., variables or constants). In effect, the sentences of $\varXi_1$ cannot distinguish a given tuple in S from an identical one in S'. It is easy to see that $\varXi_1$ is information monotone (since it is a subset of WFF($\mathbf{E}_1, \exists+$)). Furthermore, both of the solutions $M'_{01} = M'_{00} \cup \{R(a_2, b_2, c_2), S(c_2, d_2)\}$ and $M''_{01} = M'_{00} \cup \{R(a_2, b_2, c_2), S'(c_2, d_2)\}$ are optimal under this measure.

By choosing a suitable update classifier, rather broad notions of equivalence are hence achievable, so there is a tradeoff between the generality of the update classifier and how "equivalent" the various optimal solutions really are. In the example sketched above, the solutions are not isomorphic in any reasonable sense. On the other hand, for StdUCP($\mathbf{E}_0, M_{00}$), all optimal solutions are naturally isomorphic, a nontrivial result which requires some work to establish; the rest of this section is devoted to that task.

Definition 3.7 (Constant endomorphisms). An *endomorphism* on $\mathcal{D}$ is a function $h : \mathsf{Const}(\mathcal{D}) \rightarrow \mathsf{Const}(\mathcal{D})$ which preserves attribute types, in the precise sense that for each $A \in \mathcal{A}_\mathcal{D}$ and each $a \in \mathsf{Const}_\mathcal{D}(A)$, $h(a) \in \mathsf{Const}_\mathcal{D}(A)$. If h is additionally a bijection, then it is called an *automorphism* of $\mathcal{D}$. For $S \subseteq \mathsf{Const}(\mathcal{D})$, call h S-invariant if $h(a) = a$ for all $a \in S$.

Given a database schema $\mathbf{D}$, an endomorphism on $\mathcal{D}$ induces a mapping from GrAtoms($\mathbf{D}$) to itself given by sending $t \in$ GrAtoms($\mathbf{D}$) to the tuple t' with $t'[\mathsf{RName}] = t[\mathsf{RName}]$ and $t'[A] = t[h(A)]$ for all $A \in \mathsf{Ar}_{t[\mathsf{RName}]}$. This mapping on atoms will also be represented by h, as will the induced mapping from DB($\mathbf{D}$) to itself given by $M \mapsto \{h(t) \mid t \in M\}$.

Definition 3.8 (Armstrong models in an information-monotone context). Let $\Psi \subseteq$ WFS($\mathbf{D}$) and let $\Phi \subseteq \Psi$. Informally, an Armstrong model for Φ relative to Ψ is a model of Φ which satisfies only those constraints of Ψ which are implied by Φ. More formally, an *Armstrong model* for Φ relative to Ψ is an $M \in \mathsf{Mod}_{\mathcal{I}}(\Phi)$ with the property that for any $\psi \in \Psi$, if $M \in \mathsf{Mod}_{\mathcal{I}}(\psi)$, then $\mathsf{Mod}_{\mathcal{I}}(\Phi) \subseteq \mathsf{Mod}_{\mathcal{I}}(\psi)$. Armstrong models have been studied extensively for database dependencies; see, for example, [8] and [15]. In the current context, it will be shown that if (M_1, M_2) is a $\mathsf{StdUCP}(\mathbf{D}, M_1)$-optimal reflection of the update request (M_1, N_2), then M_2 is a minimal Armstrong model with respect to $\mathsf{StdUCP}(\mathbf{D}, M_1)$. It will furthermore be shown that if (M_1, M_2') is another such optimal reflection, there is an an automorphism h which is constant on $\mathsf{Const}_{\mathcal{D}}(M)$ with $M_2' = h(M_2)$.

Definition 3.9 (Representation of $\exists\wedge+$-sentences as sets of D-atoms). There is an alternative syntactic representation for formulas in $\mathsf{WFS}(\mathbf{D}, \exists\wedge+)$ which will be used in that which follows. Specifically, for $\varphi \in \mathsf{WFS}(\mathbf{D}, \exists\wedge+)$ define $\mathsf{AtRep}(\varphi)$ to be the set of all atoms which occur as conjuncts in φ. For example, if $\varphi = (\exists x_1)(\exists x_2)(\exists x_3)(R(x_1, a)\wedge R(x_1, b)\wedge S(x_2, a)\wedge T(x_2, x_3))$ then $\mathsf{AtRep}(\varphi) = \{R(x_1, a), R(x_1, b), S(x_2, a), T(x_2, x_3)\}$.

This representation is dual to that used in theorem-proving contexts in classical artificial intelligence [13, 4.1]. Here the variables are existentially quantified and the atoms are conjuncts of one another; in the AI setting the atoms are disjuncts of one another and the variables are universally quantified.

Definition 3.10 (Substitutions). Let $V = \{v_1, v_2, \ldots, v_n\} \subseteq \mathsf{GenVars}(\mathcal{D})$. A *(constant) substitution* for V (in $\mathcal{D}$) is a function $s : V \to \mathsf{Const}(\mathcal{D})$. If $s(x_i) = a_i$ for $i \in \{1, 2, \ldots, n\}$, following (somewhat) standard notation this substitution is often written $\{a_1/x_1, a_2/x_2, \ldots, a_n/x_n\}$ (although some authors [13, 4.2] write $\{x_1/a_1, x_2/a_2, \ldots, x_n/a_n\}$ instead).

Let $\varphi \in \mathsf{WFS}(\mathbf{D}, \exists\wedge+)$ with $\mathsf{Vars}(\varphi) \subseteq V$. Call s *correctly typed* for φ if for each $t \in \mathsf{AtRep}(\varphi)$ and each $A \in \mathsf{Ar}_{\mathbf{D}}(t[\mathsf{RName}])$, if $t[A] \in \mathsf{Vars}(\mathbf{D})$ then $s(t[A]) \in \mathsf{Const}_{\mathcal{D}}(A)$. Define $\mathsf{Subst}(\varphi, s)$ to be the set of ground atoms obtained by substituting $s(x_i)$ for x_i in $\mathsf{AtRep}(\varphi)$. For example, with $s = \{a_1/x_1, a_2/x_2, a_3/x_3\}$ and $\mathsf{AtRep}(\varphi) = \{R(x_1, a), R(x_1, b), S(x_2, a), T(x_2, x_3)\}$, $\mathsf{Subst}(\varphi, s) = \{R(a_1, a), R(a_1, b), S(a_2, a), T(a_2, a_3)\}$.

Now let $\Phi \subseteq \mathsf{WFS}(\mathbf{D}, \exists\wedge+)$ be a finite set. A *substitution set* for Φ is a Φ-indexed set $S = \{s_\varphi \mid \varphi \in \Phi\}$ of substitutions, with s_φ a substitution for $\mathsf{Vars}(\varphi)$. S is *free for* Φ if each s_φ is correctly typed for φ, injective, and, furthermore, for any distinct $\varphi_1, \varphi_2 \in \Phi$, $s_{\varphi_1}(\mathsf{Vars}(\varphi_1)) \cap s_{\varphi_2}(\mathsf{Vars}(\varphi_2)) = \emptyset$.

For a free substitution S, the *canonical model* defined by (Φ, S) is obtained by applying the substitution s_φ to φ for each $\varphi \in \Phi$. Formally, $\mathsf{CanMod}\langle\Phi, S\rangle = \bigcup\{\mathsf{Subst}(\varphi, s_\varphi) \mid \varphi \in \Phi\}$.

For an illustrative example, let $\varphi_1 = (\exists x_1)(\exists x_2)(\exists x_3)(R(x_1, x_2)\wedge R(x_2, x_3))$, $\varphi_2 = (\exists x_1)(\exists x_2)(\exists x_3)(R(a_1, x_1)\wedge S(x_1, a_2)\wedge T(x_2, x_3)\wedge T(x_2, a_3))$, $\varphi_3 = R(a_1, a_2)$, and $\varphi_4 = S(a_2, a_3)$, with the a_i's all distinct constants, and let $\Phi = \{\varphi_1, \varphi_2, \varphi_3, \varphi_4\}$. Put $s_{\varphi_1} = \{b_{11}/x_1, b_{12}/x_2, b_{13}/x_3\}$ $s_{\varphi_2} = \{b_{21}/x_1, b_{22}/x_2, b_{23}/x_3\}$, and $s_{\varphi_3} = s_{\varphi_4} = \emptyset$, with all of the b_{ij}'s distinct constants. Then $S = \{s_{\varphi_1}, s_{\varphi_2}, s_{\varphi_3}, s_{\varphi_4}\}$ is free for Φ, with $\mathsf{Subst}(\varphi_1, s_{\varphi_1}) = \{R(b_{11}, b_{12}), S(b_{12}, b_{13})\}$, $\mathsf{Subst}(\varphi_2, s_{\varphi_2}) = \{R(a_1, b_{21}), S(b_{21}, a_2), T(b_{22}, b_{23}), T(b_{22}, a_3)\}$, $\mathsf{Subst}(\varphi_3, s_{\varphi_3}) = \{R(a_1, a_2)\}$, and $\mathsf{Subst}(\varphi_4, s_{\varphi_4}) = \{S(a_2, a_3)\}$. Unfortunately, while $\mathsf{CanMod}\langle\Phi, S\rangle = \{\mathsf{Subst}(\varphi_i, s_{\varphi_i}) \mid 1 \leq i \leq 4\}$ is an Armstrong model for

$\{\varphi_i \mid 1 \leq i \leq 4\}$ with respect to $\mathsf{WFF}(\mathbf{D}, \exists\wedge+)$, it is not not minimal. The problem is that there is redundancy within Φ, which results in redundancy in the canonical model. For example, it is easy to see that φ_1 is a logical consequence of φ_2, and so $\mathsf{Subst}(\varphi_1, s_{\varphi_1})$ can be removed from $\mathsf{CanMod}\langle\Phi, S\rangle$ entirely, with the result still an Armstrong model. While this problem could be resolved by choosing the substitutions more cleverly, it is more straightforward to normalize the the the set of sentences before applying the construction of the canonical model, as developed next.

Definition 3.11 (Reduction steps). To construct a minimal Armstrong model from a set $\Phi \subseteq \mathsf{WFS}(\mathbf{D}, \exists\wedge+)$, it is first necessary to normalize Φ by applying three simple reduction rules, defined as follows.

Decomposition: For $\varphi \in \Phi$, if $\{X_1, X_2\}$ partitions $\mathsf{AtRep}(\varphi)$ into disjoint sets, and $\mathsf{Mod}_{\mathcal{I}}(\mathsf{AtRep}^{-1}(X_1)) \cap \mathsf{Mod}_{\mathcal{I}}(\mathsf{AtRep}^{-1}(X_2)) = \mathsf{Mod}_{\mathcal{I}}(\varphi)$, then remove φ from Φ and add both $\mathsf{AtRep}^{-1}(X_1)$ and $\mathsf{AtRep}^{-1}(X_2)$.

Collapsing: For $\varphi \in \Phi$, if $\{X_1, X_2\}$ partitions $\mathsf{AtRep}(\varphi)$ into disjoint sets, and $\mathsf{Mod}_{\mathcal{I}}(\mathsf{AtRep}^{-1}(X_1)) \subseteq \mathsf{Mod}_{\mathcal{I}}(\mathsf{AtRep}^{-1}(X_2))$, then remove φ from Φ and add $\mathsf{AtRep}^{-1}(X_1)$.

Minimization: If $\mathsf{Mod}_{\mathcal{I}}(\Phi \setminus \{\varphi\}) = \mathsf{Mod}_{\mathcal{I}}(\Phi)$, then remove φ from Φ.

It is clear that each of these steps preserves $\mathsf{Mod}_{\mathcal{I}}(\Phi)$, and that they may only be applied a finite number of times before none is applicable. Call Φ *reduced* if none of these steps is applicable.

A simple example will help illustrate how these rules work. Let Φ be as in Definition 3.10 above. Using the decomposition rule, $\mathsf{AtRep}(\varphi_2) = \{R(a_1, x_1), R(x_1, a_2), T(x_2, x_3), T(x_2, a_3)\}$ may be replaced with $\{R(a_1, x_1), S(x_1, a_2)\}$ and $\{T(x_2, x_3), T(x_2, a_3)\}$, since these two sets have no variables in common. Next, $\{T(x_2, x_3), T(x_2, a_3)\}$ may be replaced with $\{T(x_2, a_3)\}$ using the collapsing rule. Finally, φ_1 may be removed using the minimization rule, since it is a consequence of $\varphi_3\wedge\varphi_4$. The final reduced version of Φ is thus $\{(\exists x_1)(R(a_1, x_1)\wedge S(x_1, a_2)), (\exists x_2)(T(x_2, a_3)), R(a_1, a_2), S(a_2, a_3)\}$. Note that $(\exists x_1)(R(a_1, x_1)\wedge S(x_1, a_2))$ is not a consequence of $R(a_1, a_2)$ and $S(a_2, a_3)$, and so it cannot be removed by this procedure. A minimal Armstrong model is obtained by substituting a distinct new constant for each variable: $\{R(a_1, b_1), R(b_1, a_2), T(b_2, a_3), R(a_1, a_2), S(a_2, a_3)\}$. Furthermore, this model is obtained from the one of Definition 3.10 above via the endomorphism which maps $b_{11} \mapsto a_1$, $b_{12} \mapsto a_2$, $b_{13} \mapsto a_3$, $b_{21} \mapsto b_1$, $b_{22} \mapsto b_2$, $b_{23} \mapsto a_3$, and is the identity on everything else. To establish this result in a completely formal fashion requires a bit of work, and is presented below.

Theorem 3.12 (Characterization of minimal Armstrong models). *Let $\Phi \subseteq \mathsf{WFS}(\mathbf{D}, \exists\wedge+)$ be a finite set of constraints, and assume furthermore that Φ is reduced in the sense of Definition 3.11 above. Let S be a substitution set which is free for Φ. Then the following hold.*

(a) $\mathsf{CanMod}\langle\Phi, S\rangle$ *is a minimal Armstrong model for Φ relative to* $\mathsf{WFF}(\mathbf{D}, \mathsf{ConstSym}(\Phi), \exists\wedge+)$.

(b) *For any $M \in \mathsf{Mod}_{\mathcal{I}}(\varPhi)$, there is a $\mathsf{ConstSym}(\varPhi)$-invariant endomorphism h on $\mathcal{D}$ with $h(\mathsf{CanMod}\langle \varPhi, S \rangle) \subseteq M$.*

(c) *If M is any other minimal Armstrong model for $\varPhi$ relative to $\mathsf{WFS}(\mathbf{D}, \mathsf{ConstSym}(\varPhi), \exists \wedge +)$, then there is a $\mathsf{ConstSym}(\varPhi)$-invariant automorphism h on $\mathcal{D}$ with $h(\mathsf{CanMod}\langle \varPhi, S \rangle) = M$.*

Proof. It is immediate that $\mathsf{CanMod}\langle \varPhi, S \rangle$ is a model of $\varPhi$. It is furthermore easy to see that it is minimal; if any tuple is deleted, the $\varphi \in \varPhi$ associated with the tuple in $\mathsf{CanMod}\langle \varPhi, S \rangle$ is no longer satisfied, since $\varPhi$ is assumed to be minimized, as defined in Definition 3.11 above. To show that it is an Armstrong model, let $\psi \in \mathsf{WFS}(\mathbf{D}, \exists \wedge +)$ for which $\mathsf{Mod}_{\mathcal{I}}(\varPhi) \subseteq \mathsf{Mod}_{\mathcal{I}}(\psi)$ does not hold, and let S' be a substitution set, free for $\varPhi \cup \{\psi\}$, which is built from S by adding a substitution associated with ψ. Let the resulting set of constraints by the reduction steps of Definition 3.11 from $\varPhi \cup \{\psi\}$ be denoted by $\varPhi'$. For the reduction steps of Definition 3.11, it suffices to note that ψ cannot be removed by minimization. Hence $\mathsf{CanMod}\langle \varPhi', S' \rangle \subseteq \mathsf{CanMod}\langle \varPhi, S \rangle$ cannot hold, and so $\mathsf{CanMod}\langle \varPhi, S \rangle$ cannot be a model of ψ, whence $\mathsf{CanMod}\langle \varPhi, S \rangle$ is an Armstrong model of $\varPhi$.

To establish (b), let $M \in \mathsf{Mod}_{\mathcal{I}}(\varPhi)$, and for each $\varphi \in \varPhi$, let M_φ be a minimal subset of M with $M_\varphi \in \mathsf{Mod}_{\mathcal{I}}(\varphi)$. Let V_φ denote the set of variables of $s_\varphi \in S$. It is easy to see that there must be a substitution s'' with $\mathsf{Vars}(s'') = V_\varphi$ and $\mathsf{Subst}(\varphi, s'') = M_\varphi$. Indeed, there is trivially a substitution with $\mathsf{Subst}(\varphi, s'') \subseteq M_\varphi$, but if the subset inclusion were proper, M_φ would not be minimal.

Now define $h : s_\varphi(V_\varphi) \to s''(V_\varphi)$ by $a \mapsto s''(s_\varphi^{-1}(a))$. Since s_φ is injective, h is well defined. Since $s_{\varphi_1}(\mathsf{Vars}(\varphi_1)) \cap s_{\varphi_2}(\mathsf{Vars}(\varphi_2)) = \emptyset$ for distinct $\varphi_1, \varphi_2 \in \varPhi$, there are no conflicts in this definition of h. Finally, extend h to be the identity on all $a \in \mathsf{Const}(\mathcal{D})$ which are not covered by the above definition. The result is a endomorphism on $\mathcal{D}$ which satisfies $h(\mathsf{CanMod}\langle \varPhi, S \rangle) \subseteq M$.

To show (c), let M be any other minimal Armstrong model for $\varPhi$ relative to $\mathsf{WFS}(\mathbf{D}, \mathsf{ConstSym}(\varPhi)) \exists \wedge +$. In the above construction for the proof of (b), the resulting h must be surjective (else M would not be minimal), and it must be injective (since there must also be an endomorphism in the opposite direction, and both $\mathsf{CanMod}\langle \varPhi, S \rangle$ and M are finite, by assumption). Hence, h is an automorphism. □

The desired result, that any two optimal realizations are isomorphic up to a renaming via an automorphism, follows directly as a corollary.

Corollary 3.13 (Optimal updates are unique up to constant automorphism). *Let (M_1, N_2) be an update request from $\varGamma$ to $\mathbf{D}$, and let (M_1, M_2) and (M_1, M_2') be $\langle \mathsf{StdUCP}(\mathbf{D}, M_1), \mathsf{UpdRealiz}\langle M_1, N_2, \varGamma \rangle \rangle$-optimal realizations of (M_1, N_2). Then there is a $\mathsf{ConstSym}_{\mathcal{D}}^+(M)$-invariant automorphism h on $\mathcal{D}$ with $M_2' = h(M_2)$.* □

In some ways, the construction given above is similar to the construction of the universal solutions of [16, Def. 2.4], in that both are based upon similar notions of endomorphism (there termed *homomorphism*). However, those universal solutions are not required to be minimal. On the other hand, they are not limited to positive sentences, but rather apply to XEIDs, as developed in the next section.

4   Optimal Insertion in the Context of XEIDs

In this section, it is shown that in the context of database constraints which are extended embedded implicational dependencies (XEIDs), and views which are of class $\exists\wedge+$, all admissible realizations of an insertion request are optimal. In other words, there cannot be non-isomorphic minimal realizations of an update request which is an insertion. To establish this isomorphism, it is necessary to rule out the kind of non-isomorphic alternatives which are illustrated in Example 3.6. The logical formulation which formalizes this idea is splitting of disjunctions. Informally, disjunction splitting [8, Thm. 3.1(c)] stipulates that nondeterminism in logical implication cannot occur. If a set Ψ_1 of sentences implies the disjunction of all sentences in Ψ_2, then it must in fact imply some sentence in Ψ_2. Since Ψ_2 may be infinite, the notion of disjunction must be formulated carefully.

Notation 4.1 (Notational convention).   Throughout this section, unless stated explicitly to the contrary, take Σ to be an update classifier for $\mathbf{D}$.

Definition 4.2 (Splitting of disjunctions over finite databases).   The family $\Phi \subseteq \mathsf{WFS}(\mathbf{D})$ *splits disjunctions over finite databases* if whenever $\Psi_1, \Psi_2 \subseteq \Phi$ with Ψ_2 nonempty have the property that $\mathsf{RNeMod}_\mathcal{I}(\Psi_1) \subseteq \bigcup\{\mathsf{RNeMod}_\mathcal{I}(\psi') \mid \psi' \in \Psi_2\}$, then there is a $\psi \in \Psi_2$ with $\mathsf{RNeMod}_\mathcal{I}(\Psi_1) \subseteq \mathsf{RNeMod}_\mathcal{I}(\psi)$. The limitation to relationwise-nonempty databases is a technical one which will ultimately be necessary. The definition can, of course, be made without this restriction, and even Theorem 4.3 below is true without it, but the critical result Theorem 4.9 would fail. Since requiring databases to be relationwise nonempty is not much of restriction, it is easiest to require it throughout.

Define $\mathsf{SDConstr}\langle \mathbf{D}, \Gamma, \Sigma \rangle = \mathsf{Constr}(\mathbf{D}) \cup \mathsf{GrAtoms}(\mathbf{D}) \cup \{\mathsf{Subst}\langle \gamma, t \rangle \mid t \in \mathsf{GrAtoms}(\mathbf{V})\} \cup \Sigma$. Basically, $\mathsf{SDConstr}\langle \mathbf{D}, \Gamma, \Sigma \rangle$ is the set of all sentences which can arise in the construction of updates on $\mathbf{D}$ induced via updates on the view Γ. $\mathsf{Constr}(\mathbf{D})$ is (a basis for) the set of all constraints on $\mathbf{D}$, $\mathsf{GrAtoms}(\mathbf{D})$ is the set of all ground atoms which can occur in a database of $\mathbf{D}$, $\{\mathsf{Subst}\langle \gamma, t \rangle \mid t \in \mathsf{GrAtoms}(\mathbf{V})\}$ is the set of all sentences on $\mathbf{D}$ which can arise by reflecting an atom of the view Γ to the main schema, and Σ is the set of all sentences which are used in measuring information content. If all of these together split disjunctions, the constructions will work. Formally, say that the triple $(\mathbf{D}, \Gamma, \Sigma)$ *supports disjunction splitting over finite databases* if $\mathsf{SDConstr}\langle \mathbf{D}, \Gamma, \Sigma \rangle$ splits disjunctions over finite databases.

Theorem 4.3 (Disjunction splitting implies that admissible insertions are optimal).   *Assume that $(\mathbf{D}, \Gamma, \Sigma)$ supports disjunction splitting over finite databases, and let (M_1, N_2) be an insertion request from Γ to $\mathbf{D}$ with the property that M_1 is relationwise nonempty. Then all $\langle \Sigma, \uparrow \rangle$-admissible realizations of (M_1, N_2) are $\langle \Sigma, \uparrow \rangle$-optimal.*

Proof. First of all, observe that $\Psi_1 = \mathsf{Constr}(\mathbf{D}) \cup \{\mathsf{Subst}\langle\gamma, t\rangle \mid t \in N_2\} \cup M_1$ is precisely the set of constraints which the updated database of $\mathbf{D}$ must satisfy; $(M_1, M_2) \in \mathsf{InsRealiz}\langle M_1, N_2, \Gamma\rangle$ iff $M_2 \in \mathsf{Mod}_\mathcal{I}(\Psi_1)$. Furthermore, since $\Psi_1 \subseteq \mathsf{SDConstr}\langle \mathbf{D}, \Gamma, \Sigma\rangle$, it splits disjunctions over finite databases.

Now, let S denote the set of all $M_2 \in \mathsf{LDB}(\mathbf{D})$ for which (M_1, M_2) is a $\langle\Sigma, \uparrow\rangle$-admissible realization of (M_1, N_2), and assume that S is nonempty. Let Ψ_2 denote the set of all $\psi \in \Sigma$ with the property that $M_2 \in \mathsf{Mod}_\mathcal{I}(\psi)$ for some, but not all, $M_2 \in S$. If $\Psi_2 = \emptyset$, then all members of S are Σ-equivalent, and so all are least with respect to Σ and hence $\langle\Sigma, \uparrow\rangle$-optimal. If $\Psi_2 \neq \emptyset$, then for each $M_2 \in S$, there must be some $\psi \in \Psi_2$ with the property that $M_2 \in \mathsf{Mod}_\mathcal{I}(\psi)$. Otherwise, $\mathsf{Info}\langle M_2, \Sigma\rangle \subsetneq \mathsf{Info}\langle M_2', \Sigma\rangle$ for all $M_2' \in S \cap \mathsf{Mod}_\mathcal{I}(\psi)$, which would contradict the Σ-admissibility of any such M_2'. Thus $\mathsf{RNeMod}_\mathcal{I}(\Psi_1) \subseteq \bigcup\{\mathsf{RNeMod}_\mathcal{I}(\psi') \mid \psi' \in \Psi_2\}$. Since M_1 is relationwise nonempty, so too is each $M_2 \in S$. Now, using the fact that Ψ_1 splits disjunctions, there is some $\psi \in \Psi_2$ with the property that $\mathsf{RNeMod}_\mathcal{I}(\Psi_1) \subseteq \mathsf{RNeMod}_\mathcal{I}(\psi)$; i.e., that $M_2 \in \mathsf{Mod}_\mathcal{I}(\psi)$ for all $M_2 \in S$. This is a contradiction, and so $\Psi_2 = \emptyset$. Thus all $\langle\Sigma, \uparrow\rangle$-admissible realizations of (M_1, N_2) are are $\langle\Sigma, \uparrow\rangle$-optimal. $\square$

Definition 4.4 (Extended tuple databases). The results which follow use heavily the framework developed by Fagin in [8]. It is necessary in particular to be able to construct infinite products of databases. This leads to two complications. First of all, the databases of this paper are finite, while such products may be infinite. Second, the databases of this paper here have a fixed bijective correspondence between the domain of the interpretation and constant symbols which cannot be preserved completely under products. Fortunately, such products are not really used as databases; rather they are just artefacts which arise in the proof to show that a certain context supports the splitting of disjunctions. The solution is to embed the $\mathbf{D}$-tuples of this paper into a larger set, called the extended $\mathbf{D}$-tuples, and to carry out the infinite-product constructions on databases of these extended tuples. Since every tuple database in the sense of this paper is also an extended database, the results will follow.

Formally, an *extended tuple database* M over $\mathbf{D}$ consists of the following:

(xtdb-i) A set $\mathsf{Dom}(M)$, called the *domain* of M.

(xtdb-ii) An injective function $\iota_M : \mathsf{Dom}_\mathcal{I} \to \mathsf{Dom}(M)$.

(xtdb-iii) A (not necessarily finite) set $\mathsf{XTuples}(M)$ of extended $\mathbf{D}$-tuples over $(\mathsf{Dom}(M), \iota_M)$.

For $R \in \mathsf{Rels}(\mathbf{D})$, an *extended R-tuple* t over $(\mathsf{Dom}(M), \iota_M)$ is a function $t : \{\mathsf{RName}\} \cup \mathsf{Ar}_\mathbf{D}(R) \to \mathsf{Dom}(M) \cup \mathsf{Rels}(\mathbf{D})$ with the property that $t[\mathsf{RName}] = R$ and, for all $A \in \mathcal{A}_\mathcal{D}$, if $t[A] \in \iota_M(\mathsf{Dom}_\mathcal{I})$, then $\iota_M^{-1}(t[A]) \in \mathsf{Dom}_\mathcal{I}(A)$. An *extended $\mathbf{D}$-tuple* over $(\mathsf{Dom}(M), \iota_M)$ is an extended R-tuple over that same pair for some $R \in \mathsf{Rels}(\mathbf{D})$. $\mathsf{XTuples}(M)$ denotes $\bigcup\{\mathsf{XTuples}(R, M) \mid R \in \mathsf{Rels}(\mathbf{D})\}$. The collection of all extended tuple databases over $\mathbf{D}$ is denoted $\mathsf{XTDB}(\mathbf{D})$, with $\mathsf{RNeXTDB}(\mathbf{D})$ denoting the subcollection consisting of all relationwise-nonempty members (obvious definition). As a slight abuse of notation, $t \in M$ will be used as shorthand for $t \in \mathsf{XTuples}(M)$.

Note that every $M \in \mathsf{TDB}(\mathbf{D})$ may be regarded as an extended tuple database by setting $\mathsf{Dom}(M) = \mathsf{Dom}_{\mathcal{I}}$ and taking ι_M to be the identity function. In an extended $\mathbf{D}$-tuple, domain elements which are not in $\iota_M(\mathsf{Dom}_{\mathcal{I}})$ are not associated with any constant symbol.

For $\varphi \in \mathsf{WFS}(\mathbf{D})$, define $\mathsf{XMod}_{\mathcal{I}}(\varphi)$ to be the set of $M \in \mathsf{XTDB}(\mathbf{D})$ which interpret the constant symbols according to $\mathsf{IntFn}_{\mathcal{I}}$, and which are models (in the usual logical sense) of both φ and $\mathsf{UNA}(\mathcal{D})$. For $\varPhi \subseteq \mathsf{WFS}(\mathbf{D})$, $\mathsf{XMod}_{\mathcal{I}}(\varPhi) = \bigcap\{\mathsf{XMod}_{\mathcal{I}}(\varphi) \mid \varphi \in \varPhi\}$. The relationwise-nonempty versions, $\mathsf{RNeXMod}_{\mathcal{I}}(\varphi)$ and $\mathsf{RNeXMod}_{\mathcal{I}}(\varPhi)$, are defined analogously. Note that $\mathsf{Mod}_{\mathcal{I}}(\varphi) \subseteq \mathsf{XMod}_{\mathcal{I}}(\varphi)$ and $\mathsf{RNeMod}_{\mathcal{I}}(\varphi) \subseteq \mathsf{RNeXMod}_{\mathcal{I}}(\varphi)$ under these definitions; i.e., ordinary models are extended models.

Definition 4.5 (Products of extended tuple databases). Let $P = \{M_j \mid j \in J\}$ be an indexed set of nonempty extended tuple databases over $\mathbf{D}$. The $\mathcal{D}$-product of P, denoted $\otimes^{\mathcal{D}}(P)$, is the extended tuple database defined as follows:

(i) $\mathsf{Dom}(\otimes^{\mathcal{D}}(P)) = \prod_{j \in J} \mathsf{Dom}(M_j)$.

(ii) $\iota_{\otimes^{\mathcal{D}}(P)} : x \mapsto \langle \iota_{M_j}(x) \rangle_{j \in J}$ (the J-tuple whose j^{th} entry is $\iota_{M_j}(x)$).

(iii) $\mathsf{XTuples}(R, \otimes^{\mathcal{D}}(P)) = \{\otimes\langle t_j \rangle_{j \in J} \mid t_j \in \mathsf{XTuples}(R, M_j)\}$.

In the above, $t' = \otimes\langle t_j \rangle_{j \in J}$ is the extended R-tuple with $t'[A] = \langle t_j[A] \rangle_{j \in J}$ for each $A \in \mathsf{Ar}_{\mathbf{D}}(R)$.

Call $\otimes^{\mathcal{D}}(P)$ *lossless* if each M_j can be recovered from it. Note that $\otimes^{\mathcal{D}}(P)$ is lossless if each $M_j \in P$ is in $\mathsf{RNeXTDB}(\mathbf{D})$; however, given $R \in \mathsf{Rels}(\mathbf{D})$, if some M_j contains no R-tuples, then the entire product will contain no R-tuples. Since it is essential to be able to recover P from $\otimes^{\mathcal{D}} D$, the condition that each $M_j \in P$ be relationwise nonempty will be enforced.

Definition 4.6 (Splitting of disjunctions over extended databases). Definition 4.2 can be extended in the obvious fashion to extended databases. Specifically, the family $\varPhi \subseteq \mathsf{WFS}(\mathbf{D})$ *splits disjunctions over extended databases* if whenever $\varPsi_1 \subseteq \varPhi$ and $\varPsi_2 \subseteq \varPhi$ with $\varPsi_2$ nonempty have the property that $\mathsf{RNeXMod}_{\mathcal{I}}(\varPsi_1) \subseteq \bigcup\{\mathsf{RNeXMod}_{\mathcal{I}}(\psi') \mid \psi' \in \varPsi_2\}$, then there is a $\psi \in \varPsi_2$ with $\mathsf{RNeXMod}_{\mathcal{I}}(\varPsi_1) \subseteq \mathsf{RNeXMod}_{\mathcal{I}}(\psi)$. Similarly, the triple $\mathsf{SDConstr}\langle \mathbf{D}, \varGamma, \varSigma \rangle$ *supports disjunction splitting over extended databases* if $\mathsf{SDConstr}\langle \mathbf{D}, \varGamma, \varSigma \rangle$ splits disjunctions over extended databases.

Because ordinary tuples may be interpreted as extended tuples, splitting of disjunctions over extended databases trivially implies splitting of disjunction over ordinary finite databases. Due to its importance, this fact is recorded formally.

Observation 4.7. *If the family $\varPhi \subseteq \mathsf{WFS}(\mathbf{D})$ splits disjunctions over extended databases, then it splits disjunctions over finite databases as well. In particular, if the triple $\mathsf{SDConstr}\langle \mathbf{D}, \varGamma, \varSigma \rangle$ supports disjunction splitting over extended databases, then it supports disjunction splitting over finite databases.* □

Definition 4.8 (Faithful sentences). Informally, a sentence $\varphi \in \mathsf{WFS}(\mathbf{D})$ is faithful [8] if it is preserved under the formation of products and under the projection of factors from products. Formally, $\varphi \in \mathsf{WFS}(\mathbf{D})$ is said to be *faithful* if whenever $P = \{M_j \mid j \in J\} \subseteq \mathsf{RNeXTDB}(\mathbf{D})$ is a nonempty (indexed) set, $\otimes^{\mathcal{D}}(P) \in \mathsf{XMod}_{\mathcal{I}}(\varphi)$ iff $M_j \in \mathsf{XMod}_{\mathcal{I}}(\varphi)$ for each $M_j \in P$. The family $\Phi \subseteq \mathsf{WFS}(\mathbf{D})$ is *faithful* precisely in the case that each $\varphi \in \Phi$ is.

Theorem 4.9 (Faithful $\equiv$ disjunction splitting). *Let $\Phi \subseteq \mathsf{WFS}(\mathbf{D})$. Then Φ is faithful iff it splits disjunctions over extended databases.*

Proof. See [8, Thm. 3.1]. □

Definition 4.10 (XEIDs). The *extended embedded implicational dependencies (XEIDs)* form a very general class which includes most types of dependencies which have been studied, including functional dependencies, multivalued dependencies, (embedded) join dependencies, and inclusion dependencies. Formally, an *XEID* [8, Sec. 7] is a sentence in $\mathsf{WFS}(\mathbf{D})$ of the form

$$(\forall x_1)(\forall x_2)\ldots(\forall x_n)((A_1 \wedge A_2 \wedge \ldots \wedge A_n) \Rightarrow (\exists y_1)(\exists y_2)\ldots(\exists y_r)(B_1 \wedge B_2 \wedge \ldots \wedge B_s))$$

such that each A_i is a relational atom for the same relation, i.e., the left-hand side is unirelational, each B_i is a relational atom or an equality, each x_i occurs in some A_j, the left-hand side is typed in the sense that no variable is used for more than one attribute. In the original definition of Fagin, it is also required that $n \geq 1$. However, this is an inessential constraint which may easily be dropped, as long as at least one of the $B_i's$ is a relational atom (and not an equality). Indeed, let $\varphi = (\exists y_1)(\exists y_2)\ldots(\exists y_r)(B_1 \wedge B_2 \wedge \ldots \wedge B_s)) \in \mathsf{WFS}(\mathbf{D}, \exists \wedge +)$, with B_i, say, a relational atom for relation symbol R. Let $A = R(x_1, x_2, .., x_n)$ be a relational atom for R with variables as arguments. Then $\varphi' = (\forall x_1)(\forall x_2)\ldots(\forall x_n)(A \Rightarrow (\exists y_1)(\exists y_2)\ldots(\exists y_r)(A \wedge B_1 \wedge B_2 \wedge \ldots \wedge B_s)) \in \mathsf{WFS}(\mathbf{D})$ is equivalent to φ, but with $n \geq 1$. Thus, without loss of generality, in this paper sentences in $\mathsf{WFS}(\mathbf{D}, \exists \wedge +)$ will also be regarded as XEIDs.

The set of all XEIDs on $\mathbf{D}$ is denoted $\mathsf{XEID}(\mathbf{D})$, while those XEIDs involving only the constant symbols in $S \subseteq \mathsf{Const}(\mathcal{D})$ is denoted $\mathsf{XEID}(\mathbf{D}, S)$.

The reason that XEIDS are of interest here is the following.

Proposition 4.11. *Every $\Phi \subseteq \mathsf{XEID}(\mathbf{D})$ is faithful.*

Proof. This is essentially [8, Thm. 7.2]. The only complication is the constant symbols, which are not part of the framework of [8], so the integrity of $\mathsf{UNA}(\mathcal{D})$ (not a set of XEIDs) must be verified. To this end, simply note that a domain value in an extended model is associated with constant a iff *each* of its projections is the domain value $\mathsf{IntFn}_I(a)$, so $\mathsf{UNA}(\mathcal{D})$ is enforced by construction. □

Lemma 4.12 (XEIDs support disjunction splitting). *Let $\mathsf{Constr}(\mathbf{D}) \subseteq \mathsf{XEID}(\mathbf{D})$, Γ a view of class $\exists \wedge +$, and Σ an update classification pair for $\mathbf{D}$ with $\Sigma \subseteq \mathsf{XEID}(\mathbf{D})$. Then $\mathsf{SDConstr}\langle \mathbf{D}, \Gamma, \Sigma \rangle$ supports disjunction splitting over extended databases.*

Proof. By construction $\mathsf{SDConstr}\langle \mathbf{D}, \Gamma, \Sigma \rangle \subseteq \mathsf{XEID}(\mathbf{D})$, and so the result follows from Theorem 4.9 and Proposition 4.11. □

Finally, the main result on the existence of optimal reflections may be established.

Theorem 4.13 (XEIDs imply optimal insertions). *Let* Constr($\mathbf{D}$) $\subseteq$ XEID($\mathbf{D}$), Γ *a view of class* $\exists \wedge +$, (M_1, N_2) *an insertion request from* Γ *to* $\mathbf{D}$ *with* $M \in$ RNeLDB($\mathbf{D}$), *and* Σ *an update classification pair for* $\mathbf{D}$ *with* $\Sigma \subseteq$ XEID($\mathbf{D}$). *Then every* $\langle \Sigma, \uparrow \rangle$*-minimal realization of* (M_1, N_2) *is* $\langle \Sigma, \uparrow \rangle$*-optimal.*

Proof. Combine Theorem 4.3, Lemma 4.12, and Observation 4.7. □

Discussion 4.14 (Dependencies which guarantee minimal realizations). The above result states that whenever an admissible realization exists, it must be optimal. However, it says nothing about existence, and, indeed it is possible to construct views for which no Σ-admissible update exists. For example, let the schema $\mathbf{E}_2$ have three relational symbols $R[A]$, $S[AB]$, and $T[AB]$ with the inclusion dependencies $R[A] \subseteq S[A]$, $S[A] \subseteq T[A]$, and $T[B] \subseteq S[B]$. Let $M_1 = \{R(a_0), S(a_0, b_0), T(a_0, b_0)\}$. Consider the view $\Pi_{R[A]} = (R[A], \pi_{R[A]})$, which preserves R but discards S and T. Let the current state of this view be $N_1 = \{R(a_0)\}$; consider updating it to $N_2 = N_1 \cup \{R(a_1)\}$. For $\Sigma = $ StdUCP($\mathbf{E}_2, M_1$), there is no Σ-admissible realization of (M_1, N_2). Indeed, a tuple of the form $S(a_1, b_1)$ must be inserted, and this then implies that one of the form $T(a_2, b_1)$ must be inserted as well, which in turn implies that one of the form $S(a_2, b_3)$ must be inserted, and so forth. It is easy to see that if this sequence is terminated by forcing an equality (say, by replacing b_3 with b_2), then the resulting insertion is not Σ-admissible. In other words, relative to WFS($\mathbf{D}$, ConstSym$_{\mathcal{D}}^+(M_1)$, $\exists \wedge +$), there are no admissible solutions. In the vernacular of traditional dependency theory, this is a situation in which the chase inference process does not terminate with a finite solution [16, Def. 3.2]. To ensure termination, attention may be restricted to the subclass of XEIDs consisting of the *weakly acyclic tuple generating dependencies (TGDs)* together with the *equality generating dependencies (EGDs)*. See [16, Thm. 3.9] for details.

5   Conclusions and Further Directions

A strategy for the optimal reflection of view updates has been developed, based upon the concept of least information change. It has been shown in particular that optimal insertions are supported in a reasonable fashion — they are unique up to a renaming of the newly-inserted constants. Nonetheless, a number of issues remain for future investigation. Among the most important are the following.

Optimization of tuple modification: Although the general formulation applies to all types of updates, the results focus almost entirely upon insertions. Due to space limitation, deletions have not been considered in this paper; however, since deletions introduce no new constants or tuples, their analysis is relatively unremarkable within this context. Modification of single tuples ("updates" in SQL), on the other hand, are of fundamental importance. With the standard update classification pair of Example 3.6, only very special cases admit optimal solutions. The difficulty arises from the fact that the framework, which is based entirely upon information content, cannot distinguish between the process of modifying a tuple and that of deleting it and then inserting a new one. Consequently, both appear as admissible updates, but neither is optimal relative to the other. Further work must therefore look for a way to recapture the distinction between tuple modification and a delete-insert pair.

Application to database components: This investigation began as an effort to understand better how updates are propagated between database components, as forwarded in [17, Sec. 4], but then took on a life of its own as it was discovered that the component-based problems were in turn dependent upon more fundamental issues. Nevertheless, it is important to return to the roots of this investigation — database components. This includes not only the purely autonomous case, as sketched in [17, Sec. 4], but also the situation in which users cooperate to achieve a suitable reflection, as introduced in [18]

Relationship to work in logic programming: The problem of view update has also been studied extensively in the context of deductive databases. Often, only tuple minimality is considered as an admissibility criterion, and the focus then becomes one of identifying efficient algorithms for identifying all such admissible updates [19]. However, some recent work has introduced the idea of using active constraints to establish a preference order on admissible updates [20]. Thus, rather than employing a preference based upon information content, one based upon explicit rules is employed. The relationship between such approaches and that of this paper warrants further investigation. Also, there has been a substantial body of work on updates to disjunctive deductive databases [21], in which the extensional database itself consists of a collection of alternatives. The approach of minimizing information change in the disjunctive context deserves further attention as well.

Acknowledgment. Much of this research was carried out while the author was a visitor at the Information Systems Engineering Group at the University of Kiel. He is indebted to Bernhard Thalheim and the members of the group for the invitation and for the many discussions which led to this work. Also, the anonymous reviewers made numerous suggestions which (hopefully) have led to a more readable presentation.

References

1. Bancilhon, F., Spyratos, N.: Update semantics of relational views. ACM Trans. Database Systems 6, 557–575 (1981)
2. Hegner, S.J.: An order-based theory of updates for database views. Ann. Math. Art. Intell. 40, 63–125 (2004)
3. Dayal, U., Bernstein, P.A.: On the correct translation of update opeartions on relational views. ACM Trans. Database Systems 8, 381–416 (1982)

4. Keller, A.M.: Updating Relational Databases through Views. PhD thesis, Stanford University (1985)
5. Langerak, R.: View updates in relational databases with an independent scheme. ACM Trans. Database Systems 15, 40–66 (1990)
6. Bentayeb, F., Laurent, D.: Inversion de l'algèbre relationnelle et mises à jour. Technical Report 97-9, Université d'Orléans, LIFO (1997)
7. Bentayeb, F., Laurent, D.: View updates translations in relational databases. In: Quirchmayr, G., Bench-Capon, T.J.M., Schweighofer, E. (eds.) DEXA 1998. LNCS, vol. 1460, pp. 322–331. Springer, Heidelberg (1998)
8. Fagin, R.: Horn clauses and database dependencies. J. Assoc. Comp. Mach. 29, 952–985 (1982)
9. Paredaens, J., De Bra, P., Gyssens, M., Van Gucht, D.: The Structure of the Relational Database Model. Springer, Heidelberg (1989)
10. Abiteboul, S., Hull, R., Vianu, V.: Foundations of Databases. Addison-Wesley, Reading (1995)
11. Ceri, S., Gottlog, G., Tanca, L.: Logic Programming and Databases. Springer, Heidelberg (1989)
12. Monk, J.D.: Mathematical Logic. Springer, Heidelberg (1976)
13. Genesereth, M.R., Nilsson, N.J.: Logical Foundations of Artificial Intelligence. Morgan Kaufmann, San Francisco (1987)
14. Jacobs, B.E., Aronson, A.R., Klug, A.C.: On interpretations of relational languages and solutions to the implied constraint problem. ACM Trans. Database Systems 7, 291–315 (1982)
15. Fagin, R., Vardi, M.Y.: Armstrong databases for functional and inclusion dependencies. Info. Process. Lett. 16, 13–19 (1983)
16. Fagin, R., Kolaitis, P.G., Miller, R.J., Popa, L.: Data exchange: Semantics and query answering. Theoret. Comput. Sci. 336, 89–124 (2005)
17. Hegner, S.J.: A model of database components and their interconnection based upon communicating views. In: Jakkola, H., Kiyoki, Y., Tokuda, T. (eds.) Information Modelling and Knowledge Systems XXIV. Frontiers in Artificial Intelligence and Applications. IOS Press, Amsterdam (in press)
18. Hegner, S.J., Schmidt, P.: Update support for database views via cooperation. In: Ioannis, Y., Novikov, B., Rachev, B. (eds.) ADBIS 2007. LNCS, vol. 4690, pp. 98–113. Springer, Heidelberg (2007)
19. Behrend, A., Manthey, R.: Update propagation in deductive databases using soft stratification. In: Benczúr, A.A., Demetrovics, J., Gottlob, G. (eds.) ADBIS 2004. LNCS, vol. 3255, pp. 22–36. Springer, Heidelberg (2004)
20. Greco, S., Sirangelo, C., Trubitsyna, I., Zumpano, E.: Preferred repairs for inconsistent databases. In: IDEAS 2003. 7th International Database Engineering and Applications Symposium, Hong Kong, China, pp. 202–211. IEEE Computer Society, Los Alamitos (2003)
21. Fernández, J.A., Grant, J., Minker, J.: Model theoretic approach to view updates in deductive databases. J. Automated Reasoning 17, 171–197 (1996)

Merging First-Order Knowledge
Using Dilation Operators

Nikos Gorogiannis and Anthony Hunter

Department of Computer Science,
University College London,
Gower Street, WC1E 6BT, London, UK
{n.gkorogiannis,a.hunter}@cs.ucl.ac.uk

Abstract. The area of knowledge merging is concerned with merging conflicting information while preserving as much as possible. Most proposals in the literature work with knowledge bases expressed in propositional logic. We propose a new framework for merging knowledge bases expressed in (subsets of) first-order logic. Dilation operators (a concept originally introduced by Bloch and Lang) are employed and developed, and by combining them with the concept of comparison orderings we obtain a framework that is driven by model-based intuitions but that can be implemented in a syntax-based manner. We demonstrate specific dilation operators and comparison orderings for use in applications. We also show how postulates from the literature on knowledge merging translate into our framework and provide the conditions that dilation operators and comparison orderings must satisfy in order for the respective merging operators to satisfy the new postulates.

1 Introduction

The dynamic nature of many information sources makes the appearance of inconsistencies a natural occurrence, either due to the need to consider multiple disparate sources, or because a single source is continually updated with new information. The need to study and handle these inconsistencies is well recognised and has spawned numerous research directions in non-monotonic logics, paraconsistent logics, belief revision and knowledge merging, to name a few. Logic-based knowledge merging is concerned with combining a set of potentially conflicting sources expressed in a logical language (*knowledge bases*), producing a single consistent knowledge base that maintains as much information content of the original knowledge bases as possible.

A *merging operator* is typically a function from tuples of knowledge bases (*profiles*) to knowledge bases. In recent years, the area of knowledge merging has seen significant research results, both in proposals for specific merging operators [1,2,3,4,5] as well as in proposals for unifying frameworks, where specific merging operators can be seen as instantiations of these frameworks [6,7,8,9].

Significantly, most of these results concern merging knowledge bases expressed in propositional logic (PL). If logic-based merging is to succeed in applications, then it is clear that more expressive languages need to be employed.

S. Hartmann and G. Kern-Isberner (Eds.): FoIKS 2008, LNCS 4932, pp. 132–150, 2008.

Although there exist proposals for knowledge merging with richer logics (mostly consistency-based, e.g. [1,10] but see also [11] for merging on infinite propositional logics), there is still a range of unresolved issues. In addition, we believe that computational issues should remain at the core of such investigations. In this paper, we examine the problems that arise when trying to construct a framework for merging knowledge expressed in first-order logic (FOL). We present a general framework that can be used when the language is (a subset of) FOL, and which makes use of the notion of a *dilation operator*, corresponding to the notion of distance employed in the knowledge merging frameworks such as the ones of [8,9]. The concept of a *comparison ordering* is then employed to order the result of applying the dilation operator to the original profile. The merging operators of our framework are, then, defined by minimising over these comparison orderings. The resulting framework has similarities to the work by Booth [12] in that repeated weakenings of the original profile are employed with the aim of reaching consensus; in contrast to that work, our framework is not limited to propositional logic, and does not explicitly use sets of models, making it computationally more appealing for richer logics. In addition, we instantiate this framework with specific dilation operators for use in different applications.

The outline of this paper is as follows. First, we cover some of the background literature in Section 2. Then, we introduce our approach in Section 3 and cover preliminary definitions in Section 4.1. We propose concrete dilation operators in Section 4.2 and comparison orderings in Section 4.3. We define the notion of a merging operator in Section 4.4 and present worked examples of concrete merging operators in Section 4.5. The general properties of merging operators in terms of merging postulates are examined in Section 4.6. Finally, we look at the issues that remain unaddressed and suggest further research topics in Section 5.

2   Background on Propositional Knowledge Merging

Merging operators from the literature can be broadly categorised in two groups: the *model-based*, or *semantic*, operators are defined using orderings over the models of PL, expressing how close models are with respect to the knowledge bases being merged [2,3,4,5]; and the *syntax-based*, or *consistency-based*, operators, where the merging process is defined in terms of consistent unions of subsets of the original knowledge bases [1,10,13]. We believe that the semantic approach offers several advantages such as syntax-independence and generality (e.g. sometimes syntax-based operators can be expressed as semantic ones), hence our approach is geared towards a semantic framework. In fact, we draw our intuitions from an important subset of the model-based operators, those that can be defined using a notion of *distance* between models (see, e.g. [9]).

We briefly cover the core concepts behind model-based merging operators. Let $\mathcal{L}$ be a finite PL language (i.e. a logic with a finite number of atoms), with $\mathcal{M}$ as its set of models and the usual operations mod $: \mathcal{L} \rightarrow 2^{\mathcal{M}}$ (which returns the set of models of a formula) and form $: 2^{\mathcal{M}} \rightarrow \mathcal{L}$ (which returns a formula whose set of models is the one provided). Knowledge bases are represented by formulae (equivalent to the conjunction of the formulae in the knowledge base). The set of all profiles (tuples of consistent formulae of $\mathcal{L}$) is denoted by $\mathcal{E}$. The concatenation of two profiles E_1, E_2 is written $E_1 \sqcup E_2$ and the length of E_1 as $|E_1|$. Two profiles $E_1 = \langle \phi_1, \ldots, \phi_k \rangle, E_2$ are equivalent, written $E_1 \leftrightarrow E_2$, iff $|E_1| = |E_2|$ and there is a bijection f from E_1 to E_2 such that $\phi_i \leftrightarrow f(\phi_i)$ for all $i \leq |E_1|$. The conjunction of all formulae in a tuple E is written $\bigwedge E$. We will abbreviate $E \sqcup \{\phi\}$ with $E \sqcup \phi$, and $\phi \sqcup \ldots \sqcup \phi$ (n times) with ϕ^n. The lexicographic ordering on tuples of integers is defined as follows. For two tuples of integers $A = \langle a_1, \ldots, a_k \rangle, B = \langle b_1, \ldots, b_k \rangle$ we write $A <_{\text{lex}} B$ iff there exists $i \leq |A|$ such that $a_i < b_i$ and for all j such that $1 \leq j < i$, $a_j = b_j$; and we write $A \leq_{\text{lex}} B$ iff $A <_{\text{lex}} B$ or $A = B$. We denote by $sort^a$ the function that takes a tuple of integers and sorts it in ascending order and by $sort^d$ the function that takes a tuple of integers and sorts it in descending order.

For most model-based merging operators the concept of a *distance function* is central. A distance function $d : \mathcal{M} \times \mathcal{M} \to \mathbb{N}$ evaluates how "close" two models are, and has the properties that for any $\omega_1, \omega_2 \in \mathcal{M}$, $d(\omega_1, \omega_1) = 0$ (identity of indiscernibles) and $d(\omega_1, \omega_2) = d(\omega_2, \omega_1)$ (symmetry). On top of this distance, a notion of distance between a formula $\phi \in \mathcal{L}$ and a model $\omega \in \mathcal{M}$ can be defined by setting $d(\phi, \omega) = \min\{ d(\omega_1, \omega) \mid \omega_1 \in \text{mod}(\phi) \}$. Most model-based operators use this notion of distance to construct another distance between a whole profile and a model. Then, the merging operator is defined by minimising this distance over $\mathcal{M}$. Some examples from the literature are given below.

The operator $\triangle_{\text{Max}}$ **[6].** The distance from a profile to a model for this operator is simply the maximum distance of each formula-to-model distance (where $E = \langle \phi_1, \ldots, \phi_k \rangle$):

$$d_{\text{Max}}(E, \omega) = \max\{ d(\phi_1, \omega), \ldots, d(\phi_k, \omega) \}$$
$$\triangle_{\text{Max}}(E) = \text{form}(\min\{ \omega \in \mathcal{M} \mid d_{\text{Max}}(E, \omega) \text{ is minimal} \})$$

The operator $\triangle_{\Sigma}$ **[6].** Here, the distance from a profile to a model is the sum of the individual formula-to-model distances:

$$d_{\Sigma}(E, \omega) = \sum_{i=1}^{k} d(\phi_i, \omega)$$
$$\triangle_{\Sigma}(E) = \text{form}(\min\{ \omega \in \mathcal{M} \mid d_{\Sigma}(E, \omega) \text{ is minimal} \})$$

The operator $\triangle^{\text{Gmax}}$ **[6].** In this case, the tuple of formula-to-model distances is sorted in ascending order and compared lexicographically.

$$d_{\text{Gmax}}(E, \omega) = sort^a \langle d(\phi_1, \omega), \ldots, d(\phi_k, \omega) \rangle$$
$$\triangle^{\text{Gmax}}(E) = \text{form}(\min\{ \omega \in \mathcal{M} \mid d_{\text{Gmax}}(E, \omega) \text{ is minimal w.r.t. } \leq_{\text{lex}} \})$$

The operator $\triangle^{\text{Gmin}}$ **[5].** Not dissimilarly with $\triangle^{\text{Gmax}}$, the tuple of formula-to-model distances is sorted in *descending* order and compared lexicographically.

$$d_{\text{Gmin}}(E, \omega) = sort^d \langle d(\phi_1, \omega), \ldots, d(\phi_k, \omega) \rangle$$
$$\triangle^{\text{Gmin}}(E) = \text{form}(\min\{ \omega \in \mathcal{M} \mid d_{\text{Gmin}}(E, \omega) \text{ is minimal w.r.t. } \leq_{\text{lex}} \})$$

A1. $\triangle(E)$ is consistent.

A2. If E is consistent, then $\triangle(E) = \bigwedge E$.

A3. If $E_1 \leftrightarrow E_2$, then $\vdash \triangle(E_1) \leftrightarrow \triangle(E_2)$.

A4. If $\phi \wedge \psi$ is inconsistent, then $\triangle(\phi \sqcup \psi) \nvdash \phi$.

A5. $\triangle(E_1) \wedge \triangle(E_2) \vdash \triangle(E_1 \sqcup E_2)$.

A6. If $\triangle(E_1) \wedge \triangle(E_2)$ is consistent, then $\triangle(E_1 \sqcup E_2) \vdash \triangle(E_1) \wedge \triangle(E_2)$.

Fig. 1. Merging postulates

There is a set of postulates, first proposed in [6] and listed in Figure 1, that merging operators are expected to satisfy. Postulate $A1$ simply means that a merging operator should produce consistent results. Postulate $A2$ states that in the case where a profile is already consistent, the operator will not alter it. Postulate $A3$ expresses the condition for syntax-independence. Postulate $A4$ is a fairness condition, requiring that the merging operator will not give preference to one knowledge base over another. Postulate $A5$ requires that if the two profiles, when merged separately, agree on a compromise then merging the concatenation of the profiles should include that compromise. Postulate $A6$ states that if there is agreement between two merged profiles, then the merged aggregate profile should entail all the consequences the two merged profiles agree on. Note that this list has been enhanced in later publications, such as [7,8], to include integrity constraints, i.e. formulae that must be consistent with the result of the merging; for simplicity of exposition, we have used the simpler version that does not employ intergrity constraints.

Bloch and Lang [14] explore how some operations from mathematical morphology can be translated and used in logic. One of the core concepts is that of a dilation operator, a function from formulae to formulae, $D : \mathcal{L} \to \mathcal{L}$.

$$D(\phi) = \text{form}\left(\{\, \omega \in \mathcal{M} \mid d(\phi, \omega) \leq 1 \,\}\right)$$

Intuitively, dilating a formula yields another formula, the models of which are at a distance at most one from the models of the original formula. Dilations can be iterated: $D^{n+1}(\phi) = D(D^n(\phi))$ and $D^0(\phi) = \phi$. In [14], it is shown that $\triangle_{\text{Max}}(E)$ is equivalently defined by use of dilations:

$$\triangle_{\text{Max}}(E) = D^n(\phi_1) \wedge \cdots \wedge D^n(\phi_k)$$

where n is the least number such that this conjunction is consistent.

Several syntax-based merging operators have been proposed [1,10,13]. These operators generally work by producing the maximal consistent subsets of the profile (maximising either with respect to subset inclusion or using set cardinality); and then taking the disjunction of (potentially, a subset of) the maximal consistent subsets. Although these operators are useful, they fail to satisfy some of the postulates for merging operators from the literature, and have been criticised for sacrificing too much information.

3 Motivation for Dilation-Based Merging

It should be clear that using such model-based operators in a FOL setting can be problematic. The number of models of even the simplest FOL formulae is usually infinite and, therefore, several results from the literature that make use of the fact that the number of models of (finite) PL is finite will fail. Also, it may be that the set of models resulting from merging a profile cannot be represented by a formula, or even worse, not even by an infinite set of formulae (even though there always exists a set of formulae whose models are a superset of a given set X, e.g. the theory of that set, $\text{Th}(X)$).[1] Although there are ways to address this issue, e.g. by using maximal consistent sets of formulae instead of sets of models, as for example in [16], such approaches are not amenable computationally. Therefore, a compromise between model-based and syntax-based approaches that aims to address these issues is desirable. We will now look at how developing the idea of dilations can help achieve these aims.

The merging operators described previously can be expressed in an equivalent form using only dilations.

- $\triangle_{\text{Max}}(E) \leftrightarrow D^n(\phi_1) \wedge \cdots \wedge D^n(\phi_k)$, where n is the least number such that this conjunction is consistent [14].
- $\triangle_\Sigma(E) \leftrightarrow \bigvee_{c_1+\cdots+c_k=n} D^{c_1}(\phi_1) \wedge \cdots \wedge D^{c_k}(\phi_k)$, where n is the least number such that this disjunction is consistent [17].
- $\triangle^{\text{Gmax}}(E) \leftrightarrow \bigvee_{\langle c_1,\ldots,c_k \rangle \in perm(T)} D^{c_1}(\phi_1) \wedge \cdots \wedge D^{c_k}(\phi_k)$, where T is the lexicographically-least tuple of integers that is sorted in ascending order for which this disjunction is consistent [17] ($perm(T)$ is the set of all tuples that are permutations of T).
- $\triangle^{\text{Gmin}}(E) \leftrightarrow \bigvee_{\langle c_1,\ldots,c_k \rangle \in perm(T)} D^{c_1}(\phi_1) \wedge \cdots \wedge D^{c_k}(\phi_k)$, where T is a tuple of integers that is sorted in *descending* order and is lexicographically least, such that this disjunction is consistent [17].

Example 1. Consider the profile $E = \langle p \wedge q, \neg p \wedge \neg q \rangle$. Clearly, $\bigwedge E$ is inconsistent. Using the dilation operator defined using Dalal's notion of distance [18], we obtain:

$$D^1(p \wedge q) \leftrightarrow (p \wedge q) \vee (p \wedge \neg q) \vee (\neg p \wedge q)$$
$$D^2(p \wedge q) \leftrightarrow \top$$
$$D^1(\neg p \wedge \neg q) \leftrightarrow (\neg p \wedge \neg q) \vee (\neg p \wedge q) \vee (p \wedge \neg q)$$
$$D^2(\neg p \wedge \neg q) \leftrightarrow \top$$

Clearly, the following conjunctions are consistent:

$$D^0(p \wedge q) \wedge D^2(\neg p \wedge \neg q)$$
$$D^1(p \wedge q) \wedge D^1(\neg p \wedge \neg q)$$
$$D^2(p \wedge q) \wedge D^0(\neg p \wedge \neg q)$$

---

[1] See [15] for the notion of *definability preservation* in this context.

Therefore, the following hold:

$$\triangle_{\text{Max}}(E) \leftrightarrow D^1(p \wedge q) \wedge D^1(\neg p \wedge \neg q) \leftrightarrow (p \wedge \neg q) \vee (\neg p \wedge q)$$
$$\triangle_{\Sigma}(E) \leftrightarrow ((D^0(p \wedge q) \wedge D^2(\neg p \wedge \neg q)) \vee$$
$$(D^1(p \wedge q) \wedge D^1(\neg p \wedge \neg q)) \vee$$
$$(D^2(p \wedge q) \wedge D^0(\neg p \wedge \neg q))) \leftrightarrow \top$$
$$\triangle^{\text{Gmax}}(E) \leftrightarrow D^1(p \wedge q) \wedge D^1(\neg p \wedge \neg q) \leftrightarrow (p \wedge \neg q) \vee (\neg p \wedge q)$$
$$\triangle^{\text{Gmin}}(E) \leftrightarrow (D^0(p \wedge q) \wedge D^2(\neg p \wedge \neg q)) \vee (D^2(p \wedge q) \wedge D^0(\neg p \wedge \neg q)) \leftrightarrow$$
$$(p \wedge q) \vee (\neg p \wedge \neg q)$$

The advantage of the above expressions is that, assuming a syntactic formulation for dilation is provided, they are entirely syntax-based and are, therefore, appealing in the context of a FOL merging framework without compromising the model-based properties of the original operators. In addition, a number of observations are in order.

- A dilation operator encapsulates a concept of distance in the sense that if a model is in $\text{mod}(D^n(\phi))$ and not in $\text{mod}(D^{n-1}(\phi))$, then it follows that the distance of the model to ϕ is n. In other words, given a dilation operator we can recover the corresponding formula-to-model distance. Also, a dilation operator can be seen as a *weakening* operator in that $D(\phi)$ represents a weakened version of ϕ, in the sense that $\phi \vdash D(\phi)$. This is especially appealing for knowledge merging, as the process of merging can be understood as a succession of compromises on the original points of view, until consensus is reached.
- In some sense, the merging operators presented above work by manipulating conjunctions of dilations of each of the formulae in a profile ($D^{c_1}(\phi_1) \wedge \cdots \wedge D^{c_k}(\phi_k)$). This is important because it exemplifies that models are not the smallest unit necessary for building merging operators, and therefore, that in our move from PL to FOL we need not use models explicitly.
- All of these merging operators can be expressed as a disjunction of conjunctions of dilations of profile formulae (for $\triangle_{\text{Max}}$ it suffices to use the fact that $\phi \vdash D(\phi)$ to see this). The set of conjunctions is determined by an ordering which is used for minimisation. This ordering effectively works on tuples of numbers. In other words, for a given profile, a tuple of dilations $\langle D^{c_1}(\phi_1), \ldots, D^{c_k}(\phi_k) \rangle$ corresponds to the *distance tuple* $\langle c_1, \ldots, c_k \rangle$. In order to produce the merged knowledge base we only need to know how to find the minimal distance tuples among all those that correspond to consistent conjunctions of dilations of profile formulae. Therefore, orderings of distance tuples (which we call *comparison orderings*) encapsulate the nature of the merging operator independently of the dilation operator (or distance concept) used.

Whilst it is not impossible to use the model-based operators presented earlier in a FOL setting, there are shortcomings in doing so. An approach demonstrated in several papers [2,19,4] assumes a function-free FOL language and modifies Dalal's distance to one that counts the number of tuples that belong to the symmetric difference of the interpretations of a predicate. The models of the knowledge bases are, then, produced for *the minimum domain size* that is admitted by the number of constants in the language. The merging operator works by minimising over these sets of models, producing a subset of models at that domain size. This approach, we feel, restricts the expressive power of FOL too much as it is geared primarily towards databases, and so leads to problems when used more generally as illustrated by the following example. Suppose that there is one predicate, P, and that the profile we want to merge is $\langle \exists x P(x), \exists x \neg P(x) \rangle$. Since there are no constants, the merge will happen at domain size 1, where these formulae are obviously inconsistent, yielding a tautology, an unsatisfactory result. We can avoid this pitfall by skolemising, but at the potential cost of introducing functions, which the notion of distance proposed is unable to handle. Also, after such a merge, one is left with a set of models of a particular domain size, and therefore one that entails many more facts than were implicit in the original profile. In addition, we feel that using such a version of Dalal's distance introduces a level of granularity that can be too fine-grained. Our framework, which is based on dilations, is a way around these problems.

An additional consideration is that the most frequently used notion of distance d between models in PL is the Dalal (or Hamming) distance [18,20]. In FOL it is much harder to come up with a single, all-encompassing notion of distance (indeed, even in PL, the Dalal distance suffers from well-known issues [21], e.g. some propositional letters may be more relevant to the merging process than others). To address this problem our framework admits several dilation operators, corresponding to different notions of distance.

Furthemore, PL distances like Dalal's enjoy a certain property of *commensurateness* in the sense that the Dalal dilation of a formula changes it by a degree that is in a sense equal to the degree of change that results from the application of Dalal's dilation to another formula (this condition corresponds to the symmetry condition for the distance function employed). This is especially hard to maintain in a FOL setting, as we will demonstrate in Section 4.2. Our approach, first, relaxes this constraint for dilation operators by letting them be relations as opposed to functions; and, second, changes the type of merging operators from a function that returns a single knowledge base, to a function that returns a set of alternative knowledge bases. The user, then, would assess the resulting set of candidate knowledge bases and act depending on the application. In other words, in order to get around the problem of lacking commensurateness, we increase the freedom of choice over what the merged profile should be.

4   Framework for Dilation-Based Merging

4.1   Preliminaries

We will work with subsets of FOL, although the proposed framework should be usable with other classical logics such as modal logics. So, let $L = \langle \mathcal{L}, \vdash \rangle$

be a sound and complete logic, where $\mathcal{L}$ is the set of formulae of the logic, and $\vdash \subseteq \mathcal{L} \times \mathcal{L}$ is the (transitive and reflexive) entailment relation. Two formulae are equivalent, $\phi \equiv \psi$, iff $\phi \vdash \psi$ and $\psi \vdash \phi$. We assume that L has the connectives $\wedge$, $\vee$ and $\neg$ and that they behave classically. A formula $\phi \in \mathcal{L}$ is consistent iff $\phi \nvdash \bot$.

We use the notation for profiles introduced previously with the following additions. The notation for concatenation of tuples of formulae, $\sqcup$, repeated concatenation, $\cdot^n$, and the length of a tuple of formulae, $|\cdot|$, will also be used for tuples of numbers. A profile is a tuple of consistent formulae of $\mathcal{L}$ and denoted by E. The set of all profiles of finite length is denoted by $\mathcal{E}$. Two profiles $E_1 = \langle \phi_1, \ldots, \phi_k \rangle$, E_2 are equivalent, written $E_1 \equiv E_2$, iff $|E_1| = |E_2|$ and there is a bijection f from E_1 to E_2 such that $\phi_i \equiv f(\phi_i)$ for all $i \leq |E_1|$. We also use the $\equiv$ notation for sets of formulae with the corresponding meaning.

If R is a binary relation over a set S and $x \in S$ then we denote the image of x through R as $R(x)$, i.e. $R(x) = \{ y \in S \mid \langle x, y \rangle \in R \}$. We extend this notation over sets of elements, i.e. if $X \subseteq S$ then $R(X) = \{ y \in S \mid \exists x \in X, \langle x, y \rangle \in R \}$. We define $R^n(x)$ inductively as $R^1(x) = R(x)$ and $R^{n+1}(x) = R^n(R(x))$ and use the natural extension of this notation over sets, e.g. $R^n(X)$.

If $A = \langle n_1, \ldots, n_k \rangle$ is a tuple of natural numbers, then $\sum A = \sum_{i=1}^k n_i$ and $\max A = \max \{ n_i \mid 1 \leq i \leq k \}$.

4.2 Dilation Operators

Similarly to the literature on propositional knowledge merging, we will list some postulates that the concepts we define may satisfy. However, we do not make these postulates a part of the definitions as we want to highlight the dependencies of the results as well as to allow more freedom as to which postulates a particular instantiation of the definitions happens to satisfy. Note that a further modification to the notion of a dilation is that we set it to depend on the profile in question. We do this to allow a dilation operator to perform a guided weakening to the formulae involved; given the difficulty of ensuring commensurateness for dilation, this may allow for more effective definitions of concrete dilation operators.

Definition 1. A dilation operator *is a relation* $D_E \subseteq \mathcal{L} \times \mathcal{L}$, *where* $E \in \mathcal{E}$.

A dilation operator D may satisfy the following postulates.

D1. If $\phi \equiv \psi$ then $D_E(\phi) \equiv D_E(\psi)$ (syntax independence).

D2. If $\chi \in D_E(\phi)$ then $\phi \vdash \chi$ (weakening).

D3. Let $E = \langle \phi, \psi \rangle$. For all $a \in \mathbb{N}$ and $\phi' \in D_E^a(\phi)$ such that $\phi' \wedge \psi$ is consistent, there exists $b \leq a$ and $\psi' \in D_E^b(\psi)$ such that $\phi \wedge \psi' \nvdash \bot$ (commensurateness).

Postulate $D1$ is a standard syntax-independence postulate. Postulate $D2$ is important because it demands that the dilation operator is, in effect, a subset of the entailment relation. Postulate $D3$ corresponds to the symmetry condition on distances in the sense that if a certain number of dilations of ϕ is enough to make it consistent with ψ then it should be the case that an equal or lower number of dilations of ψ will make it consistent with ϕ.

We list, below, several concrete dilation operators. We begin with a trivial dilation operator that corresponds to the *drastic distance* from the PL literature.

Definition 2. *Let $\mathcal{L}$ be a FOL fragment. Then for all $\phi \in \mathcal{L}$,*

$$D_E^{\mathrm{drastic}}(\phi) = \{\top\}.$$

This operator can be used to reconstruct some of the consistency-based operators from the literature and we will show specific examples of that in Section 4.5. It is easy to see that postulates $D1$, $D2$ and $D3$ trivially hold.

Another simple dilation operator is obtained by taking the disjunction of the formula being dilated with some other formula from the profile.

Definition 3. *Let $\mathcal{L}$ be a FOL fragment which is closed under disjunction. Then, if $E = \langle \psi_1, \ldots, \psi_k \rangle$,*

$$D_E^{\mathrm{disj}}(\phi) = \{\, \phi \vee \psi_i \mid 1 \leq i \leq k \text{ and } \psi_i \nvdash \phi \,\}$$

Again, it is trivial to see that $D1$, $D2$ and $D3$ are satisfied. As an example, if $E = \langle a \to b, a, \neg b, c, d \rangle$ then,

$$D_E^{\mathrm{disj}}(a \to b) = \{(a \to b) \vee a, (a \to b) \vee \neg b, (a \to b) \vee c, (a \to b) \vee d\}$$
$$\equiv \{\top, (a \to b) \vee c, (a \to b) \vee d\}$$
$$D_E^{\mathrm{disj}}(D_E^{\mathrm{disj}}(a \to b)) \equiv \{\top, (a \to b) \vee c, (a \to b) \vee d, (a \to b) \vee c \vee d\}$$

Finally, the following definition concerns a dilation operator that weakens formulae by changing universal quantifiers to existential. Its effect on the input formulae is rather profound, and as such it may be suitable only for specific applications. We present a more illuminating worked example in Section 4.4.

Definition 4. *Let $\mathcal{L}$ be a (prenex) FOL fragment and ψ a formula of the form $\psi = Q_1 x_1 \cdots Q_n x_n \phi$ where Q_i is a quantifier and ϕ is quantifier-free.*

$$D_E^Q(\psi) = \left\{ Q_1' x_1 \cdots Q_n' x_n \phi \;\middle|\; \begin{array}{l} \text{there exists } j \leq n \text{ such that} \\ Q_j = \forall \text{ and } Q_j' = \exists \text{ and} \\ \text{for all } i \leq n \text{ such that } i \neq j, Q_i' = Q_i \end{array} \right\}$$

As an example, for some profile E,

$$D_E^Q(\forall x \forall y\, P(x, y)) = \{\forall x \exists y\, P(x, y), \exists x \forall y\, P(x, y)\}$$
$$D_E^Q(D_E^Q(\forall x \forall y\, P(x, y))) = \{\exists x \exists y\, P(x, y)\}.$$

Postulate $D2$ is trivially satisfied. Syntax independence (postulate $D1$) fails, since $\forall x \forall y\, P(x, y) \equiv \forall y \forall x\, P(x, y)$ but, $D_E^Q(\forall y \forall x\, P(x, y)) \not\equiv D_E^Q(\forall x \forall y\, P(x, y))$ as can be seen below.

$$D_E^Q(\forall y \forall x\, P(x, y)) = \{\forall y \exists x\, P(x, y), \exists y \forall x\, P(x, y)\}$$

Commensurateness ($D3$) also fails. Consider the formulae $\phi = \forall x \forall y \, P(x,y)$ and $\psi = \exists x \forall y \, \neg P(x,y)$. Clearly, $\phi \wedge \psi$ is inconsistent and there exists a formula in $D_E^Q(\phi)$, namely $\chi = \exists x \forall y \, P(x,y)$ such that $\chi \wedge \psi$ is consistent. However, the only formula in $D_E^Q(\exists x \forall y \, \neg P(x,y))$ is $\exists x \exists y \neg P(x,y)$ which is not consistent with ϕ.

If we restrict the logical language to propositional logic, then we can define a dilation operator that exactly captures Dalal's dilation. Supposing that ϕ is written in Disjunctive Normal Form (DNF), i.e. $\phi = \bigvee_{i=1}^{l} \bigwedge_{j=1}^{m_i} L_{i,j}$ with $L_{i,j}$ being literals, then it can be shown (e.g. [14]) that

$$
D_E^{\mathrm{Dalal}}(\psi) = \left\{ \bigvee_{i=1}^{l} \bigvee_{j=1}^{m_i} \bigwedge_{\substack{k=1 \\ k \neq j}}^{m_i} L_{i,k} \right\}.
$$

This dilation operator allows us to subsume the PL merging operators from the literature within our framework.

4.3   Comparison Orderings

The notion of a comparison ordering is meant to capture the nature of a merging operator independently of the particular dilation operator used in its definition. Since dilating a formula constitutes a discrete step, the purpose of a comparison ordering is to evaluate how much compromise is involved in making a profile consistent by applying the dilation operator to the profile formulae a number of times. In other words, the comparison ordering is meant to order distance tuples in how far they are from the original profile. This concept is partly related to the orderings over models used in frameworks such as the one in [8] (syncretic assignments) but also to the aggregation functions of [9].

Definition 5. *A comparison ordering is a collection of total preorders over tuples of natural numbers, one for each length k, $\sqsubseteq \subseteq \mathbb{N}^k \times \mathbb{N}^k$.*

Comparison orderings may satisfy the following postulates.

$C1$. For any $\langle n_1, \ldots, n_k \rangle \in \mathbb{N}^k$, $\langle 0, \ldots, 0 \rangle \sqsubseteq \langle n_1, \ldots, n_k \rangle$. Furthermore, if there exists $i \leq k$ such that $n_i > 0$, then $\langle 0, \ldots, 0 \rangle \sqsubset \langle n_1, \ldots, n_k \rangle$

$C2$. If A_1, A_2 and B_1, B_2 are pairs of tuples of equal respective lengths, then $A_1 \sqsubseteq A_2$ and $B_1 \sqsubseteq B_2$ implies $A_1 \sqcup B_1 \sqsubseteq A_2 \sqcup B_2$.

$C3$. If A_1, A_2 and B_1, B_2 are pairs of tuples of equal respective lengths, then $A_1 \sqsubset A_2$ and $B_1 \sqsubseteq B_2$ implies $A_1 \sqcup B_1 \sqsubset A_2 \sqcup B_2$.

$C4$. For all $a, b \in \mathbb{N}$, if $a \leq b$, then $\langle a, 0 \rangle \sqsubseteq \langle 0, b \rangle$.

Postulate $C1$ corresponds to the conditions of *non-decreasingness* and *minimality* from [9] and to the postulate $A2$ in a way that will become obvious after we define what a merging operator is, in Definition 11. Postulates $C2$ and $C3$ correspond to the fifth and sixth conditions for syncretic assignments respectively [8].

Note that $C3$ entails $C2$. These two postulates allow us to infer facts about the relation between two concatenations of tuples when the corresponding relations are known, but there is no way to infer facts about concatenations in different permutations. Postulate $C4$ fills in this gap and is related to commensurateness.

The PL merging operators listed in Section 2 can be translated into our framework as follows.

Definition 6. *The comparison ordering* $\sqsubseteq_{\mathrm{Max}}$ *is defined as follows.*

$$\langle a_1, \ldots, a_k \rangle \sqsubseteq_{\mathrm{Max}} \langle b_1, \ldots, b_k \rangle \text{ iff } \max_{1 \le i \le k} a_i \le \max_{1 \le i \le k} b_i$$

It is easy to see that the $\sqsubseteq_{\mathrm{Max}}$ satisfies $C1$. It also satisfies $C2$ since if $\max A_1 \le \max A_2$ and $\max B_1 \le \max B_2$ then it will be the case that $\max A_1 \sqcup A_2 \le \max B_1 \sqcup B_2$. Note, however, that $\sqsubseteq_{\mathrm{Max}}$ does not satisfy $C3$, because if $\max A_1 < \max A_2$ but $\max B_1 = \max B_2$ we cannot conclude that $\max A_1 \sqcup A_2 < \max B_1 \sqcup B_2$. Finally, it is easy to see that $\sqsubseteq_{\mathrm{Max}}$ satisfies $C4$.

Definition 7. *The comparison ordering* $\sqsubseteq_\Sigma$ *is defined as follows.*

$$\langle a_1, \ldots, a_k \rangle \sqsubseteq_\Sigma \langle b_1, \ldots, b_k \rangle \text{ iff } \sum \langle a_1, \ldots, a_k \rangle \le \sum \langle b_1, \ldots, b_k \rangle$$

The ordering $\sqsubseteq_\Sigma$ clearly satisfies $C1$. It is also easy to see that it satisfies $C3$ (and therefore $C2$ as well): if $\sum A_1 < \sum A_2$ and $\sum B_1 \le \sum B_2$ then $\sum A_1 \sqcup B_1 = \sum A_1 + \sum B_1 < \sum A_2 + \sum B_1 \le \sum A_2 + \sum B_2 = \sum A_2 \sqcup B_2$. It is clear that $\sqsubseteq_\Sigma$ satisfies $C4$ as well.

Definition 8. *The comparison ordering* $\sqsubseteq_{\mathrm{Gmax}}$ *is defined as follows.*

$$\langle a_1, \ldots, a_k \rangle \sqsubseteq_{\mathrm{Gmax}} \langle b_1, \ldots, b_k \rangle \text{ iff } sort^d \langle a_1, \ldots, a_k \rangle \le_{\mathrm{lex}} sort^d \langle b_1, \ldots, b_k \rangle$$

It is clear that $\triangle^{\mathrm{Gmax}}$ satisfies $C1$. That it satisfies $C3$ follows from a result presented in [6], namely that if A_1, A_2 and B_1, B_2 are tuples of numbers of equal lengths, and $sort^d A_1 <_{\mathrm{lex}} sort^d A_2$ and $sort^d B_1 \le_{\mathrm{lex}} sort^d B_2$ then $sort^d A_1 \sqcup B_1 <_{\mathrm{lex}} sort^d A_2 \sqcup B_2$. It is also easy to check that $\sqsubseteq_{\mathrm{Gmax}}$ satisfies $C4$.

Definition 9. *The comparison ordering* $\sqsubseteq_{\mathrm{Gmin}}$ *is defined as follows.*

$$\langle a_1, \ldots, a_k \rangle \sqsubseteq_{\mathrm{Gmin}} \langle b_1, \ldots, b_k \rangle \text{ iff } sort^a \langle a_1, \ldots, a_k \rangle \le_{\mathrm{lex}} sort^a \langle b_1, \ldots, b_k \rangle$$

Again, it is obvious that $\sqsubseteq_{\mathrm{Gmin}}$ satisfies $C1$. In similar manner to that of the result on $\sqsubseteq_{\mathrm{Gmax}}$ we can show that $\sqsubseteq_{\mathrm{Gmin}}$ satisfies $C3$ as well. Also, it is clear that $C4$ holds of $\sqsubseteq_{\mathrm{Gmin}}$ as well.

4.4   Merging Operators

As noted in Section 3, the smallest unit that we can distinguish using a profile and a dilation operator is the conjunction of dilations of the profile formulae. This set of candidate tuples of dilations is captured by the next definition. Note that we discard tuples that yield inconsistent conjunctions as they represent a set of compromises that has not reached consensus, and therefore cannot form part of the merging.

Definition 10. *The function* $\mathcal{C}_D : \mathcal{E} \to 2^{\mathcal{E}}$ *generates the set of all consistent combinations of dilations of a profile, using the dilation operator* D_E. *Let* $E = \langle \phi_1, \ldots, \phi_k \rangle$.

$$\langle \psi_1, \ldots, \psi_k \rangle \in \mathcal{C}_D(E) \text{ iff } \begin{cases} \psi_1 \wedge \cdots \wedge \psi_k \text{ is consistent and} \\ \text{for all } i \leq k \text{ there exists } n_i, \text{ such that } \psi_i \in D_E^{n_i}(\phi_i) \end{cases}$$

In effect, $\mathcal{C}_D$ generates the set of all possible tuples that may play a role in the merging process.

For a profile $E = \langle \phi_1, \ldots, \phi_k \rangle$ of length k, the function $dt : \mathcal{C}_D(E) \to \mathbb{N}^k$ (*distance tuple*) returns the tuple of the (minimum) numbers of dilations of each profile formula:

$$dt(\langle \psi_1, \ldots, \psi_k \rangle) = \langle n_1, \ldots, n_k \rangle \text{ iff } \begin{cases} \text{for all } i \leq k, n_i \text{ is the least number} \\ \text{such that } \psi_i \in D_E^{n_i}(\phi_i) \end{cases}$$

We can now give the definition of a merging operator, using the notions of a dilation operator and comparison ordering.

Definition 11. *A merging operator is a function* $\triangle_{D,\sqsubseteq} : \mathcal{E} \to 2^{\mathcal{L}}$ *defined in terms of a dilation operator* D *and a comparison ordering* $\sqsubseteq$ *in the following manner.*

$$\bigwedge A \in \triangle_{D,\sqsubseteq}(E) \quad \text{iff} \quad \begin{cases} A \in \mathcal{C}_D(E) \text{ and} \\ dt(A) \sqsubseteq dt(B) \text{ for all } B \in \mathcal{C}_D(E) \end{cases}$$

Note that a merging operator according to this definition is a function that returns a set of formulae instead of a single formula. This is because the dilation operators are now relations and this opens up the question of how to treat the resulting conjunctions. In PL, merging operators form the union of the corresponding sets of models. This is not appropriate here, since the union may discard too much information. As an example suppose that for any $\psi \in D(\phi)$, $\psi \not\equiv \top$, but $\bigvee D(\phi) \equiv \top$. To address this, instead of taking the disjunction of the formulae in the result of the merging, the user is free to choose which operation to apply according to what makes sense in the context of the application in question. For example, the original behaviour of PL merging operators can be recovered by taking the disjunction of the members of this set.

In the following, we will abuse notation slightly and say that a tuple $A \in \mathcal{C}_D(E)$ is $\sqsubseteq$-minimal when $dt(A)$ is actually $\sqsubseteq$-minimal. We will also omit $\sqsubseteq$ and D as subscripts of a merging operator when they are clear from the context.

4.5   Examples of Dilation-Based Merging

In this section we demonstrate some concrete merging operators by combining specific dilation operators with the comparison orderings presented in Section 4.3. We begin with the drastic dilation and show how some consistency-based merging operators can be subsumed by our framework.

Example 2. Suppose $E = \langle a \to b, a, \neg b, c \rangle$. It is easy to see that $\bigwedge E$ is inconsistent and that any two formulae of $\{a \to b, a, \neg b\}$ are consistent with c. Hence,

$$\mathcal{C}_{D_E^{\text{drastic}}}(E) = \left\{ \begin{array}{l} \langle \ \top, \quad a, \ \neg b, \ c \ \rangle, \\ \langle \ a \to b, \ \top, \ \neg b, \ c \ \rangle, \\ \langle \ a \to b, \ a, \quad \top, \ c \ \rangle \\ \qquad\qquad\vdots \end{array} \right\}.$$

Using this dilation operator and the $\sqsubseteq_{\text{Max}}$ comparison ordering we obtain the following result.

$$\triangle_{\sqsubseteq_{\text{Max}}}(E) = \left\{ \bigwedge A \ \middle| \ A \in \mathcal{C}_{D_E^{\text{drastic}}}(E) \right\}$$

Obviously, this operator is very coarse in the comparisons it makes and that is a direct result of the deficiencies of $\sqsubseteq_{\text{Max}}$. PL merging operators work by taking the disjunction of the returned set (e.g. $\bigvee \psi$ for all $\psi \in \triangle_{\sqsubseteq_{\text{Max}}}(E)$), which in our case would yield a tautology.

Using $\sqsubseteq_\Sigma, \sqsubseteq_{\text{Gmax}}$ or $\sqsubseteq_{\text{Gmin}}$ we obtain the same intuitive result corresponding to the maximal consistent subsets of the profile E:

$$\triangle_{\sqsubseteq_\Sigma}(E) = \triangle_{\sqsubseteq_{\text{Gmax}}}(E) = \triangle_{\sqsubseteq_{\text{Gmin}}}(E) = \left\{ \begin{array}{l} a \wedge \neg b \wedge c, \\ a \to b \wedge \neg b \wedge c, \\ a \to b \wedge a \wedge c \end{array} \right\}.$$

The disjunction of the members of this set is equivalent to $(b \to a) \wedge c$.

It is easy to see that, when using the drastic dilation operator, the three comparison orderings $\sqsubseteq_\Sigma, \sqsubseteq_{\text{Gmax}}$ or $\sqsubseteq_{\text{Gmin}}$ yield the same merging operator, which selects the cardinality-maximal consistent subsets of the original profile. This is, effectively, one of the consistency-based merging operators found in [1,10,13].

Example 3. Suppose $E = \langle \forall x \forall y \, P(x,y) \leftrightarrow Q(y,x), \forall x \, P(x,x) \leftrightarrow \neg Q(x,x) \rangle$. The first formula effectively forces the relation Q to be the transpose of the relation P. The second formula prescribes that, if we view P and Q as incidence matrices, that they disagree on their diagonals. Clearly, E is inconsistent, since the first formula entails the negation of the second. Using the dilation operator D_E^Q, we obtain:

$$D_E^Q(\forall x \forall y \, P(x,y) \leftrightarrow Q(y,x)) = \left\{ \begin{array}{l} \forall x \exists y \, P(x,y) \leftrightarrow Q(y,x), \\ \exists x \forall y \, P(x,y) \leftrightarrow Q(y,x) \end{array} \right\}$$

$$D_E^Q(D_E^Q(\forall x \forall y \, P(x,y) \leftrightarrow Q(y,x))) = \{\exists x \exists y \, P(x,y) \leftrightarrow Q(y,x)\}$$

$$D_E^Q(\forall x \, P(x,x) \leftrightarrow \neg Q(x,x)) = \{\exists x \, P(x,x) \leftrightarrow \neg Q(x,x)\}.$$

Choosing any comparison ordering out of $\sqsubseteq_\Sigma, \sqsubseteq_{\text{Gmax}}, \sqsubseteq_{\text{Gmin}}$ will yield the same merged set, i.e.

$$\triangle_\Sigma(E) = \triangle^{\text{Gmax}}(E) = \triangle^{\text{Gmin}}(E) = \left\{ \bigwedge \left\{ \begin{array}{l} \forall x \exists y \, P(x,y) \leftrightarrow Q(y,x), \\ \forall x \, P(x,x) \leftrightarrow \neg Q(x,x) \end{array} \right\} \right\}$$

The result entails that for each pair of elements in P there is a transposed pair in Q and that the two relations disagree on their diagonals.

It is clear that a merging operator based on D_E^Q will have a very strong effect on the input profile and, therefore, may be used in limited contexts.

Next, we look at a dilation operator that weakens universal formulae by adding exceptions using the constants of the language.

Definition 12. *Assume $\mathcal{L}$ is a (prenex) universal FOL fragment with a finite number of constants $c_1, \ldots, c_n$. Let $\psi \in \mathcal{L}$ be of the form $\forall x_1 \cdots \forall x_k \phi$.*

$$exc_{i,j}(\phi) = \forall x_1 \cdots \forall x_k (x_i \neq c_j \rightarrow \phi)$$
$$D_E^{exc}(\phi) = \{ \, exc_{i,j}(\phi) \mid 1 \leq i \leq k, 1 \leq j \leq n \text{ and } exc_{i,j}(\phi) \nvdash \bot \, \}$$

It should be easy to see that postulates $D1$ and $D2$ clearly hold.

Example 4. Let us assume a language with three constants, a, b, c and the following profile $E = \langle \forall y\, P(c,y), \forall y\, P(a,y), \forall x\, \neg P(x,b) \rangle$. The dilation operator applied to the profile formula $\forall y\, P(c,y)$ gives:

$$D_E^{exc}(\forall y\, P(c,y)) = \left\{ \begin{array}{c} \forall y\, (y \neq a \rightarrow P(c,y)), \\ \forall y\, (y \neq b \rightarrow P(c,y)), \\ \forall y\, (y \neq c \rightarrow P(c,y)) \end{array} \right\}.$$

Similar results are obtained by dilating the rest of the profile formulae. By using $\sqsubseteq_\Sigma, \sqsubseteq_{Gmax}, \sqsubseteq_{Gmin}$ we obtain the following results:

$$\triangle_{\sqsubseteq_{Gmax}}(E) = \left\{ \bigwedge \left\{ \begin{array}{c} \forall y\, (y \neq b \rightarrow P(c,y)), \\ \forall y\, (y \neq b \rightarrow P(a,y)), \\ \forall x\, \neg P(x,b) \end{array} \right\} \right\}$$

$$\triangle_{\sqsubseteq_{Gmin}}(E) \equiv \left\{ \bigwedge \left\{ \begin{array}{c} \forall y\, P(c,y), \\ \forall y\, P(a,y), \\ \forall x\, (x \neq a \wedge x \neq c) \rightarrow \neg P(x,b) \end{array} \right\} \right\}$$

$$\triangle_{\sqsubseteq_\Sigma}(E) \equiv \triangle_{\sqsubseteq_{Gmax}}(E) \cup \triangle_{\sqsubseteq_{Gmin}}(E)$$

The operator $\triangle_{\sqsubseteq_\Sigma}$ is selecting dilated profiles based on the total number of dilation operations applied (here the minimum for a consistent result is 2), $\triangle_{\sqsubseteq_{Gmax}}$ sorts the distance tuples in descending order and therefore selects the dilated profiles that have (any permutation of) $\langle 1, 1, 0 \rangle$ as their distance tuple. Finally, $\triangle_{\sqsubseteq_{Gmin}}$ sorts in ascending order and, thus, selects the dilated profiles that correspond to the distance tuple $\langle 2, 0, 0 \rangle$ and its permutations.

We note that if the language $\mathcal{L}$ does not allow function symbols and the input profile does not contain existential quantifiers, then this merging operator can be equivalently defined using the notion of Dalal's distance on the symmetrical set differences of the interpretation of the predicates, as explained in Section 3 in relation to the papers [2,19,4]. However, if we allow function symbols in the language and/or existential quantifiers in the input profile, then our dilation-based merging operator becomes distinct from the approach described in the papers cited above.

4.6   Properties of Merging Operators

Obviously, the list of postulates presented in Figure 1 is not appropriate for operators defined through Definition 11. To address this, we present a list in Figure 2 that we believe captures the intention of the original postulates and fits with our framework. We comment on the modified postulates below. Note that most postulates are essentially quantified versions of the originals, over the set returned by the merging operator.

$B1$. For any $\phi \in \triangle(E)$, ϕ is consistent.

$B2$. If $\bigwedge E$ is consistent, then $\triangle(E) \equiv \{\bigwedge E\}$.

$B3$. If $E_1 \equiv E_2$, then $\triangle(E_1) \equiv \triangle(E_2)$.

$B4$. If $\phi \wedge \psi$ is inconsistent, then there exists $\chi \in \triangle(\phi \sqcup \psi)$ such that $\chi \nvdash \phi$.

$B5$. If $\phi_1 \in \triangle(E_1)$, $\phi_2 \in \triangle(E_2)$ and $\phi_1 \wedge \phi_2 \nvdash \bot$, then $\phi_1 \wedge \phi_2 \in \triangle(E_1 \sqcup E_2)$.

$B6$. If there exist $\phi_1 \in \triangle(E_1)$, $\phi_2 \in \triangle(E_2)$ such that $\phi_1 \wedge \phi_2$ is consistent, then for any $\chi \in \triangle(E_1 \sqcup E_2)$ there are $\psi_1 \in \triangle(E_1)$ and $\psi_2 \in \triangle(E_2)$ such that $\chi = \psi_1 \wedge \psi_2$.

Fig. 2. Merging postulates for our framework

Postulate $B1$ is a straightforward translation of $A1$ requiring that any member of $\triangle(E)$ is consistent. Similarly, postulates $B2$, $B3$ correspond directly to $A2$ and $A3$. Postulate $B4$ corresponds to $A4$ but requires some explanation; since the PL merging operators can be expressed as disjunctions of conjunctions of dilations, in order to satisfy $A4$ it suffices to have a conjunction of dilations that does not entail ϕ and this is why $B4$ is existentially quantified. In order to translate $A5$ and $A6$ we interpret the conjunction of two merged profiles as an operation that produces all consistent conjunctions from the two sets $\triangle(E_1)$ and $\triangle(E_2)$. In other words, we translate $\triangle(E_1) \wedge \triangle(E_2)$ into the set $\{\phi_1 \wedge \phi_2 \mid \phi_1 \in \triangle(E_1), \phi_2 \in \triangle(E_2), \phi_1 \wedge \phi_2 \nvdash \bot\}$, over which the postulates $B5$ and $B6$ are quantified.

Using Definitions 1, 10, 5 and 11 we can now explore the conditions under which a merging operator will satisfy the postulates listed in Figure 2.

Proposition 1. *Any operator $\triangle$ defined through Definition 11 in terms of a dilation operator D and a comparison ordering $\sqsubseteq$ will satisfy the following postulates, under the listed conditions.*

- *If D satisfies $D2$ then $\triangle$ satisfies $B1$.*
- *If $\sqsubseteq$ satisfies $C1$ then $\triangle$ satisfies $B2$.*
- *If D satisfies $D1$ then $\triangle$ satisfies $B3$.*
- *If D satisfies $D3$ and $\sqsubseteq$ satisfies $C1$, $C2$ and $C4$, then $\triangle$ satisfies $B4$.*
- *If $\sqsubseteq$ satisfies $C2$ then $\triangle$ satisfies $B5$.*
- *If $\sqsubseteq$ satisfies $C3$ then $\triangle$ satisfies $B6$.*

The proof for Proposition 1 is included in the appendix.

So, returning to Example 2, we can see that the merging operators $\triangle_{\sqsubseteq_\Sigma}(E)$, $\triangle_{\sqsubseteq_{\mathrm{Gmax}}}(E)$ and $\triangle_{\sqsubseteq_{\mathrm{Gmin}}}(E)$, when based on the drastic dilation operator, will satisfy all the above postulates. On the other hand, in Example 4, the operators $\triangle_{\sqsubseteq_\Sigma}(E)$, $\triangle_{\sqsubseteq_{\mathrm{Gmax}}}(E)$ and $\triangle_{\sqsubseteq_{\mathrm{Gmin}}}(E)$, when based on D_E^{exc}, will satisfy all the postulates apart from $B4$ since it is an open question as to whether D_E^{exc} satisfies $D3$.

5   Conclusions

We have outlined a dilation-based framework for knowledge merging that is appropriate for more expressive logics than propositional logic. This framework is flexible, admitting different notions of compromise in the form of dilation operators, as well as allowing different ways of minimising the extent of compromise in the form of comparison orderings. In addition, it can subsume important proposals from the literature, such as some kinds of consistency-based merging, as well as the PL merging operators based on Dalal's distance.

We believe that the framework and its specific instantiations presented, avoid the problems described in Section 3. Artefacts that result from using variants of Dalal's distance on the interpretations of predicates, such as the exceedingly fine-grained nature of the resulting knowledge base or the inability to allow function symbols in the language, are avoided. In addition, merging operators that return a multiplicity of weakened versions of a profile are allowed, enabling the user of the merging operator to make choices that may depend on extra-logical information, or information that is not available to the merging operator. Furthermore, the user can compute the disjunction of the resulting set of profiles, thus recovering the behaviour of many of the merging operators in the literature.

There are many possibilities for additional dilation operators. One that we will explore in the future is the following. The intuition behind it is to try and capture some of the characteristics of the Dalal dilation, but in a FOL setting.

Definition 13. *Let $\mathcal{L}$ be a (prenex) universal FOL fragment and $\psi \in \mathcal{L}$ a formula of the form $Q_1 x_1 \cdots Q_n x_n \phi$ where Q_i is a quantifier and ϕ is quantifier-free and in DNF form. This means that $\phi = \bigvee_{i=1}^{l} \bigwedge_{j=1}^{m_i} L_{i,j}$ with $L_{i,j}$ being possibly negated atomic formulae.*

$$D_E^{\mathrm{Dalal}}(\psi) = \left\{ Q_1 x_1 \cdots Q_n x_n \bigvee_{i=1}^{l} \bigvee_{j=1}^{m_i} \bigwedge_{\substack{k=1 \\ k \neq j}}^{m_i} L_{i,k} \right\}$$

Several questions remain open. For example, are there conditions that can capture the essence behind arbitration and majority operators, as defined in the literature (see, e.g. [6])? Also, the properties we have outlined are sufficient to force a merging operator to satisfy the postulates listed in Figure 2, but can this result can be strengthened towards a characterisation theorem?

In addition, although we believe that the format of our definitions lends itself nicely to implementation, decidability is obviously a deeper question, especially with regards to the exact logical language chosen. For example, our definition of dilation allows for an infinity of weakened versions of a formula to be returned, something that would obviously complicate computation, even though all the examples of dilation operators we presented do not suffer from this issue. Also related to this is the possibility of infinite chains of weakenings that never produce consensus in the original profile (this is interrelated to the existence of infinitely-descending chains of models in frameworks that employ distances or orderings). While this is not a definitional problem of the framework ($\mathcal{C}_D(E)$ and $\triangle(E)$ will both simply be empty), it would be interesting to designate the conditions under which this does not happen.

Other open issues include whether we can define a distance function between models based on a syntactic construction such as a dilation operator, similar to the work done by Lehmann et al on belief revision [15]. Finally, even though the use of integrity constraints is very appealing for purposes like expressing background knowledge, for simplicity of exposition, we have omitted their treatment. Extending the framework to allow integrity constraints to be expressed and handled appropriately would be another possible research direction.

References

1. Baral, C., Kraus, S., Minker, J.: Combining multiple knowledge bases. IEEE Transactions on Knowledge and Data Engineering 3(2), 208–220 (1991)
2. Revesz, P.Z.: On the semantics of arbitration. Journal of Algebra and Computation 7(2), 133–160 (1997)
3. Liberatore, P., Schaerf, M.: Arbitration (or how to merge knowledge bases). IEEE Transactions on Knowledge and Data Engineering 10(1), 76–90 (1998)
4. Lin, J., Mendelzon, A.O.: Merging databases under constraints. International Journal of Cooperative Information Systems 7(1), 55–76 (1998)
5. Everaere, P., Konieczny, S., Marquis, P.: Quota and Gmin merging operators. In: IJCAI 2005. Nineteenth International Joint Conference on Artificial Intelligence, pp. 424–429 (2005)
6. Konieczny, S., Pérez, R.P.: On the logic of merging. In: KR 1998. Sixth International Conference on Principles of Knowledge Representation and Reasoning, pp. 488–498 (1998)
7. Konieczny, S., Pérez, R.P.: Merging with integrity constraints. In: Hunter, A., Parsons, S. (eds.) ECSQARU 1999. LNCS (LNAI), vol. 1638, pp. 233–244. Springer, Heidelberg (1999)
8. Konieczny, S., Pérez, R.P.: Merging Information Under Constraints: A Logical Framework. Journal of Logic and Computation 12(5), 773–808 (2002)

9. Konieczny, S., Lang, J., Marquis, P.: DA^2 merging operators. Artificial Intelligence 157(1–2), 49–79 (2004)
10. Baral, C., Kraus, S., Minker, J., Subrahmanian, V.S.: Combining knowledge bases consisting of first-order theories. Computational Intelligence 8, 45–71 (1992)
11. Chacón, J.L., Pérez, R.P.: Merging operators: Beyond the finite case. Information Fusion 7(1), 41–60 (2006)
12. Booth, R.: Social contraction and belief negotiation. In: KR 2002. Proceedings of the Eighth International Conference on Principles of Knowledge Representation and Reasoning, pp. 375–384. Morgan Kaufmann, San Francisco (2002)
13. Konieczny, S.: On the difference between merging knowledge bases and combining them. In: Cohn, A.G., Giunchiglia, F., Selman, B. (eds.) KR 2000. International Conference on Principles of Knowledge Representation and Reasoning, pp. 135–144. Morgan Kaufmann, San Francisco (2000)
14. Bloch, I., Lang, J.: Towards Mathematical Morpho-Logics. In: Technologies for Constructing Intelligent Systems, vol. 2, pp. 367–380. Springer, Heidelberg (2002)
15. Lehmann, D., Magidor, M., Schlechta, K.: Distance semantics for belief revision. Journal of Symbolic Logic 66(1), 295–317 (2001)
16. Levesque, H.J., Lakemeyer, G.: The Logic of Knowledge Bases. MIT Press, Cambridge (2000)
17. Gorogiannis, N., Hunter, A.: Implementing semantic merging operators using binary decision diagrams. Technical report, University College London (2007)
18. Dalal, M.: Investigations into a theory of knowledge base revision. In: AAAI 1988. Proceedings of the Seventh National Conference on Artificial Intelligence, vol. 2, pp. 475–479 (1988)
19. Grahne, G., Mendelzon, A.O., Revesz, P.Z.: Knowledgebase transformations. Journal of Computer and System Sciences 54(1), 98–112 (1997)
20. Hamming, R.W.: Error detecting and error correcting codes. Bell System Technical Journal 29, 147–160 (1950)
21. Lafage, C., Lang, J.: Propositional distances and compact preference representation. European Journal of Operational Research 160(3), 741–761 (2005)

A   Proof of Proposition 1

– If D satisfies $D2$ then $\triangle$ satisfies $B1$.

Let $E = \langle \phi_1, \ldots, \phi_k \rangle$. From $D2$ and the fact that for all i, ϕ_i is consistent we can deduce that for all i, j, any member of $D_E^j(\phi_i)$ is consistent. This fact in conjunction with the restriction that $\mathcal{C}_D(E)$ consists of consistent tuples of dilations concludes the proof.

– If $\sqsubseteq$ satisfies $C1$ then $\triangle$ satisfies $B2$.

If $\bigwedge E$ is consistent, then it will be the case that $E \in \mathcal{C}_D(E)$, and $dt(E) = \langle 0, \ldots, 0 \rangle$. From the first half of $C1$ we conclude that for all tuples $E' \in \mathcal{C}_D(E)$, it will be the case that $dt(E) \sqsubseteq dt(E')$, and as such through Definition 11 we conclude that $\bigwedge E \in \triangle(E)$. From the second half of $C1$ we obtain that for any tuple $E' \in \mathcal{C}_D(E)$ such that $E \neq E'$ it will be the case that $dt(E) \sqsubset dt(E')$. This concludes the proof that $\triangle(E) \equiv \{\bigwedge E\}$.

– If D satisfies $D1$ then $\triangle$ satisfies $B3$.

Follows from $D1$ and Definition 11.

- If D satisfies $D3$ and $\sqsubseteq$ satisfies $C4$ and $C2$, then $\triangle$ satisfies $B4$.

  We assume that $\chi \in \triangle(\phi \sqcup \psi)$. If $\chi \nvdash \phi$ then we are done. So we assume that for all $\chi \in \triangle(\phi \sqcup \psi)$ it is the case that $\chi \vdash \phi$. By Definition 11, it follows that there exist $a_1, b_1 \in \mathbb{N}$ and $\phi_1 \in D_E^{a_1}(\phi)$, $\psi_1 \in D_E^{b_1}(\psi)$ such that $\langle \phi_1, \psi_1 \rangle$ is minimal in $\mathcal{C}_D(E)$. From $\chi \vdash \phi$ we get that $\phi_1 \wedge \psi_1 \vdash \phi$, therefore $\phi_1 \wedge \psi_1 \vdash \phi \wedge \psi_1$. By Definition 10, $\phi_1 \wedge \psi_1$ is consistent and, therefore, so is $\phi \wedge \psi_1$. By applying $D3$, we obtain $a_2 \in \mathbb{N}$ and $\phi_2 \in D_E^{a_2}(\phi)$ such that $\phi_2 \wedge \psi$ is consistent and $a_2 \leq b_1$. Using $C4$, we obtain $\langle a_2, 0 \rangle \sqsubseteq \langle 0, b_1 \rangle$. However, from $C1$ and $C2$ we know that $\langle 0, b_1 \rangle \sqsubseteq \langle a_1, b_1 \rangle$ and by transitivity we conclude that $\langle a_2, 0 \rangle \sqsubseteq \langle a_1, b_1 \rangle$ which means that, since $\phi_2 \wedge \psi$ is consistent, $\langle \phi_2, \psi \rangle$ is minimal in $\mathcal{C}_D(\phi \sqcup \psi)$, or equivalently, that $\phi_2 \wedge \psi \in \triangle(\phi \sqcup \psi)$. But then it must be the case that $\phi_2 \wedge \psi \vdash \phi$, which gives $\phi_2 \wedge \psi \vdash \phi \wedge \psi$ which is a contradiction, since by assumption $\phi \wedge \psi$ is inconsistent.

- If $\sqsubseteq$ satisfies $C2$ then $\triangle$ satisfies $B5$.

  The assumptions mean that there exist $A_1 \in \mathcal{C}_D(E_1)$ and $A_2 \in \mathcal{C}_D(E_2)$ such that $\phi_1 = \bigwedge A_1$, $\phi_2 = \bigwedge A_2$, and that $\bigwedge A_1 \wedge \bigwedge A_2$ is consistent. Moreover, A_1 and A_2 are $\sqsubseteq$-minimal, i.e. for all $B_1 \in \mathcal{C}_D(E_1)$, $dt(A_1) \sqsubseteq dt(B_1)$ and for all $B_2 \in \mathcal{C}_D(E_2)$, $dt(A_2) \sqsubseteq dt(B_2)$. Since $\bigwedge A_1 \wedge \bigwedge A_2$ is consistent, it follows that $A_1 \sqcup A_2 \in \mathcal{C}_D(E_1 \sqcup E_2)$. In addition, from $C2$ and the minimalities of A_1 and A_2 as above, it follows that for all $C \in \mathcal{C}_D(E_1 \sqcup E_2)$, $dt(A_1 \sqcup A_2) \sqsubseteq dt(C)$ and thus, $\bigwedge A_1 \sqcup A_2 \in \triangle(E_1 \sqcup E_2)$.

- If $\sqsubseteq$ satisfies $C3$ then $\triangle$ satisfies $B6$.

  Assume that $\chi \in \triangle(E_1 \sqcup E_2)$, which is to say that there is a $\sqsubseteq$-minimal tuple $A \in \mathcal{C}_D(E_1 \sqcup E_2)$ and $\chi = \bigwedge A$. Since χ is consistent, there exist tuples $B_1 \in \mathcal{C}_D(E_1)$, $B_2 \in \mathcal{C}_D(E_2)$ such that $A = B_1 \sqcup B_2$. Also, we know that there are $\sqsubseteq$-minimal tuples in $\mathcal{C}_D(E_1)$ and $\mathcal{C}_D(E_2)$ and that there are consistent pairs of those.

  If both B_1 and B_2 are $\sqsubseteq$-minimal in $\mathcal{C}_D(E_1)$ and $\mathcal{C}_D(E_2)$ respectively then we have nothing to prove.

  Let us assume that B_1 is *not* $\sqsubseteq$-minimal in $\mathcal{C}_D(E_1)$. This means that for any $\sqsubseteq$-minimal tuple $B_1' \in \mathcal{C}_D(E_1)$ it will be the case that $dt(B_1') \sqsubset dt(B_1)$ (on account of the totality of the preorders). Irrespective of the $\sqsubseteq$-minimality status of B_2, it will also be the case that for any $\sqsubseteq$-minimal tuple B_2' in $\mathcal{C}_D(E_2)$, $dt(B_2') \sqsubseteq dt(B_2)$.

  Since the choice of B_1' and B_2' is arbitrary among the minimal tuples of $\mathcal{C}_D(E_1)$ and $\mathcal{C}_D(E_2)$ we will choose them so that $\bigwedge B_1' \wedge \bigwedge B_2'$ is consistent, as per assumption. By using $C3$ we obtain $dt(B_1' \sqcup B_2') \sqsubset dt(B_1 \sqcup B_2)$. But then, $B_1 \sqcup B_2$ cannot be minimal in $\mathcal{C}_D(E_1 \sqcup E_2)$ anymore, a contradiction. Thus, there exist $\psi_1 = \bigwedge B_1$ and $\psi_2 = \bigwedge B_2$ such that $\psi_1 \wedge \psi_2 = \chi$ and $\psi_1 \in \triangle^{\mathrm{Gmax}}(E_1)$ and $\psi_2 \in \triangle^{\mathrm{Gmax}}(E_2)$.

On the Existence of Armstrong Instances with Bounded Domains

Attila Sali[1,*] and László Székely[2,**]

[1] Alfréd Rényi Institute of Mathematics, Hungarian Academy of Sciences
Budapest, P.O.B. 127, H-1364 Hungary
sali@renyi.hu

[2] Department of Mathematics University of South Carolina
Columbia, SC 29208 USA
szekely@math.sc.edu

Abstract. The existence of Armstrong-instances of bounded domains is investigated for specific key systems. This leads to the concept of Armstrong(q, k, n)-codes. These are q-ary codes of length n, minimum distance $n - k + 1$ and have the property that for any possible $k - 1$ coordinate positions there are two codewords that agree exactly there. We derive upper and lower bounds on the length of the code as function of q and k. The upper bounds use geometric arguments and bounds on spherical codes, the lower bounds are probabilistic.

Keywords: Armstrong instance, bounded domain, functional dependency, Armstrong-code, spherical code, Lovász' Local Lemma.

1 Introduction

Arguably the most important database constraint is the collection of functional dependencies that a relational schema satisfies, in particular, the key dependencies. If R denotes the set of attributes, then $K \subseteq R$ is a *key*, if the functional dependency $K \to R$ holds. In what follows we use the terminology of the book [1].

It is interesting from the point of view of schema design that given a collection Σ of functional dependencies, what other dependencies hold in a database instance that satisfies Σ. A way of solving this problem is the construction of an *Armstrong instance* for Σ, that is a database that satisfies a functional dependency $X \to Y$ if and only if $\Sigma \models X \to Y$. Silva and Melkanoff [19] developed a design aid that for a collection of functional and multivalued dependencies as input presents an Armstrong instance for that set. The existence of Armstrong instance for a set of functional dependencies was proved by Armstrong [3] and

---

[*] This research was done while the first author visited the Department of Mathematics, University of South Carolina. Research was supported in part by Hungarian National Research Fund (OTKA) grants T037846 and AT048826.

[**] Research was supported in part by NSF DMS grants 0302307 and 0701111.

S. Hartmann and G. Kern-Isberner (Eds.): FoIKS 2008, LNCS 4932, pp. 151–157, 2008.
© Springer-Verlag Berlin Heidelberg 2008

Demetrovics [4]. Later Fagin [11] gave a necessary and sufficient condition for general dependencies.

Further investigations concentrated on the minimum size of an Armstrong instance, since it is a good measure of the complexity of the collection of dependencies or system of minimal keys in question [5,6,7,8,9,10,12].

All papers cited above assumed that the *domain* of each attribute is unbounded, countably infinite. However, in the study of *Higher Order Datamodel* [14,16,17,18] the question of bounded domains arises naturally. In fact, if a minimal key system contains only *counter attributes*, then the possible number of tuples in an Armstrong instance is bounded from above. Another reason to consider bounded domains comes from real life databases. In many cases the domain of an attribute is a well defined finite set, for example in car rental, the class of cars can take values from the set {subcompact, compact, mid-size, full-size, SUV, sports-car, van}. Same kind of finiteness may occur in the case of job assignments, schedules, etc.

Thalheim [20] investigated the maximum number of minimal keys in the case of bounded domains and showed that having restrictions on the sizes of domains makes a significant difference.

It is natural to ask what can be said about Armstrong instances if attribute A_i has a domain of size q. The main question investigated in this paper was introduced in [18]. Let $\mathcal{K}_n^k$ denote the collection of all k-subsets of an n-element attribute set $\mathbf{R}$.

Definition 1. *Let $q > 1$ and $k > 1$ be given natural numbers. Let $f(q,k)$ be the maximum such n that there exists an Armstrong instance for $\mathcal{K}_n^k$ being the system of minimal keys.*

It is clear that for a meaningful Armstrong instance we need at least two distinct symbols, so $q > 1$ is necessary. On the other hand the minimal Armstrong instance for $\mathcal{K}_n^1$ uses only two symbols for arbitrary n [7], hence $f(q,k)$ is well defined only for $k > 1$.

Definition 2. *Let $\mathcal{K}$ be a Sperner system of minimal keys.*

$$\mathcal{K}^{-1} = \{A \subset \mathbf{R} : \nexists K \in \mathcal{K} \text{ such that } K \subseteq A \text{ and } A \text{ is maximal subject to this condition}\}$$

is the collection of maximal antikeys *corresponding to $\mathcal{K}$.*

The following basic fact is known [7].

Proposition 1. $\mathbf{A}$ *is an Armstrong instance for $\mathcal{K}$ iff the following two properties hold:*

(K) There are no two rows of $\mathbf{A}$ that agree in all positions for any $K \in \mathcal{K}$ and
(A) For every $A \in \mathcal{K}^{-1}$ there exist two rows of $\mathbf{A}$ that agree in all positions of A.

It is helpful to view an Armstrong instance for $\mathcal{K}_n^k$ as minimal key system using at most q symbols as a q-ary code $\mathcal{C}$ of length n, where codewords are the tuples, or rows of the instance. Using $(\mathcal{K}_n^k)^{-1} = \mathcal{K}_n^{k-1}$ we obtain (**md**) $\mathcal{C}$ has minimum Hamming-distance at least $n - k + 1$ by (**K**).

(**di**) For any set of $k - 1$ coordinates there exist two codewords that agree exactly there by (**A**).

A $k-1$-set of coordinate can be considered as a 'direction', so in $\mathcal{C}$ the minimum distance is *attained in all directions*. Such a code $\mathcal{C}$ is called *Armstrong-instance type code* of parameters (q, k, n), or *Armstrong(q, k, n)-code* for short. For example, the rows of the $k + 1 \times k + 1$ identity matrix form an Armstrong$(2, k, k+1)$-code.

Remark 1. Let $q > 1$ and $k > 1$ be given natural numbers. Then $f(q, k)$ is the maximum n such that there exists an Armstrong(q, k, n)-code.

In the following q is considered to be fixed, while k is let to increase without bound. That is, we consider the size of the domain as a fixed finite number, while the sizes of keys increase.

$f(q, k)$ was investigated in [13]. It was shown that

$$f(q, k) \leq q(k - 1) \tag{1}$$

with the following exceptions: $(k, q) = (5, 2), (5, 3), (5, 4), (5, 5), (6, 2)$. It was believed that the bound in (1) is of the right order of magnitude. The goal of the present paper is to show that this is not true by improving the coefficient of k to $q - \log q$ for large enough k. Also, the lower bound of [13] is improved to $\frac{1}{e}\sqrt{q}k$ by replacing the greedy construction by a probabilistic argument using Lovász' Local Lemma.

2   Upper Bound

The idea is to embed an Armstrong(q, k, n)-code into an $n' = (q-1)n$-dimensional space as a spherical code and use existing bounds for the size spherical codes of given minimum distance. On the other hand, an old result of Demetrovics and Katona [7] gives a lower bound for the size of an Armstrong(q, k, n)-code. Comparing the two estimates results in the lower bound for c, where $k - 1 = cn$.

Theorem 1. *For $k > k_0(q)$ we have $f(q, k) < (q - \log q)k$.*

Proof. It is not hard to see that if k is fixed and an Armstrong(q, k, n)-code exists for some $k < n$, then Armstrong(q, k, n')-codes also exist for all $k < n' < n$. Let $\mathcal{C}$ be an Armstrong(q, k, n)-code of size $m = |\mathcal{C}|$. Let $\ell = k - 1$. Using (**di**) and the argument of [7],

$$\binom{n}{\ell} \leq \binom{m}{2} \tag{2}$$

is obtained. Let $s: \{0, 1, \ldots, q - 1\} \to \mathbb{R}^{q-1}$ be a bijective mapping of the q symbols to the vertices of a regular simplex centered at the origin. Extend this mapping to codewords by juxtaposition of coordinates of vectors that are images of symbols of codewords under s. Thus each codeword of $\mathcal{C}$ is mapped to a vector from $\mathbb{R}^{(q-1)n}$ and we normalize them so they are unit vectors. Let $\mathcal{D}$ be the spherical code obtained. Using the minimum distance of $\mathcal{C}$ we obtain that $\mathcal{D}$ has minimum angle ϕ with $\cos\phi = \frac{\ell q - n}{(q-1)n}$ and $\sin(\frac{\phi}{2}) = \sqrt{\frac{q(n-k+1)}{2(q-1)n}}$. By (2) and the upper bound of Rankin [15]

$$A(n,\phi) \le \sqrt{\frac{\pi}{2}n^3 \cos\phi} \left(\sqrt{2}\sin(\frac{\phi}{2})\right)^{-n} (1+o(1)) \tag{3}$$

on the maximum size of a spherical code in n dimension with minimum angle ϕ. Applying that $m \le A((q-1)n, \phi)$ we obtain

$$\sqrt{2\binom{n}{\ell}} < m \le \sqrt{\frac{\pi}{2}(q-1)^3 n^3 \frac{\ell q - n}{(q-1)n}} \left(\sqrt{\frac{q(n-k+1)}{(q-1)n}}\right)^{-(q-1)n} (1+o(1)). \tag{4}$$

Writing $\ell = cn$ and using the approximation of $\binom{n}{cn}$ (4) yields

$$\sqrt{2}\left(\frac{1}{c^c(1-c)^{1-c}}\right)^{\frac{n}{2}} < \sqrt{\frac{\pi}{2}(q-1)n\sqrt{(cq-1)n}} \left(\sqrt{\frac{q-1}{q(1-c)}}\right)^{(q-1)n}. \tag{5}$$

Now, (5) can only hold for large enough n if

$$\frac{1}{c^c(1-c)^{1-c}} < \left(\frac{q-1}{q(1-c)}\right)^{q-1}. \tag{6}$$

It is easy to see that for $c = \frac{1}{q}$ LHS>1 and RHS=1 in (6). However, that only gives the upper bound established in [13].

With considerably more effort it can be shown that LHS>RHS in (6) for $c = \frac{1}{q-\log q}$, as well. Indeed, let $x = q - \log q$, $c = \frac{1}{x}$ and consider x as a function of variable q. LHS>RHS in (6) means

$$\frac{1}{\left(\frac{1}{x}\right)^{\frac{1}{x}}\left(\frac{x-1}{x}\right)^{\frac{x-1}{x}}} > \left(1-\frac{1}{q}\right)^{q-1}\left(\frac{x-1}{x}\right)^{1-q}. \tag{7}$$

Now, (7) can be written as

$$\frac{x}{(x-1)^{\frac{x-1}{x}}} > \left[\left(1-\frac{1}{q}\right)\left(1+\frac{1}{x-1}\right)\right]^{q-1}. \tag{8}$$

taking the logarithm of both sides of (8)

$$\log x - \frac{x-1}{x}\log(x-1) > (q-1)\left[\log\left(1-\frac{1}{q}\right) + \log\left(1+\frac{1}{x-1}\right)\right] \tag{9}$$

is obtained. Both sides of (9) is differentiated with respect to q to show that the difference LHS−RHS is an increasing function of q. For small values of q the validity of (6) is checked by computer. The derivative of LHS is

$$\frac{1}{x}\cdot x' - \frac{x-1}{x}\cdot\frac{1}{x-1}\cdot x' - \frac{-1}{(x-1)^2}\cdot x'\cdot\log(x-1), \tag{10}$$

where $x' = 1 - \frac{1}{q}$. Similarly, the derivative of the RHS is

$$\log\left(1 - \frac{1}{q}\right) + \log\left(1 + \frac{1}{x-1}\right) + (q-1)\frac{1}{(1 - \frac{1}{q})}\cdot\frac{-1}{q^2} + (q-1)\frac{1}{1 + \frac{1}{x-1}}\cdot\frac{-1}{(x-1)^2}\cdot x'.$$

(11)

Simplifying the inequality $(10) - (11) \geq 0$

$$\frac{1}{(x-1)^2}\cdot\frac{q-1}{q}\cdot\log(x-1) + \frac{1}{q} + \frac{(q-1)^2}{q(x-1)x} \geq \log\left(1 - \frac{1}{q}\right) + \log\left(1 + \frac{1}{x-1}\right)$$

(12)

is obtained. The right hand side of (12) can be estimated as

$$\log(q-1) - \log q + \log x - \log(x-1) = -\int_{q-1}^{q}\frac{dy}{y} + \int_{x-1}^{x}\frac{dy}{y} < \frac{-1}{q} + \frac{1}{x-1}.$$

(13)

Considering that the left hand side of (12) is $\frac{1}{q}$ plus two positive numbers, it is enough to see that

$$\frac{1}{q} \geq \frac{-1}{q} + \frac{1}{x-1}.$$

(14)

However, (14) holds if $q \geq 2\log q + 2$, which is true as long as $q \geq 8$.

Solving for c the equality of LHS and RHS in (6) numerically in the range of $q = 2, 3, \ldots, 200$ suggests that the limitation of the method above is really $\frac{1}{c} = q - \log q$.

3   Lower Bound

We will use the Lovász Local Lemma, that is a probabilistic construction will be given. For a comprehensive introduction of the method consult the book of Alon and Spencer [2].

For each $|K| = k - 1$ subset of coordinate positions a pair of codewords (A_1^K, A_2^K) is chosen randomly that agree exactly at those positions. That is, in each coordinate position each symbol is chosen with probability $\frac{1}{q}$, and the choices are pairwise independent for distinct positions. Consider events $v(A_i^K, A_j^L)$ where $i, j \in \{1, 2\}$ and $K \neq L$ are $k - 1$-sets of coordinate positions, that the two codewords agree in at least k coordinates. Two such events $v(A_i^K, A_j^L)$ and $v(A_i^{K'}, A_j^{L'})$ are independent if $\{K, L\} \cap \{K', L'\} = \emptyset$. Define the dependency graph $G = (V, E)$ by V being the set of events $v(A_i^K, A_j^L)$, and $v(A_i^K, A_j^L)$ and $v(A_i^{K'}, A_j^{L'})$ are connected by an edge if $\{K, L\} \cap \{K', L'\} \neq \emptyset$. Thus the degree of $v(A_i^K, A_j^L)$ in the dependency graph is $4\binom{n}{k-1} - 4$. On the other hand,

$$\mathrm{Prob}(v(A_i^K, A_j^L)) = \sum_{\ell=k}^{n}\binom{n}{\ell}\left(\frac{1}{q}\right)^{\ell}\left(\frac{q-1}{q}\right)^{n-\ell} = B(k, n, \frac{1}{q}).$$

(15)

By the well-known Chernoff bound $B(k,n,p) \leq \left(\frac{np}{k}\right)^k e^{k-np}$ if $k > np$ that in our case means $k > \frac{n}{q}$, which can be assumed without loss of generality. If $4\binom{n}{k-1}B(k,n,\frac{1}{q}) < \frac{1}{e}$, then

$$\text{Prob}\left(\bigcap \overline{v(A_i^K, A_j^L)}\right) > 0 \tag{16}$$

is obtained using Lovász' Local Lemma, which means that an Armstrong(q,k,n)-code code exists with these parameters. Writing $n = ck$, $4\binom{n}{k-1}B(k,n,\frac{1}{q}) < \frac{1}{e}$ becomes

$$\left[\frac{1}{\left(\frac{1}{c}\right)^{\frac{1}{c}}\left(1 - \frac{1}{c}\right)^{1-\frac{1}{c}}}\right]^{ck} \left(\frac{c}{q}\right)^k e^{k\left(1-\frac{c}{q}\right)} < \frac{1}{4e}. \tag{17}$$

The LHS of (17) is a complete kth power, thus if

$$\frac{1}{\frac{1}{c}\left(1-\frac{1}{c}\right)^{c-1}}\left(\frac{c}{q}\right)e^{\left(1-\frac{c}{q}\right)} < 1, \tag{18}$$

then (17) holds for $k > k_0$. Since $\left(\frac{c}{c-1}\right)^{c-1} < e$, (18) holds if $\frac{c^2}{q}e^{2-\frac{c}{q}} < 1$. This latter one is certainly true for $c < \frac{\sqrt{q}}{e}$. Thus, we have proved.

Theorem 2. *An Armstrong(q,k,n)-code code exists for $n < \frac{\sqrt{q}}{e}k$ if $k > k_0$, that is*

$$f(q,k) \geq \frac{\sqrt{q}}{e}k$$

for $k > k_0$.

4   Conclusions

We have proved general lower and upper bounds on $f(q,k)$. However there is a significant gap in them. If the lower bound can be improved, then it would require some design-theory like construction of codes. However, in the binary case one can prove that for small k, such as $k = 3, 4, 5$, the best construction is the $k + 1 \times k + 1$ identity. In particular, neither the Fano plane nor any of its extensions work. This makes it hard to find a "pattern" to generalize. On the other hand, the upper bound might also be improved, since the extremal spherical codes cannot be translated back to Armstrong-codes.

A more practical setting would be allowing the domain size to depend on the attribute, say $|\mathbf{dom}(A_i)| \leq q_i$. However, this destroys the coding theory connection.

References

1. Abiteboul, S., Hull, R., Vianu, V.: Foundations of Databases. Addison-Wesley, Reading (1995)
2. Alon, N., Spencer, J.: The Probabilistic Method. John Wiley and Sons, Chichester (2002)

3. Armstrong, W.W.: Dependency structures of database relationships. Information Processing, 580–583 (1974)
4. Demetrovics, J.: On the equivalence of candidate keys with Sperner systems. Acta Cybernetica 4, 247–252 (1979)
5. Demetrovics, J., Füredi, Z., Katona, G.O.H.: Minimum matrix reperesentation of closure operetions. Discrete Applied Mathematics 11, 115–128 (1985)
6. Demetrovics, J., Gyepesi, G.: A note on minimum matrix reperesentation of closure operetions. Combinatorica 3, 177–180 (1983)
7. Demetrovics, J., Katona, G.: Extremal combinatorial problems in relational data base. In: Gecseg, F. (ed.) FCT 1981. LNCS, vol. 117, pp. 110–119. Springer, Heidelberg (1981)
8. Demetrovics, J., Katona, G., Sali, A.: The characterization of branching dependencies. Discrete Applied Mathematics 40, 139–153 (1992)
9. Demetrovics, J., Katona, G., Sali, A.: Design type problems motivated by database theory. Journal of Statistical Planning and Inference 72, 149–164 (1998)
10. Demetrovics, J., Katona, G.O.H.: A survey of some combinatorial results concerning functional dependencies in databases. Annals of Mathematics and Artificial Intelligence 7, 63–82 (1993)
11. Fagin, R.: Horn clauses and database dependencies. Journal of the Association for Computing Machinery 29(4), 952–985 (1982)
12. Füredi, Z.: Perfect error-correcting databases. Discrete Applied Mathematics 28, 171–176 (1990)
13. Katona, G.O.H., Sali, A., and Schewe, K.-D.: Codes that attain minimum distance in all possible directions. Central European J. Math. (2007)
14. Hartmann, S., Link, S., Schewe, K.-D.: Weak functional dependencies in higher-order datamodels. In: Seipel, D., Turull-Torres, J.M. (eds.) FoIKS 2004. LNCS, vol. 2942. Springer, Heidelberg (2004)
15. Rankin, R.: The closest packing of spherical caps in n dimensions. Proceedings of the Glagow Mathematical Society 2, 145–146 (1955)
16. Sali, A.: Minimal keys in higher-order datamodels. In: Seipel, D., Turull-Torres, J.M.a (eds.) FoIKS 2004. LNCS, vol. 2942. Springer, Heidelberg (2004)
17. Sali, A., Schewe, K.-D.: Counter-free keys and functional dependencies in higher-order datamodels. Fundamenta Informaticae 70, 277–301 (2006)
18. Sali, A., Schewe, K.-D.: Keys and Armstrong databases in trees with restructuring. Acta Cybernetica (2007)
19. Silva, A., Melkanoff, M.: A method for helping discover the dependencies of a relation. In: Gallaire, H., Minker, J., Nicolas, J.-M. (eds.) Advances in Data Base Theory, vol. 1. Plenum Publishing, New York (1981)
20. Thalheim, B.: The number of keys in relational and nested relational databases. Discrete Applied Mathematics 40, 265–282 (1992)

Reasoning on Data Models in Schema Translation

Paolo Atzeni*, Giorgio Gianforme**, and Paolo Cappellari* * *

Dipartimento di Informatica e Automazione
Università Roma Tre, Italy
{atzeni,cappellari}@dia.uniroma3.it, giorgio.gianforme@gmail.com

Abstract. We refer to the problem of translating schemas from a model to another, in a model independent framework. Specifically, we consider an approach where translations are specified as Datalog programs. In this context we show how it is possible to reason on models and schemas involved as input and output for a translation. The various notions are formalized: (i) concise descriptions of models in terms of sets of constructs, with associated propositional formulas; (ii) a notion of signature for translation rules (with the property that signatures can be automatically computed out of rules); (iii) the "application" of signatures to models. The main result is that the target model of a translation can be completely characterized given the description of the source model and the signatures of the rules. This result is being exploited in the framework of a tool that implements model generic translations.

1   Introduction

The translation of schemas from a data model to another has received attention for decades [1,13,14,15]. An ambitious goal is to consider translations in a model generic setting [5,6], where the major problem can be formulated as follows: "given two data models M_1, M_2 (from a set of models of interest) and a schema S_1 of M_1, translate S_1 into a schema S_2 of M_2 that properly represents S_1."

The goal of this paper is to give formal grounds to the notion of model and to the management of translations of schemas with reference to a recent approach to this problem, the MIDST proposal [3]. MIDST assumes that there is a set of *generic* constructs, each with a number of possible variations. Constructs with the "same" meaning in different models are defined in terms of the same generic construct; for example, "entity" in an ER model and "class" in an object model both correspond to the *abstract* construct. Then, a data model is defined by specifying the constructs it includes. The notion of a *supermodel* is introduced: it is a model that includes all constructs (each in its most general version) and so generalizes all the other models (modulo renaming of constructs).

* Supported by MIUR, within the FIRB-MAIS project, and by an IBM Faculty Award.
** Supported by Microsoft Research through its European PhD Scholarship Programme.
* * * Supported by MIUR, within the FIRB-MAIS project. Currently with IBM.

S. Hartmann and G. Kern-Isberner (Eds.): FoIKS 2008, LNCS 4932, pp. 158–177, 2008.

In this framework, the set of models can become very large: each construct has many variations, with many possible combinations. For example, binary relationships (in the family of ER models) can be many-to-many or be limited to one-to-many; they might allow attributes or not, and so on. With more constructs (for example nested structures, as appearing in object models or in XML contexts) it is easy to reach hundreds or thousands of different models.[1] With this size for the space of data models, it would be hopeless to have translations between every pair of models. If we refer to the supermodel, then we would need one translation for each model (from the supermodel to it), but with many models this would still be impractical. In order to tackle this problem, MIDST (following MDM [5], and concurrently with other proposals [7,16]) has complex translations that are built as composition of elementary ones. The idea is to have "basic" translations that perform elementary steps that refer to generic constructs, and so can be reused. For example, both elements in the XSD world and entities in the ER model can be seen as "abstracts," and so both the translation from XSD and that from the ER to the relational model require the replacement of abstracts with tables. A translation from a model to another would then be obtained as a sequence of basic translations. For example, if we want to translate schemas from an ER model with generalizations and n-ary relationships to an object model with no generalizations, then we would have the following steps (where, for the sake of clarity, we use the specific names, such as "entity," instead of the generic ones, such as "abstract"):

1. replace n-ary relationships with binary ones (adding some entities);
2. eliminate many-to-many relationships (adding entities again);
3. replace one-to-many and one-to-one relationships with references;
4. eliminate generalizations (in an object model, new references are added).

With this fine-grained approach, the elementary steps can be reused, but care is needed. For example, steps 1-4 are meaningful in this order; moreover, step 2 is not needed if the source model has no many-to-many relationships. On the other hand, if step 2 is omitted and there are many-to-many relationships, then the latter are not translated and get "lost." Therefore, a major issue arises: given a set of basic translations, how do we build the actual translations we need? How do we verify that a given sequence of basic translations produces the model we are interested in? And that it does not "forget" any construct?

The contribution of this paper is a formal approach to answering these questions. We refer to our current project in this area, MIDST [3], where models can be defined in terms of constructs, as above, and translations are implemented in a variant of Datalog. Let us give an idea of the result. First, we associate with each model a "signature" that describes its constructs. We introduce a similar notion for our Datalog programs and for their applications: given a program, we can derive its signature, and we have defined the application of signatures of programs to signatures of models. Then the result is that signatures completely

---

[1] Obviously, all these models need not be defined at the same time. However, even if few models are used, they are chosen in a very large set.

describe the behavior of programs on models, in the sense that the application of signatures provides a "derivation" of models that is sound and complete with respect to the schemas generated by programs. Let us describe the main results with the help of Fig. 1. Let S_1 be a schema for a model $\mathcal{M}_1$, $\mathbf{P}$ a Datalog program

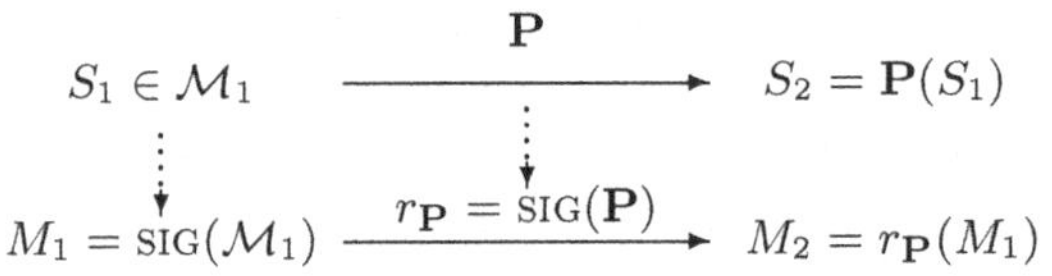

$$
\begin{array}{ccc}
 & \mathbf{P} & \\
S_1 \in \mathcal{M}_1 & \longrightarrow & S_2 = \mathbf{P}(S_1) \\
\vdots & \vdots & \vdots \\
M_1 = \text{SIG}(\mathcal{M}_1) & \xrightarrow{\ r_{\mathbf{P}} = \text{SIG}(\mathbf{P})\ } & M_2 = r_{\mathbf{P}}(M_1)
\end{array}
$$

Fig. 1.

implementing a basic translation, and S_2 the schema obtained by applying $\mathbf{P}$ to S_1. Then, let M_1 be the signature of $\mathcal{M}_1$ and $r_{\mathbf{P}}$ the signature of $\mathbf{P}$, which corresponds to a function (the "application" $r_{\mathbf{P}}()$ of $r_{\mathbf{P}}$) from signatures of models to signatures of models. Soundness and completeness (of signatures with respect to Datalog programs) can be claimed as follows, respectively:

1. the application of $\mathbf{P}$ to a schema S_1 of $\mathcal{M}_1$ produces a schema $\mathbf{P}(S_1)$ that belongs to the model whose signature $r_{\mathbf{P}}(M_1)$ is obtained by applying the signature $r_{\mathbf{P}}$ of program $\mathbf{P}$ to the signature M_1 of $\mathcal{M}_1$;
2. among the possible schemas of $\mathcal{M}_1$, there exists a schema S^* such that the application of $\mathbf{P}$ to S^* produces a schema $\mathbf{P}(S^*)$ that (beside belonging to $r_{\mathbf{P}}(M_1)$, as stated by the previous claim) does not belong to any model that is "strictly more restricted" than $r_{\mathbf{P}}(M_1)$.

The two claims together say that the model we derive by means of the application of signatures is exactly the model that allows the set of schemas that can be obtained by means of the Datalog programs. Claim (1) says that the derived model is liberal enough (soundness) and claim (2) says that it is restricted enough (completeness). This can be seen as a "syntactic" aspect of the correctness of rules to be complemented by a "semantic" one.

The contents of this paper complement those we have recently published on a tool for translations (MIDST [3,4]). In the latter papers, we have shown the overall approach to the problem and how individual translations can be defined and manually composed. Here we show how it is possible to describe the properties of translations and to reason on them. Various applications are possible for the results we have here. First of all, they can be used to perform simple checks on the behavior of Datalog programs, without the need to run them on specific schemas. Moreover, they can be the basis for the automatic generation of complex translations given a library of elementary steps.

The rest of the paper is organized as follows. The main results are in Section 5; they are preceded by an overview of MIDST and an informal presentation of the results (Section 2) and by a formalization of the approach, both for models (Section 3) and "signatures" of rules (Section 4). Then we discuss in Section 6 the interesting consequences of the results in the context of a tool. We also discuss some related work (Section 7) and draw our conclusions (Section 8).

2   Context and Goals

We set the context for our approach, by discussing the features of the MIDST [3] project that are of interest, together with a running example that will be used to illustrate the technical development.

We assume the availability of a *universe* of constructs. Each construct has a set of references (which relate its occurrences to other constructs; references are acyclic) and Boolean properties. Occurrences of constructs also have names and possibly types (for example, in the ER model each entity, relationship, and attribute has a name, and each attribute has a type), but we need not refer to them here. Let us explain the basic idea by means of the universe we will use in our running example. This universe has the following constructs (just a subset of the universe we are using in the MIDST tool, which allows to handle many versions of the XSD, ER, OO and relational models):

Construct	*References*	*Properties*
ENTITY		
ATTRIBUTEOFENT.	Entity	isKey, isNullable
RELATIONSHIP	Entity1, Entity2	isOpt1, isFunct1, isIdent, isOpt2, isFunct2
ATTRIBUTEOFREL.	Relationship	isNullable
TABLE		
COLUMN	Table	isKey, isNullable

Let us comment on the various aspects. References are rather intuitive: each attribute and column has a reference to the construct it belongs to; each relationship (binary here for the sake of simplicity) has references to the two involved entities. Properties require some explanation: for each attribute of an entity, we can specify, with isKey, whether it belongs to the primary key and, with isNullable, whether it allows null values; each relationship has five properties, which describe its cardinality and whether it contributes to the identification of an entity: isOpt1 (isOpt2) tells us whether the participation of the first (second) entity is optional or compulsory, that is, whether its minimum cardinality is 0 or 1, isFunct1 (isFunct2) tells whether its maximum cardinality is 1 or is unbounded ('N', as we usually write), isIdent tells us whether the first entity has an external identifier to which this relationship contributes (a *weak* entity [18]).

In this framework, given a set of constructs, the references are always required to build schemas for meaningful models (for example, a relationship without references to entities would make no sense), whereas properties could be restricted in some way: for example, we can think of models where all cardinalities for relationships are allowed and models where many-to-many relationships are not allowed. Therefore, we can think that models are defined by means of the constructs, each with a condition on its properties; this idea will be formalized in Section 3. Let us list a set of models to be used in the discussion:

- $\mathcal{M}_{\text{REL}}$: a relational model, with tables and columns (and no restrictions);
- $\mathcal{M}_{\text{RELNON}}$: a relational model with no null values: all columns must have a value *false* for isNullable property;

- $\mathcal{M}_{\text{ER}}$: an ER model with all the available features;
- $\mathcal{M}_{\text{ER}_{\text{SIMPLE}}}$: an ER model with no null values on attributes (all attributes have a value *false* for isNullable) and no attributes on relationships;
- $\mathcal{M}_{\text{ER}_{\text{NOM2N}}}$: an ER model with no many-to-many relationships (all relationships have a value *true* for isFunct1 or isFunct2).

As we said in the introduction, translations in this approach are specified by means of sequences of elementary steps. A translation from the ER model $\mathcal{M}_{\text{ER}}$ to the relational model $\mathcal{M}_{\text{REL}}$ could be composed of the basic translations:

- $\mathbf{P_1}$: eliminate many-to-many relationships;
- $\mathbf{P_2}$: translate (i) entities into tables, (ii) attributes and one-to-many (and one-to-one) relationships into columns.

In MIDST [3] translations are specified in a Datalog variant with OID invention, with ideas from ILOG [12]. The predicates correspond to constructs and their arguments may be (beside OIDs, names and possibly types), names of references and properties. A Datalog program for $\mathbf{P_1}$ (shown for completeness in the appendix) includes rules of two kinds: rules that copy the constructs that go unchanged from the source schema to the target one (four rules, $R_{1,1}$-$R_{1,4}$ in the appendix) and rules that actually perform the translation, as follows:

- $R_{1,5}$ generates an entity for each many-to-many relationship;
- $R_{1,6}, R_{1,7}$ generate, for each entity generated by $R_{1,5}$, relationships between it and the copies of the two entities involved in the original many-to-many relationship, respectively;
- $R_{1,8}$ generates, for each attribute of each many-to-many relationship, an attribute for the entity generated by $R_{1,5}$.

Let us show rules $R_{1,5}$ and $R_{1,6}$, in order to discuss a few issues that are important for understanding the way we use rules.

- rule $R_{1,5}$

    ENTITY(OID: #entity_1(rOid), sOID: tgt, Name: n)

    ←RELATIONSHIP(OID: rOid, sOID: src, Name: n, isFunct1: false, isFunct2: false)

- rule $R_{1,6}$

    RELATIONSHIP(OID: #relationship_1(eOid,rOid), sOID: tgt, Name: eN+rN,

        Entity1: #entity_1(rOid), isOpt1: false, isFunct1: true, isIdent: true,

        Entity2: #entity_0(eOid), isOpt2: isOpt, isFunct2: false)

    ←RELATIONSHIP(OID: rOid, sOID: src, Name: rN, Entity1: eOid, isOpt1: isOpt,

        isFunct1: false, isFunct2: false),

    ENTITY(OID: eOid, sOID: src, Name: eN)

We use a non-positional notation for rules, so we indicate the names of the fields, and omit those that are not needed (rather than using anonymous variables). Rules generate constructs for a target schema (tgt) from those in a source schema (src), and we may assume that variables tgt and src are bound to constants when the rule is executed. Each predicate has an OID argument, used for unique identification. For each schema we have different identifiers, and so, when a construct is produced by a rule, it has to have a "new" identifier, which is generated by means of a Skolem functor (denoted by the # sign in the rules). In general, there are several functors (with disjoint ranges) for each construct. Let us comment on a few issues with reference to $R_{1,5}$ and $R_{1,6}$:

- both rules are applied only if both isFunct1 and isFunct2 are `false`: so, if the source model has no many-to-many relationships (for example, $\mathcal{M}_{\text{ERNoM2N}}$), these rules have no effect;
- $R_{1,6}$ always generates relationships with isFunct1 = `true`; so, it does not generate many-to-many relationships; constants in the head tell us some restrictions on the construct that is generated;
- the repeated variable isOpt in $R_{1,6}$ transfers the value of isOpt1 in the source to the value of isOpt2 in the target; if isOpt1 were constrained in the source, so would isOpt2 be in the target.

Therefore, if we apply program $\mathbf{P}_1$ to a schema in model $\mathcal{M}_{\text{ERNoM2N}}$, we obtain the same schema,[2] and so, if we want to translate a schema from $\mathcal{M}_{\text{ERNoM2N}}$ to $\mathcal{M}_{\text{REL}}$, then it suffices to apply program $\mathbf{P}_2$. In this paper we show that it is possible to formalize these ideas, by means of the notions of signatures of models and programs mentioned in the introduction. In fact, if, for each of the models listed above, M_α is the signature of model $\mathcal{M}_\alpha$ and $r_{\mathbf{P}_1}$ and $r_{\mathbf{P}_2}$ are the signatures of programs $\mathbf{P}_1$ and $\mathbf{P}_2$, respectively, then:

(i) $r_{\mathbf{P}_1}(M_{\text{ERNoM2N}}) = M_{\text{ERNoM2N}}$;
(ii) $r_{\mathbf{P}_1}(M_{\text{ER}}) = M_{\text{ERNoM2N}}$;
(iii) $r_{\mathbf{P}_2}(r_{\mathbf{P}_1}(M_{\text{ERSIMPLE}})) = M_{\text{RelNoN}}$.

The results in Section 5 guarantee that these "syntactic" observations correspond indeed to the "semantics": for example, from (iii) we will be able to know that the application of the sequence of programs $\mathbf{P}_1$ and $\mathbf{P}_2$ to a schema of $\mathcal{M}_{\text{ERSIMPLE}}$ produces a schema of $\mathcal{M}_{\text{RELNoN}}$. Moreover, the technical development will allow us to say that if we apply program $\mathbf{P}_2$ to a schema of $\mathcal{M}_{\text{ERSIMPLE}}$ or $\mathcal{M}_{\text{ER}}$, then some constructs are ignored and so get lost: these models include many-to-many relationships, which are not translated.

Before concluding the section, let us mention that we have a number of restrictions on our rules, for which we give only the main ideas (used in the following as hypotheses for proving the main results of this paper), omitting details. It is important to observe that all these assumptions correspond to features we have in the MIDST approach and that they can be checked automatically. First of all, we have the standard "safety" requirements [18]. Second, we assume that boolean variables for properties cannot be repeated in the head nor in the body (the only allowed repetition is: once in the body and once in the head). Third, repeated variables for oids are allowed only in fields that are subject to referential integrity constraints and in comparison atoms, which may contain only inequality comparisons (equalities are handled by means of repeated variables).

---

[2] More precisely, we obtain an isomorphic copy of the source schema.

It is also important to observe that we refer to rules without negation. Indeed, negations arise only in a few cases in the translation process, with a restriction over positive conditions which can be reduced to the positive case (possibly with some growth in the number of rules).

Moreover, our programs are coherent with respect to referential constraints: if there is a rule that produces a construct N that refers to a construct N', then there is another rule that generates a suitable N' that guarantees the satisfaction of the constraint. In the example, rule $R_{1,6}$ is acceptable because rules $R_{1,1}$ and $R_{1,5}$ produce entities and so guarantee that references of relationships generated by $R_{1,6}$ are not dangling.

A final comment on recursion is useful. Most of our rules, such as $R_{1,6}$ above, are recursive according to the standard definition. However, recursion is only "apparent": a literal occurs in both the head and the body, but the construct generated by an application of the rule belongs to the target schema, so the rule cannot be applied to it again, as the body refers to the source schema. A really recursive application happens only for rules that have atoms that refer to the target schema also in their body. In the following, we will use the term **strongly recursive** for these rules. Indeed, in our experiments with MIDST we have needed few strongly recursive rules, the most notable of which are for resolving unbounded chains of external identification (a department number is unique within a company, a sub-department number is unique within a department, and so on) and for unnesting complex structures.

3   Signatures of Constructs and Models

In this section we formalize the description of models. We define models in terms of their signatures, blurring the distinction between a model and its signature, as signatures are sufficient for the purpose of this paper.

As we illustrated in the previous section, each construct has a number of properties and references. References are tightly bound to the construct, with no variations. Properties, instead, can be subject to restrictions. Therefore we can give a synthetic description of a model by listing the constructs it involves, each with a propositional formula over its properties.

In detail, we fix a **universe** of constructs, each with a set of associated properties: $\mathcal{U} = \{N_1(P_1), \ldots, N_u(P_u)\}$. In the examples we will refer to the constructs in the previous section, in abbreviated form as follows:

Construct	Properties	Abbreviation
ENTITY		$E()$
ATTRIBUTEOFENT	isKey,isNullable	$A(K, N)$
RELATIONSHIP	isOpt1,isFunct1,isIdent,isOpt2,isFunct2	$R(O_1, F_1, I, O_2, F_2)$
ATTRIBUTEOFREL	isNullable	$AR(N)$
TABLE		$T()$
COLUMN	isKey,isNullable	$C(K, N)$

Then, **(the signature of) a model** is a mapping that associates a proposition with each construct in the universe, denoted as follows: $M = \{N_1(f_1), \ldots, N_u(f_u)\}$. We will use the term **construct signature** to refer to $N(f)$, where N is the name of a construct and f a proposition (over the associated properties).

In the definition above, we have that all constructs are mentioned in every model, possibly with a *false* proposition, which would mean that the construct does not belong to the model. In practice, we describe a model by listing only the constructs that really belong to it—those that have a satisfiable proposition; in this way, the models discussed in Section 2 would be as follows (with the propositions that specify the restrictions we have informally described there):

- $M_{\text{REL}} = \{\text{T}(true), \text{C}(true)\}$
- $M_{\text{RELNoN}} = \{\text{T}(true), \text{C}(\neg\text{N})\}$
- $M_{\text{ER}} = \{\text{E}(true), \text{A}(true), \text{R}(\text{F}_1 \vee \neg\text{F}_2), \text{AR}(true)\}^3$
- $M_{\text{ERSIMPLE}} = \{\text{E}(true), \text{A}(\neg\text{N}), \text{R}(\text{F}_1 \vee \neg\text{F}_2)\}$
- $M_{\text{ERNoM2N}} = \{\text{E}(true), \text{A}(true), \text{R}(\text{F}_1), \text{AR}(true)\}$

We can define a partial order on models, as follows:

- $M_1 \sqsubseteq M_2$ (read M_1 is **more restricted** than M_2) if for every $N \in \mathcal{U}$ it is the case that $f_1 \wedge f_2$ is equivalent to f_1 (that is, f_1 implies f_2), where $N(f_1) \in M_1$ and $N(f_2) \in M_2$

It can be shown that $\sqsubseteq$ is a partial order (modulo equivalence of propositions), as it is reflexive, antisymmetric and transitive.

If models are described only in terms of the constructs that have satisfiable properties, then the partial order can be rewritten as:

- $M_1 \sqsubseteq M_2$ if for every $N(f_1) \in M_1$ there is $N(f_2) \in M_2$ such that $f_1 \wedge f_2$ is equivalent to f_1

In words, $M_1 \sqsubseteq M_2$ means that M_2 has at least the constructs of M_1 and, for those in M_1, it allows at least the same variants. For the example models:

- $M_{\text{RELNoN}} \sqsubseteq M_{\text{REL}}$ (and $M_{\text{REL}} \not\sqsubseteq M_{\text{RELNoN}}$): they have the same constructs, but M_{RELNoN} has a more restrictive condition on construct C than M_{REL};
- $M_{\text{ERSIMPLE}} \sqsubseteq M_{\text{ER}}$ (and $M_{\text{ER}} \not\sqsubseteq M_{\text{ERSIMPLE}}$): the constructs in M_{ERSIMPLE} are a proper subset of those in M_{ER} and, for each of them, the condition in M_{ERSIMPLE} is at least as restrictive as the respective one in M_{ER};
- $M_{\text{ERNoM2N}} \not\sqsubseteq M_{\text{ERSIMPLE}}$, $M_{\text{ERSIMPLE}} \not\sqsubseteq M_{\text{ERNoM2N}}$: in fact M_{ERNoM2N} has AR which is not in M_{ERSIMPLE}, but has a more restrictive condition on R.

We can define two binary operators on the space of models as follows:

$$M_1 \sqcup M_2 = \{N(f_1 \vee f_2)|N(f_1) \in M_1 \text{ and } N(f_2) \in M_2\}$$
$$M_1 \sqcap M_2 = \{N(f_1 \wedge f_2)|N(f_1) \in M_1 \text{ and } N(f_2) \in M_2\}$$

---

[3] Without loss of generality, we assume that in a one-to-many relationship, it is the first entity that has a functional role, and so $\text{F}_1 = true$ and $\text{F}_2 = false$.

It can be shown that the space of models forms a lattice with respect to these two operators (modulo equivalence of propositions). The proofs of the claims that guarantee the lattice structure follow the definitions and the fact that the boolean operators in propositional logic form a lattice. The supermodel (the fictitious most general model mentioned in the Introduction) is the top element of the lattice (with the true proposition for every construct). It is worth noting that models obtained as the result of these operations, especially ⊓, could have, in some extreme cases, little practical meaning. For example, the bottom element of the lattice is the (degenerate) empty model, which has all false propositions (or, in other words, no constructs).

4 Signatures of Datalog Rules and Their Application

In order to handle rules and to reason on them, in an effective way, we introduce the notion of "signature" of a Datalog rule. The definition gives a unique construction, so the signature can be automatically computed for each rule.

As a preliminary step, let us define the **signature of an atom** in a Datalog rule. Given an atom $N(\mathrm{ARGS})$, consider the fields in ARGS that correspond to properties (ignoring the others); let them be $p_1{:}v_1, \ldots, p_k{:}v_k$; each v_i is either a variable or a boolean constant *true* or *false*. Then, the signature of $N(\mathrm{ARGS})$ is a construct signature $N(f)$, where the proposition f is the conjunction of literals corresponding to the properties in $p_1, \ldots, p_k$ that are associated with a constant; each of them is positive if the constant is *true* and negated if it is *false*. If there are no constants, then the proposition is *true*. For example, the signatures of the two atoms in the body of rule $R_{1,6}$ are $\mathrm{R}(\neg\mathrm{F}_1 \wedge \neg\mathrm{F}_2)$ and $\mathrm{E}(true)$, respectively.

Let us now define the **signature of a Datalog rule**. Let R be a rule, with a head $N(\mathrm{ARGS})$ and a body with a list of atoms referring to constructs which need not be distinct $\langle N_{j_1}(\mathrm{ARGS}_1), \ldots, N_{j_h}(\mathrm{ARGS}_h)\rangle$; comparison terms (with inequalities, according to our hypotheses) do not affect the signature, and so we can ignore them. The signature r_R of R is composed of three parts (B, H, MAP):

- B (the **body of** r_R) describes the applicability of the rule, by referring to the constructs in the body of R; B is a list of signatures of constructs, $\langle N_{j_1}(f_1), \ldots, N_{j_h}(f_h)\rangle$, where $N_{j_i}(f_i)$ is the signature of the atom $N_{j_i}(\mathrm{ARGS}_i)$.
- H (the **head of** r_R) indicates the conditions that definitely hold on the result of the application of R, because of constants in its head; H is defined as the signature $N(f)$ of the atom $N(\mathrm{ARGS})$ in the head.
- MAP (the **mapping of** r_R) is a partial function that describes where values of properties in the head originate from. It is defined as follows. Its domain is the set of properties of the construct in the head; MAP is defined for the properties that are associated, in the head, with a variable. For our assumptions, each variable in the head appears also in the body, and only once. If a variable appears for a property p' in the head and for a property p of a construct N_{j_k} in the body, then MAP is defined on p' and $\mathrm{MAP}(p') = N_{j_k}(p)$.

Let us see the definition on rule $R_{1,6}$ in our running example. The body is $B_{1,6} = \langle R(\neg F_1 \wedge \neg F_2), E(true) \rangle$; indeed, the rule is applicable only to many-to-many relationships, that is, if both F_1 and F_2 are *false*.

The head of the signature of rule $R_{1,6}$ is $H_{1,6} = R(\neg O_1 \wedge F_1 \wedge I \wedge \neg F_2)$: the relationships produced by the rule all have O_1 and F_2 equal to *false* and F_1 and I equal to *true*.

The mapping for rule $R_{1,6}$ is $MAP_{1,6} = \langle O_2 : R(O_1) \rangle$ (we denote the function as a list of pairs, including only the properties on which it is defined). The name of the construct in the head is not mentioned, because it is known, but let us note that it might be different from the one in the body; this is the case for $R_{1,8}$ (in the appendix) where $MAP = \langle N : AR(N) \rangle$ and the first N is a property of A, the construct in the head.

Then, we can define the "application" of the signature of a rule to a model. We need two preliminary notions. First, we say that the signature $r_R = (B, H, MAP)$ of a rule R is **applicable** to a model M if, for each $N_{j_i}(f_i)$ in B, there is $N_{j_i}(f_{j_i}^M) \in M$ such that $f_{j_i}^M \wedge f_i$ is satisfiable. In words, each construct in the body has to appear in the model, and the two propositions must not contradict one another. For example, $R_{1,6}$ is not applicable to $M_{ERNoM2N}$ because we have $R(\neg F_1 \wedge \neg F_2)$ in the body of the rule and $R(F_1)$ in the model: the conjunction of $\neg F_1 \wedge \neg F_2$ and F_1 is not satisfiable.

Second, let us define the **transformation** $\mu_{MAP}()$ induced by mapping MAP on literals. Let l be a literal for a property p of an atom $N_{j_i}(\ldots)$ in the body of rule R. Then, if $N_{j_i}(p)$ belongs to the range of MAP, with $MAP(p') = N_{j_i}(p)$, we have that $\mu_{MAP}(l)$ is a literal for the property p' with the same sign as l; if $N_{j_i}(p)$ does not belong to the range of MAP, then $\mu_{MAP}(l) = true$. Let us define μ_{MAP} also on constants: $\mu_{MAP}(true) = true$ and $\mu_{MAP}(false) = false$. The notion can be extended to disjunctions and conjunctions: (i) $\mu_{MAP}(f_1 \wedge f_2) = \mu_{MAP}(f_1) \wedge \mu_{MAP}(f_2)$ (if $f_1 \wedge f_2$ is satisfiable, otherwise $\mu_{MAP}(f_1 \wedge f_2) = false$); (ii) $\mu_{MAP}(f_1 \vee f_2) = \mu_{MAP}(f_1) \vee \mu_{MAP}(f_2)$. In plain words, we use μ_{MAP} to "transfer" constraints on literals over properties in the body to literals in the head according to the MAP of the rule.

We are now ready for the definition of $r_R(M)$, the **application** of the signature of a rule R to a model M. If r_R is not applicable to M, then we define $r_R(M) = \{\}$. The interesting case is when r_R is applicable to M. Let the signatures of the body and of the head of R be $B = \langle N_{j_1}(f_1), N_{j_2}(f_2), \ldots, N_{j_h}(f_h) \rangle$ and $H = N(f)$, respectively. For every atom $N_{j_i}(f_i)$ in the body, let $f_{j_i}^M$ be the proposition associated with N_{j_i} in the model.

Let us first give the definition in the special case where all the constructs in the source model M have propositions that are just conjunctions of literals. In this case, $r_R(M) = \{N(f')\}$, where: $f' = f \wedge (\bigwedge_{i=1}^{h} \mu_{MAP}(f_{j_i}^M \wedge f_i))$. Let us note that $(f_{j_i}^M \wedge f_i)$ is satisfiable, since the rule is applicable, and that it is just a conjunction of literals, because this is the case for f_i, by construction, and for $f_{j_i}^M$, by hypothesis. In words, the condition in the result is obtained as the conjunction of the proposition in the head, f, with those obtained, by means of

MAP, from those in the source model (the $f_{j_i}^M$'s) and those in the body of the rule (the f_i's).

If the $f_{j_i}^M$'s include disjunctions, then let us rewrite $f_{j_i}^M \wedge f_i$ in disjunctive normal form $g_{i,1} \vee \ldots \vee g_{i,q_i}$. Then f' is built as the conjunction of the disjunctions of the applications of μ_{MAP} to the disjuncts: $f' = f \wedge (\bigwedge_{i=1}^{h} (\bigvee_{t=1}^{q_i} \mu_{\text{MAP}}(g_{i,t})))$.

Let us see three cases of applications of a rule in our running example. First, we have that $r_{R_{1,6}}(M_{\text{ERNoM2N}}) = \{\}$, as the rule is not applicable to the model (as we already saw).

Second, we have $r_{R_{1,6}}(M_{\text{ER}}) = \{\text{R}(\neg O_1 \wedge F_1 \wedge I \wedge \neg F_2)\}$. The rule is applicable since only construct R has an associated proposition and $(F_1 \vee \neg F_2) \wedge (\neg F_1 \wedge \neg F_2)$ is satisfiable, as it is equivalent to $\neg F_1 \wedge \neg F_2$. Then, applying the definition $f' = f \wedge (\bigwedge_{i=1}^{h} (\bigvee_{t=1}^{q_i} \mu_{\text{MAP}}(g_{i,t})))$ we have that the conjunction of the disjunctions $\mu_{\text{MAP}}(\ldots)$ is *true*, since the only property in the body mapped to the head is O_1, which does not appear in the argument of μ_{MAP}. Therefore, f' equals the condition f in the head of the signature: $f' = f = \neg O_1 \wedge F_1 \wedge I \wedge \neg F_2$.

As a third example, to see MAP and μ_{MAP} really in action, let us apply $R_{1,6}$ to model $M = \{\text{E}(\textit{true}), \text{R}((F_1 \vee \neg F_2) \wedge \neg O_1)\}$. The rule is applicable and we have $r_{R_{1,6}}(M) = \{\text{R}(\neg O_1 \wedge F_1 \wedge I \wedge \neg F_2 \wedge \neg O_2)\}$ as

$$
\begin{aligned}
f' &= f \wedge \mu_{\text{MAP}}(((F_1 \vee \neg F_2) \wedge \neg O_1) \wedge (\neg F_1 \wedge \neg F_2)) \wedge \mu_{\text{MAP}}(\textit{true} \wedge \textit{true}) \\
&= f \wedge (\mu_{\text{MAP}}(F_1 \wedge \neg O_1 \wedge \neg F_1 \wedge \neg F_2) \vee \mu_{\text{MAP}}(\neg F_2 \wedge \neg O_1 \wedge \neg F_1 \wedge \neg F_2)) \wedge \textit{true} \\
&= f \wedge (\mu_{\text{MAP}}(\textit{false}) \vee (\mu_{\text{MAP}}(\neg F_2) \wedge \mu_{\text{MAP}}(\neg O_1) \wedge \mu_{\text{MAP}}(\neg F_2))) \\
&= f \wedge (\textit{false} \vee (\textit{true} \wedge \neg O_2 \wedge \textit{true})) = \neg O_1 \wedge F_1 \wedge I \wedge \neg F_2 \wedge \neg O_2.
\end{aligned}
$$

We are now ready for our final definition, that of the application of the signature of a Datalog program (implementing a basic translation) to a model. Let us first consider programs with no strongly recursive rules; we will remove this assumption before the end of the section. Given a program $\mathbf{P}$ consisting of a set of Datalog rules $R_1, \ldots, R_n$, the **application** of $\mathbf{P}$ to a model M is the least upper bound of the applications of the R_i's to M: $r_{\mathbf{P}}(M) = \bigsqcup_{i=1}^{n} r_{R_i}(M)$. In this way, we have a construct for each applicable rule and, if a construct is generated by more than one rule, the associated condition is the disjunction of the conditions of the various rules.

If we apply the program $\mathbf{P}_1$ in our running example to the ER model M_{ER}, then all constructs get copied and maintain the *true* proposition, except relationships, for which rules $R_{1,3}$, $R_{1,6}$ and $R_{1,7}$ generate, respectively, $\text{R}(F_1)$, $\text{R}(\neg O_1 \wedge F_1 \wedge I \wedge \neg F_2)$ (as we saw above) and $\text{R}(\neg O_1 \wedge F_1 \wedge I \wedge \neg F_2)$. Therefore, as the disjunction of the three formulas is F_1, the target model will have $\text{R}(F_1)$ and so we can say that the application of the signature of the program to M_{ER} produces M_{ERNoM2N}. The results in Section 5 will tell us that, as a consequence, the application of $\mathbf{P}_1$ to schemas of M_{ER} produce schemas of M_{ERNoM2N}.

Let us now consider also strongly recursive rules, that is, according to our definition, rules whose body includes atoms referring to the target schema. These rules may be applied on the basis of constructs generated by previous applications of other constructs. As a consequence, the application of signatures is also defined recursively, as a minimum fixpoint. We can redefine $r_\mathbf{P}$ to have two arguments, the source model and the target one, and its recursive application to a model M_0 is the fixpoint of the recursive expression $M = r_\mathbf{P}(M_0, M)$. Since it turns out that $r_\mathbf{P}()$ is monotonic, then, by Tarski's theorem [17], we have that the minimum fixpoint exists and can be obtained by computing $M_1 = r_\mathbf{P}(M_0, \perp)$ (where $\perp$ is the empty model), $M_{i+1} = r_\mathbf{P}(M_0, M_i)$ and stopping when $M_{i+1} = M_i$.

5   Inferring Models from Rules

In this section we show the main results of this paper, namely the fact that we can characterize the models obtained by applying Datalog rules by means of the syntactical notion of application of the signature of a rule to a model.

We need a few preliminary concepts. A **pseudoschema** is a set of ground atoms (called **ground constructs** hereinafter) each of which has the form $N(\text{OID}: o,\ p_1: v_1,\ \ldots,\ p_k: v_k,\ r_1: o_1,\ \ldots,\ r_z: o_z)$, where N is the name of a construct that has exactly the properties $p_1, \ldots, p_k$ and the references $r_1, \ldots, r_z$, each v_i is a boolean constant and each o_i is an identifier. A **schema** is a pseudoschema that satisfies the referential constraints defined over constructs. That is, if a schema includes a ground construct $N(\ldots)$ with a reference $r_i: o_i$ and r_i is subject to a referential constraint to a construct N', then the schema has to include a ground atom of the form $N'(\text{OID}: o_i, \ldots)$.

Given a pseudoschema S_0, a schema S that contains all ground constructs in S_0 is called a **closure** of S_0. It can be shown that a closure of a pseudoschema can be obtained by applying procedure similar to the chase for inclusion dependencies [9], which has a finite result in our case as we have acyclic referential constraints, and then replacing variables with "new," distinct values.

A schema **belongs to a model** if its predicate symbols (that is, its constructs) belong to the model and, for each ground atom $N(\ldots)$, the boolean values for its properties satisfy the proposition associated with N in the model.

Given a ground construct $c = N(\text{OID}: o,\ p_1: v_1,\ \ldots,\ p_k: v_k,\ r_1: o_1,\ \ldots,\ r_z: o_z)$, we define the **signature** of c, denoted with $\text{SIG}(c)$, as $N(f)$ where $f = l_1 \wedge \ldots \wedge l_k$, and each l_j is a literal with the symbol p_j, positive if $v_j = true$ and negated if $v_j = false$.

As an example, $\text{R}(\text{OID}: o,\ \text{O}_1: false,\ \text{F}_1: false,\ \text{I}: false,\ \text{O}_2: true,\ \text{F}_2: false,\ \ldots)$ is a ground construct (with references omitted as not relevant) describing a many-to-many relationship, optional on one side and not optional on the other and without external identification. Its signature is $\text{R}(\neg \text{O}_1 \wedge \neg \text{F}_1 \wedge \neg \text{I} \wedge \text{O}_2 \wedge \neg \text{F}_2)$.

The notion of signature can be extended to schemas: given a schema S, we define $\text{SIG}(S) = \bigsqcup_{c \in S} \{\text{SIG}(c)\}$. It is interesting to note (even if we will not use this property) that $\text{SIG}(S) = \sqcap \{M | S \text{ belongs to } M\}$ (that is, $\text{SIG}(S)$ is the greatest lower bound of the models to which S belongs). Therefore, $\text{SIG}(S) \sqsubseteq M$ if and only if S belongs to M.

We are now ready to show our results. Because of space limitations, we give only sketches of the proofs.

Lemma 1. Let M be a model and R a Datalog rule. For each ground construct c in the pseudoschema $R(S)$ produced by the application of R to a schema S of M, it is the case that $\{\text{SIG}(c)\} \sqsubseteq r_R(M)$.

Proof sketch. Let $c = N(\text{OID} : o, p_1 : v_1, \ldots, p_k : v_k, r_1 : o_1, \ldots, r_z : o_z)$ be a ground construct generated by R.

The proof proceeds by first showing that r_R is applicable to M: the idea is that if the rule generates a new ground construct, then, for each of its body literals there is some ground construct in the schema that unifies with it, and therefore the schema element satisfies both the condition in the body and that in the model, and so their conjunction is satisfiable.

Then, since r_R is applicable to M, we have that $r_R = N(f')$, with $f' = f \wedge (\bigwedge_{i=1}^{h}(\bigvee_{t=1}^{q_i} \mu_{\text{MAP}}(g_{i,t})))$. Then, the proof shows that the assignment $\phi = (p_1 = v_1, \ldots, p_k = v_k)$ (that is, the one with the constants in c) satisfies f', by showing that it satisfies both f (and this is easy, as the constants in the head of the rule appear also in c) and $\bigwedge_{i=1}^{h}(\bigvee_{t=1}^{q_i} \mu_{\text{MAP}}(g_{i,t}))$; this latter part follows the structure of the propositional formula and the definition of $\mu_{\text{MAP}}()$. $\square$

Lemma 2. Let M be a model and R a Datalog rule. If s is a construct signature such that $\{s\} \sqsubseteq r_R(M)$ then there is a schema S of M such that the application of R to S produces a pseudoschema $R(S)$ that contains exactly one construct c such that $\text{SIG}(c) = s$.

Proof sketch. The proof proceeds by considering a construct c with signature $s = \text{SIG}(c)$ and showing that there is a set of ground constructs corresponding to the atoms in the body out of which c can be produced.

If the rule R has the form $N \leftarrow N_{j_1}(\ldots), \ldots, N_{j_h}(\ldots)$, then consider a pseudoschema S_0 that contains h ground constructs (one for each atom), with repeated oids for repeated variables and distinct ones elsewhere. The boolean properties in these ground constructs are traced back, from the properties in c, by means of MAP (which gives no ambiguity, as there are no repeated boolean values in the head). Then, consider a closure S of S_0, which is a schema. The proof is completed by showing that the application of R to S produces a result that is indeed a schema and includes the construct c and nothing else. $\square$

Let us briefly comment on the latter lemma. Given our example model M_{ER} and rule $R_{1,6}$, we have that (as we saw) $r_{R_{1,6}}(M_{\text{ER}}) = \{\text{R}(\neg O_1 \wedge F_1 \wedge I \wedge \neg F_2)\}$. The lemma says that all construct signatures $s \sqsubseteq r_{R_{1,6}}(M_{\text{ER}})$ can be obtained as a result of $R_{1,6}$ applied to some schema of M_{ER}. For example, signature $s = \{\text{R}(\neg O_1 \wedge F_1 \wedge I \wedge \neg O_2 \wedge \neg F_2)\}$, which satisfies $s \sqsubseteq r_{R_{1,6}}(M_{\text{ER}})$ can be obtained by applying rule $R_{1,6}$ to a schema which includes (together with other constructs) at least a many-to-many relationship, and all of them with $O_1 = \textit{false}$; this follows from $\text{MAP}_{1,6} = \langle O_2 : \text{R}(O_1) \rangle$.

Lemmas 1 and 2 can be synthesized as the following theorem, which describes the behavior of individual Datalog rules with respect to models and signatures of schemas. It says that using the signature of a rule we can characterize the signatures of the constructs that can be generated out of a given model.

Theorem 1. Let M be a model and R a Datalog rule. Then $\{s\} \sqsubseteq r_R(M)$ if and only if there is a schema S of M such that $R(S)$ contains exactly one construct c such that $\text{SIG}(c) = s$.

Let us now extend the results to Datalog programs.

Lemma 3. Let M be a model and $\mathbf{P}$ a Datalog program. The application of $\mathbf{P}$ to a schema S of M produces a schema $\mathbf{P}(S)$ that belongs to $r_{\mathbf{P}}(M)$.

Proof sketch. The result of the application of $\mathbf{P}$ to a schema S produces a pseudoschema that is indeed a schema because the program is coherent with respect to referential integrity constraints.

Then, the fact this schema belongs to $r_{\mathbf{P}}(M)$ is a consequence of Lemma 1 and of the definition of $r_{\mathbf{P}}(M)$ as $\bigsqcup_{i=1}^{n} r_{R_i}(M)$: if a ground construct c is generated by a rule $R_i \in \mathbf{P}$, then, by Lemma 1, $\{\text{SIG}(c)\} \sqsubseteq r_{R_i}(M)$, and so, since $r_{\mathbf{P}}(M) = \bigsqcup_{i=1}^{n} r_{R_i}(M)$, it also satisfies $r_{\mathbf{P}}(M)$.   □

Lemma 4. Let M be a model and $\mathbf{P}$ a Datalog program. If S' is a schema that belongs to $r_{\mathbf{P}}(M)$, then there is a schema S of M such that $\text{SIG}(S') \sqsubseteq \text{SIG}(\mathbf{P}(S))$.

Proof sketch. Let us consider first programs with no strongly recursive rules. The proof is essentially based on Lemma 2: for every ground construct c in S', we have (partly by hypothesis and partly by definition), that $\{\text{SIG}(c)\} \sqsubseteq \text{SIG}(S') \sqsubseteq r_{\mathbf{P}}(M)$; so, there is a construct signature $N(f)$ in $r_{\mathbf{P}}(M)$ such that c has the predicate symbol N and satisfies f. Now, by definition of $r_{\mathbf{P}}(M)$, we have that f is obtained as the disjunction of the formulas associated with the various rules that have N in the head, and therefore c (since it satisfies f) has to satisfy one of them. If R is such a rule, it turns out that $\{\text{SIG}(c)\} \sqsubseteq r_R(M) \sqsubseteq r_{\mathbf{P}}(M)$ and so, by Lemma 2, we have that there is a schema S_c of M such that $R(S_c)$ contains only constructs c' such that $\text{SIG}(c) = \text{SIG}(c')$. Then, let us consider the "union" of such schemas for the various constructs $S = \bigcup_{c \in S'}(S_c)$: it can be shown that S is a schema for model M and $\text{SIG}(S') \sqsubseteq \text{SIG}(\mathbf{P}(S))$.

The proof for strongly recursive rules proceeds by induction on the number steps needed to reach the fixpoint, with the individual step based on the arguments above.   □

Theorem 2. Let M be a model and $\mathbf{P}$ a Datalog program. Then a schema S' belongs to $r_{\mathbf{P}}(M)$ if and only if there is a schema S of M such that $\text{SIG}(S') \sqsubseteq \text{SIG}(\mathbf{P}(S))$ modulo equivalence of propositions.

Theorem 2 synthesizes Lemmas 3 and 4 in a direct way. However, there is another point of view, which is more interesting, as follows.

Theorem 3. Let M be a model and $\mathbf{P}$ a Datalog program. Then,

1. for every schema S of M, it is the case that $\text{SIG}(\mathbf{P}(S)) \sqsubseteq r_{\mathbf{P}}(M)$
2. there is a schema S of M such that $\text{SIG}(\mathbf{P}(S)) = r_{\mathbf{P}}(M)$

Proof sketch

Claim 1 is essentially Lemma 3.

Claim 2 follows from the application of Lemma 4 to the extreme case of a schema S' whose signature $\text{SIG}(S')$ is exactly $r_{\mathbf{P}}(M)$ (this is possible because the only constraints we have on our schemas are referential ones): by Lemma 4 we have that there is a schema S such that $\text{SIG}(S') \sqsubseteq \text{SIG}(\mathbf{P}(S))$, that is, $r_{\mathbf{P}}(M) \sqsubseteq \text{SIG}(\mathbf{P}(S))$ and by Lemma 3 we have that $\text{SIG}(\mathbf{P}(S)) \sqsubseteq r_{\mathbf{P}}(M)$. □

Theorem 3 is our main result. It states that the derivation of model signatures by means of the application of the signatures of Datalog programs is sound and complete with respect to the models generated by the program: a Datalog program can generate schemas with all and only the signatures described by the application of the signature of rules. In other words, signatures completely characterize the models that can be generated by means of a Datalog program.

6   Applications of the Results

The technical development of the previous sections can be used in various ways to support the activities of an actual tool for schema translation, such as the MIDST tool [3,4] we are developing.

A simple use of the result is the possibility offered to check which is the model obtained as the result of a Datalog program: the results in Section 5 allow the "rule designer" to know the target model without running (or inspecting) Datalog rules, but simply generating signatures and running their applications.

A related use, still in rule specification, is the possibility to check whether a Datalog program takes into consideration all the constructs of a given source model. Let us say that the **domain** of a rule with respect to a model is the set of constructs of M that are considered by the rule; formally, given R and M, if R is applicable to M, then the domain $\text{DOM}(r_R, M)$ of R with respect to M is composed of the constructs that unify with the atoms of the body B of the signature of R; if R is not applicable to M, then $\text{DOM}(r_R, M)$ is the empty model. We can extend this notion to Datalog programs. Given a program $\mathbf{P}$ and a model M, the **domain** of $\mathbf{P}$ with respect to M, $\text{DOM}(r_{\mathbf{P}}, M)$ is $\bigsqcup_{R \in \mathbf{P}'} \text{DOM}(r_R, M)$ where $\mathbf{P}'$ denotes the program that includes only the rules in $\mathbf{P}$ that are applicable to M. Now, constructs (or variant of them) **ignored** by a program $\mathbf{P}$ when applied to a model M are those in the difference between M and $\text{DOM}(r_{\mathbf{P}}, M)$, where the difference between two models is defined in a natural way, given our framework: $M_2 - M_1 = \{N(f_2 \wedge \neg f_1) \mid N(f_1) \in M_1 \text{ and } N(f_2) \in M_2\}$.

A more interesting use of signatures is possible in order to handle the combination of Datalog programs to form complex translations. In a "manual" approach, the designer has a library of translations at his/her disposal and could use signatures in order to verify what is the result for a complex translation and which are the constructs that are ignored in the various steps. In practice, there could be different alternatives, and the designer could choose among them.

A more ambitious goal would involve the automatic selection of rules for the generation of complex translations out of a library. A general approach for this, followed also by other authors in similar contexts [7,16], would be based on the generation of a search tree and on the adoption of heuristics (for example based on A*-type algorithms) which need not even terminate in general (as the need could arise for multiple application of rules, and no bounds) or produces a translation plan driven for a specific criterion (which may differ from the optimal one). Our approach allows the automatic generation of concise description of translation steps, and then could be the basis for the application of such algorithm or, under suitable hypotheses, of more effective and efficient ones. In fact, it turns out to be the case that most rules are essentially **reductions**, so that, given a source model M, we have $\mathbf{P}(M) \sqsubseteq M$. This is due to the fact that in most cases we have variants of constructs within a "family" of models, and we need to simplify them. Under suitable hypotheses, for the discussion of which we do not have space here, we have that, if there is a complex transformation from a model to another, then there exist a transformation which is composed of a sequence of reductions followed by just one transformation that is not a reduction and another sequence of reductions.

On the basis of this property, we can build simple algorithms that automatically find the sequence of programs (a number of reductions followed by a nonreduction transformation and another sequence of reductions) that allows to go from a source to a target model if and only if the set of available program admits such a transformation. Our hypotheses prune the search space significantly, and so an exhaustive search become feasible: they are guaranteed to terminate and to find a suitable sequence if and only of it exists.

7   Related Work

To the best of our knowledge, there is no approach in the literature that tackles the problem we are considering here. There are pieces of work that consider the translation of schemas in heterogeneous frameworks [3,5,7,16], but none has techniques for inferring high level descriptions of translations from their specification. They all propose some way of generating a complex translation plan but they either handle very simple descriptions of models or have to rely on a hard coding of knowledge of behaviour of transformations in terms of pattern of constructs removed and introduced. [7,16] both use some form of signature to implement an A* algorithm that produces the shortest transformation plan (if it exists) between the source and the target model in terms of number of transformations.

Translations of schemas by means of Datalog variants have been proposed by various authors [1,8,10], but no explicit reference to models and to the possibility of reasoning on models has been proposed. The latter work includes some reasoning on constraints, but without reference to the features of models.

Various pieces of work exist on the correctness of transformations of schemas, with reference to the well known notion of information capacity dominance and equivalence [2,11,15]. Here we are not studying the correctness of the individual translation steps, but the correctness of complex translations, assumed that the elementary steps are correct, following an "axiomatic" approach [5].

8   Conclusions

We have given the definition of a formal system to infer the model $r_{\mathbf{P}}(M)$ obtained out of a model M by applying the signature $r_{\mathbf{P}}$ of a Datalog program **P**, and have shown that the derivation is sound and complete: the application of **P** to a schemas S of M produces only schemas that belongs to $r_{\mathbf{P}}(M)$ and potentially all of them. The techniques developed here can be the basis for a formal support to a tool for the automatic generation of translations, because signatures can be obtained directly out of programs.

References

1. Abiteboul, S., Cluet, S., Milo, T.: Correspondence and translation for heterogeneous data. Theor. Comput. Sci. 275(1–2), 179–213 (2002)
2. Abiteboul, S., Hull, R.: Restructuring hierarchical database objects. Theor. Comput. Sci. 62(1–2), 3–38 (1988)
3. Atzeni, P., Cappellari, P., Bernstein, P.A.: Model-independent schema and data translation. In: EDBT, pp. 368–385 (2006)
4. Atzeni, P., Cappellari, P., Gianforme, G.: MIDST: Model independent schema and data translation. In: Chan, C.Y., Ooi, B.C., Zhou, A. (eds.) SIGMOD Conference, pp. 1134–1136. ACM, New York (2007)
5. Atzeni, P., Torlone, R.: Management of multiple models in an extensible database design tool. EDBT 0, 79–95 (1996)
6. Bernstein, P.A.: Applying model management to classical meta data problems. In: CIDR, pp. 209–220 (2003)
7. Bernstein, P.A., Melnik, S., Mork, P.: Interactive schema translation with instance-level mappings. In: VLDB, pp. 1283–1286 (2005)
8. Bowers, S., Delcambre, L.M.L.: The Uni-Level Description: A uniform framework for representing information in multiple data models. In: Song, I.-Y., Liddle, S.W., Ling, T.-W., Scheuermann, P. (eds.) ER 2003. LNCS, vol. 2813, pp. 45–58. Springer, Heidelberg (2003)
9. Cosmadakis, S., Kanellakis, P.: Functional and inclusion dependencies - a graph theoretical approach. In: Kanellakis, P., Preparata, F. (eds.) Advances in Computing Research, vol. 3, pp. 163–184. JAI Press, Greenwich (1986)
10. Davidson, S.B., Kosky, A.: Wol: A language for database transformations and constraints. In: ICDE, pp. 55–65 (1997)
11. Hull, R.: Relative information capacity of simple relational schemata. SIAM J. Comput. 15(3), 856–886 (1986)
12. Hull, R., Yoshikawa, M.: ILOG: Declarative creation and manipulation of object identifiers. In: VLDB 1990. Sixteenth International Conference on Very Large Data Bases, Brisbane, pp. 455–468 (1990)

13. Markowitz, V.M., Shoshani, A.: On the correctness of representing extended entity-relationship structures in the relational model. In: SIGMOD, pp. 430–439 (1989)
14. McGee, W.C.: A contribution to the study of data equivalence. In: IFIP Working Conference Data Base Management, pp. 123–148 (1974)
15. Miller, R.J., Ioannidis, Y.E., Ramakrishnan, R.: The use of information capacity in schema integration and translation. In: VLDB, pp. 120–133 (1993)
16. Papotti, P., Torlone, R.: Heterogeneous data translation through XML conversion. J. Web Eng. 4(3), 189–204 (2005)
17. Tarski, A.: A lattice-theorethic Fixpoint Theorem and its applications. Pacific Journal of Mathematics 5, 285–309 (1955)
18. Ullman, J.D., Widom, J.: A First Course in Database Systems. Prentice-Hall, Englewood Cliffs (1997)

Appendix A: Rules

This appendix lists the complete set of rules for the running example used in the paper.

Rule $R_{1,1}$: *copy entities*

ENTITY(OID: #entity_0(eOid), sOID: tgt, Name: n)
←ENTITY(OID: eOid, sOID: src, Name: n)
Signature:
   $H_{1,1} = \langle \text{E}(true) \rangle$
   $B_{1,1} = \langle \text{E}(true) \rangle$
   MAP$_{1,1} = \langle \rangle$

Rule $R_{1,2}$: *copy attributes of entities*

ATTRIBUTEOFENT(OID: #attribute_0(aOid), sOID: tgt, Name: n, IsKey: isK,
      IsNullable: isN, EntityOID: #entity_0(eOid))
←ATTRIBUTEOFENT(OID: aOid, sOID: src, Name: n, IsKey: isK,
      IsNullable: isN, EntityOID: eOid)
Signature:
   $H_{1,2} = \langle \text{A}(true) \rangle$
   $B_{1,2} = \langle \text{A}(true) \rangle$
   MAP$_{1,2} = \langle \text{K} : \text{A}(\text{K}), \text{N} : \text{A}(\text{N}) \rangle$

Rule $R_{1,3}$: *copy one-to-one and one-to-many relationships*

RELATIONSHIP(OID: #relationship_0(rOid), sOID: tgt, Name: n,
      Entity1: #entity_0(eOid1), isOpt1: isO1, isFunct1: true, isIdent: isId,
      Entity2: #entity_0(eOid2), isOpt2: isO2, isFunct2: isF2)
←RELATIONSHIP(OID: rOid, sOID: src, Name: n, Entity1: eOid1, isOpt1: isO1,
      isFunct1: true, isIdent: isId Entity2: eOid2, isOpt2: isO2, isFunct2: isF2)
Signature:
   $H_{1,3} = \langle \text{R}(\text{F}_1) \rangle$
   $B_{1,3} = \langle \text{R}(\text{F}_1) \rangle$
   MAP$_{1,3} = \langle \text{O}_1 : \text{R}(\text{O}_1), \text{I} : \text{R}(\text{I}), \text{O}_2 : \text{R}(\text{O}_2), \text{F}_2 : \text{R}(\text{F}_2) \rangle$

Rule $R_{1,4}$: *copy attributes of one-to-one and one-to-many relationships*

ATTRIBUTEOFREL(OID:#attributeOfRel_0(arOid), sOID: tgt, Name: n,
    IsNullable: isN, RelationshipOID: #relationship_0(rOid))
←ATTRIBUTEOFREL(OID: arOid, sOID: src, Name: n,
    IsNullable: isN, RelationshipOID: rOid),
    RELATIONSHIP(OID: rOid, sOID: src, isFunct1: **true**)

Signature:
$H_{1,4} = \langle \text{AR}(true) \rangle$
$B_{1,4} = \langle \text{AR}(true), \text{R}(\text{F}_1) \rangle$
$\text{MAP}_{1,4} = \langle \text{N} : \text{AR}(\text{N}) \rangle$

Rule $R_{1,5}$: *generate an entity for each many-to-many relationship*

ENTITY(OID: #entity_1(rOid), sOID: tgt, Name: n)
←RELATIONSHIP(OID: rOid, sOID: src, Name: n, isFunct1: **false**, isFunct2: **false**)

Signature:
$H_{1,5} = \langle \text{E}(true) \rangle$
$B_{1,5} = \langle \text{R}(\neg \text{F}_1 \wedge \neg \text{F}_2) \rangle$
$\text{MAP}_{1,5} = \langle \rangle$

Rule $R_{1,6}$: *for each entity generated by $R_{1,5}$, generate a relationship between it and the copy of the first entity involved in the many-to-many relationship*

RELATIONSHIP(OID: #relationship_1(eOid,rOid), sOID: tgt, Name: eN+rN,
    Entity1: #entity_1(rOid), isOpt1: **false**, isFunct1: **true**, isIdent: **true**,
    Entity2: #entity_0(eOid), isOpt2: isOpt, isFunct2: **false**)
←RELATIONSHIP(OID: rOid, sOID: src, Name: rN, Entity1: eOid, isOpt1: isOpt,
    isFunct1: **false**, isFunct2: **false**),
    ENTITY(OID: eOid, sOID: src, Name: eN)

Signature:
$H_{1,6} = \langle \text{R}(\neg \text{O}_1 \wedge \text{F}_1 \wedge \text{I} \wedge \neg \text{F}_2) \rangle$
$B_{1,6} = \langle \text{R}(\neg \text{F}_1 \wedge \neg \text{F}_2), \text{E}(true) \rangle$
$\text{MAP}_{1,6} = \langle \text{O}_2 : \text{R}(\text{O}_1) \rangle$

Rule $R_{1,7}$: *for each entity generated by $R_{1,5}$, generate a relationship between it and the copy of the second entity involved in the many-to-many relationship*

RELATIONSHIP(OID: #relationship_1(eOid,rOid), sOID: tgt, Name: eN+rN,
    Entity1: #entity_1(rOid), isOpt1: **false**, isFunct1: **true**, isIdent: **true**,
    Entity2: #entity_0(eOid), isOpt2: isOpt, isFunct2: **false**)
←RELATIONSHIP(OID: rOid, sOID: src, Name: rN, Entity2: eOid,
    isOpt2: isOpt, isFunct1: **false**, isFunct2: **false**),
    ENTITY(OID: eOid, sOID: src, Name: eN)

Signature:
$H_{1,7} = \langle \text{R}(\neg \text{O}_1 \wedge \text{F}_1 \wedge \text{I} \wedge \neg \text{F}_2) \rangle$
$B_{1,7} = \langle \text{R}(\neg \text{F}_1 \wedge \neg \text{F}_2), \text{E}(true) \rangle$
$\text{MAP}_{1,7} = \langle \text{O}_2 : \text{R}(\text{O}_2) \rangle$

Rule $R_{1,8}$: *for each attribute of each many-to-many relationship, generate an attribute for the entity generated by* $R_{1,5}$

ATTRIBUTEOFENT(OID: #attributeOfEnt_1(arOid), sOID: tgt, Name: n,
    IsKey: `false`, IsNullable: isN, EntityOID: #entity_1(rOid))
←ATTRIBUTEOFREL(OID: arOid, sOID: src, Name: n, IsNullable: isN,
    Relationship: rOid)
    RELATIONSHIP(OID: rOid, sOID: src, Name: n, isFunct1: `false`, isFunct2: `false`)

Signature:

$H_{1,8} = \langle A(\neg K) \rangle$
$B_{1,8} = \langle AR(true), R(\neg F_1 \wedge \neg F_2) \rangle$
$MAP_{1,8} = \langle N : AR(N) \rangle$

Tightly Integrated Probabilistic Description Logic Programs for Representing Ontology Mappings

Andrea Calì[1], Thomas Lukasiewicz[2,3], Livia Predoiu[4], and Heiner Stuckenschmidt[4]

[1] Computing Laboratory, University of Oxford, UK
andrea.cali@comlab.ox.ac.uk
[2] Dipartimento di Informatica e Sistemistica, Sapienza Università di Roma, Italy
lukasiewicz@dis.uniroma1.it
[3] Institut für Informationssysteme, Technische Universität Wien, Austria
lukasiewicz@kr.tuwien.ac.at
[4] Institut für Informatik, Universität Mannheim, Germany
{livia,heiner}@informatik.uni-mannheim.de

Abstract. Creating mappings between ontologies is a common way of approaching the semantic heterogeneity problem on the Semantic Web. To fit into the landscape of semantic web languages, a suitable, logic-based representation formalism for mappings is needed. We argue that such a formalism has to be able to deal with uncertainty and inconsistencies in automatically created mappings. We analyze the requirements for such a formalism, and we propose a novel approach to probabilistic description logic programs as such a formalism, which tightly combines disjunctive logic programs under the answer set semantics with both description logics and Bayesian probabilities. We define the language, and we show that it can be used to resolve inconsistencies and merge mappings from different matchers based on the level of confidence assigned to different rules. Furthermore, we explore the computational aspects of consistency checking and query processing in tightly integrated probabilistic description logic programs. We show that these problems are decidable and computable, respectively, and that they can be reduced to consistency checking and cautious/brave reasoning, respectively, in tightly integrated disjunctive description logic programs. We also analyze the complexity of consistency checking and query processing in the new probabilistic description logic programs in special cases. In particular, we present a special case of these problems with polynomial data complexity.

1 Introduction

The problem of aligning heterogeneous ontologies via semantic mappings has been identified as one of the major challenges of semantic web technologies. In order to address this problem, a number of languages for representing semantic relations between elements in different ontologies as a basis for reasoning and query answering across multiple ontologies have been proposed [27]. In the presence of real world ontologies, it is unrealistic to assume that mappings between ontologies are created manually by domain experts, since existing ontologies, e.g., in the area of medicine contain thousands of concepts and hundreds of relations. Recently, a number of heuristic methods for

S. Hartmann and G. Kern-Isberner (Eds.): FoIKS 2008, LNCS 4932, pp. 178–198, 2008.

matching elements from different ontologies have been proposed that support the creation of mappings between different languages by suggesting candidate mappings (e.g., [10]). These methods rely on linguistic and structural criteria. Evaluation studies have shown that existing methods often trade off precision and recall. The resulting mapping either contains a fair amount of errors or only covers a small part of the ontologies involved [9,11]. To leverage the weaknesses of the individual methods, it is common practice to combine the results of a number of matching components or even the results of different matching systems to achieve a better coverage of the problem [10].

This means that automatically created mappings often contain uncertain hypotheses and errors that need to be dealt with, briefly summarized as follows:

- mapping hypotheses are often oversimplifying, since most matchers only support very simple semantic relations (mostly equivalence between individual elements);
- there may be conflicts between different hypotheses for semantic relations from different matching components and often even from the same matcher;
- semantic relations are only given with a degree of confidence in their correctness.

If we want to use the resulting mappings, we have to find a way to deal with these uncertainties and errors in a suitable way. We argue that the most suitable way of dealing with uncertainties in mappings is to provide means to explicitly represent uncertainties in the target language that encodes the mappings. In this paper, we address the problem of designing a mapping representation language that is capable of representing the kinds of uncertainty mentioned above. We propose an approach to such a language, which is based on an integration of ontologies and rules under probabilistic uncertainty.

There is a large body of work on integrating ontologies and rules, which is a promising way of representing mappings between ontologies. One type of integration is to build rules on top of ontologies, that is, rule-based systems that use vocabulary from ontology knowledge bases. Another form of integration is to build ontologies on top of rules, where ontological definitions are supplemented by rules or imported from rules. Both types of integration have been realized in recent hybrid integrations of rules and ontologies, called *description logic programs* (or *dl-programs*), which have the form $KB = (L, P)$, where L is a description logic knowledge base, and P is a finite set of rules involving either queries to L in a loose integration [6,7] or concepts and roles from L as unary resp. binary predicates in a tight integration [18] (see especially [7,24,18] for detailed overviews on the different types of description logic programs).

Other works explore formalisms for *uncertainty reasoning in the Semantic Web* (an important recent forum for approaches to uncertainty in the Semantic Web is the annual *Workshop on Uncertainty Reasoning for the Semantic Web (URSW)*; there also exists a W3C Incubator Group on *Uncertainty Reasoning for the World Wide Web*). There are especially probabilistic extensions of description logics [15,19], web ontology languages [2,3], and description logic programs [20] (to encode ambiguous information, such as "John is a student with probability 0.7 and a teacher with probability 0.3", which is very different from vague/fuzzy information, such as "John is tall with degree of truth 0.7"). In particular, [20] extends the loosely integrated description logic programs of [6,7] by probabilistic uncertainty as in Poole's independent choice logic (ICL) [26]. The ICL is a powerful representation and reasoning formalism for single- and also multi-agent systems, which combines logic and probability, and which can represent a number of important uncertainty formalisms, in particular, influence diagrams, Bayesian networks, Markov decision processes, and normal form games. It also allows for natural notions of causes and explanations as in Pearl's structural causal models [13].

In this paper, we propose *tightly integrated probabilistic description logic programs under the answer set semantics* as a language for representing and reasoning with uncertain and possibly inconsistent ontology mappings. The approach is a tight integration of disjunctive logic programs under the answer set semantics, the expressive description logics $\mathcal{SHIF}(\mathbf{D})$ and $\mathcal{SHOIN}(\mathbf{D})$ (which stand behind the standard web ontology languages OWL Lite and OWL DL [16], respectively), and Bayesian probabilities. More concretely, the tight integration between ontology and rule languages of [18] is combined with probabilistic uncertainty as in the ICL [26]. The resulting language has the following useful features, which will be explained in more detail later:

- The semantics of the language is based on the tight integration between ontology and rule languages of [18], which assumes no structural separation between the vocabularies of the description logic and the logic program components. This enables us to have description logic concepts and roles in both rule bodies and rule heads. This is necessary if we want to use rules to combine ontologies.
- The rule language is quite expressive. In particular, we can have disjunctions in rule heads and nonmonotonic negations in rule bodies. This gives a rich basis for refining and rewriting automatically created mappings for resolving inconsistencies.
- The integration with probability theory provides us with a sound formal framework for representing and reasoning with confidence values. In particular, we can interpret the confidence values as error probabilities and use standard techniques for combining them. We can also resolve inconsistencies by using trust probabilities.
- Consistency checking and query processing in the new rule language are decidable resp. computable, and they can be reduced to their classical counterparts in tightly integrated disjunctive description logic programs. We also analyze the complexity of consistency checking and query processing in special cases, which turn out to be complete for the complexity classes NEXP^{NP} and co-NEXP^{NP}, respectively.
- There are tractable subsets of the language that are of practical relevance. In particular, we show later that in the case where ontologies are represented in *DL-Lite*, reasoning in the language can be done in polynomial time in the data complexity.

It is important to point out that the probabilistic description logic programs here are very different from the ones in [20] (and their recent tractable variant in [21]). First, they are based on the tight integration between the ontology component L and the rule component P of [18], while the ones in [20,21] realize the loose query-based integration between the ontology component L and the rule component P of [6]. This implies in particular that the vocabularies of L and P here may have common elements (see also Example 4.1), while the vocabularies of L and P in [20,21] are necessarily disjoint. Furthermore, the probabilistic description logic programs here behave semantically very differently from the ones in [20,21] (see Example 4.2). As a consequence, the probabilistic description logic programs here are especially useful for sophisticated probabilistic reasoning tasks involving ontologies (including representing and reasoning with ontology mappings under probabilistic uncertainty and inconsistency), while the ones in [20,21] can especially be used as query interfaces to web databases (including RDF theories). Second, differently from the programs here, the ones in [20,21] do not allow for disjunctions in rule heads. Third, differently from here, the works [20,21] do not explore the use of probabilistic description logic programs for representing and reasoning with ontology mappings under probabilistic uncertainty and inconsistency.

The rest of this paper is structured as follows. In Section 2, we analyze the requirements of an ontology mapping language. Section 3 briefly reviews description logics as a basis for representing ontologies to be connected by mappings. In Sections 4 and 5, we describe tightly integrated description logic programs as a basis for representing mappings between ontologies as logical rules and explain how the rule language supports the refinement and repair of oversimplifying or inconsistent mappings. Sections 6 and 7 present a probabilistic extension thereof and show that it can be used to represent and combine confidence values of different matchers in terms of error probabilities, and to resolve inconsistencies by using trust probabilities. Sections 8 and 9 address the computational aspects of reasoning in the novel language. In particular, Section 9 identifies a tractable subset of the language. Section 10 concludes with a summary and an outlook. Note that the proofs for the results of this paper are given in Appendix A.

2   Representation Requirements

The problem of ontology matching can be defined as follows [10]. Ontologies are theories encoded in a certain language L. In this work, we assume that ontologies are encoded in OWL DL or OWL Lite. For each ontology O in language L, we denote by $Q(O)$ the matchable elements of the ontology O. Given two ontologies O and O', the task of matching is now to determine correspondences between the matchable elements in the two ontologies. Correspondences are 5-tuples (id, e, e', r, n) such that

- id is a unique identifier for referring to the correspondence;
- $e \in Q(O)$ and $e' \in Q(O')$ are matchable elements from the two ontologies;
- $r \in R$ is a semantic relation (in this work, we consider the case where the semantic relation can be interpreted as an implication);
- n is a degree of confidence in the correctness of the correspondence.

In this paper, we develop a formal language for representing and combining correspondences that are produced by different matching components or systems. From the above general description of automatically generated correspondences between ontologies, we can derive a number of requirements for such a formal language for representing the results of multiple matchers as well as the contained uncertainties:

- *Tight integration of mapping and ontology language:* The semantics of the language used to represent the correspondences between different ontologies has to be tightly integrated with the semantics of the used ontology language (in this case OWL). This is important if we want to use the correspondences to reason across different ontologies in a semantically coherent way. In particular, this means that the interpretation of the mapped elements depends on the definitions in the ontologies.

– *Support for mappings refinement:* The language should be expressive enough to allow the user to refine oversimplifying correspondences suggested by the matching system. This is important to be able to provide a precise account of the true semantic relation between elements in the mapped ontologies. In particular, this requires the ability to describe correspondences that include several elements from the two ontologies.

– *Support for repairing inconsistencies:* Inconsistent mappings are a major problem for the combined use of ontologies because they can cause inconsistencies in the mapped ontologies. These inconsistencies can make logical reasoning impossible, since everything can be derived from an inconsistent ontology. The mapping language should be able to represent and reason about inconsistent mappings in an approximate fashion.

– *Representation and combination of confidence:* The confidence values provided by matching systems is an important indicator for the uncertainty that has to be taken into account. The mapping representation language should be able to use these confidence values when reasoning with mappings. In particular, it should be able to represent the confidence in a mapping rule and to combine confidence values on a sound formal basis.

– *Decidability and efficiency of instance reasoning:* An important use of ontology mappings is the exchange of data across different ontologies. In particular, we normally want to be able to ask queries using the vocabulary of one ontology and receive answers that do not only consist of instances of this ontology but also of ontologies connected through ontology mappings. To support this, query answering in the combined formalism consisting of ontology language and mapping language has to be decidable and there should be efficient algorithms for answering queries at least for relevant cases.

Throughout the paper, we use real data from the Ontology Alignment Evaluation Initiative (OAEI)[1] to illustrate the different aspects of mapping representation. In particular, we use examples from the benchmark and the conference data set. The benchmark dataset consists of five OWL ontologies (tests 101 and 301–304) describing scientific publications and related information. The conference dataset consists of about 10 OWL ontologies describing concepts related to conference organization and management. In both cases, we give examples of mappings that have been created by the participants of the 2006 evaluation campaign. In particular, we use mappings created by state-of-the-art ontology matching systems like falcon, hmatch, and coma++.

3   Description Logics

In this section, we recall the description logics $\mathcal{SHIF}(\mathbf{D})$ and $\mathcal{SHOIN}(\mathbf{D})$, which stand behind the web ontology languages OWL Lite and OWL DL [16], respectively. Intuitively, description logics model a domain of interest in terms of concepts and roles, which represent classes of individuals and binary relations between classes of

---

[1] http://oaei.ontologymatching.org/2006/

individuals, respectively. A description logic knowledge base encodes especially subset relationships between concepts, subset relationships between roles, the membership of individuals to concepts, and the membership of pairs of individuals to roles.

Syntax. We first describe the syntax of $\mathcal{SHOIN}(\mathbf{D})$. We assume a set of *elementary datatypes* and a set of *data values*. A *datatype* is either an elementary datatype or a set of data values (*datatype oneOf*). A *datatype theory* $\mathbf{D} = (\varDelta^{\mathbf{D}}, \cdot^{\mathbf{D}})$ consists of a *datatype domain* $\varDelta^{\mathbf{D}}$ and a mapping $\cdot^{\mathbf{D}}$ that assigns to each elementary datatype a subset of $\varDelta^{\mathbf{D}}$ and to each data value an element of $\varDelta^{\mathbf{D}}$. The mapping $\cdot^{\mathbf{D}}$ is extended to all datatypes by $\{v_1, \ldots\}^{\mathbf{D}} = \{v_1^{\mathbf{D}}, \ldots\}$. Let $\mathbf{A}$, $\mathbf{R}_A$, $\mathbf{R}_D$, and $\mathbf{I}$ be pairwise disjoint (denumerable) sets of *atomic concepts*, *abstract roles*, *datatype roles*, and *individuals*, respectively. We denote by $\mathbf{R}_A^-$ the set of *inverses* R^- of all $R \in \mathbf{R}_A$.

A *role* is any element of $\mathbf{R}_A \cup \mathbf{R}_A^- \cup \mathbf{R}_D$. *Concepts* are inductively defined as follows. Every $\phi \in \mathbf{A}$ is a concept, and if $o_1, \ldots, o_n \in \mathbf{I}$, then $\{o_1, \ldots, o_n\}$ is a concept (*oneOf*). If ϕ, ϕ_1, and ϕ_2 are concepts and if $R \in \mathbf{R}_A \cup \mathbf{R}_A^-$, then also $(\phi_1 \sqcap \phi_2)$, $(\phi_1 \sqcup \phi_2)$, and $\neg\phi$ are concepts (*conjunction*, *disjunction*, and *negation*, respectively), as well as $\exists R.\phi$, $\forall R.\phi$, $\geqslant nR$, and $\leqslant nR$ (*exists*, *value*, *atleast*, and *atmost restriction*, respectively) for an integer $n \geqslant 0$. If D is a datatype and $U \in \mathbf{R}_D$, then $\exists U.D$, $\forall U.D$, $\geqslant nU$, and $\leqslant nU$ are concepts (*datatype exists*, *value*, *atleast*, and *atmost restriction*, respectively) for an integer $n \geqslant 0$. We write $\top$ and $\bot$ to abbreviate the concepts $\phi \sqcup \neg\phi$ and $\phi \sqcap \neg\phi$, respectively, and we eliminate parentheses as usual.

An *axiom* has one of the following forms: (1) $\phi \sqsubseteq \psi$ (*concept inclusion axiom*), where ϕ and ψ are concepts; (2) $R \sqsubseteq S$ (*role inclusion axiom*), where either $R, S \in \mathbf{R}_A \cup \mathbf{R}_A^-$ or $R, S \in \mathbf{R}_D$; (3) $\mathrm{Trans}(R)$ (*transitivity axiom*), where $R \in \mathbf{R}_A$; (4) $\phi(a)$ (*concept membership axiom*), where ϕ is a concept and $a \in \mathbf{I}$; (5) $R(a, b)$ (resp., $U(a, v)$) (*role membership axiom*), where $R \in \mathbf{R}_A$ (resp., $U \in \mathbf{R}_D$) and $a, b \in \mathbf{I}$ (resp., $a \in \mathbf{I}$ and v is a data value); and (6) $a = b$ (resp., $a \neq b$) (*equality* (resp., *inequality*) *axiom*), where $a, b \in \mathbf{I}$. A *(description logic) knowledge base* L is a finite set of axioms. For decidability, number restrictions in L are restricted to simple abstract roles [17].

The syntax of $\mathcal{SHIF}(\mathbf{D})$ is as the above syntax of $\mathcal{SHOIN}(\mathbf{D})$, but without the oneOf constructor and with the atleast and atmost constructors limited to 0 and 1.

Example 3.1. A university database may use a knowledge base L to characterize students and exams. For example, suppose that (1) every bachelor student is a student; (2) every master student is a student; (3) every student is either a bachelor student or a master student; (4) professors are not students; (5) only students give exams and only exams are given; (6) *mary* is a student, *john* is a master student, *java* is an exam, and *john* has given it. These relationships are expressed by the following axioms in L:

(1) *bachelor_student* $\sqsubseteq$ *student*; (2) *master_student* $\sqsubseteq$ *student*;
(3) *student* $\sqsubseteq$ *bachelor_student* $\sqcup$ *master_student*; (4) *professor* $\sqsubseteq$ ¬*student*;
(5) $\geqslant 1$ *given* $\sqsubseteq$ *student*; $\geqslant 1$ *given*$^{-1}$ $\sqsubseteq$ *exam*;
(6) *student(mary)*; *master_student(john)*; *exam(java)*; *given(john, java)* .

Semantics. An *interpretation* $\mathcal{I} = (\varDelta^{\mathcal{I}}, \cdot^{\mathcal{I}})$ relative to a datatype theory $\mathbf{D} = (\varDelta^{\mathbf{D}}, \cdot^{\mathbf{D}})$ consists of a nonempty (*abstract*) *domain* $\varDelta^{\mathcal{I}}$ disjoint from $\varDelta^{\mathbf{D}}$, and a mapping $\cdot^{\mathcal{I}}$ that assigns to each atomic concept $\phi \in \mathbf{A}$ a subset of $\Delta^{\mathcal{I}}$, to each individual $o \in \mathbf{I}$ an element of $\Delta^{\mathcal{I}}$, to each abstract role $R \in \mathbf{R}_A$ a subset of $\Delta^{\mathcal{I}} \times \Delta^{\mathcal{I}}$, and to each datatype role $U \in \mathbf{R}_D$ a subset of $\Delta^{\mathcal{I}} \times \Delta^{\mathbf{D}}$. We extend $\cdot^{\mathcal{I}}$ to all concepts and roles, and we define the *satisfaction* of an axiom F in an interpretation $\mathcal{I} = (\Delta^{\mathcal{I}}, \cdot^{\mathcal{I}})$, denoted $\mathcal{I} \models F$, as usual [16]. We say $\mathcal{I}$ *satisfies* the axiom F, or $\mathcal{I}$ is a *model* of F, iff $\mathcal{I} \models F$. We say $\mathcal{I}$ *satisfies* a knowledge base L, or $\mathcal{I}$ is a *model* of L, denoted $\mathcal{I} \models L$, iff $\mathcal{I} \models F$ for all $F \in L$. We say L is *satisfiable* iff L has a model. An axiom F is a *logical consequence* of L, denoted $L \models F$, iff every model of L satisfies F.

4   Tightly Integrated Disjunctive DL-Programs

In this section, we recall the *tightly integrated* approach to *disjunctive description logic programs* (or simply *disjunctive dl-programs*) $KB = (L, P)$ under the answer set semantics from [18], where KB consists of a description logic knowledge base L and a disjunctive logic program P. Their semantics is defined in a modular way as in [6,7], but it allows for a much tighter integration of L and P. Note that we do not assume any structural separation between the vocabularies of L and P. The main idea behind their semantics is to interpret P relative to Herbrand interpretations that are compatible with L, while L is interpreted relative to general interpretations over a first-order domain. Thus, we modularly combine the standard semantics of logic programs and of description logics, which allows for building on the standard techniques and results of both areas. As another advantage, the novel disjunctive dl-programs are decidable, even when their components of logic programs and description logic knowledge bases are both very expressive. We refer especially to [18] for further details on the novel approach to disjunctive dl-programs and for a detailed comparison to related works.

Syntax. We assume a first-order vocabulary Φ with finite nonempty sets of constant and predicate symbols, but no function symbols. We use Φ_c to denote the set of all constant symbols in Φ. We also assume a set of data values $\mathbf{V}$ (relative to a datatype theory $\mathbf{D} = (\Delta^{\mathbf{D}}, \cdot^{\mathbf{D}})$) and pairwise disjoint (denumerable) sets $\mathbf{A}, \mathbf{R}_A, \mathbf{R}_D$, and $\mathbf{I}$ of atomic concepts, abstract roles, datatype roles, and individuals, respectively, as in Section 3. We assume that (i) Φ_c is a subset of $\mathbf{I} \cup \mathbf{V}$, and that (ii) Φ and $\mathbf{A}$ (resp., $\mathbf{R}_A \cup \mathbf{R}_D$) may have unary (resp., binary) predicate symbols in common.

Let $\mathcal{X}$ be a set of variables. A *term* is either a variable from $\mathcal{X}$ or a constant symbol from Φ. An *atom* is of the form $p(t_1, \ldots, t_n)$, where p is a predicate symbol of arity $n \geqslant 0$ from Φ, and $t_1, \ldots, t_n$ are terms. A *literal* l is an atom p or a default-negated atom $not\, p$. A *disjunctive rule* (or simply *rule*) r is an expression of the form

$$\alpha_1 \vee \cdots \vee \alpha_k \leftarrow \beta_1, \ldots, \beta_n, not\, \beta_{n+1}, \ldots, not\, \beta_{n+m}, \tag{1}$$

where $\alpha_1, \ldots, \alpha_k, \beta_1, \ldots, \beta_{n+m}$ are atoms and $k, m, n \geqslant 0$. We call $\alpha_1 \vee \cdots \vee \alpha_k$ the *head* of r, while the conjunction $\beta_1, \ldots, \beta_n, not\, \beta_{n+1}, \ldots, not\, \beta_{n+m}$ is its *body*. We define $H(r) = \{\alpha_1, \ldots, \alpha_k\}$ and $B(r) = B^+(r) \cup B^-(r)$, where $B^+(r) = \{\beta_1, \ldots, \beta_n\}$ and $B^-(r) = \{\beta_{n+1}, \ldots, \beta_{n+m}\}$. A *disjunctive program* P is a finite set of disjunctive rules of the form (1). We say P is *positive* iff $m = 0$ for all disjunctive rules (1) in P. We say P is a *normal program* iff $k \leqslant 1$ for all disjunctive rules (1) in P.

A *tightly integrated disjunctive description logic program* (or simply *disjunctive dl-program*) $KB = (L, P)$ consists of a description logic knowledge base L and a disjunctive program P. We say KB is *positive* iff P is positive. We say KB is a *normal dl-program* iff P is a normal program.

Example 4.1. Consider the disjunctive dl-program $KB = (L, P)$, where L is the description logic knowledge base from Example 3.1, and P is the following set of rules, which express that (1) *bill* is either a master student or a Ph.D. student (which is encoded by a rule that has the form of a disjunction of ground atoms), (2) the relation of propaedeuticity enjoys the transitive property, (3) if a student has given an exam, then he/she has given all exams that are propaedeutic to it, and (4) *unix* is propaedeutic for *java*, and *java* is propaedeutic for *programming_languages*:

    (1) $master\_student(bill) \lor phd\_student(bill)$;
    (2) $propaedeutic(X, Z) \leftarrow propaedeutic(X, Y), propaedeutic(Y, Z)$;
    (3) $given(X, Z) \leftarrow given(X, Y), propaedeutic(Z, Y)$;
    (4) $propaedeutic(unix, java); propaedeutic(java, programming\_languages)$.

The above disjunctive dl-program also shows the advantages and flexibility of the tight integration between rules and ontologies (compared to the loose integration in [6,7]): Observe that the predicate symbol *given* in P is also a role in L, and it freely occurs in both rule bodies and rule heads in P (which is both not possible in [6,7]). Moreover, we can easily use L to express additional constraints on the predicate symbols in P. For example, we may use the two axioms $\geqslant 1 \, propaedeutic \sqsubseteq exam$ and $\geqslant 1 \, propaedeutic^{-1} \sqsubseteq exam$ in L to express that *propaedeutic* in P relates only exams.

Semantics. We now define the answer set semantics of disjunctive dl-programs as a generalization of the answer set semantics of ordinary disjunctive logic programs. In the sequel, let $KB = (L, P)$ be a disjunctive dl-program.

A *ground instance* of a rule $r \in P$ is obtained from r by replacing every variable that occurs in r by a constant symbol from Φ_c. We denote by $ground(P)$ the set of all ground instances of rules in P. The *Herbrand base* relative to Φ, denoted HB_Φ, is the set of all ground atoms constructed with constant and predicate symbols from Φ. We use DL_Φ to denote the set of all ground atoms in HB_Φ that are constructed from atomic concepts in $\mathbf{A}$, abstract roles in $\mathbf{R}_A$, and datatype roles in $\mathbf{R}_D$.

An *interpretation* I is any subset of HB_Φ. Informally, every such I represents the Herbrand interpretation in which all $a \in I$ (resp., $a \in HB_\Phi - I$) are true (resp., false). We say an interpretation I is a *model* of a description logic knowledge base L, denoted $I \models L$, iff $L \cup I \cup \{\neg a \mid a \in HB_\Phi - I\}$ is satisfiable. We say I is a *model* of a ground atom $a \in HB_\Phi$, or I *satisfies* a, denoted $I \models a$, iff $a \in I$. We say I is a *model* of a ground rule r, denoted $I \models r$, iff $I \models \alpha$ for some $\alpha \in H(r)$ whenever $I \models B(r)$, that is, $I \models \beta$ for all $\beta \in B^+(r)$ and $I \not\models \beta$ for all $\beta \in B^-(r)$. We say I is a *model* of a set of rules P iff $I \models r$ for every $r \in ground(P)$. We say I is a *model* of a disjunctive dl-program $KB = (L, P)$, denoted $I \models KB$, iff I is a model of both L and P.

We now define the answer set semantics of disjunctive dl-programs by generalizing the ordinary answer set semantics of disjunctive logic programs. We generalize the definition via the FLP-reduct [12], which is equivalent to the standard definition via the

Gelfond-Lifschitz reduct [14]. Given a dl-program $KB = (L, P)$, the *FLP-reduct* of KB relative to $I \subseteq HB_\Phi$, denoted KB^I, is the disjunctive dl-program (L, P^I), where P^I is the set of all $r \in ground(P)$ with $I \models B(r)$. Note that the *Gelfond-Lifschitz reduct* of KB relative to $I \subseteq HB_\Phi$ is the positive disjunctive dl-program $(L, \hat{P}^I)$, where $\hat{P}^I$ is obtained from $ground(P)$ by (i) deleting every rule r such that $I \models \beta$ for some $\beta \in B^-(r)$ and (ii) deleting the negative body from each remaining rule. An interpretation $I \subseteq HB_\Phi$ is an *answer set* of KB iff I is a minimal model of KB^I. A dl-program KB is *consistent* (resp., *inconsistent*) iff it has an (resp., no) answer set.

We finally define the notion of *cautious* (resp., *brave*) *reasoning* from disjunctive dl-programs under the answer set semantics as follows. A ground atom $a \in HB_\Phi$ is a *cautious* (resp., *brave*) *consequence* of a disjunctive dl-program KB under the answer set semantics iff every (resp., some) answer set of KB satisfies a.

Semantic Properties. We now summarize some important semantic properties of disjunctive dl-programs under the above answer set semantics. In the ordinary case, every answer set of a disjunctive program P is also a minimal model of P, and the converse holds when P is positive. This result holds also for disjunctive dl-programs.

As another important semantic property, the answer set semantics of disjunctive dl-programs faithfully extends its ordinary counterpart. That is, the answer set semantics of a disjunctive dl-program with empty description logic knowledge base coincides with the ordinary answer set semantics of its disjunctive program.

Furthermore, the answer set semantics of disjunctive dl-programs also faithfully extends (from the perspective of answer set programming) the first-order semantics of description logic knowledge bases. That is, a ground atom $\alpha \in HB_\Phi$ is true in all answer sets of a positive disjunctive dl-program $KB = (L, P)$ iff α is true in all first-order models of $L \cup ground(P)$. In particular, a ground atom $\alpha \in HB_\Phi$ is true in all answer sets of $KB = (L, \emptyset)$ iff α is true in all first-order models of L. Note that this result holds also when α is a ground formula constructed from HB_Φ using the operators $\wedge$ and $\vee$.

The tight integration of ontologies and rules semantically behaves very differently from the loose integration. This makes the former more (and the latter less) suitable for representing ontology mappings. The following example illustrates this difference.

Example 4.2. The normal dl-program $KB = (L, P)$, where

$$L = \{person(a), person \sqsubseteq male \sqcup female\} \text{ and}$$
$$P = \{client(X) \leftarrow male(X), client(X) \leftarrow female(X)\}$$

implies $client(a)$, while the normal dl-program $KB' = (L', P')$ as in [6,7]

$$L' = \{person(a), person \sqsubseteq male \sqcup female\} \text{ and}$$
$$P' = \{client(X) \leftarrow DL[male](X), client(X) \leftarrow DL[female](X)\}$$

does *not* imply $client(a)$, since the two queries are evaluated independently from each other, and neither $male(a)$ nor $female(a)$ follows from L'. To obtain the conclusion $client(a)$ in [6,7], one has to directly use the rule $client(X) \leftarrow DL[male \sqcup female](X)$.

5 Representing Ontology Mappings

In this section, we show how tightly integrated disjunctive dl-programs $KB = (L, P)$ can be used for representing (possibly inconsistent) mappings (without confidence values) between two ontologies. Intuitively, L encodes the union of the two ontologies, while P encodes the mappings between the ontologies, where disjunctions in rule heads and nonmonotonic negations in rule bodies in P can be used to resolve inconsistencies.

Tightly integrated disjunctive dl-programs $KB = (L, P)$ naturally represent two heterogeneous ontologies O_1 and O_2, and mappings between O_1 and O_2 as follows. The description logic knowledge base L is the union of two independent description logic knowledge bases L_1 and L_2, which encode the ontologies O_1 and O_2, respectively. Here, we assume that L_1 and L_2 have signatures $\mathbf{A}_1$, $\mathbf{R}_{A,1}$, $\mathbf{R}_{D,1}$, $\mathbf{I}_1$ and $\mathbf{A}_2$, $\mathbf{R}_{A,2}$, $\mathbf{R}_{D,2}$, $\mathbf{I}_2$, respectively, such that $\mathbf{A}_1 \cap \mathbf{A}_2 = \emptyset$, $\mathbf{R}_{A,1} \cap \mathbf{R}_{A,2} = \emptyset$, $\mathbf{R}_{D,1} \cap \mathbf{R}_{D,2} = \emptyset$, and $\mathbf{I}_1 \cap \mathbf{I}_2 = \emptyset$. Note that this can easily be achieved for any pair of ontologies by a suitable renaming. A mapping between elements e_1 and e_2 from L_1 and L_2, respectively, is then represented by a simple rule $e_2(\mathbf{x}) \leftarrow e_1(\mathbf{x})$ in P, where $e_1 \in \mathbf{A}_1 \cup \mathbf{R}_{A,1} \cup \mathbf{R}_{D,1}$, $e_2 \in \mathbf{A}_2 \cup \mathbf{R}_{A,2} \cup \mathbf{R}_{D,2}$, and $\mathbf{x}$ is a suitable variable vector. Informally, such a rule encodes that every instance of (the concept or role) e_1 in O_1 is also an instance of (the concept or role) e_2 in O_2. Note that demanding the signatures of L_1 and L_2 to be disjoint guarantees that the rule base that represents mappings between different ontologies is stratified as long as there are no cyclic mappings.

Example 5.1. Taking an example from the conference data set of the OAEI challenge 2006, we find e.g. the following mappings that have been created by the hmatch system for mapping the CRS Ontology (O_1) on the EKAW Ontology (O_2):

$$EarlyRegisteredParticipant(X) \leftarrow Participant(X);$$
$$LateRegisteredParticipant(X) \leftarrow Participant(X).$$

Informally, these two mapping relationships express that every instance of the concept *Participant* of the ontology O_1 is also an instance of the concepts *EarlyRegisteredParticipant* and *LateRegisteredParticipant*, respectively, of the ontology O_2.

We now encode the two ontologies and the mappings by a tightly integrated disjunctive dl-program $KB = (L, P)$, where L is the union of two description logic knowledge bases L_1 and L_2 encoding the ontologies O_1 resp. O_2, and P encodes the mappings. However, we cannot directly use the two mapping relationships as two rules in P, since this would introduce an inconsistency in KB. More specifically, recall that a model of KB has to satisfy both L and P. Here, the two mapping relationships interpreted as rules in P would require that if there is a participant Alice ($Participant(alice)$) in the ontology O_1, an answer set of KB contains $EarlyRegisteredParticipant(alice)$ and $LateRegisteredParticipant(alice)$ at the same time. Such an answer set, however, is invalidated by the ontology O_2, which requires the concepts $EarlyRegistered$-$Participant$ and $LateRegisteredParticipant$ to be disjoint. Therefore, these mappings are useless, since they do not actively participate in the creation of any model of KB.

In [23], we present a method for detecting such inconsistent mappings. There are different approaches for resolving this inconsistency. The most straightforward one is to drop mappings until no inconsistency is present any more. Peng and Xu [25] have proposed a more suitable method for dealing with inconsistencies in terms of a relaxation of the mappings. In particular, they propose to replace a number of conflicting mappings by a single mapping that includes a disjunction of the conflicting concepts. In the example above, we would replace the two mapping rules by the following one:

$$EarlyRegisteredParticipant(X) \lor LateRegisteredParticipant(X) \leftarrow Participant(X).$$

This new mapping rule can be represented in our framework and resolves the inconsistency. More specifically, for a particular participant Alice ($Participant(alice)$) in the ontology O_1, it imposes the existence of two answer sets

$$\{EarlyRegisteredParticipant(alice), Participant(alice)\};$$
$$\{LateRegisteredParticipant(alice), Participant(alice)\}.$$

None of these answer sets is invalidated by the disjointness constraints imposed by the ontology O_2. However, we can deduce only $Participant(alice)$ cautiously, the other atoms can be deduced bravely. More generally, with such rules, instances that are only available in the ontology O_1 cannot be classified with certainty.

We can solve this issue by refining the rules again and making use of nonmonotonic negation. In particular, we can extend the body of the original mappings with the following additional requirement:

$$EarlyRegisteredParticipant(X) \leftarrow Participant(X) \land RegisterdbeforeDeadline(X);$$
$$LateRegisteredParticipant(X) \leftarrow Participant(X) \land not\ RegisteredbeforeDeadline(X).$$

This refinement of the mapping rules resolves the inconsistency and also provides a more correct mapping because background information has been added. A drawback of this approach is the fact that it requires manual post-processing of mappings because the additional background information is not obvious. In the next section, we present a probabilistic extension of tightly integrated disjunctive dl-programs that allows us to directly use confidence estimations of matching engines to resolve inconsistencies and to combine the results of different matchers.

6   Tightly Integrated Probabilistic DL-Programs

In this section, we present a *tightly integrated* approach to *probabilistic disjunctive description logic programs* (or simply *probabilistic dl-programs*) under the answer set semantics. Differently from [20] (in addition to being a tightly integrated approach), the probabilistic dl-programs here also allow for disjunctions in rule heads. Similarly to the probabilistic dl-programs in [20], they are defined as a combination of dl-programs with Poole's ICL [26], but using the tightly integrated disjunctive dl-programs of [18] (see Section 4), rather than the loosely integrated dl-programs of [6,7]. Poole's ICL is based on ordinary acyclic logic programs P under different "choices", where every choice along with P produces a first-order model, and one then obtains a probability distribution over the set of all first-order models by placing a probability distribution over the different choices. We use the tightly integrated disjunctive dl-programs under the answer set semantics of [18], instead of ordinary acyclic logic programs under their canonical semantics (which coincides with their answer set semantics). We first introduce the syntax of probabilistic dl-programs and then their answer set semantics.

Syntax. We now define the syntax of probabilistic dl-programs and of probabilistic queries to them. We first introduce choice spaces and probabilities on choice spaces.

A *choice space* C is a set of pairwise disjoint and nonempty sets $A \subseteq HB_{\Phi} - DL_{\Phi}$. Any $A \in C$ is an *alternative* of C and any element $a \in A$ an *atomic choice* of C. Intuitively, every alternative $A \in C$ represents a random variable and every atomic choice $a \in A$ one of its possible values. A *total choice* of C is a set $B \subseteq HB_{\Phi}$ such that $|B \cap A| = 1$ for all $A \in C$ (and thus $|B| = |C|$). Intuitively, every total choice B of C represents an assignment of values to all the random variables. A *probability* μ on a choice space C is a probability function on the set of all total choices of C. Intuitively, every probability μ is a probability distribution over the set of all variable assignments. Since C and all its alternatives are finite, μ can be defined by (i) a mapping $\mu \colon \bigcup C \to [0, 1]$ such that $\sum_{a \in A} \mu(a) = 1$ for all $A \in C$, and (ii) $\mu(B) = \Pi_{b \in B} \mu(b)$ for all total choices B of C. Intuitively, (i) defines a probability over the values of each random variable of C, and (ii) assumes independence between the random variables.

A *tightly integrated probabilistic disjunctive description logic program* (or simply *probabilistic dl-program*) $KB = (L, P, C, \mu)$ consists of a disjunctive dl-program (L, P), a choice space C such that no atomic choice in C coincides with the head of any rule in $ground(P)$, and a probability μ on C. Intuitively, since the total choices of C select subsets of P, and μ is a probability distribution on the total choices of C, every probabilistic dl-program is the compact representation of a probability distribution on a finite set of disjunctive dl-programs. Observe here that P is fully general and not necessarily stratified or acyclic. We say KB is *normal* iff P is normal. A *probabilistic query* to KB has the form $\exists (c_1(\boldsymbol{x}) \vee \cdots \vee c_n(\boldsymbol{x}))[r, s]$, where $\boldsymbol{x}, r, s$ is a tuple of variables, $n \geqslant 1$, and each $c_i(\boldsymbol{x})$ is a conjunction of atoms constructed from predicate and constant symbols in Φ and variables in $\boldsymbol{x}$. Note that the above probabilistic queries can also be easily extended to conditional expressions as in [20].

Example 6.1. Consider $KB = (L, P, C, \mu)$, where L and P are as in Examples 3.1 and 4.1, respectively, except that the following two (probabilistic) rules are added to P:

$$given(X, operating\_systems) \leftarrow master\_student(X), given(X, unix), choice_m ;$$
$$given(X, operating\_systems) \leftarrow bachelor\_student(X), given(X, unix), choice_b .$$

Let $C = \{\{choice_m, not\_choice_m\}, \{choice_b, not\_choice_b\}\}$, and let the probability μ on C be given by $\mu \colon choice_m, not\_choice_m, choice_b, not\_choice_b \mapsto 0.9, 0.1, 0.7, 0.3$. Here, the new (probabilistic) rules express that if a master (resp., bachelor) student has given the exam *unix*, then there is a probability of 0.9 (resp., 0.7) that he/she has also given *operating_systems*. Note that probabilistic facts can be encoded by rules with only atomic choices in their body. Our wondering about the entailed tight interval for the probability that *john* has given an exam on *java* can be expressed by the probabilistic query $\exists (given(john, java))[R, S]$. Our wondering about which exams *john* has given with which tight probability interval can be encoded by $\exists (given(john, E))[R, S]$.

Semantics. We now define an answer set semantics of probabilistic dl-programs, and we introduce the notions of consistency, consequence, tight consequence, and correct and tight answers for probabilistic queries to probabilistic dl-programs. Note that the semantics is based on subjective probabilities defined on a set of possible worlds.

Given a probabilistic dl-program $KB = (L, P, C, \mu)$, a *probabilistic interpretation* Pr is a probability function on the set of all $I \subseteq HB_\Phi$. We say Pr is an *answer set* of KB iff (i) every interpretation $I \subseteq HB_\Phi$ with $Pr(I) > 0$ is an answer set of $(L, P \cup \{p \leftarrow \mid p \in B\})$ for some total choice B of C, and (ii) $Pr(\bigwedge_{p \in B} p) = \sum_{I \subseteq HB_\Phi,\, B \subseteq I} Pr(I) = \mu(B)$ for every total choice B of C. Informally, Pr is an answer set of $KB = (L, P, C, \mu)$ iff (i) every interpretation $I \subseteq HB_\Phi$ of positive probability under Pr is an answer set of the dl-program (L, P) under some total choice B of C, and (ii) Pr coincides with μ on the total choices B of C. We say KB is *consistent* iff it has an answer set Pr.

We define the notions of consequence and tight consequence as follows. Given a probabilistic query $\exists(q(\boldsymbol{x}))[r, s]$, the *probability* of $q(\boldsymbol{x})$ in a probabilistic interpretation Pr under a variable assignment σ, denoted $Pr_\sigma(q(\boldsymbol{x}))$ is defined as the sum of all $Pr(I)$ such that $I \subseteq HB_\Phi$ and $I \models_\sigma q(\boldsymbol{x})$. We say $(q(\boldsymbol{x}))[l, u]$ (where $l, u \in [0, 1]$) is a *consequence* of KB, denoted $KB \|\!\sim (q(\boldsymbol{x}))[l, u]$, iff $Pr_\sigma(q(\boldsymbol{x})) \in [l, u]$ for every answer set Pr of KB and every variable assignment σ. We say $(q(\boldsymbol{x}))[l, u]$ (where $l, u \in [0, 1]$) is a *tight consequence* of KB, denoted $KB \|\!\sim_{tight} (q(\boldsymbol{x}))[l, u]$, iff l (resp., u) is the infimum (resp., supremum) of $Pr_\sigma(q(\boldsymbol{x}))$ subject to all answer sets Pr of KB and all σ. A *correct* (resp., *tight*) *answer* to a probabilistic query $\exists(c_1(\boldsymbol{x}) \vee \cdots \vee c_n(\boldsymbol{x}))[r, s]$ is a ground substitution θ (for the variables $\boldsymbol{x}, r, s$) such that $(c_1(\boldsymbol{x}) \vee \cdots \vee c_n(\boldsymbol{x}))[r, s]\,\theta$ is a consequence (resp., tight consequence) of KB.

Example 6.2. Consider again $KB = (L, P, C, \mu)$ of Example 6.1. The tight answer for $\exists(given(john, java))[R, S]$ to KB is given by $\theta = \{R/1, S/1\}$, while some tight answers for $\exists(given(john, E))[R, S]$ to KB are given by $\theta = \{E/java, R/1, S/1\}$, $\theta = \{E/unix, R/1, S/1\}$ and $\theta = \{E/operating\_systems, R/0.9, S/0.9\}$.

7   Representing Ontology Mappings with Confidence Values

We now show how tightly integrated probabilistic dl-programs $KB = (L, P, C, \mu)$ can be used for representing (possibly inconsistent) mappings with confidence values between two ontologies. Intuitively, L encodes the union of the two ontologies, while P, C, and μ encode the mappings between the ontologies, where confidence values can be encoded as error probabilities, and inconsistencies can also be resolved via trust probabilities (in addition to using disjunctions and nonmonotonic negations in P).

The probabilistic extension of tightly integrated disjunctive dl-programs $KB = (L, P)$ to tightly integrated probabilistic dl-programs $KB' = (L, P, C, \mu)$ provides us with a means to explicitly represent and use the confidence values provided by matching systems. In particular, we can interpret the confidence value as an *error probability* and state that the probability that a mapping introduces an error is $1 - n$. Conversely, the probability that a mapping correctly describes the semantic relation between elements of the different ontologies is $1 - (1 - n) = n$. This means that we can use the confidence value n as a probability for the correctness of a mapping. The indirect formulation is chosen, because it allows us to combine the results of different matchers in a meaningful way. In particular, if we assume that the error probabilities of two matchers are independent, we can calculate the joint error probability of two matchers that have found the same mapping rule as $(1 - n_1) \cdot (1 - n_2)$. This means that we can get a new probability for the correctness of the rule found by two matchers which is

$1 - (1 - n_1) \cdot (1 - n_2)$. This way of calculating the joint probability meets the intuition that a mapping is more likely to be correct if it has been discovered by more than one matcher because $1 - (1 - n_1) \cdot (1 - n_2) \geqslant n_1$ and $1 - (1 - n_1) \cdot (1 - n_2) \geqslant n_2$.

In addition, when merging inconsistent results of different matching systems, we weigh each matching system and its result with a (user-defined) *trust probability*, which describes our confidence in its quality. All these trust probabilities sum up to 1. For example, the trust probabilities of the matching systems m_1, m_2, and m_3 may be 0.6, 0.3, and 0.1, respectively. That is, we trust most in m_1, medium in m_2, and less in m_3.

Example 7.1. We illustrate this approach using an example from the benchmark data set of the OAEI 2006 campaign. In particular, we consider the case where the publication ontology in test 101 (O_1) is mapped on the ontology of test 302 (O_2). Below we show some mappings that have been detected by the matching system hmatch that participated in the challenge. The mappings are described as rules in P, which contain a conjunct indicating the matching system that has created it and a number for identifying the mapping. These additional conjuncts are atomic choices of the choice space C and link probabilities (which are specified in the probability μ on the choice space C) to the rules (where the common concept *Proceedings* of both ontologies O_1 and O_2 is renamed to the concepts *Proceedings*$_1$ and *Proceedings*$_2$, respectively):

$$Book(X) \leftarrow Collection(X) \wedge hmatch_1 \, ;$$
$$Proceedings_2(X) \leftarrow Proceedings_1(X) \wedge hmatch_2 \, .$$

We define the choice space according to the interpretation of confidence described above. The resulting choice space is $C = \{\{hmatch_i, not\_hmatch_i\} \mid i \in \{1,2\}\}$. It comes along with the probability μ on C, which assigns the corresponding confidence value n (from the matching system) to each atomic choice $hmatch_i$ and the complement $1 - n$ to the atomic choice not_hmatch_i. In our case, we have $\mu(hmatch_1) = 0.62$, $\mu(not\_hmatch_1) = 0.38$, $\mu(hmatch_2) = 0.73$, and $\mu(not\_hmatch_2) = 0.27$.

The benefits of this explicit treatment of uncertainty becomes clear when we now try to merge this mapping with the result of another matching system. Below are two examples of rules that describe correspondences for the same ontologies that have been found by the falcon system:

$$InCollection(X) \leftarrow Collection(X) \wedge falcon_1 \, ;$$
$$Proceedings_2(X) \leftarrow Proceedings_1(X) \wedge falcon_2 \, .$$

Here, the confidence encoding yields the choice space $C' = \{\{falcon_i, not\_falcon_i\} \mid i \in \{1,2\}\}$ along with the probabilities $\mu'(falcon_1) = 0.94$, $\mu'(not\_falcon_1) = 0.06$, $\mu'(falcon_2) = 0.96$, and $\mu'(not\_falcon_2) = 0.04$.

Note that directly merging these two mappings as they are would not be a good idea for two reasons. The first one is that we might encounter an inconsistency problem like shown in Section 5. For example, in this case, the ontology O_2 imposes that the concepts *InCollection* and *Book* are to be disjoint. Thus, for each publication *pub* belonging to the concept *Collection* in the ontology O_1, the merged mappings infer $Book(pub)$ and $InCollection(pub)$. Therefore, the first rule of each of the mappings cannot contribute to a model of the knowledge base. The second reason is that a simple merge does not account for the fact that the mapping between the $Proceedings_1$ and $Proceedings_2$ concepts has been found by both matchers and should therefore be strengthened. Here, the mapping rule has the same status as any other rule in the mapping and each instance of the rule has two probabilities at the same time.

Suppose we associate with hmatch and falcon the trust probabilities 0.55 and 0.45, respectively. Based on the interpretation of confidence values as error probabilities, and on the use of trust probabilities when resolving inconsistencies between rules, we can now define a merged mapping set that consists of the following rules:

$$Book(X) \leftarrow Collection(X) \wedge hmatch_1 \wedge sel\_hmatch_1 ;$$
$$InCollection(X) \leftarrow Collection(X) \wedge falcon_1 \wedge sel\_falcon_1 ;$$
$$Proceedings_2(X) \leftarrow Proceedings_1(X) \wedge hmatch_2 ;$$
$$Proceedings_2(X) \leftarrow Proceedings_1(X) \wedge falcon_2 .$$

The new choice space C'' and the new probability μ'' on C'' are obtained from $C \cup C'$ and $\mu \cdot \mu'$ (which is the product of μ and μ', that is, $(\mu \cdot \mu')(B \cup B') = \mu(B) \cdot \mu'(B')$ for all total choices B of C and B' of C'), respectively, by adding the alternative $\{sel\_hmatch_1, sel\_falcon_1\}$ and the two probabilities $\mu''(sel\_hmatch_1) = 0.55$ and $\mu''(sel\_falcon_1) = 0.45$ for resolving the inconsistency between the first two rules.

It is not difficult to verify that, due to the independent combination of alternatives, the last two rules encode that the rule $Proceedings_2(X) \leftarrow Proceedings_1(X)$ holds with the probability $1 - (1 - \mu''(hmatch_2)) \cdot (1 - \mu''(falcon_2)) = 0.9892$, as desired. Informally, any randomly chosen instance of $Proceedings$ of the ontology O_1 is also an instance of $Proceedings$ of the ontology O_2 with the probability 0.9892. In contrast, if the mapping rule would have been discovered only by falcon or hmatch, respectively, such an instance of $Proceedings$ of the ontology O_1 would be an instance of $Proceedings$ of the ontology O_2 with the probability 0.96 or 0.73, respectively.

A probabilistic query Q asking for the probability that a specific publication pub in the ontology O_1 is an instance of the concept $Book$ of the ontology O_2 is given by $Q = \exists(Book(pub))[R, S]$. The tight answer θ to Q is given by $\theta = \{R/0, S/0\}$, if pub is not an instance of the concept $Collection$ in the ontology O_1 (since there is no mapping rule that maps another concept than $Collection$ to the concept $Book$). If pub is an instance of the concept $Collection$, however, then the tight answer to Q is given by $\theta = \{R/0.341, S/0.341\}$ (as $\mu''(hmatch_1) \cdot \mu''(sel\_hmatch_1) = 0.62 \cdot 0.55 = 0.341$). Informally, pub belongs to the concept $Book$ with the probabilities 0 resp. 0.341. Note that we may obtain real intervals when there are total choices with multiple answer sets.

8   Algorithms and Complexity

In this section, we characterize the consistency and the query processing problem in probabilistic dl-programs in terms of the consistency and the cautious/brave reasoning problem in disjunctive dl-programs (which are all decidable [18]). These characterizations show that the consistency and the query processing problem in probabilistic dl-programs are decidable resp. computable, and they directly reveal algorithms for solving these problems. We also give a precise picture of the complexity of deciding consistency and correct answers when the choice space C is bounded by a constant.

Algorithms. The following theorem shows that a probabilistic dl-program $KB = (L, P, C, \mu)$ is consistent iff $(L, P \cup \{p \leftarrow | p \in B\})$ is consistent, for every total choice B of C with $\mu(B) > 0$. This implies that deciding whether a probabilistic dl-program is consistent can be reduced to deciding whether a disjunctive dl-program is consistent.

Theorem 8.1. *A probabilistic dl-program $KB = (L, P, C, \mu)$ is consistent iff $(L, P \cup \{p \leftarrow | p \in B\})$ is consistent for every total choice B of C with $\mu(B) > 0$.*

The next theorem shows that computing tight answers for $\exists (q)[r, s]$ to KB, where $q \in HB_\Phi$, can be reduced to brave and cautious reasoning from disjunctive dl-programs. Informally, to obtain the tight lower (resp., upper) bound, we have to sum up all $\mu(B)$ such that q is a cautious (resp., brave) consequence of $(L, P \cup \{p \leftarrow | p \in B\})$. The theorem holds also when q is a ground formula constructed from HB_Φ. Note that this result implies also that tight query processing in probabilistic dl-programs KB can be done by an anytime algorithm (along the total choices of KB).

Theorem 8.2. *Let $KB = (L, P, C, \mu)$ be a consistent probabilistic dl-program, and let q be a ground atom from HB_Φ. Then, l (resp., u) such that $KB \|\!\sim_{tight} (q)[l, u]$ is the sum of all $\mu(B)$ such that (i) B is a total choice of C and (ii) q is true in all (resp., some) answer sets of $(L, P \cup \{p \leftarrow | p \in B\})$.*

Complexity. The following theorem shows that deciding whether a probabilistic dl-program is consistent is complete for NEXP^{NP} (and so has the same complexity as deciding consistency in ordinary disjunctive logic programs) when the size of its choice space is bounded by a constant. Here, the lower bound follows from the NEXP^{NP}-hardness of deciding whether an ordinary disjunctive logic program has an answer set [5].

Theorem 8.3. *Given Φ and a probabilistic dl-program $KB = (L, P, C, \mu)$, where L is defined in $\mathcal{SHIF}(\mathbf{D})$ or $\mathcal{SHOIN}(\mathbf{D})$, and the size of C is bounded by a constant, deciding whether KB is consistent is complete for NEXP^{NP}.*

The next theorem shows that deciding correct answers for probabilistic queries $\exists (q)[r, s]$, where $q \in HB_\Phi$, to a probabilistic dl-program is complete for co-NEXP^{NP} when the size of the choice space is bounded by a constant. The theorem holds also when q is a ground formula constructed from HB_Φ.

Theorem 8.4. *Given Φ, a probabilistic dl-program $KB = (L, P, C, \mu)$, where L is defined in $\mathcal{SHIF}(\mathbf{D})$ or $\mathcal{SHOIN}(\mathbf{D})$, and the size of C is bounded by a constant, a ground atom q from HB_Φ, and $l, u \in [0, 1]$, deciding whether $(q)[l, u]$ is a consequence of KB is complete for co-NEXP^{NP}.*

9 Tractability Results

In this section, we describe a special class of probabilistic dl-programs for which deciding consistency and query processing can both be done in polynomial time in the data complexity. These programs are normal, stratified, and defined relative to *DL-Lite* [4], which allows for deciding knowledge base satisfiability in polynomial time.

We first recall *DL-Lite*. Let $\mathbf{A}$, $\mathbf{R}_A$, and $\mathbf{I}$ be pairwise disjoint sets of atomic concepts, abstract roles, and individuals, respectively. A *basic concept in DL-Lite* is either an atomic concept from $\mathbf{A}$ or an exists restriction on roles $\exists R.\top$ (abbreviated as $\exists R$), where $R \in \mathbf{R}_A \cup \mathbf{R}_A^-$. A *literal in DL-Lite* is either a basic concept b or the negation of a basic concept $\neg b$. *Concepts in DL-Lite* are defined by induction as follows. Every basic concept in *DL-Lite* is a concept in *DL-Lite*. If b is a basic concept in *DL-Lite*, and ϕ_1 and ϕ_2 are concepts in *DL-Lite*, then $\neg b$ and $\phi_1 \sqcap \phi_2$ are also concepts in *DL-Lite*. An *axiom in DL-Lite* is either (1) a concept inclusion axiom $b \sqsubseteq \phi$, where b is a basic concept in *DL-Lite*, and ϕ is a concept in *DL-Lite*, or (2) a *functionality axiom* (funct R), where $R \in \mathbf{R}_A \cup \mathbf{R}_A^-$, or (3) a concept membership axiom $b(a)$, where b is a basic concept in *DL-Lite* and $a \in \mathbf{I}$, or (4) a role membership axiom $R(a, c)$, where $R \in \mathbf{R}_A$ and $a, c \in \mathbf{I}$. A *knowledge base in DL-Lite L* is a finite set of axioms in *DL-Lite*.

Every knowledge base in *DL-Lite L* can be transformed into an equivalent one in *DL-Lite trans(L)* in which every concept inclusion axiom is of form $b \sqsubseteq \ell$, where b (resp., ℓ) is a basic concept (resp., literal) in *DL-Lite* [4]. We then define $trans(P) = P \cup \{b'(X) \leftarrow b(X) \mid b \sqsubseteq b' \in trans(L),\ b'\text{ is a basic concept}\} \cup \{\exists R(X) \leftarrow R(X, Y) \mid R \in \mathbf{R}_A \cap \Phi\} \cup \{\exists R^-(Y) \leftarrow R(X, Y) \mid R \in \mathbf{R}_A \cap \Phi\}$. Intuitively, we make explicit all the relationships between the predicates in P that are implicitly encoded in L.

We define stratified normal dl- and stratified normal probabilistic dl-programs as follows. A normal dl-program $KB = (L, P)$ is *stratified* iff (i) L is defined in *DL-Lite* and (ii) $trans(P)$ is locally stratified. A probabilistic dl-program $KB = (L, P, C, \mu)$ is *normal* iff P is normal. A normal probabilistic dl-program $KB = (L, P, C, \mu)$ is *stratified* iff every of KB's represented dl-programs is stratified.

The following result shows that stratified normal probabilistic dl-programs allow for consistency checking and query processing with a polynomial data complexity. It follows from Theorems 8.1 and 8.2 and that consistency checking and reasoning in stratified normal dl-programs can be done in polynomial time in the data complexity [18].

Theorem 9.1. *Given Φ and a stratified normal probabilistic dl-program KB, (a) deciding if KB has an answer set, and (b) computing $l, u \in [0, 1]$ for a given ground atom q such that $KB \mathrel{|\!\!\sim}_{tight}(q)[l, u]$ can be done in polynomial time in the data complexity.*

10   Conclusion

We have presented tightly integrated probabilistic (disjunctive) dl-programs as a rule-based framework for representing ontology mappings that supports the resolution of inconsistencies on a symbolic and a numeric level. While the use of disjunction and nonmonotonic negation allows the rewriting of inconsistent rules, the probabilistic extension of the language allows us to explicitly represent numeric confidence values as error probabilities, to resolve inconsistencies by using trust probabilities, and to reason about these on a numeric level. While being expressive and well-integrated with description logic ontologies, the language is still decidable and has data-tractable subsets that make it particularly interesting for practical applications.

Note that probabilistic queries in tightly integrated probabilistic dl-programs can syntactically and semantically easily be generalized to contain conditionals of disjunctions of conjunctions of atoms, rather than only disjunctions of conjunctions of atoms. The characterization in Theorem 8.2 can be generalized to such probabilistic queries, and the completeness for co-NEXP^{NP} of Theorem 8.4 also carries over to them. Furthermore, note that tightly integrated probabilistic dl-programs are also a natural approach to combining languages for reasoning about actions with both description logics and Bayesian probabilities (which is especially directed towards Web Services) [1].

We leave for future work the implementation of tightly integrated probabilistic dl-programs. Another interesting topic for future work is to explore whether the tractability results can be extended to an even larger class of tightly integrated probabilistic dl-programs. One way to achieve this could be to approximate the answer set semantics through the well-founded semantics (which may be defined similarly as in [21]). Furthermore, it would be interesting to investigate whether one can develop an efficient top-k query technique (as in [28,22]) for tightly integrated probabilistic dl-programs: Rather than computing the tight probability interval for a given ground atom, such a technique returns the k most probable ground instances of a given non-ground atom.

Acknowledgments. Andrea Calì is supported by the STREP FET project TONES (FP6-7603) of the EU. Thomas Lukasiewicz is supported by the German Research Foundation (DFG) under the Heisenberg Programme and by the Austrian Science Fund (FWF) under the project P18146-N04. Heiner Stuckenschmidt and Livia Predoiu are supported by an Emmy-Noether Grant of the German Research Foundation (DFG).

References

1. Calì, A., Lukasiewicz, T.: Tightly integrated probabilistic description logic programs for the Semantic Web. In: Dahl, V., Niemelä, I. (eds.) ICLP 2007. LNCS, vol. 4670, pp. 428–429. Springer, Heidelberg (2007)
2. da Costa, P.C.G.: Bayesian Semantics for the Semantic Web. Doctoral Dissertation, George Mason University, Fairfax, VA, USA (2005)
3. da Costa, P.C.G., Laskey, K.B.: PR-OWL: A framework for probabilistic ontologies. In: Proc. FOIS-2006, pp. 237–249. IOS Press, Amsterdam (2006)
4. Calvanese, D., De Giacomo, G., Lembo, D., Lenzerini, M., Rosati, R.: DL-Lite: Tractable description logics for ontologies. In: Proc. AAAI-2005, pp. 602–607. AAAI Press / MIT Press (2005)
5. Dantsin, E., Eiter, T., Gottlob, G., Voronkov, A.: Complexity and expressive power of logic programming. ACM Comput. Surv. 33(3), 374–425 (2001)
6. Eiter, T., Lukasiewicz, T., Schindlauer, R., Tompits, H.: Combining answer set programming with description logics for the Semantic Web. In: Proc. KR-2004, pp. 141–151. AAAI Press, California (2004)
7. Eiter, T., Ianni, G., Lukasiewicz, T., Schindlauer, R., Tompits, H.: Combining answer set programming with description logics for the Semantic Web. Technical Report INFSYS RR-1843-07-04, Institut für Informationssysteme, TU Wien (March 2007)
8. Eiter, T., Ianni, G., Schindlauer, R., Tompits, H.: Effective integration of declarative rules with external evaluations for semantic-web reasoning. In: Sure, Y., Domingue, J. (eds.) ESWC 2006. LNCS, vol. 4011, pp. 273–287. Springer, Heidelberg (2006)
9. Euzenat, J., Mochol, M., Shvaiko, P., Stuckenschmidt, H., Svab, O., Svatek, V., van Hage, W.R., Yatskevich, M.: First results of the ontology alignment evaluation initiative 2006. In: Proc. ISWC-2006 Workshop on Ontology Matching (2006)

10. Euzenat, J., Shvaiko, P.: Ontology Matching. Springer, Heidelberg (2007)
11. Euzenat, J., Stuckenschmidt, H., Yatskevich, M.: Introduction to the ontology alignment evaluation 2005. In: Proc. K-CAP-2005 Workshop on Integrating Ontologies (2005)
12. Faber, W., Leone, N., Pfeifer, G.: Recursive aggregates in disjunctive logic programs: Semantics and complexity. In: Alferes, J.J., Leite, J.A. (eds.) JELIA 2004. LNCS (LNAI), vol. 3229, pp. 200–212. Springer, Heidelberg (2004)
13. Finzi, A., Lukasiewicz, T.: Structure-based causes and explanations in the independent choice logic. In: Proc. UAI-2003, pp. 225–232. Morgan Kaufmann, San Francisco (2003)
14. Gelfond, M., Lifschitz, V.: Classical negation in logic programs and disjunctive databases. New Generation Comput. 9(3/4), 365–386 (1991)
15. Giugno, R., Lukasiewicz, T.: P-$\mathcal{SHOQ}(\mathbf{D})$: A probabilistic extension of $\mathcal{SHOQ}(\mathbf{D})$ for probabilistic ontologies in the Semantic Web. In: Flesca, S., Greco, S., Leone, N., Ianni, G. (eds.) JELIA 2002. LNCS (LNAI), vol. 2424, pp. 86–97. Springer, Heidelberg (2002)
16. Horrocks, I., Patel-Schneider, P.F.: Reducing OWL entailment to description logic satisfiability. In: Fensel, D., Sycara, K.P., Mylopoulos, J. (eds.) ISWC 2003. LNCS, vol. 2870, pp. 17–29. Springer, Heidelberg (2003)
17. Horrocks, I., Sattler, U., Tobies, S.: Practical reasoning for expressive description logics. In: Ganzinger, H., McAllester, D., Voronkov, A. (eds.) LPAR 1999. LNCS, vol. 1705, pp. 161–180. Springer, Heidelberg (1999)
18. Lukasiewicz, T.: A novel combination of answer set programming with description logics for the Semantic Web. In: Franconi, E., Kifer, M., May, W. (eds.) ESWC 2007. LNCS, vol. 4519, pp. 384–398. Springer, Heidelberg (2007)
19. Lukasiewicz, T.: Expressive probabilistic description logics. Artif. Intell. (in press)
20. Lukasiewicz, T.: Probabilistic description logic programs. Int. J. Approx. Reason. 45(2), 288–307 (2007)
21. Lukasiewicz, T.: Tractable probabilistic description logic programs. In: Prade, H., Subrahmanian, V.S. (eds.) SUM 2007. LNCS (LNAI), vol. 4772, pp. 143–156. Springer, Heidelberg (2007)
22. Lukasiewicz, T., Straccia, U.: Top-k retrieval in description logic programs under vagueness for the Semantic Web. In: Prade, H., Subrahmanian, V.S. (eds.) SUM 2007. LNCS (LNAI), vol. 4772, pp. 16–30. Springer, Heidelberg (2007)
23. Meilicke, C., Stuckenschmidt, H., Tamilin, A.: Repairing ontology mappings. In: Proc. AAAI-2007, pp. 1408–1413. AAAI Press, California (2007)
24. Motik, B., Horrocks, I., Rosati, R., Sattler, U.: Can OWL and logic programming live together happily ever after? In: Cruz, I., Decker, S., Allemang, D., Preist, C., Schwabe, D., Mika, P., Uschold, M., Aroyo, L. (eds.) ISWC 2006. LNCS, vol. 4273, pp. 501–514. Springer, Heidelberg (2006)
25. Wang, P., Xu, B.: Debugging ontology mapping: A static method. Computation and Intelligence (to appear, 2007)
26. Poole, D.: The independent choice logic for modelling multiple agents under uncertainty. Artif. Intell. 94(1/2), 7–56 (1997)
27. Serafini, L., Stuckenschmidt, H., Wache, H.: A formal investigation of mapping languages for terminological knowledge. In: Proc. IJCAI-2005, pp. 576–581 (2005)
28. Straccia, U.: Towards top-k query answering in description logics: The case of DL-Lite. In: Fisher, M., van der Hoek, W., Konev, B., Lisitsa, A. (eds.) JELIA 2006. LNCS (LNAI), vol. 4160, pp. 439–451. Springer, Heidelberg (2006)

Appendix A: Proofs

Proof of Theorem 8.1. Recall first that KB is consistent iff KB has an answer set Pr, which is a probabilistic interpretation Pr such that (i) every interpretation $I \subseteq HB_\Phi$

with $Pr(I) > 0$ is an answer set of the disjunctive dl-program $(L, P \cup \{p \leftarrow \mid p \in B\})$ for some total choice B of C, and (ii) $Pr(\bigwedge_{p \in B} p) = \mu(B)$ for each total choice B of C.

($\Rightarrow$) Suppose that KB is consistent. We now show that the disjunctive dl-program $(L, P \cup \{p \leftarrow \mid p \in B\})$ is consistent, for every total choice B of C with $\mu(B) > 0$. Towards a contradiction, suppose the contrary. That is, $(L, P \cup \{p \leftarrow \mid p \in B\})$ is not consistent for some total choice B of C with $\mu(B) > 0$. So, it follows that $Pr(\bigwedge_{p \in B} p) = 0$. But this contradicts $Pr(\bigwedge_{p \in B} p) = \mu(B) > 0$. This shows that $(L, P \cup \{p \leftarrow \mid p \in B\})$ is consistent, for every total choice B of C with $\mu(B) > 0$.

($\Leftarrow$) Suppose that the disjunctive dl-program $(L, P \cup \{p \leftarrow \mid p \in B\})$ is consistent, for every total choice B of C with $\mu(B) > 0$. That is, there exists some answer set I_B of $(L, P \cup \{p \leftarrow \mid p \in B\})$, for every total choice B of C with $\mu(B) > 0$. Let the probabilistic interpretation Pr be defined by $Pr(I_B) = \mu(B)$ for every total choice B of C with $\mu(B) > 0$ and by $Pr(I) = 0$ for all other $I \subseteq HB_\Phi$. Then, Pr is an interpretation that satisfies (i) and (ii). That is, Pr is an answer set of KB. Thus, KB is consistent. $\square$

Proof of Theorem 8.2. The statement of the theorem follows from the observation that the probability $\mu(B)$ of all total choices B of C such that q is true in all (resp., some) answer sets of $(L, P \cup \{p \leftarrow \mid p \in B\})$ contributes (resp., may contribute) to the probability $Pr(q)$, while the probability $\mu(B)$ of all total choices B of C such that q is false in all answer sets of $(L, P \cup \{p \leftarrow \mid p \in B\})$ does not contribute to $Pr(q)$. $\square$

Proof of Theorem 8.3. We first show membership in $\mathrm{NEXP}^{\mathrm{NP}}$. By Theorem 8.1, we check whether $(L, P \cup \{p \leftarrow \mid p \in B\})$ is consistent, for every total choice B of C with $\mu(B) > 0$. Since C is bounded by a constant, the number of all total choices B of C with $\mu(B) > 0$ is also bounded by a constant. As shown in [18], deciding whether a disjunctive dl-program has an answer set is in $\mathrm{NEXP}^{\mathrm{NP}}$. In summary, this shows that deciding whether KB is consistent is in $\mathrm{NEXP}^{\mathrm{NP}}$.

Hardness for $\mathrm{NEXP}^{\mathrm{NP}}$ follows from the $\mathrm{NEXP}^{\mathrm{NP}}$-hardness of deciding whether a disjunctive dl-program has an answer set [18], since by Theorem 8.1 a disjunctive dl-program $KB = (L, P)$ has an answer set iff the probabilistic dl-program $KB = (L, P, C, \mu)$ has answer set, for the choice space $C = \{\{a\}\}$, the probability function $\mu(a) = 1$, and any ground atom $a \in HB_\Phi$ that does not occur in $ground(P)$. $\square$

Proof of Theorem 8.4. We first show membership in co-$\mathrm{NEXP}^{\mathrm{NP}}$. We show that deciding whether $(q)[l, u]$ is not a consequence of KB is in $\mathrm{NEXP}^{\mathrm{NP}}$. By Theorem 8.2, $(q)[l, u]$ is not a consequence of KB iff there exists a set $\mathcal{B}$ of total choices B of C such that either (a.1) q is true in some answer set of $(L, P \cup \{p \leftarrow \mid p \in B\})$, for every $B \in \mathcal{B}$, and (a.2) $\sum_{B \in \mathcal{B}} \mu(B) > u$, or (b.1) q is false in some answer set of $(L, P \cup \{p \leftarrow \mid p \in B\})$, for every $B \in \mathcal{B}$, and (a.2) $\sum_{B \in \mathcal{B}} \mu(B) < l$. As shown in [18], deciding whether q is true in some answer set of a disjunctive dl-program is in $\mathrm{NEXP}^{\mathrm{NP}}$. It thus follows that deciding whether $(q)[l, u]$ is not a consequence of KB is in $\mathrm{NEXP}^{\mathrm{NP}}$, and thus deciding whether $(q)[l, u]$ is a consequence of KB is in co-$\mathrm{NEXP}^{\mathrm{NP}}$.

Hardness for co-$\mathrm{NEXP}^{\mathrm{NP}}$ follows from the co-$\mathrm{NEXP}^{\mathrm{NP}}$-hardness of deciding whether a ground atom q is true in all answer sets of a disjunctive dl-program [18], since by Theorem 8.2 a ground atom q is true in all answer sets of a disjunctive dl-program $KB = (L, P)$ iff $(q)[1, 1]$ is a consequence of the probabilistic dl-program $KB = (L, P, C, \mu)$ under the answer set semantics, for the choice space $C = \{\{a\}\}$, the probability function $\mu(a) = 1$, and any $a \in HB_{\Phi}$ that does not occur in $ground(P)$.   □

Proof of Theorem 9.1. As shown in [18], deciding the existence of (and computing) the answer set of a stratified normal dl-program has a polynomial data complexity. Observe then that in the case of data complexity, the choice space C is fixed. By Theorems 8.1 and 8.2, it thus follows that the problems of (a) deciding whether KB has an answer set, and (b) computing $l, u \in [0, 1]$ for a given ground atom q such that $KB \mathrel{|\!\!\sim}_{tight} (q)[l, u]$, respectively, can both be solved in polynomial time in the data complexity.   □

Using Transversals for Discovering XML Functional Dependencies

Thu Trinh

Information Research Centre, Department of Information Systems
Massey University, Palmerston North, New Zealand
t.trinh@massey.ac.nz

Abstract. Dependency discovery is an important technique for relational database design due to numerous applications like normalisation and query optimisation. Similar applications are likely to exist in the context of XML. To date, there are few investigations into dependency discovery in the context of XML. We propose a transversal approach for discovering a class of subgraph-based XML functional dependencies with pre-image semantics (XFDs). We cover all aspects of the discovery process, from determining agree sets to the extraction of an XFD cover. An inherent challenge is the large and complex search space of valid XFDs due to the tree-structure of XML data.

1 Introduction

The discovery of satisfied functional dependencies from given data instances has long been acknowledged as an important problem for relational databases. This is due to the myriad of applications for functional dependencies, including database design, reverse engineering and query optimisation. Similar applications have already been shown to exist for XML functional dependencies, for example for identifying redundancies [13] and normalisation of XML schemas [1]. XML databases are designed in an ad-hoc manner without much consideration for functional dependencies, so tools for dependency discovery will become important for unlocking the potentials of such applications.

For the relational data model there is a single widely-accepted notion for functional dependencies, but in contrast, there are numerous proposals for XML functional dependencies, for example [1,3,10,11,13]. The different notions of XML functional dependencies vary in terms of expressiveness but are all justified by their natural occurrence in XML data. Dependency discovery is also a useful tool for understanding the expressiveness of different dependencies and the nature in which they arise in practice.

A vast majority of proposals for XML functional dependencies feature paths [1,10,11]. Examples of such XML functional dependencies often bear close similarity to relational functional dependencies because a flat tuple-like semantic is adopted. Consider an XML document about purchases at different outlets of

S. Hartmann and G. Kern-Isberner (Eds.): FoIKS 2008, LNCS 4932, pp. 199–218, 2008.
© Springer-Verlag Berlin Heidelberg 2008

a supermarket, with details about items bought and discounts received as depicted by the tree in Figure 1. We can observe that every outlet in the tree has a unique name and unique address location. This can be expressed by two functional dependencies: two outlets with the same name have the same address, and conversely two outlets with the same address also have the same name. The data also suggests that an item's description and price determine the item's discount amount.

The tree-structure of XML data cannot always adequately specified using sets of paths. Suppose items in our purchase example are identifiable by description and price. Then purchases at nodes [2] and [5] are for different sets of items. However both purchases share the same multiset of item descriptions, and similarly the same multiset of item prices. Therefore, with functional dependencies featuring paths we cannot express the business rule stating two purchases for the same multiset of items receive the same amount of savings.

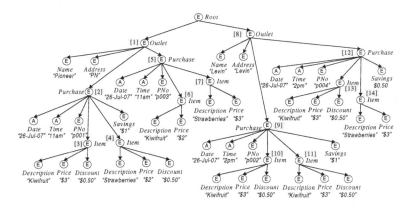

Fig. 1. XML data tree of purchases

The subgraph-based XML functional dependencies with pre-image semantics (XFDs) proposed in [3] can express functional relationships between tree structures rather than just paths and have a multiset-based semantics. XFDs are capable of expressing business rules like the ones above. A nice feature of the class is that we can efficiently reason about XFDs [4]. In this paper, we address the problem of discovering XFDs which holds for a given XML data tree.

Organisation. The paper is organised as follows. In Section 2, we present some basic terminology and definitions relating to the tree-based XML data model and XFDs with pre-image semantics. Section 3 identifies some related work. Details of the dependency discovery approach are given in Section 4-5. Specifically, we propose a transversal-based approach for discovering XFDs, followed by a discussion of how to determine the difference sets that are needed as input to the transversal-based approach. We provide a summary of the key features of the approach in Section 6.

2   Preliminary

2.1   XML Schema Tree and XML Data Tree

We use the simple XML graph model from [3]. An *XML tree* is a rooted tree T with node set V_T, arc set A_T, root r_T , and mappings $name : V_T \to Names$ and $kind : V_T \to \{E, A\}$. The symbols E and A indicate elements and attributes, with attributes only appearing as leaf nodes. A *data tree* is an XML tree T' with a mapping *valuation* assigning string values to leaves. A *schema tree* is an XML tree T with frequencies $?, 1, *, +$ assigned to its arcs, where no two siblings have the same name and kind. A data tree T' is T-*compatible* whenever there is a homomorphism $\phi : V_{T'} \to V_T$ (i.e. root-preserving, name-preserving, kind-preserving and arc-preserving mapping) such that for every vertex v' of T' and every arc $a = (\phi(v'), w)$ of T, the number of arcs $a' = (v', w')$ mapped to a is at most one if a has frequency label $?$, exactly one if a has frequency label 1, at least one if a has frequency label $+$, and arbitrarily many if a has frequency label $*$. A homomorphism $\phi : V_{T'} \to V_T$ is an *isomorphism* if ϕ is bijective and ϕ^{-1} is a homomorphism.

The purchase data tree from Figure 1 is compatible (but not isomorphic) with the schema tree depicted in Figure 2.

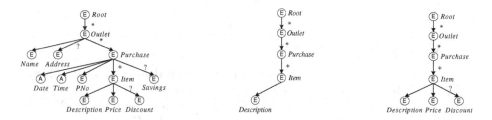

Fig. 2. An XML schema tree for purchases and some of its subgraphs: walk `De` and subgraph $\{$`De`, `Pr`, `Di`$\}$ (for convenience, each walk is denoted by the first two letters of its leaf label)

2.2   Subgraph Terminology

Given a set L of leaves of T, a *walk* of T is a path from the root of T to a member of L. For a node v of T, every walk containing v is a v-*walk*. A *subgraph* of T is a (possibly empty) set of walks of T. For a node v of T, a subgraph of T is a v-*subgraph* if each of its walks contains v. Clearly a v-subgraph and v-walk depict again an XML tree.

By $S_T(v)$ we denote the set of all v-subgraphs of T. It is easy to see that $S_T(v)$ contains the empty subgraph and is closed under the union, intersection and difference operators. For a subgraph $\mathcal{X}$ which consists of a single walk X we tend to write X instead of $\{X\}$ and we sometimes also refer to such singleton subgraphs as walks.

Consider two XML trees T' and T with a homomorphism between them. A subgraph U' of T' is a *subcopy* of T if U' is isomorphic to some subgraph U of T. Given a subgraph U of T, the *projection of T' to U* is the union of all subcopies of U in T', and denoted by $T'|_U$.

A *total v-subgraph* of an XML tree is its set of all v-walks. The homomorphism ϕ between a T-compatible data tree T' and schema tree T induces a mapping of the total subgraphs of T' to the total subgraphs of T. For a fixed node v of T, a *pre-image of v* is just a total w-subgraph with $\phi(w) = v$, where the node id of w will be used to identify the pre-image. By $V_{T'}(v)$ we denote the set of all pre-images of v.

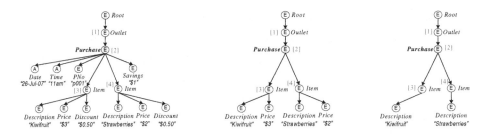

Fig. 3. Example of a pre-image of $v_{Purchase}$ from the purchase data tree together with two of its projections: to {De, Pr} and to De

2.3   XML Functional Dependencies

Two data trees T' and T are *value-equal* if and only if there is an isomorphism $\phi : V_{T'} \to V_T$ between them which is also valuation-preserving.

An *XML functional dependency (XFD)* over T is an expression $v : \mathcal{X} \to \mathcal{Y}$ where both $\mathcal{X}$ and $\mathcal{Y}$ are non-empty sets of v-subgraphs. Herein, v is referred to as the *target*, $\mathcal{X}$ is called the LHS and $\mathcal{Y}$ the RHS. An XFD with a singleton RHS is said to be *singular*.

A T-compatible data tree T' *satisfies* $v : \mathcal{X} \to \mathcal{Y}$ (or equivalently $v : \mathcal{X} \to \mathcal{Y}$ *holds* in T'), written as $\models_{T'} v : \mathcal{X} \to \mathcal{Y}$, if and only if for any two pre-images $p_1, p_2 \in V_{T'}(v)$ projections $p_1|_Y, p_2|_Y$ are value-equal for all $Y \in \mathcal{Y}$ whenever the projections $p_1|_X, p_2|_X$ are value-equal for all $X \in \mathcal{X}$.

Example 2.1. The purchase data tree of Figure 1 satisfies the following XFDs:

XFD1 - $v_{Outlet} : \{Na\} \to \{Ad\}$ XFD3 - $v_{Item} : \{De, Pr\} \to \{Di\}$
XFD2 - $v_{Outlet} : \{Ad\} \to \{Na\}$ XFD4 - $v_{Purchase} : \{\{De, Pr\}\} \to \{Sa\}$

XFD1 and XFD2 express that two outlets with the same name have the same address and vice versa. Two purchases for the same set of items must have the same savings is represented by XFD4, while XFD3 captures that items with the same description and price receives the same discount.

Refining XFDs. The number of possible LHS is exponential in the number of v-subgraphs which is already exponential in the number of v-walks. Fortunately, not all of these possible XFDs are interesting.

In the introduction, $v_{Purchase}$ nodes [2] and [5] illustrated that projections of two XML trees to two v-subgraphs X, Y may be value-equal without implying that projections to the union v-subgraph $X \cup Y$ are value-equal. On the other hand, for pre-images of v_{Item}, if projections to De and projections to Pr are value-equal then projections to $\{De, Pr\}$ are necessarily value-equal.

Two distinct v-subgraphs X, Y are called v-*reconcilable* if and only if for any proper descendant w of v which is shared by X and Y, X contains every w-walk in Y or Y contains every w-walk in X or the path from v to w does not contain an arc of frequency other than ? and 1.

Example 2.2. Subgraphs De and Pr are v_{Item}-reconcilable but not $v_{Purchase}$-reconcilable. Also De and Pr are each v_{Item} and $v_{Purchase}$-reconcilable with $\{De, Pr, Sa\}$.

Lemma 2.1. *Let X, Y be v-reconcilable v-subgraphs. For any two pre-images $p_1, p_2 \in V_{T'}(v)$ $p_1|_X, p_2|_X$ are value-equal and $p_1|_Y, p_2|_Y$ are value-equal if and only if $p_1|_{X \cup Y}, p_2|_{X \cup Y}$ are value-equal.*

Proof. One direction is trivial since the value-equality of projections to a v-subgraph X always imply the value-equality of projections to any v-subgraph contained in X. Suppose $p_1|_X, p_2|_X$ are value-equal and $p_1|_Y, p_2|_Y$ are value-equal for v-reconcilable v-subgraphs X, Y. Let ϕ_X, ϕ_Y be valuation-preserving isomorphisms for $p_1|_X, p_2|_X$ and $p_1|_Y, p_2|_Y$, respectively. Since X, Y are v-reconcilable, ϕ_X and ϕ_Y map every shared vertex to the same target. Thus ϕ_X union ϕ_Y is a function. Since ϕ_X and ϕ_Y are valuation-preserving isomorphisms, so is ϕ_X union ϕ_Y. $\qquad\qquad\square$

The previous lemma shows that v-reconcilability characterises when value-equality of projections to some v-subgraph X is implied by value-equality of projections to certain v-subgraphs which are contained in X. We proceed to refine the notion of XFDs in view of this.

Let $E_T(v)$ be the smallest subset of $S_T(v)$ such that all singleton v-subgraph consisting of a single v-walk belong to $E_T(v)$ and such that, if two v-subgraphs $X, Y \in E_T(v)$ are not v-reconcilable then $X \cup Y \in E_T(v)$. The members of $E_T(v)$ are called the *essential v-subgraphs*. For a set $\mathcal{X}$ of v-subgraphs consider the set of all essential v-subgraphs that are included in some member of $\mathcal{X}$. We use $\vartheta_v(\mathcal{X})$ to denote the set of maximal (with respect to containment) members of this set.

Example 2.3. For the set $\mathcal{X} = \{\{Da, De, Pr\}, \{Pr, Di\}, Pr\}\}$ we get $\vartheta_{v_{Purchase}}(\mathcal{X})$ $= \{Da, \{De, Pr\}, \{Pr, Di\}\}$. Essential $v_{Purchases}$-subgraphs included in some member of $\mathcal{X}$ but not in $\vartheta_{v_{Purchase}}(\mathcal{X})$ are: De, Pr, Di. De and Pr are contained in the second member of $\vartheta_{v_{Purchase}}(\mathcal{X})$ while Di is contained in the third member of $\vartheta_{v_{Purchase}}(\mathcal{X})$.

As shown by the next lemma, we can, without loss of generality, refine XFDs to be of the form $v : \mathcal{X} \to \mathcal{Y}$ with $\mathcal{X} = \vartheta_v(\mathcal{X})$ and $\mathcal{Y} = \vartheta_v(\mathcal{Y})$.

Lemma 2.2. *A T-compatible data tree T' satisfies $v : \mathcal{X} \to \mathcal{Y}$ if and only if T' satisfies $v : \vartheta_v(\mathcal{X}) \to \vartheta_v(\mathcal{Y})$.*

Proof. The lemma immediately follows from the following observation: for any two pre-images $p_1, p_2 \in V_{T'}(v)$ projections $p_1|_X, p_2|_X$ are value-equal for all $X \in \mathcal{X}$ if and only if projections $p_1|_Y, p_2|_Y$ are value-equal for all $Y \in \vartheta_v(\mathcal{X})$.

One direction of the proof is trivial since every $Y \in \vartheta_v(\mathcal{X})$ is contained in some $X \in \mathcal{X}$. For the other direction, suppose $p_1|_Y, p_2|_Y$ are value-equal for all $Y \in \vartheta_v(\mathcal{X})$ but $p_1|_X, p_2|_X$ are not value-equal for some $X \in \mathcal{X}$. Clearly the elements of $\vartheta_v(X) = \{X_1, \ldots, X_n\}$ are pair-wised v-reconcilable and $\vartheta_v(X)$ is a subset of $\vartheta_v(\mathcal{X})$. Thus projections $p_1|_{X_i}, p_2|_{X_i}$ are value-equal (for $i = 1, \ldots, n$). This implies $p_1|_X, p_2|_X$ are value-equal (Lemma 2.1), a contradiction. $\square$

A v-*unit* is a v-subgraph U such that: (1) U consists of a single v-walk in which every proper descendant of v has an incoming arc of frequency ? or 1; or (2) U consists of all w-walks where w is a proper descendant of v whose incoming arc is the only arc in the path from v to w with frequency other than ? and 1. Observe that the set of all v-units induces a (total) partition on the set of all v-walks.

We can use v-units to identify essential v-subgraphs in the following way: An essential v-subgraph is a non-empty v-subgraph that is contained in some v-unit. The proof for this observation has been omitted due to to space restriction.

3   Related Work

One of the earliest investigations of functional dependency (FD) discovery appeared in [8,9]. In these papers, Mannila and Räihä present a hypergraph-transversal method for discovering FDs in relational databases. Central to this approach is the characterisation that whenever two tuples do not agree on their projections to RHS then they must also not agree on their projections to LHS for the corresponding functional dependency to be satisfied.

Many other researchers have explored the problem of dependency discovery. In the relational context, some proposals include: [5,7,12]. More recent algorithms use partitions to capture information about value-equality. Dep-Miner and FastFDs algorithms (see [7,12] respectively) are hypergraph-based approaches involving partitions. A popular alternative to the hypergraph-transversal method is a level-wise examination of syntactically valid FDs (e.g. [5]). For such approaches, it is important to identify the order in which valid FDs are examined and pruning criteria that would reduce the search space.

In the context of XML, few works have addressed dependency discovery. Discovery of XML keys have been discussed in [2]. Zhou has investigated approaches for discovering XML functional dependencies with tuple-like semantics [14]. Several algorithms were proposed for discovering XML functional dependencies from relational representation of XML data. The algorithms are adapted from procedures for discovering relational FDs. Yu and Jagadish have presented a level-wised procedure for discovering a class of path-based XML functional dependencies evaluated over generalised tree tuples [13]. The approach computes partitions from XML data that are stored as a set of relations. The partitions are then used to determine the satisfiability of valid XML functional dependencies. Our approach differs in several ways. Firstly, the classes of XML functional dependencies differ in terms of expressibility. Secondly, by considering subgraphs as well as walks, we face a larger and more intricate search space of valid dependencies. Systematic examination of syntactically valid XFDs is not as straightforward. Fortunately, our transversal approach for discovering XFDs is quite natural and preserves the basic ideas of the hypergraph-transversal approach that has been proposed for the relational data model.

4   Transversal Approach to Mining XFDs

In this section, we describe the core of our XFD discovery approach. Specifically we focus on discovering XFDs for some fixed target. The correctness of the approach is proven in two steps: we show firstly that canonical XFD covers can be used to represent all satisfied XFDs, and secondly that computation of canonical XFD covers relates to finding minimal transversals. Ultimately, the XFD discovery problem is reduced to the problem of finding minimal transversals.

We want to use a notion of canonical XFD cover to represent the set of all satisfied XFDs. As such we want this set to be as small as possible. For this, several natural implications of XFDs are explored.

Firstly, observe that $v : \mathcal{X} \to \mathcal{Y}$ and $v : \mathcal{X} \to \mathcal{Z}$ implies $v : \mathcal{X} \to \mathcal{Y} \cup \mathcal{Z}$ (analogous to the *union rule* from relational data model). This means, for the discovery we can restrict ourselves to singular XFDs.

Secondly, we can exclude XFDs which are trivially satisfied, that is those which hold for every T-compatible data tree. We say $v : \mathcal{X} \to \{Y\}$ is *trivial* if Y is contained in some $X \in \mathcal{X}$ or v is simple, where v is said to be *simple* if the path from the root to v includes only arcs of frequency ? or 1. In the first case the value-equality of projections to RHS is implied by value-equality of projections to LHS and, in the second case, where the target v is simple, there can be at most one pre-image of v.

Finally, we can identify reduced XFDs such that all satisfied XFDs are reduced or implied by an XFD which is reduced. Unlike in the relational data model, we have an ordering of both possible LHSs and possible RHSs, and the notion of reduced XFDs should consider both.

For comparing possible LHSs, we have to contend with the subset relationship among sets of v-subgraphs, as well as the containment relationship amongst members of these sets. The following subsumption ordering incorporates both aspects. Let $\mathcal{X}$ and $\mathcal{Y}$ be two sets of v-subgraphs. We use $\mathcal{X} \leq \mathcal{Y}$ to denote that for every v-subgraph $X \in \mathcal{X}$ there is some $Y \in \mathcal{Y}$ such that X is contained in Y. We say that $\mathcal{Y}$ *subsumes* $\mathcal{X}$ or equivalently $\mathcal{X}$ *is subsumed by* $\mathcal{Y}$. If $\mathcal{X} \leq \mathcal{Y}$ but

$\mathcal{Y} \not\leq \mathcal{X}$ then we say $\mathcal{Y}$ *strictly subsumes* $\mathcal{X}$ and write $\mathcal{X} < \mathcal{Y}$. We can easily verify that subsumption is in fact a partial order over the family of possible LHSs.

An XFD $v : \mathcal{X} \to \{Y\}$ is *left-reduced* with respect to T' if and only if $\models_{T'} v : \mathcal{X} \to \{Y\}$ and $\mathcal{W} < \mathcal{X}$ implies that $v : \mathcal{W} \to \{Y\}$ does not hold in T'.

For possible RHSs, we only have the containment relationship. We say that $v : \mathcal{X} \to \{Y\}$ is *right-maximal* with respect to T' if and only if $\models_{T'} v : \mathcal{X} \to \{Y\}$ implies that $v : \mathcal{X} \to \{Z\}$ does not hold in T' for any essential v-subgraph Z which properly contains Y.

An XFD $v : \mathcal{X} \to \{Y\}$ is said to be *reduced* with respect to T' if it is both left-reduced and right-maximal with respect to T'.

Lemma 4.1. *If $\models_{T'} v : \mathcal{X} \to \{Y\}$ then $v : \mathcal{X} \to \{Y\}$ is implied by some XFD (of the form $v : \mathcal{W} \to \{Z\}$ with $\mathcal{W} \leq \mathcal{X}$ and Z is an essential v-subgraph containing Y) which is reduced with respect to T'.*

Proof Sketch. The following observations can be easily verified:

- If XFD $v : \mathcal{X} \to \{Y\}$ is not left-reduced then it is implied by some left-reduced $v : \mathcal{W} \to \{Y\}$ with $\mathcal{W} < \mathcal{X}$.
- If XFD $v : \mathcal{X} \to \{Y\}$ is not right-maximal then it is implied by some right-maximal $v : \mathcal{X} \to \{Z\}$ where essential v-subgraph Z properly contains Y.

The lemma holds if the XFD is reduced so assume it is not reduced. Particularly, this means that the XFD is not left-reduced and/or not right-maximal.

If $v : \mathcal{X} \to \{Y\}$ is not left-reduced then it is implied by some left-reduced $v : \mathcal{W} \to \{Y\}$ with $\mathcal{W} < \mathcal{X}$ (first observation above). If $v : \mathcal{W} \to \{Y\}$ is not right-maximal then it is implied by some right-maximal $v : \mathcal{W} \to \{Z\}$ with Y being properly contained in Z (second observation above). Now suppose $v : \mathcal{W} \to \{Z\}$ is not left-reduced. This means there is some left-reduced $v : \mathcal{W}' \to \{Z\}$ with $\mathcal{W}' < \mathcal{W}$ which implies it. But from this we can infer $v : \mathcal{W}' \to \{Y\}$ which contradicts $v : \mathcal{W} \to \{Y\}$ being left-reduced. Hence $v : \mathcal{W} \to \{Z\}$ is both right and left-reduced, and therefore reduced. Implication of $v : \mathcal{X} \to \{Y\}$ from $v : \mathcal{W} \to \{Z\}$ is implicit in the proof of the two statements above. A similar argument can be presented, supposing $v : \mathcal{X} \to \{Y\}$ is not right-maximal. □

In summary of our discussion thus far, the *canonical XFD cover* for target v and T-compatible data tree T' is defined by

$$\mathfrak{C}_v(T') = \{v : \mathcal{X} \to \{Y\} \mid v : \mathcal{X} \to \{Y\} \text{ is not trivial and}$$
$$v : \mathcal{X} \to \{Y\} \text{ is reduced with respect to } T'\}$$

Theorem 4.1. *If $\models_{T'} v : \mathcal{X} \to \{Y\}$ then $v : \mathcal{X} \to \{Y\}$ is trivial or implied by some XFD which is in the canonical XFD cover $\mathfrak{C}_v(T')$.*

Proof. The statement follows from $v : \mathcal{X} \to \{Y\}$ being trivial, therefore assume it is not. Furthermore, if $v : \mathcal{X} \to \{Y\}$ is also reduced then it belongs to $\mathfrak{C}_v(T')$ and the statement again follows. So suppose $v : \mathcal{X} \to \{Y\}$ is not reduced and not trivial. Lemma 4.1 states that $v : \mathcal{X} \to \{Y\}$ is implied by some reduced

XFD $v : W \rightarrow \{Z\}$ where $W \leq X$ and Y is contained in Z. It must be that $v : W \rightarrow \{Z\}$ is trivial, otherwise it would belong to $\mathfrak{C}_v(T')$ and the statement follows. This means the path from the root to v is simple or, Z is contained in some $W \in \mathcal{W}$. The first case makes $v : X \rightarrow \{Y\}$ trivial as well, a contradiction. In the second case, Y is contained in W. Since $W \leq X$, there is some $X \in \mathcal{X}$ which contains W and consequently contains Y. Again we contradict $v : X \rightarrow \{Y\}$ not being trivial. All cases have been considered, thus the proof is complete.   □

The previous theorem translates the XFD discovery problem for some fixed target v to the problem of finding the canonical XFD cover $\mathfrak{C}_v(T')$. In the following, we introduce v-difference sets and transversals and discuss how the problem of finding the canonical XFD cover relates to the problem of finding transversals. There are two obvious cases where all syntactically valid XFDs are trivial: (1) if the target is simple and, (2) if there is only one walk containing the target in the schema tree. Case (2) is exists because we consider LHSs to be non-empty. In these cases the following discussion is not needed.

For two pre-images $p_1, p_2 \in V_{T'}(v)$, the v-*difference set of* p_1 *and* p_2, written $\mathfrak{D}_v(p_1, p_2)$, consists of all essential v-subgraphs X where the projections $p_1|_X, p_2|_X$ are not value-equal. We use $\mathfrak{D}_v(T') = \{\mathfrak{D}_v(p_1, p_2) \mid p_1, p_2 \in V_{T'}(v)$ and $p_1 \neq p_2$ and $\mathfrak{D}_v(p_1, p_2) \neq \emptyset\}$ to denote the family of all v-*difference sets of* T'. For a given essential v-subgraph Y, the family of v-*difference sets of* T' *modulo* Y is defined by $\mathfrak{D}_v^Y(T') = \{D \in \mathfrak{D}_v(T') \mid Y \in D\}$. In other words, $\mathfrak{D}_v^Y(T')$ contains all v-difference sets for pairs of pre-images $p_1, p_2 \in V_{T'}(v)$ whose projections $p_1|_Y, p_2|_Y$ are not value-equal. Note that v-difference sets are upward closed with respect to essential v-subgraphs. Thus, for simplicity, we will write every v-difference set as its subset of minimal essential v-subgraphs.

For a family $\mathfrak{D}_v$ of v-difference sets, a *transversal* of $\mathfrak{D}_v$ is a non-empty set X of essential v-subgraphs such that for every $D \in \mathfrak{D}_v$ we have $X \cap D \neq \emptyset$. Moreover, a *minimal transversal* is a transversal $\mathcal{T}$ such that no $\mathcal{T}' < \mathcal{T}$ is a transversal of $\mathfrak{D}_v$. For $\mathfrak{D}_v = \emptyset$, the minimal transversals of $\mathfrak{D}_v$ are just the sets consisting of individual v-walks of T.

Example 4.1. Continuing with our purchase example, consider the target $v_{Purchase}$. There are four pre-images of $v_{Purchase}$ in the purchase data tree of Figure 1: these are total $v_{Purchase}$-subgraphs rooted at vertices [2], [5], [9] and [12], respectively. The family of $v_{Purchase}$-difference sets is as follows:

$$\mathfrak{D}_{v_{Purchase}}([2], [5]) = \{\mathtt{Pn}, \mathtt{Di}, \{\mathtt{De}, \mathtt{Pr}\}, \mathtt{Sa}\}$$
$$\mathfrak{D}_{v_{Purchase}}([2], [9]) = \{\mathtt{Ti}, \mathtt{Pn}, \mathtt{De}, \mathtt{Pr}\}$$
$$\mathfrak{D}_{v_{Purchase}}([2], [12]) = \{\mathtt{Ti}, \mathtt{Pn}, \mathtt{Pr}, \mathtt{Di}, \mathtt{Sa}\} = \mathfrak{D}_{v_{Purchase}}([5], [12])$$
$$\mathfrak{D}_{v_{Purchase}}([5], [9]) = \{\mathtt{Ti}, \mathtt{Pn}, \mathtt{De}, \mathtt{Pr}, \mathtt{Di}, \mathtt{Sa}\}$$
$$\mathfrak{D}_{v_{Purchase}}([9], [12]) = \{\mathtt{Pn}, \mathtt{De}, \mathtt{Di}, \mathtt{Sa}\}$$

The family $\mathfrak{D}_{v_{Purchase}}^{\{\mathtt{De}, \mathtt{Pr}, \mathtt{Di}\}}(T')$ of $v_{Purchase}$-difference sets modulo subgraph $\{\mathtt{De}, \mathtt{Pr}, \mathtt{Di}\}$ includes all of the $v_{Purchase}$-difference sets above. There are eight minimal transversals for $\mathfrak{D}^{\{De,Pr,Di\}}_{v_{Purchase}}(T')$: $\{Pn\}$, $\{\{De, Pr\}\}$, $\{De, Di\}$, $\{De, Sa\}$, $\{Di, Ti\}$, $\{Di, Pr\}$, $\{Sa, Ti\}$ and $\{Sa, Pr\}$.

Lemma 4.2. $\mathcal{X}$ *is a transversal of* $\mathfrak{D}^Y_v(T')$ *if and only if* $\models_{T'} v : \mathcal{X} \to \{Y\}$.

Proof. The proof is trivial. $\square$

Corollary 4.1. *An XFD* $v : \mathcal{X} \to \{Y\}$ *is left-reduced if and only if* $\mathcal{X}$ *is a minimal transversal of* $\mathfrak{D}^Y_v(T')$.

Thus we can find all left-reduced XFDs by considering the v-difference sets of T' modulo each essential v-subgraph. If a resulting left-reduced XFD is in fact non-trivial and right-maximal, then it belongs to the canonical XFD cover. Clearly not all essential v-subgraphs may occur as RHS for some XFD belonging to $\mathfrak{C}_v(T')$. Observe that if $\mathfrak{D}^Y_v(T') = \mathfrak{D}^Z_v(T')$ with Y containing Z, then all satisfied XFDs with Z for its RHS are not right-maximal. There is a simple characterisation for this scenario which we exploit for skipping essential v-subgraphs which are definitely not RHS for any member of $\mathfrak{C}_v(T')$.

Lemma 4.3. *Let* $\mathcal{X}$ *be a minimal transversal of* $\mathfrak{D}^Y_v(T')$ *with* $\mathcal{X} \leq \{Y\}$ *and, let* X *be a* v-subgraph containing $\bigcup \mathcal{X}$ *which is itself contained in* Y. *We have* $\mathfrak{D}^X_v(T') = \mathfrak{D}^Y_v(T')$.

Proof. ($\subseteq$) Consider some arbitrary $\mathfrak{D}_v(p_1, p_2) \in \mathfrak{D}^X_v(T')$. That is, $\mathfrak{D}_v(p_1, p_2)$ includes X and $p_1|_X, p_2|_X$ are not value-equal. It follows from X being contained in Y that $p_1|_Y, p_2|_Y$ are not value-equal. This means $Y \in \mathfrak{D}_v(p_1, p_2)$ and so $\mathfrak{D}_v(p_1, p_2) \in \mathfrak{D}^Y_v(T')$.

($\supseteq$) Suppose $\mathfrak{D}_v(p_1, p_2) \in \mathfrak{D}^Y_v(T')$. Since $\mathcal{X}$ is a minimal transversal of $\mathfrak{D}^Y_v(T')$, $\mathcal{X}$ contains some v-subgraph W which belongs to $\mathfrak{D}_v(p_1, p_2)$ and $p_1|_W, p_2|_W$ are not value-equal. Further, we know that X contains W since X contains $\bigcup \mathcal{X}$. This implies $p_1|_X, p_2|_X$ are not value-equal. Hence X is a subgraph in $\mathfrak{D}_v(p_1, p_2)$ and $\mathfrak{D}_v(p_1, p_2) \in \mathfrak{D}^X_v(T')$. $\square$

We call an essential v-subgraph Z a *candidate RHS* if we cannot say for certain that no XFD belonging to the canonical XFD cover will have Z as the RHS. We propose to generate the set of all candidate RHSs recursively. We first provide some intuitions and give a formal definition later on.

To start with, our notion of right-maximality results in maximal essential v-subgraphs (i.e. the v-units) being candidate RHSs. Let $\mathcal{T}r(Y)$ be the set of all minimal transversals of $\mathfrak{D}^Y_v(T')$, and $\mathtt{rel}\mathcal{T}r(Y) = \{\mathcal{X} \mid \mathcal{X} \in \mathcal{T}r(Y)$ with $\mathcal{X} \leq \{Y\}\}$. By applying Lemma 4.3 we infer that a v-subgraph Z which is properly contained in Y is a candidate RHS if it does not contain $\bigcup \mathcal{X}$ for any $\mathcal{X} \in \mathtt{rel}\mathcal{T}r(Y)$ and, is maximal and non-empty with this property. Specifically, we say that Z is a *candidate RHS generated from* Y. We obtain such Z by removing from Y a subgraph W which contains exactly one v-walk from each $\mathcal{X} \in \mathtt{rel}\mathcal{T}r(Y)$. We find W using hypergraph theory.

The following are some notations from hypergraph theory [6] (pg. 239): "A *hypergraph* $\mathcal{H} = (\mathcal{V}, \mathcal{E})$ is a finite collection $\mathcal{E}$ of sets over a finite set $\mathcal{V}$. The elements of $\mathcal{V}$ are called *nodes* while the elements of $\mathcal{E}$ are called *hyperedges*. A *transversal* (or *hitting set*) of $\mathcal{H}$ is a set $\mathcal{T} \subseteq \mathcal{V}$ that has a non-empty intersection with every hyperedge of $\mathcal{H}$. A transversal $\mathcal{T}$ is *minimal* if no proper subset of $\mathcal{T}$ is a hitting set of $\mathcal{H}$. The collection of all minimal transversals of $\mathcal{H}$, denoted by $Tr(\mathcal{H})$, is called the *transversal hypergraph* of $\mathcal{H}$." To avoid confusion with our previous notion of transversal we will use the term "hitting set" to refer to transversals of hypergraphs.

Consider the hypergraph $\mathcal{H}_{relTr(Y)}$ whose nodes are v-walks of Y and hyperedges are $\bigcup \mathcal{X}$ with $\mathcal{X} \in relTr(Y)$. Then the subgraph W which we want to remove from a candidate RHS Y to generate a candidate RHS Z is a minimal hitting set of $\mathcal{H}_{relTr(Y)}$.

Example 4.2. The $v_{Purchase}$-unit $\{De, Pr, Di\}$ is a candidate RHS. Recall the eight minimal transversals of $\mathfrak{D}_{v_{Purchase}}^{\{De,Pr,Di\}}(T')$ from Example 4.1. Three of them are subsumed by $\{\{De, Pr, Di\}\}$: $\{\{De, Pr\}\}$, $\{\{De, Di\}\}$ and $\{\{Di, Pr\}\}$. So, we find the minimal hitting sets of the family consisting of $\{De, Pr\}$, $\{De, Di\}$ and $\{Di, Pr\}$, which are actually these three sets. Hence three candidate RHSs are generated from $\{De, Pr, Di\}$: $\{De, Pr, Di\} - \{De, Pr\} = Di$, $\{De, Pr, Di\} - \{De, Di\} = Pr$ and $\{De, Pr, Di\} - \{Di, Pr\} = De$.

The set of all candidate RHS for v, denoted by $\mathtt{candRHS}(v)$, is defined inductively as follows: every v-unit belongs to $\mathtt{candRHS}(v)$ and, for each $Y \in \mathtt{candRHS}(v)$ every candidate RHS generated from Y belongs to $\mathtt{candRHS}(v)$. The following lemma supports that this choice of $\mathtt{candRHS}(v)$ is adequate.

Lemma 4.4. *If $Y \notin \mathtt{candRHS}(v)$ then no XFD with RHS $\{Y\}$ is right-maximal.*

Proof. Since $Y \notin \mathtt{candRHS}(v)$ then Y is not a v-unit nor is it generated by any v-essential v-subgraph containing Y. Therefore we must be able to find v-subgraphs Y' and Y'' such that: $Y' \subset Y \subset Y''$ with $Y'' \in \mathtt{candRHS}(v)$ and, Y' is empty or a candidate RHS generated from Y''. Particularly, Y includes at least one more v-walk from Y'' than Y'. From how it is generated, Y' does not contain one walk from each minimal transversal belonging to $relTr(Y'')$. It follows that $\bigcup \mathcal{X}$ is contained in Y for some minimal transversal $\mathcal{X}$ of $\mathfrak{D}_v^Y(T')$. By Lemma 4.3, $\mathfrak{D}_v^Y(T') = \mathfrak{D}_v^{Y''}(T')$ and particularly, the set of minimal transversals for these two difference sets will be the same. Therefore no XFD of the form $v : \mathcal{X} \to \{Y\}$ is right-maximal due to the satisfiability of $v : \mathcal{X} \to \{Y''\}$. $\square$

The previous lemma imposes that a right-maximal XFD must have a RHS from $\mathtt{candRHS}(v)$. The converse however may not be true because two different families of v-difference sets may have some of the same minimal transversals. We have also not eliminated all trivial XFDs. Specifically the target v should not be simple and we should further exclude trivial XFDs of the form $v : \{Y\} \to \{Y\}$.

Theorem 4.2. *Let v not be simple and $LHS(Y) = Tr(Y) - \{\{Y\}\} - \bigcup Tr(Z)$ for all $Z \in \mathtt{candRHS}(v)$ which properly contain Y. We have $v : \mathcal{X} \to \{Y\} \in \mathfrak{C}_v(T')$ if and only if $Y \in \mathtt{candRHS}(v)$ and $\mathcal{X} \in LHS(Y)$.*

Proof. (If) Assume $Y \in \texttt{candRHS}(v)$ and $\mathcal{X} \in \texttt{LHS}(Y)$. By Corollary 4.1, $v : \mathcal{X} \to \{Y\}$ is left-reduced. Since v is also not simple and no minimal transversal of $\mathfrak{D}_v^Y(T')$ includes a v-subgraph containing Y, $v : \mathcal{X} \to \{Y\}$ is not trivial. Suppose $v : \mathcal{X} \to \{Y\}$ is not right-maximal, then it is implied by some reduced $v : \mathcal{X} \to \{Z\}$ where Y is a properly contained in the essential v-subgraph Z (see proof of Lemma 4.1). It follows that $\mathcal{X}$ is a transversal of $\mathfrak{D}_v^Z(T')$ (see Lemma 4.2). In fact $\mathcal{X}$ is a minimal transversal since $v : \mathcal{X} \to \{Z\}$ is left-reduced. In other words $\mathcal{X} \in \mathcal{T}r(Z)$. If $Z \notin \texttt{candRHS}(v)$ then we contradict $v : \mathcal{X} \to \{Z\}$ being right-maximal (Lemma 4.4). Therefore $Z \in \texttt{candRHS}(v)$ which contradicts $\mathcal{X} \in \texttt{LHS}(Y)$. Therefore $v : \mathcal{X} \to \{Y\}$ is right-maximal and, more specifically reduced. By definition $v : \mathcal{X} \to \{Y\} \in \mathfrak{C}_v(T')$.

(Only If) Assume $v : \mathcal{X} \to \{Y\} \in \mathfrak{C}_{T'}(v)$. Then the XFD is reduced. $Y \in \texttt{candRHS}(v)$, otherwise $v : \mathcal{X} \to \{Y\}$ is not right-maximal according to Lemma 4.4. Since it is left-reduced and not trivial $\mathcal{X} \in \mathcal{T}r(Y) - \{Y\}$.

If $\mathcal{X} \in \mathcal{T}r(Z)$ for some $Z \in \texttt{candRHS}(v)$ that properly contains Y then $\models_{T'} v : \mathcal{X} \to Z$ (see Lemma 4.2). But this would mean $v : \mathcal{X} \to \{Y\}$ is not right-maximal, contradicting our assumption. Therefore, $\mathcal{X} \in \texttt{LHS}(Y)$. $\qquad\square$

Summary of XFD Discovery Approach Using Transversals

Algorithm 4.1 summarises the approach for computing the canonical XFD cover $\mathfrak{C}_v(T')$. The set $\texttt{candRHS}$ of candidate RHSs is generated recursively and $\texttt{candRHS}$ will always be augmented by v-subgraphs which are strictly smaller than the one currently being considered. Step 7 applies the result of Theorem 4.2 for finding new XFDs to add to $\mathfrak{C}_v(T')$. Particularly, we need only re-examine considered RHS for extracting XFDs because the largest v-subgraph amongst the elements of $\texttt{candRHS}$ was chosen each time. In step 8 of the algorithm, we identify any candidate RHS generated by Y following the approach illustrated in Example 4.2. Actually, this step can be skipped whenever $\mathfrak{D}_v^Y(T') = \emptyset$ or Y consists of a single v-walk. Recall that in the former case each v-walk is a minimal transversal.

Algorithm 4.1 computeCanonicalCover

1: $\mathfrak{C}_{T'}(v) = \emptyset$
2: $candRHS = $ family of v-units
3: **while** $candRHS \neq \emptyset$ **do**
4:   $Y = $ largest v-subgraph from $candRHS$ not yet considered
5:   Find family $\mathfrak{D}_v^Y(T')$ of difference sets of T' modulo Y
6:   Compute family $\mathcal{T}r(Y)$ of all minimal transversals of $\mathfrak{D}_v^Y(T')$
7:   Update $\mathfrak{C}_{T'}(v)$ with any XFD extracted from $\mathcal{T}r(Y)$
8:   Update $candRHS$ with any potential RHSs generated by Y
9: **end while**

Example 4.3. Let us compute the canonical XFD cover for target $v_{Purchase}$. Recall we have five $v_{Purchase}$-units: $\{\texttt{Da}\}$, $\{\texttt{Ti}\}$, $\{\texttt{Pn}\}$, $\{\texttt{De},\texttt{Pr},\texttt{Di}\}$ and $\{\texttt{Sa}\}$. For the largest $v_{Purchase}$-unit, $\{\texttt{De}, \texttt{Pr}, \texttt{Di}\}$ there are eight minimal transversals (see Example 4.1). Since no essential v-subgraph can properly contain a $v_{Purchase}$-unit, and $\{\texttt{De}, \texttt{Pr}, \texttt{Di}\}$ is not one of the minimal transversals, we extract eight XFDs:

$$v_{Purchase} : \{\texttt{Pn}\} \quad \rightarrow \{\{\texttt{De}, \texttt{Pr}, \texttt{Di}\}\} \qquad v_{Purchase} : \{\texttt{Di}, \texttt{Ti}\} \rightarrow \{\{\texttt{De}, \texttt{Pr}, \texttt{Di}\}\}$$
$$v_{Purchase} : \{\{\texttt{De}, \texttt{Pr}\}\} \rightarrow \{\{\texttt{De}, \texttt{Pr}, \texttt{Di}\}\} \qquad v_{Purchase} : \{\texttt{Di}, \texttt{Pr}\} \rightarrow \{\{\texttt{De}, \texttt{Pr}, \texttt{Di}\}\}$$
$$v_{Purchase} : \{\texttt{De}, \texttt{Di}\} \quad \rightarrow \{\{\texttt{De}, \texttt{Pr}, \texttt{Di}\}\} \qquad v_{Purchase} : \{\texttt{Sa}, \texttt{Ti}\} \rightarrow \{\{\texttt{De}, \texttt{Pr}, \texttt{Di}\}\}$$
$$v_{Purchase} : \{\texttt{De}, \texttt{Sa}\} \quad \rightarrow \{\{\texttt{De}, \texttt{Pr}, \texttt{Di}\}\} \qquad v_{Purchase} : \{\texttt{Sa}, \texttt{Pr}\} \rightarrow \{\{\texttt{De}, \texttt{Pr}, \texttt{Di}\}\}$$

Three candidate RHSs are generated from $\{\texttt{De}, \texttt{Pr}, \texttt{Di}\}$: $\texttt{Di}, \texttt{Pr}, \texttt{De}$ (see Example 4.2). Since the candidate RHS still to be examined are all walks, we can consider them in any order and no further candidate RHS can be generated . The $v_{Purchase}$-difference sets modulo $\texttt{Di}$ consists of: $\{\texttt{Pn}, \texttt{Di}, \{\texttt{De}, \texttt{Pr}\}, \texttt{Sa}\}$, $\{\texttt{Ti}, \texttt{Pn}, \texttt{Pr}, \texttt{Di}, \texttt{Sa}\}$, $\{\texttt{Ti}, \texttt{Pn}, \texttt{De}, \texttt{Pr}, \texttt{Di}, \texttt{Sa}\}$ and $\{\texttt{Pn}, \texttt{De}, \texttt{Di}, \texttt{Sa}\}$. The four minimal transversals for this family of sets are: $\{\texttt{Pn}\}$, $\{\{\texttt{De}, \texttt{Pr}\}\}$, $\{\texttt{Sa}\}$, and $\{\texttt{Di}\}$. $\{\texttt{Di}\}$ yields a trivial XFD while $\{\texttt{Pn}\}$ and $\{\{\texttt{De}, \texttt{Pr}\}\}$ are minimal transversals of the $v_{Purchase}$-difference sets modulo $\{\texttt{De}, \texttt{Pr}, \texttt{Di}\}$. Thus, we extract one new XFD $v_{Purchase} : \{\texttt{Sa}\} \rightarrow \{\texttt{Di}\}$. A summary of the rest of the computation is shown in Table 1.

Table 1. Partial computation of the canonical XFD cover for target $v_{Purchase}$

pot. RHS	difference sets modulo RHS	minimal transversals	extracted XFDs
Pr	$\{\texttt{Ti}, \texttt{Pn}, \texttt{De}, \texttt{Pr}\}$, $\{\texttt{Ti}, \texttt{Pn}, \texttt{De}, \texttt{Pr}, \texttt{Di}, \texttt{Sa}\}$, and $\{\texttt{Ti}, \texttt{Pn}, \texttt{Pr}, \texttt{Di}, \texttt{Sa}\}$	$\{\texttt{Ti}\}, \{\texttt{Pn}\}$, $\{\texttt{De}, \texttt{Di}\}, \{\texttt{De}, \texttt{Sa}\}$ and $\{\texttt{Pr}\}$	$v_{Purchase} : \{\texttt{Ti}\} \rightarrow \{\{\texttt{De}, \texttt{Pr}, \texttt{Di}\}\}$
De	$\{\texttt{Ti}, \texttt{Pn}, \texttt{De}, \texttt{Pr}\}$, $\{\texttt{Ti}, \texttt{Pn}, \texttt{De}, \texttt{Pr}, \texttt{Di}, \texttt{Sa}\}$ and $\{\texttt{Pn}, \texttt{De}, \texttt{Di}\}$	$\{\texttt{Pn}\}, \{\texttt{De}\}$, $\{\texttt{Di}, \texttt{Ti}\}$ and $\{\texttt{Di}, \texttt{Pr}\}$	
Da	$\emptyset$	$\{\texttt{Da}\}, \{\texttt{Ti}\}, \{\texttt{Pn}\}$ $\{\texttt{De}\}, \{\texttt{Pr}\}, \{\texttt{Di}\}$ and $\{\texttt{Sa}\}$	$v_{Purchase} : \{\texttt{Ti}\} \rightarrow \{\texttt{Da}\}$ $v_{Purchase} : \{\texttt{Pn}\} \rightarrow \{\texttt{Da}\}$ $v_{Purchase} : \{\texttt{De}\} \rightarrow \{\texttt{Da}\}$ $v_{Purchase} : \{\texttt{Pr}\} \rightarrow \{\texttt{Da}\}$ $v_{Purchase} : \{\texttt{Di}\} \rightarrow \{\texttt{Da}\}$ $v_{Purchase} : \{\texttt{Sa}\} \rightarrow \{\texttt{Da}\}$
Ti	$\{\texttt{Ti}, \texttt{Pn}, \texttt{De}, \texttt{Pr}\}$, $\{\texttt{Ti}, \texttt{Pn}, \texttt{Pr}, \texttt{Di}, \texttt{Sa}\}$, and $\{\texttt{Ti}, \texttt{Pn}, \texttt{De}, \texttt{Pr}, \texttt{Di}, \texttt{Sa}\}$	$\{\texttt{Ti}\}, \{\texttt{Pn}\}, \{\texttt{Pr}\}$, $\{\texttt{De}, \texttt{Di}\}$, and $\{\texttt{De}, \texttt{Sa}\}$	$v_{Purchase} : \{\texttt{Pn}\} \rightarrow \{\texttt{Ti}\}$ $v_{Purchase} : \{\texttt{Pr}\} \rightarrow \{\texttt{Ti}\}$ $v_{Purchase} : \{\texttt{De}, \texttt{Di}\} \rightarrow \{\texttt{Ti}\}$ $v_{Purchase} : \{\texttt{De}, \texttt{Sa}\} \rightarrow \{\texttt{Ti}\}$
Pn	same as for RHS $\{\texttt{De}, \texttt{Pr}, \texttt{Di}\}$	same as for RHS $\{\texttt{De}, \texttt{Pr}, \texttt{Di}\}$	$v_{Purchase} : \{\{\texttt{De}, \texttt{Pr}\}\} \rightarrow \{\texttt{Pn}\}$ $v_{Purchase} : \{\texttt{De}, \texttt{Di}\} \rightarrow \{\texttt{Pn}\}$ $v_{Purchase} : \{\texttt{De}, \texttt{Sa}\} \rightarrow \{\texttt{Pn}\}$ $v_{Purchase} : \{\texttt{Di}, \texttt{Ti}\} \rightarrow \{\texttt{Pn}\}$ $v_{Purchase} : \{\texttt{Di}, \texttt{Pr}\} \rightarrow \{\texttt{Pn}\}$ $v_{Purchase} : \{\texttt{Sa}, \texttt{Ti}\} \rightarrow \{\texttt{Pn}\}$ $v_{Purchase} : \{\texttt{Sa}, \texttt{Pr}\} \rightarrow \{\texttt{Pn}\}$
Sa	$\{\texttt{Pn}, \texttt{Di}, \{\texttt{De}, \texttt{Pr}\}, \texttt{Sa}\}$, $\{\texttt{Ti}, \texttt{Pn}, \texttt{Pr}, \texttt{Di}, \texttt{Sa}\}$, and $\{\texttt{Ti}, \texttt{Pn}, \texttt{De}, \texttt{Pr}, \texttt{Di}, \texttt{Sa}\}$	$\{\texttt{Pn}\}, \{\texttt{Di}\}$, $\{\{\texttt{De}, \texttt{Pr}\}\}$, and $\{\texttt{Sa}\}$	$v_{Purchase} : \{\texttt{Pn}\} \rightarrow \{\texttt{Sa}\}$ $v_{Purchase} : \{\texttt{Di}\} \rightarrow \{\texttt{Sa}\}$ $v_{Purchase} : \{\{\texttt{De}, \texttt{Pr}\}\} \rightarrow \{\texttt{Sa}\}$

5   Finding Difference Sets of T'

For FDs in the relational data model, this step involves computing a natural dual of difference sets in the form of agree sets. For XFDs, we can find a similar dual to v-difference sets in the form of v-agree sets.

For two distinct pre-images $p_1, p_2 \in V_{T'}(v)$, the v-agree set $ag_v(p_1, p_2)$ consists of all essential v-subgraphs X where the projections $p_1|_X, p_2|_X$ are value-equal. We use $ag_v(T') = \{ag_v(p_1, p_2) \mid p_1, p_2 \in V_{T'}(v)$ and $p_1 \neq p_2\}$ to denote the family of v-agree sets.

5.1   Finding Difference Sets from Agree Sets

For two pre-images $p_1, p_2 \in V_{T'}(v)$, an essential v-subgraph must belong to either $\mathfrak{D}_v(p_1, p_2)$ or $ag_v(p_1, p_2)$. Therefore $\mathfrak{D}_v(p_1, p_2)$ is just the set difference between the set of all essential v-subgraphs and $ag_v(p_1, p_2)$. For computational reasons one might rather use an alternative conversion of $ag_v(p_1, p_2)$ into $\mathfrak{D}_v(p_1, p_2)$ that involves transversals.

Lemma 5.1. *Let $\mathfrak{U}$ be the set of all v-units of T.*

$$\mathfrak{D}_v(p_1, p_2) = \bigcup_{U \in \mathfrak{U}} \mathfrak{D}_v(p_1, p_2)|_U$$

where $\mathfrak{D}_v(p_1, p_2)|_U$ is the set of all essential v-subgraphs X such that projections $p_1|_X, p_2|_X$ are not value-equal and X is a v-subgraph of U.

Proof. ($\supseteq$) Suppose $X \in \bigcup_{U \in \mathfrak{U}} \mathfrak{D}_v(p_1, p_2)|_U$. Then for some $U \in \mathfrak{U}$ we have $X \in \mathfrak{D}_v(p_1, p_2)|_U$. This means X is an essential v-subgraph and projections $p_1|_X, p_2|_X$ are not value-equal, which results in $X \in \mathfrak{D}_v(p_1, p_2)$.

($\subseteq$) Assume there is some $X \in \mathfrak{D}_v(p_1, p_2)$. Therefore X is an essential v-subgraph and projections $p_1|_X, p_2|_X$ are not value-equal. Because X is essential, it is contained in some v-unit U. Therefore $X \in \mathfrak{D}_v(p_1, p_2)|_U$ and consequently $X \in \bigcup_{U \in \mathfrak{U}} \mathfrak{D}_v(p_1, p_2)|_U$.   □

The previous lemma states that for finding $\mathfrak{D}_v(p_1, p_2)$ it is sufficient to consider one v-unit at a time. It follows that an essential v-subgraph X belongs to $\mathfrak{D}_v(p_1, p_2)|_U$ if and only if it is contained in some v-unit U but not a member of $ag_v(p_1, p_2)$. We encounter three possible scenarios:

1) U is a member of $ag_v(p_1, p_2)$,
2) No member of $ag_v(p_1, p_2)$ is contained in U,
3) X is not contained in any maximal member $W \in ag_v(p_1, p_2)$ which is a v-subgraph of U.

In the first scenario, $\mathfrak{D}_v(p_1, p_2)|_U$ is empty. In the second scenario, every non-empty v-subgraph of U belongs to $\mathfrak{D}_v(p_1, p_2)|_U$. Otherwise X contains some v-walk in $U - W$ for every maximal member $W \in ag_v(p_1, p_2)$ which is a contained in U. It is possible to show that the (minimal) members of $\mathfrak{D}_v(p_1, p_2)|_U$ for the second and third scenarios are (minimal) hitting sets of hypergraph $\mathcal{H}_{\mathfrak{D}_v(p_1,p_2)|_U}$ whose nodes are the v-walks contained in U and whose hyperedge(s) is/are:

$$\begin{cases} U, & scenario2 \\ U - W \neq \emptyset \text{ for each maximal } W \in ag_v(p_1,p_2)|_U, & scenario3 \end{cases}$$

where $ag_v(p_1,p_2)|_U$ is the set of all member of $ag_v(p_1,p_2)$ which are contained in U.

Example 5.1. Consider the $v_{Purchase}$-agree set $ag_{v_{Purchase}}([2],[5])=\{\mathsf{Da},\mathsf{Ti},\mathsf{De},\mathsf{Pr}\}$. Note that all $v_{Purchase}$-subgraphs in this agree set are maximal. The $v_{Purchase}$-units Da and Ti are elements of the agree set. Therefore $\mathfrak{D}_{v_{Purchase}}([2],[5])|_{\mathsf{Da}} = \emptyset = \mathfrak{D}_{v_{Purchase}}([2],[5])|_{\mathsf{Ti}}$. No member of the agree set is a subgraph of the units Pn or Sa, and so $\mathfrak{D}_{v_{Purchase}}([2],[5])|_{\mathsf{Pn}} = \{\mathsf{Pn}\}$ and $\mathfrak{D}_{v_{Purchase}}([2],[5])|_{\mathsf{Sa}} = \{\mathsf{Sa}\}$. Lastly, for unit $\{\mathsf{De},\mathsf{Pr},\mathsf{Di}\}$ we have hyperedges $\{\mathsf{Di},\mathsf{Pr}\}$ and $\{\mathsf{De},\mathsf{Di}\}$. The minimal hitting sets for this set of hyperedges are: $\{\mathsf{Di}\}$ and $\{\mathsf{De},\mathsf{Pr}\}$. Consequently $\mathfrak{D}_{v_{Purchase}}([2],[5])|_{\{\mathsf{De},\mathsf{Pr},\mathsf{Di}\}} = \{\mathsf{Di},\{\mathsf{De},\mathsf{Pr}\}\}$ Altogether, we obtain $\mathfrak{D}_{v_{Purchase}}([2],[5]) = \{\mathsf{Pn},\mathsf{Sa},\mathsf{Di},\{\mathsf{De},\mathsf{Pr}\}\}$.

5.2 Finding Agree Sets

A brute force approach to computing v-agree sets is to systematically compare every pair of distinct pre-images of v with respect to each essential v-subgraph. Comparisons for value-equality are not trivial because we need to consider isomorphism of tree-structures. More recent investigations make use of partitions to represent value-equality and for computing agree sets. Our approach to computing v-agree sets will also use a notion of partitions for representing value-equality. In particular we try to limit the use of the given data for computing the partitions. Instead only partitions induced by v-walks utilise given data while partitions induced by further sets of v-subgraph are computed by composition of partitions. We define partitions and based on this present a recursive approach for computing v-agree sets.

Partitions Terminology. Given a set M, a *partition* π on M is a family of mutually disjoint, non-empty subsets of M. The members of π are the *partition classes* and their union is the *support* of π.

We call M the *context* of π, and the *exterior* of π is the difference between its context and its support. We call a partition *total* if its exterior is empty. We obtain the *stripped partition* $\widehat{\pi}$ of π by considering only the non-singleton partition classes of π.

A set $\mathcal{X}$ of v-subgraphs induces a total partition $\Pi_{\mathcal{X}}(v)$ on $V_{T'}(v)$ where two pre-images $p_1, p_2 \in V_{T'}(v)$ belong to the same partition class if and only if for all $X \in \mathcal{X}$ their projections $p_1|_X, p_2|_X$ are value-equal. The *null class* $\perp_{\mathcal{X}}(v)$ is the set of all pre-images $p \in V_{T'}(v)$ whose projection $p|_X$ is empty for all $X \in \mathcal{X}$. Clearly, $\perp_{\mathcal{X}}(v)$ is a partition class of $\Pi_{\mathcal{X}}(v)$ if it is non-empty. Removing $\perp_{\mathcal{X}}(v)$ from the total partition $\Pi_{\mathcal{X}}(v)$ yields the *non-null partition* $\mathfrak{N}_{\mathcal{X}}(v)$.

Example 5.2. $\Pi_{\{\mathtt{Di}\}}(v_{Item}) = \{\{[3], [4], [10], [11], [13]\}, \{[6], [7], [14]\}\}$.
$\perp_{\{\mathtt{Di}\}}(v_{Item}) = \{[6], [7], [14]\}$. $\mathfrak{N}_{\{\mathtt{Di}\}}(v_{Item}) = \{\{[3], [4], [10], [11], [13]\}\}$.

We distinguish between value-equal projections which are empty from those which are non-empty because both enjoy different properties. Firstly, projections $p_1|_X, p_2|_X$ are value-equal and empty if and only if projections $p_1|_B, p_2|_B$ for each v-walk $B \in X$ are value-equal and empty, for any two pre-images $p_1, p_2 \in V_{T'}(v)$. This observation is not true in general for projections which are value-equal but non-empty. We have already seen an example with pre-images [2] and [5] in Figure 1. Secondly, value-equality of non-empty projections for pre-images of v is partly dependent on value-equality of non-empty projections of pre-images of nodes w that are proper descendants of v. This is not the case for value-equal empty projections. For example, any ancestor of the pre-images [3], [4], [10], [11], [13] must also have non-empty projections to $\mathtt{Di}$. But knowing that the pre-images [6], [7], [14] of v_{Item} have empty projections to $\mathtt{Di}$ does not tell us anything about whether the pre-images of any ancestor of v_{Item} have empty projections to $\mathtt{Di}$. These properties will be exploited for computing partitions.

The *product* of two partitions π_1, π_2 defined on some set M is the greatest partition on M which refines both π_1 and π_2. In other words, the product of π_1, π_2 is the set of all non-empty intersections between some partition class in π_1 and some partition class in π_2. The product of π_1, π_2 is again a partition on M. A *residual* of π_1 modulo π_2 consists of all non-empty sets obtained by removing the support of π_2 from some partition class $g \in \pi_1$. The *composition* of π_1, π_2 consists of the product of π_1, π_2 together with the residual of each partition π_1, π_2 modulo their product.

Example 5.3. In addition to $\mathfrak{N}_{\{\mathtt{Di}\}}(v_{Item}) = \{\{[3], [4], [10], [11], [13]\}\}$ (from Example 5.2), we also have the following total partitions:

- $\Pi_{\{\mathtt{De}\}}(v_{Item}) = \mathfrak{N}_{\{\mathtt{De}\}}(v_{Item}) = \{\{[3], [6], [10], [11], [13]\}, \{[4], [7], [14]\}$
- $\Pi_{\{\mathtt{Pr}\}}(v_{Item}) = \mathfrak{N}_{\{\mathtt{Pr}\}}(v_{Item}) = \{\{[3], [7], [10], [11], [13], [14]\}, \{[4], [6]\}\}$

The composition (and product) of $\mathfrak{N}_{\{\mathtt{De}\}}(v_{Item})$ and $\mathfrak{N}_{\{\mathtt{Pr}\}}(v_{Item})$ results in the total partition $\{\{[3], [10], [11], [13]\}, \{[7], [14]\}, \{[6]\}, \{[4]\}\}$.

For $\mathfrak{N}_{\{\mathtt{De}\}}(v_{Item})$ and $\mathfrak{N}_{\{\mathtt{Di}\}}(v_{Item})$, their product results in $\{\{[3], [10], [11], [13]\}, \{[4]\}\}$. The residual of $\mathfrak{N}_{\{\mathtt{De}\}}(v_{Item})$ modulo the product is $\{\{[6]\}, \{[7], [14]\}\}$ while the residual of $\mathfrak{N}_{\{\mathtt{Di}\}}(v_{Item})$ modulo the product is empty. Altogether the composition of $\mathfrak{N}_{\{\mathtt{De}\}}(v_{Item})$ and $\mathfrak{N}_{\{\mathtt{Di}\}}(v_{Item})$ yields $\{\{[3], [10], [11], [13]\}, \{[4]\}, \{[6]\}, \{[7], [14]\}\}$

Let v be a node of T and w a descendant of v. For a pre-image w' of w in T', we use $\alpha_v^w(w')$ to denote the unique pre-image of v in T' that contains the pre-image of w'. When we consider pre-images in terms of their identifying root node, then $\alpha_v^w(w')$ denote the unique pre-image of v that is an ancestor of w'. Further, for a set $P_w \subseteq V_{T'}(w)$ of pre-images we put $\alpha_v^w(P_w) = \{\alpha_v^w(w') \mid w' \in P_w\}$.

Consider two sets $P_v \subseteq V_{T'}(v)$ and $P_w \subseteq V_{T'}(w)$ of pre-images. We define a partition $\pi^c(P_v, P_w)$ on P_v such that two members of P_v belong to the same partition class if and only if they have the same number of descendants in P_w, and call it the *card-partition* of P_v *modulo* P_w.

Remark. To compute $\pi^c(P_v, P_w)$ we can use α_v^w to translate set P_v into a bag B_v. Then partition classes of $\pi^c(P_v, P_w)$ simply collect together elements of B_v having the same multiplicity.

Example 5.4. Consider non-null partition $\mathfrak{N}_{\{Pr\}}(v_{Item})$ with two partition classes $g_1 = \{[3], [7], [10], [11], [13], [14]\}$ and $g_2 = \{[4], [6]\}$.

From the data tree in Figure 1 we find $\alpha_{v_{Purchase}}^{v_{Item}}$. This yields $\alpha_{v_{Purchase}}^{v_{Item}}(g_1) = \{\{[2], [5], [9], [12]\}$ and $\alpha_{v_{Purchase}}^{v_{Item}}(g_2) = \{[2], [5]\}$.

Replacing each member of the partition classes by its $v_{Purchase}$ ancestor we get bags: $\langle [2], [5], [9], [9], [12], [12] \rangle$ and $\langle [2], [5] \rangle$. Thus $\pi^c(\alpha_{v_{Purchase}}^{v_{Item}}(g_1), g_1) = \{\{[2], [5]\}, \{[9], [12]\}\}$ and $\pi^c(\alpha_{v_{Purchase}}^{v_{Item}}(g_2), g_2) = \{\{[2], [5]\}\}$.

Finding v-Agree Sets from Partitions. Computations include three phases:

1. Compute non-null partitions for individual essential v-subgraphs and null classes for individual v-walks,
2. Compute non-null partitions for sets of essential v-subgraphs,
3. Extract $ag_v(T')$ from the computed partitions.

In the first phase, null classes are only computed for v-walks because null classes induced by further v-subgraphs are easily computed by intersecting null classes for v-walks. Moreover, null classes are easily computed once we have the corresponding non-null partitions: we simply remove the support of the non-null partition from the set of all pre-images of v. On the other hand, computation of non-null partitions requires more attention. We employ a recursive approach as outlined in Algorithm 5.2.

Note that we only refer to the given data in one situation: where every proper descendant of v has incoming arc with frequency ? or 1. This makes it possible to adopt a relational representation of the XML data. Let there be a table R_w for every lowest proper descendant w of v with incoming arc of frequency other than ? and 1. Each table R_w includes an attribute column B for every walk B having w as the lowest node with incoming arc of frequency other than ? and 1. Every pre-image $p \in V_{T'}(w)$ can be translated into one tuple t in the R_w-relation r_w where, for every column B, $t[B]$ has the value of the leaf node in projection $p|_B$ or "null" if the projection is empty. Computation of non-null partitions from given data using these relations is analogous to the approach for the relational data model (see for example [5]).

It remains to show that the rest of the algorithm yields correct partitions. Firstly we show the correctness of using composition of card-partitions in Line 11.

Proposition 5.1. *Let X be a w-subgraph with w being a proper descendant of v whose incoming arc is the only arc on the path from v to w with frequency other than ? and 1. Then $\mathfrak{N}_X(v)$ is the composition of the card-partitions $\pi^c(\alpha_v^w(g), g)$ with $g \in \mathfrak{N}_X(w)$.*

Algorithm 5.2 computeIndividualNonNullPartitions

1. **for all** v-unit U in T **do**

2.   **if** U is a singleton such that no proper descendant of v has incoming arc with frequency other than ? and 1 **then**
3.   Compute $\mathfrak{N}_{\{U\}}(v)$ from given data

4.   **else**
5.   Let w be a proper descendant of v whose incoming arc is the only arc on the path from v to w with frequency other than ? and 1
6.   Compute non-null partitions for all essential w-subgraphs
7.   **for all** v-subgraph X which are contained in U **do**
8.   **if** X is not an essential w-subgraph **then**
9. $\mathfrak{N}_{\{X\}}(w) =$ composition of non-null partitions $\mathfrak{N}_{\{X_i\}}$ with $X_i \in \vartheta_w(\{X\})$
10.   **end if**
11. $\mathfrak{N}_{\{X\}}(v) =$ composition of card-partitions $\pi^c(\alpha_v^w(g), g)$ with $g \in \mathfrak{N}_{\{X\}}(w)$
12.   **end for**
13.   **end if**
14. **end for**

Proof Sketch. An important observation is that if w is a child of v then $\mathfrak{N}_{\{X\}}(v)$ is the composition of the card-partitions $\pi^c(\alpha_v(g), g)$ with $g \in \mathfrak{N}_{\{X\}}(w)$. As a corollary, we can then observe that if the path from v to w contain no arc of frequency other than ? and 1 then $\mathfrak{N}_{\{X\}}(v)$ consists of the sets $\alpha_v(g)$ with $g \in \mathfrak{N}_{\{X\}}(w)$.

Let u be the parent of w. In accordance with the observation above $\mathfrak{N}_{\{X\}}(u)$ is the composition of card-partitions $\pi^c(\alpha_u^w(g), g)$ with $g \in \mathfrak{N}_{\{X\}}(w)$. Then by the corollary to the observation, $\mathfrak{N}_{\{X\}}(v)$ consists of the sets $\alpha_v^u(h)$ where $h \in \mathfrak{N}_{\{X\}}(u)$. In fact $\alpha_v^w(g) = \alpha_v^u(\alpha_u^w(g))$ with $g \in \mathfrak{N}_{\{X\}}(w)$. Since α_v^u is injective and it is easy to see that we get the same result if we compose the card-partitions after applying α_v^u. $\square$

Finally we show that the non-null partitions of the set of pre-images of w induced by an essential v-subgraph X which is not an essential w-subgraph can be computed as per Line 9. The approach is possible because we can find $\vartheta_w(\{X\})$ whose members are mutually w-reconcilable.

Lemma 5.2. $\mathfrak{N}_{\mathcal{X}}(v)$ *is the composition of the non-null partitions* $\mathfrak{N}_{\{X_i\}}(v)$ *with* $X_i \in \mathcal{X}$.

Proof. By definition, $\Pi_{\mathcal{X}}(v)$ is the product of the total partitions $\Pi_X(v)$ with $X \in \mathcal{X}$, and $\perp_{\mathcal{X}}(v)$ is the intersection of the null classes $\perp_X(v)$ with $X \in \mathcal{X}$. $\square$

Proposition 5.2. *Let* $\mathcal{X}$ *consist of mutually* v-reconcilable v-subgraphs. Then $\mathfrak{N}_{\cup \mathcal{X}}(v)$ *is the composition of the non-null partitions* $\mathfrak{N}_X(v)$ *with* $X \in \mathcal{X}$.

Proof. Observe that if X, Y, Z are mutually v-reconcilable v-subgraphs then so are X and $Y \cup Z$. Hence it remains to verify that the claim holds for two-element sets $\mathcal{X}$. By Lemma 5.2, the composition of the non-null partitions $\mathfrak{N}_X(v)$ with $X \in \mathcal{X}$ results in $\mathfrak{N}_{\mathcal{X}}(v)$. It remains to show that $\mathfrak{N}_{\mathcal{X}}(v) = \mathfrak{N}_{\cup \mathcal{X}}(v)$.

As $\mathfrak{N}_{\mathcal{X}}(v)$ is obtained by removing $\perp_{\mathcal{X}}(v)$ from $\prod_{\mathcal{X}}(v)$, two pre-images $p_1, p_2 \in V_{T'}(v)$ belong to the same partition class if and only if for all $X \in \mathcal{X}$ their projections $p_1|_X, p_2|_X$ are value-equal. Then from Lemma 2.1 we infer that $p_1, p_2 \in V_{T'}(v)$ belong to the same partition class in $\prod_{\mathcal{X}}(v)$ if and only if they belong to the same partition class in $\prod_{\cup \mathcal{X}}(v)$. Observe that a pre-image $p \in V_{T'}(v)$ has empty projection $p|_X$ for all $X \in \mathcal{X}$ if and only if it has empty projection $p|_{\cup \mathcal{X}}$. This yields $\perp_{\mathcal{X}}(v) = \perp_{\cup \mathcal{X}}(v)$. Therefore, $\mathfrak{N}_{\mathcal{X}}(v) = \prod_{\mathcal{X}}(v) - \{\perp_{\mathcal{X}}(v)\} = \prod_{\cup \mathcal{X}}(v) - \{\perp_{\cup \mathcal{X}}(v)\} = \mathfrak{N}_{\cup \mathcal{X}}(v)$. □

v-agree sets are downward closed with respect to containment and so we represent each v-agree set by its set of maximal members. In the second phase, we identify the largest sets of maximal essential v-subgraphs on which at least one pair of pre-images agrees. For this, we can apply Lemma 5.2 to compute the partitions for sets of essential v-subgraphs. In particular, we consider sets which are smaller according to the subsumption ordering $\leq$ first, and only compute non-null partitions for larger sets if the partitions for their proper subsets are non-empty when stripped.

Extracting v-agree sets from the computed partitions is straightforward. We only examine non-singleton null partition classes and non-empty stripped non-null partitions. From the partition classes we extract pairs of pre-images; these pairs of pre-images agree on at least one essential v-subgraphs. The v-agree set for each pair of pre-images $p_1, p_2 \in V_{T'}(v)$ includes the largest set $\mathcal{X}$ of v-subgraphs for which p_1, p_2 belong to the same partition class of $\mathfrak{N}_{\mathcal{X}}(v)$ together with $\vartheta_v(\{Y\})$ where Y contains all v-walks B such that p_1, p_2 both belong to the null class $\perp_B(v)$.

We have presented very basic approaches for phase 2 and 3. Alternative approaches and further tunings are possible for improving computational efficiency. For example, an alternative characterisation of v-agree sets similar to [7] addresses the last two phases simultaneously. Due to space limitation this will not be discussed here.

6   Conclusion

In this paper, we have proposed an approach for discovering reduced non-trivial XFDs which have pre-image semantics. The class of XML functional dependencies under examination is interesting because it incorporates both structure components and multiset-based semantics which are inherent characteristics of XML data.

The approach is similar to hypergraph-transversal methods proposed in [7,8,9,12] for the relational data model. We have defined analogous notions of difference sets, transversals and canonical XFD covers for the XML context. The discovery of satisfied XFDs is reduced to the problem of finding transversals of families of v-difference sets. We have also considered how to compute v-difference sets. The approach presented uses dual v-agree sets which are represented by partitions. An important feature of our approach for computing partitions is the relatively few reference to given data and re-use of previous computations. With such a recursive computation of partitions, the approach can be extended to scale well for the discovery of XFDs for multiple targets. Details, however need to be addressed in future research.

References

1. Arenas, M., Libkin, L.: A normal form for XML documents. ACM Trans. Database Syst. 29(1), 195–232 (2004)
2. Grahne, G., Zhu, J.: Discovering approximate keys in XML data. In: CIKM 2002, pp. 453–460. ACM, New York (2002)
3. Hartmann, S., Link, S.: More functional dependencies for XML. In: Kalinichenko, L.A., Manthey, R., Thalheim, B., Wloka, U. (eds.) ADBIS 2003. LNCS, vol. 2798, pp. 355–369. Springer, Heidelberg (2003)
4. Hartmann, S., Link, S., Trinh, T.: Efficient reasoning about XFDs. In: Kotagiri, R., et al. (eds.) DASFAA 2007. LNCS, vol. 4443, p. 1070. Springer, Heidelberg (2007)
5. Huhtala, Y., Kärkkäinen, J., Porkka, P., Toivonen, H.: Tane: An efficient algorithm for discovering functional and approximate dependencies. Comput. J. 42(2), 100–111 (1999)
6. Kavvadias, D.J., Stavropoulos, E.C.: An efficient algorithm for the transversal hypergraph generation. J. Graph Algorithms Appl. 9(2), 239–264 (2005)
7. Lopes, S., Petit, J.-M., Lakhal, L.: Efficient discovery of functional dependencies and armstrong relations. In: Zaniolo, C., Grust, T., Scholl, M.H., Lockemann, P.C. (eds.) EDBT 2000. LNCS, vol. 1777, pp. 350–364. Springer, Heidelberg (2000)
8. Mannila, H., Räihä, K.-J.: Dependency inference. In: Stocker, P.M., Kent, W., Hammersley, P. (eds.) VLDB 1987, pp. 155–158. Morgan Kaufmann, San Francisco (1987)
9. Mannila, H., Räihä, K.-J.: Algorithms for inferring functional dependencies from relations. Data Knowl. Eng. 12(1), 83–99 (1994)
10. Vincent, M., Liu, J.: Strong functional dependencies and a redundancy free normal form for XML. In: 7th World Multi-Conference on Systemics, Cybernetics and Informatics (2003)
11. Wang, J., Topor, R.W.: Removing XML data redundancies using functional and equality-generating dependencies. In: ADC 2005, pp. 65–74 (2005)
12. Wyss, C., Giannella, C., Robertson, E.L.: Fastfds: A heuristic-driven, depth-first algorithm for mining functional dependencies from relation instances - extended abstract. In: Kambayashi, Y., Winiwarter, W., Arikawa, M. (eds.) DaWaK 2001. LNCS, vol. 2114, pp. 101–110. Springer, Heidelberg (2001)
13. Yu, C., Jagadish, H.V.: Efficient discovery of XML data redundancies. In: Dayal, U., Whang, K.-Y., Lomet, D.B., Alonso, G., Lohman, G.M., Kersten, M.L., Cha, S.K., Kim, Y.-K. (eds.) VLDB 2006, pp. 103–114. ACM, New York (2006)
14. Zhou, Z.: Algorithms and implementation of functional dependency discovery for XML. Master thesis, Massey University, Palmerston North, New Zealand, pages 81 (2006)

Visibly Pushdown Transducers for Approximate Validation of Streaming XML

Alex Thomo, S. Venkatesh, and Ying Ying Ye

University of Victoria, Victoria, Canada
{thomo,venkat,fayye}@cs.uvic.ca

Abstract. Visibly Pushdown Languages (VPLs), recognized by Visibly Pushdown Automata (VPAs), are a nicely behaved family of context-free languages. It has been shown that VPAs are equivalent to Extended Document Type Definitions (EDTDs), and thus, they provide means for elegantly solving various problems on XML. Especially, it has been shown that VPAs are the apt device for streaming XML.

One of the important problems about XML that can be addressed using VPAs is the validation problem in which we need to decide whether an XML document conforms to the specification given by an EDTD. In this paper, we are interested in solving the approximate version of this problem, which is to decide whether an XML document can be modified by a tolerable number of edit operations to yield a valid one with respect to a given EDTD.

For this, we define Visibly Pushdown Transducers (VPTs) that give us the framework for solving this problem under two different semantics for edit operations on XML. While the first semantics is a generalization of edit operations on strings, the second semantics is new and motivated by the special nature of XML documents. Usings VPTs, we give streaming algorithms that solve the problem under both the semantics. These algorithms use storage space that only depends on the size of the EDTD and the number of tolerable errors. Furthermore, they can check approximate validity of an incoming XML document in a single pass over the document, using auxilliary stack space that is proportional to the depth of the XML document.

1 Introduction

The Extensible Markup Language (XML) is the lingua franca for data and document exchange on the Web and used in a variety of applications ranging from collaborative commerce to medical databases. One of the most important problems on XML is the validation of documents against a schema specification typically given by one of the popular schema languages, Document Type Definition (DTD), XML Schema ([19]) or Relax NG ([8]). In many applications for data and document exchange, the data is streaming in large quantities and an on-line-one-pass processing and validation of XML using limited memory is required.

S. Hartmann and G. Kern-Isberner (Eds.): FoIKS 2008, LNCS 4932, pp. 219–238, 2008.

Due to its importance, the XML validation problem has received a lot of attention (cf. [18,22,4,6,5,21]). One of the most recent developments on the problem is the use of Visibly Pushdown Automata (VPAs) for validating streaming XML ([14]). In this work, it was shown that VPAs precisely capture the languages of XML documents induced by Extended Document Type Definitions (EDTDs), introduced in [17][1].

EDTDs are essentially extended context free grammars enriched with types and can model all three popular schema formalisms mentioned above: DTD, XML Schema and Relax NG (see [17,16]). After constructing a VPA for a given EDTD, the XML validation problem reduces to the one of accepting or rejecting an XML formatted word with the constructed VPA. We note here that the correspondence of VPAs to EDTDs is with respect to the word-encoded derivation trees of EDTDs rather than the set of words they generate. We remark that, when one uses an EDTD as a specification for XML documents, it is its language of derivation trees that is relevant; namely, for an XML document to be valid, when viewed as a tree, it has to correspond to a derivation tree (after applying typing) of the given EDTD.

VPAs, which as mentioned, precisely capture XML specifications given by EDTDs, are in essence pushdown automata. Their push or pop mode can be determined by looking at the input only (hence their name). VPAs recognize Visibly Pushdown Languages (VPLs), which form a well-behaved and robust family of context-free languages. VPLs enjoy useful closure properties and several important problems for them are decidable. For example, VPLs are closed under intersection and complement, and the containment problem is decidable.

In this paper, we introduce Visibly Pushdown Transducers (VPTs), which preserve the VPL family under their transductions. That is, given a VPL L and a VPT T, the transduction of L through T is again a VPL. We give constructions to obtain transductions for VPLs, as well as to perform useful operations on VPTs. Notably, VPTs give us a framework for solving the approximate validation of XML, where the approximation is in the sense that an XML document might not conform in its current form to a given EDTD but will do so after a few edit operations. Formally, in this paper, we study the K-validation problem for streaming XML:

Given an EDTD and a positive integer K, preprocess and store the EDTD succinctly so that queries of the form "Does an XML document fit the EDTD specification after at most K edit operations?" can be answered efficiently.

Here, in line with the streaming XML model presented in [18], we distinguish two phases, which we explicitly call: the *preprocessing phase* and the *querying phase*. The preprocessing phase can be done offline and it has to be such as to facilitate the next phase of querying which in turn has to be done online and in a single pass on the streaming XML.

---

[1] [17] calls this formalism *specialized DTD* as types specialize tags. Similarly with [16], we use the term *extended DTD* to convey that this formalism is more powerful than DTD.

In this paper, we study the approximate validation problem under two different semantics for edit operations on XML. The first semantics generalizes the standard edit operations on strings to XML. These edit operations are the substitution of a symbol by another, deletion of a symbol, and insertion of a string. In the context of XML, we generalize them by the substitution, deletion and insertion of pairs of matching open and close tags.

In the *preprocessing phase* of our algorithm for the first semantics, we construct VPTs for each of the three edit operations, and then superimpose them to produce a combined VPT for all the operations. Then, we transduce the given VPA through this VPT to get a new VPA. The preprocessing phase will store this VPA as its final output. The size of the final VPA is $O(KM)$ where M is the size of the given VPA (or EDTD). Thus, our algorithm for the first semantics uses storage space that only depends polynomially on the size of the EDTD and the error parameter.

In the *querying phase*, we receive as input a streaming XML document and check if this document is accepted or not by the VPA constructed in the preprocessing phase. Our querying scheme checks membership in a single pass over the XML document using time which depends only linearly on the size of the document, and auxilliary space proportional to the depth (and not size) of the document.

In the second part of the paper, we introduce another semantics for the edit operations on XML. Under this second semantics, whenever we decide to perform an edit operation with respect to an element, we apply the operation on all the occurrences of the element. To see the usefulness of this semantics, consider the following XML document

```
<collection>
   <book> Book-One </book> ... <book> Book-One-Thousand </book>
</collection>
```

in which there are 1000 *book* elements. It is clear that if we change *book* to *livre* we need to apply this all over the board for a total of 1000 times. Nevertheless, in this example, this should semantically count as one change, not as 1000. Our second semantics does exactly that; the "same fate" happens to all the occurrences of an element and this has a cost of one. On the other hand, under the first semantics, one would need 1000 edit operations to change all the *book* elements to *livre*. Depending on the application, the user can select the first or the second semantics.

Similarly to the first semantics, the preprocessing phase of our algorithm for the new semantics constructs VPTs for each of the three edit operations and glues these VPTs together through superimposition and union to construct a combined VPT for all the operations under the second semantics. We store the transduction of the given VPA through the VPT as the final output of this phase. The edit VPT for the second semantics has a size of $R^{O(K)}$, where R is the size of the underlying alphabet, and thus, the transduction of the given VPA can have a size proportional to $R^{O(K)}M$. We believe that this exponential penalty in K is an artifact of the requirement that the queries be answered using only one pass through the XML document.

As in the case of the first semantics, the querying scheme is in fact membership testing which is done in a single pass over the XML document using time proportional to its size and auxilliary stack space proportional to the depth of the XML document.

We would like to remark that the three parameters - number of tolerable errors, size of the EDTD and the depth of the XML document are typically small in practice. Hence, our algorithms are viable for large and streaming XML.

The rest of the paper is organized as follows. In Section 2, we discuss related work. Section 3 reviews VPAs. In Section 4, we introduce VPTs, their transductions and operations on them. In Section 5 and 6, we present VPTs for edit operations under the first and second semantics respectively. Finally, Section 8 concludes the paper.

2   Related Work

The XML validation problem has received a lot of attention in the last years (cf. [18,21,14,7,20,9,4,6,5]).

The first three, [18,21,14], study the exact validation of XML in a streaming context. In [14], it was argued that VPAs are the apt device for the validation of streaming XML. Also, in the same work, it was shown that VPAs precisely correspond to EDTDs. In fact, this result could also be established based on [17] and [3]. Namely, [17] shows that the tree languages specified by EDTDs coincide with the class of regular tree languages, while [3] shows that the latter coincide with the VPL class.

The next three works, [7,20,9], consider variants of approximate XML validation, but in a non-streaming setting.

[7] presents a randomized methodology for validating and repairing XML documents. The main difference from our work is that [7] considers edit distance with *moves* and the error is relative rather than absolute. This means that the bigger the document is the bigger the error is tolerated to be. We believe that there are practical cases when an absolute tolerable error must be specified as opposed to a relative one.[2]

The methodology of [7] can be adapted to work in a streaming context with constant space for deciding the validity. However, reparing needs to build first the XML tree and then perform two passes on the tree.

[20] presents an exact algorithm, for validating and repairing XML documents. Both [7] and [20] have a similar flavor in that both have a recursive nature

---

[2] For example, suppose that there is a schema for XML documents about (people) contact information. Now, following [7], if a contact XML file has a mailing address and a phone number then we would tolerate more structural errors than for some other contact file with only the mailing address. We believe that in this case, one should use an absolute number of tolerable errors in order to not bias the tolerance of validation towards the first file.

traversing the XML tree top-down and bottom-up, thus making two passes on the document. In contrast, our (exact) algorithms do not build an XML tree, and perform only a single pass on the document considered as a word. Also, our algorithms can be easily adapted to succinctly produce all the possible repairs for an XML document.

Regarding [9], it focuses on validating and repairing XML documents under a set of integrity constraints. The general problem for [9] is undecidable, and thus, it restricts the edit operations to either deletions or insertions only. All [7,20,9] consider simple DTDs only, while we consider VPAs which computationally represent EDTDs which in turn can abstract DTD, XML Schema and Relax NG.

The other three works, [4,6,5], consider the incremental validation of XML, which is validating documents after updates are being applied on them. The challenge there is to not rescan the document from the scratch, but rather work on the relevant (updated) part of the document. Also, the validation sought is exact rather than approximate. Although these works consider operations that edit (update) documents, the studied problem is very different from the approximate validation of streaming XML.

In all [7,20,9,4,6,5], the edit operations are variants of, or can be achieved by, our edit operations under the first semantics. On the other hand, edit operations under our second semantics, although quite useful in practice, to the best of our knowledge, have not been studied by any work.

Our treatment of approximate XML validation bears some similar flavor with [10,11,12]. However, these works deal with regular languages only and revolve around a different problem, which is finding paths in graph databases that approximately spell words in a given regular language.

3   Visibly Pushdown Automata

VPAs were introduced in [3] and are a special case of pushdown automata. Thier alphabet is partitioned into three disjoint sets of call, return and local symbols, and their push or pop behavior is determined by the consumed symbol. Specifically, while scanning the input, when a call symbol is read, the automaton pushes one stack symbol onto the stack; when a return symbol is read, the automaton pops off the top of the stack; and when a local symbol is read, the automaton only moves its control state.

Formally, a *visibly pushdown automaton* (VPA) A is a 6-tuple $(Q, (\Sigma, f), \Gamma, \tau, q_0, F)$, where

1. Q is a finite set of states.
2. – Σ is the alphabet partitioned into the (sub) alphabets Σ_c, Σ_l and Σ_r of call, local and return symbols respectively.
   – f is a one-to-one mapping $\Sigma_c \to \Sigma_r$. We denote $f(a)$, where $a \in \Sigma_c$, by $\bar{a}$, which is in Σ_r.[3]

---

[3] When referring to arbitrary elements of Σ_r, we will use $\bar{a}, \bar{b}, \ldots$ in order to emphasize that these elements correspond to $a, b, \ldots$ elements of Σ_c.

3. Γ is a finite stack alphabet that (besides other symbols) contains a special "bottom-of-the-stack" symbol $\perp$.
4. q_0 is the initial state.
5. F is the set of final states.
6. $\tau = \tau_c \cup \tau_r \cup \tau_l \cup \tau_\epsilon$ is the transition relation and τ_c, τ_l, τ_r and τ_ϵ are as follows.
   - $\tau_c \subseteq Q \times \Sigma_c \times Q \times \Gamma$
   - $\tau_r \subseteq Q \times \Sigma_r \times \Gamma \times Q$
   - $\tau_l \subseteq Q \times \Sigma_l \times Q$
   - $\tau_\epsilon \subseteq Q \times \{\epsilon\} \times Q$

When reasoning about XML structure and validity, the local symbols are not important, and thus, for simplicity we will not mention local symbols in the rest of the paper. So, for the above definition, we can consider Σ being partitioned into Σ_c and Σ_r, and τ being partitioned into τ_c, τ_r and τ_ϵ only.

Any transition involves two states (not necessarily distinct). We call the first the *origin state* and the second the *destination state*.

Two transitions are called *consecutive* if the destination state of the first is the same as the origin state of the second. This definition applies regardless of whether the transitions involve a push or a pop.

A sequence of consecutive transitions is an *accepting run* if (a) the origin state of the first transition is q_0, (b) the destination state of the last transition is in F and (c) when starting with an empty stack ($\perp$) and following all the transitions in order, in the end, we get again an empty stack ($\perp$).

A word w is accepted by a VPA if there is an accepting run in the VPA which spells w. A language L is a *visibly pushdown language* (VPL) if there exists a VPA that accepts all and only the words in L. The VPL accepted by a VPA A is denoted by $L(A)$.

Example 1. Suppose that we want to build a VPA accepting XML documents about book collections. Such documents will have a *collection* element nesting any number of *book* elements in them. Each *book* element will nest a *title* element and any number of *author* elements. A VPA accepting well-formed documents of this structure is $A = (Q, (\Sigma, f), \Gamma, \tau, q_0, F)$, where

$Q = \{q_0, q_1, q_2, q_3, q_4, q_5, q_6, q_7, q_8\}$,

$\Sigma = \Sigma_c \cup \Sigma_r =$

    $\{collection, book, author, title\} \cup \{\overline{collection}, \overline{book}, \overline{author}, \overline{title}\}$,

    f maps the Σ_c elements into their "bar"-ed counterparts in Σ_r,

$\Gamma = \{\gamma_c, \gamma_b, \gamma_a, \gamma_t\} \cup \{\perp\}$,

$F = \{q_8\}$,

$\tau = \{(q_0, collection, q_1, \gamma_c), (q_1, book, q_2, \gamma_b), (q_2, author, q_3, \gamma_a),$

    $(q_3, \overline{author}, \gamma_a, q_4), (q_4, author, q_3, \gamma_a), (q_4, title, q_5, \gamma_t),$

    $(q_5, \overline{title}, \gamma_t, q_6), (q_6, \overline{book}, \gamma_b, q_7), (q_7, \overline{collection}, \gamma_c, q_8), (q_7, \epsilon, q_1)\}.$

We show this VPA in Fig. 1.

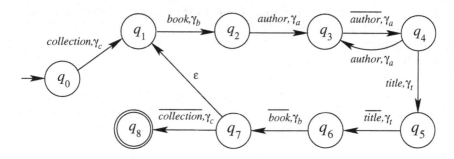

Fig. 1. Example of a VPA

Processing a document with a VPA. As mentioned in the Introduction, given a schema specification VPA $A = (Q, (\Sigma, f), \Gamma, \tau, q_0, F)$, the (exact) typechecking of an XML document (word) w amounts to accepting or rejecting w using A.

Now, the question is whether this can be done using a non-deterministic VPA A. Recall that the well-known procedure for deciding the membership of a word w to a general context-free language, is grammar-based and takes $|w|^3$ time, which is not appropriate for a streaming context.

On the other hand, as shown in [3], VPAs can be determinized, and thus allow for a linear one-pass scanning of a given word. However, there is an exponential penalty to pay for storing deterministic VPAs. Specifically, Theorem 2 in [3] shows that for a given non-deterministic VPA with M states there is an equivalent deterministic VPA with $O(2^{M^2})$ states and with stack alphabet of size $O(2^{M^2} \cdot |\Sigma_c|)$. Nevertheless, for processing a word w, we do not need the whole deterministic VPA, but rather only the single transition sequence spelling w in this automaton. Along the lines of Theorem 2 in [3], one can construct this path on the fly. The amount of space needed is $O(M^2)$, while the time for processing a symbol of w and finding for it the "next transition" in the sequence of transitions is $O((M^2 + M) \cdot |\Sigma_c|)$. In total, one needs only one pass on word w, for a time $O((M^2 + M) \cdot |\Sigma_c| \cdot |w|)$, which depends only linearly on the length of w.

4   Visibly Pushdown Transducers

A *visibly pushdown transducer* (VPT) T is a 7-tuple $(P, (I, f), (O, g), \Gamma, \tau, p_0, F)$, where

1. P is a finite set of states.
2. $-$ I is the input alphabet partitioned into the (sub) alphabets I_c and I_r of input call and return symbols.
   $-$ f is a one-to-one mapping $I_c \rightarrow I_r$. We denote $f(a)$, where $a \in I_c$, by $\bar{a}$.
3. $-$ O is the output alphabet partitioned into the (sub) alphabets O_c and O_r of output call and return symbols respectively.
   $-$ g is a one-to-one mapping $O_c \rightarrow O_r$. We denote $g(b)$, where $b \in O_c$, by $\bar{b}$.

4. Γ is a finite stack alphabet that (besides other symbols) contains a special "bottom-of-the-stack" symbol $\bot$.
5. p_0 is the initial state.
6. F is the set of final states.
7. $\tau = \tau_c \cup \tau_r \cup \tau_\epsilon$, where
   - $\tau_c \subseteq (P \times I_c \times O_c \times P \times \Gamma) \cup (P \times \{\epsilon\} \times O_c \times P \times \Gamma) \cup (P \times I_c \times \{\epsilon\} \times P \times \Gamma)$
   - $\tau_r \subseteq (P \times I_r \times O_r \times \Gamma \times P) \cup (P \times \{\epsilon\} \times O_r \times \Gamma \times P) \cup (P \times I_r \times \{\epsilon\} \times \Gamma \times P)$
   - $\tau_\epsilon \subseteq P \times \{\epsilon\} \times \{\epsilon\} \times P$.

We define an *accepting run* for T similarly as for VPAs. Now, given a word $u \in I^*$, we say that a word $w \in O^*$ is an *output of T for u* if there exists an accepting run in T spelling u as input and w as output.[4]

A transducer T might produce more than one output for a given word u. We denote the set of all outputs of T for u by $T(u)$. For a language $L \subseteq I^*$, we define the *image of L through T* as $T(L) = \bigcup_{u \in L} T(u)$.

If language L is a VPL, then we show that $T(L)$ is a VPL as well. To show this, let $A = (Q, (\Sigma^A, f^A), \Gamma^A, \tau^A, q_0, F^A)$ be a VPA accepting L, and $T = (P, (I, f^T), (O, g^T), \Gamma^T, \tau^T, p_0, F^T)$ be a VPT as above, where $I \supseteq \Sigma^A$ and f^T is an extension of f^A. Then, we present a construction to obtain a VPA B, whose accepting language is $T(L)$, showing thus that the image of L through T is again a VPL.

The construction is a Cartesian product of A and T and similar in spirit to the construction of [3] for showing the closure of VPLs under intersection.

Specifically, $B = (R, (\Sigma^B, g^B), \Gamma^B, \tau^B, r_0, F^B)$, where

1. $R = Q \times P$,
2. $\Sigma^B \subseteq O$, and g^B is a refinement of g^T
3. $\Gamma^B \subseteq (\Gamma^A \cup \dagger) \times \Gamma^T$, where $\dagger$ is a special symbol not in Γ^A and Γ^T.
4. $r_0 = (q_0, p_0)$,
5. $F^B = F^A \times F^T$,
6. $\tau^B = \tau_c^B \cup \tau_r^B$, where

$$\tau_c^B = \{(q, p), b, (q', p'), (\gamma^A, \gamma^T)) : (q, a, q', \gamma^A) \in \tau_c^A \text{ and } (p, a, b, p', \gamma^T) \in \tau_c^T\} \cup$$
$$\{((q, p), \epsilon, (q', p')) : (q, a, q', \gamma^A) \in \tau_c^A, a \neq \epsilon \text{ and } (p, a, \epsilon, p', \gamma^T) \in \tau_c^T\} \cup$$
$$\{((q, p), b, (q, p'), (\dagger, \gamma^T)) : q \in Q \text{ and } (p, \epsilon, b, p', \gamma^T) \in \tau_c^T\}$$
$$\tau_r^B = \{((q, p), \bar{b}, (\gamma^A, \gamma^T), (q', p')) : (q, \bar{a}, \gamma^A, q') \in \tau_r^A \text{ and } (p, \bar{a}, \bar{b}, \gamma^T, p') \in \tau_r^T\} \cup$$
$$\{((q, p), \epsilon, (q', p')) : (q, \bar{a}, \gamma^A, q') \in \tau_r^A, a \neq \epsilon \text{ and } (p, \bar{a}, \epsilon, \gamma^T, p') \in \tau_r^T\} \cup$$
$$\{((q, p), \bar{b}, (\dagger, \gamma^T), (q, p')) : q \in Q \text{ and } (p, \epsilon, \bar{b}, \gamma^T, p') \in \tau_r^T\}$$

Clearly, B is a VPA, and we can show that

Theorem 1. *The language accepted by B is the image of L through T, i.e. $L(B) = T(L)$.*

---

[4] In other words, we get u and w when concatenating the transitions' input and output components respectively.

Proof. A VPT T can be considered as two VPAs; the *input* VPA A_{T_I} and the output *output* VPA A_{T_O}. A_{T_I} and A_{T_O} can be obtained from T by ignoring the output and input parts, respectively, of the transitions of T. A_{T_I} and A_{T_O} have the same structure; each transition path in A_{T_I} has some corresponding transition path in A_{T_O} and vice versa.

Now, the construction of VPA B computes the Cartesian product of VPA A with VPA A_{T_I}, but instead of keeping the matched transitions, it replaces them by the corresponding transitions in A_{T_O}.

Thus, if B accepts a word w, it means that there exists a corresponding word u accepted by A and A_{T_I}, such that $w \in T(u)$. As $u \in L$, we have that $T(u) \subseteq T(L)$ and $w \in T(L)$.

On the other hand, for a word u in $T(L)$, there exists some accepting transition path in the Cartesian product of A with A_{T_I}. By the construction of B, this accepting path induces an accepting path in B as well. Let w be the word spelled out by such a path in B. We have that $w \in L(B)$, and this concludes our proof. □

Union of VPTs. In this paper, we will need to take the union of transducers. Formally given two VPTs $T_1 = (P_1, (I, f), (O, g), \Gamma_1, \tau_1, p_{01}, F_1)$ and $T_2 = (P_2, (I, f), (O, g), \Gamma_2, \tau_2, p_{02}, F_2)$, their union VPT is

$$T = (P_1 \cup P_2 \cup \{p_0\}, (I, f), (O, g), \Gamma_1 \cup \Gamma_2, \tau_0 \cup \tau_1 \cup \tau_2, p_0, F_1 \cup F_2),$$

where $p_0 \notin P_1 \cup P_2$, and $\tau_0 = \{(p_0, \epsilon, \epsilon, p_{01}), (p_0, \epsilon, \epsilon, p_{02})\}$.

Superimposition of VPTs. Given two VPTs, $T_1 = (P, (I, f), (O, g), \Gamma_1, \tau_1, p_0, F)$ and $T_2 = (P, (I, f), (O, g), \Gamma_2, \tau_2, p_0, F)$, which are the same except for the stack alphabet and transition relation, their superimposition VPT is

$$T = (P, (I, f), (O, g), \Gamma_1 \cup \Gamma_2, \tau_1 \cup \tau_2, p_0, F).$$

VPTs for edit operations. In the rest of the paper, we will work on buiding transducers for preprocessing a given VPA specification A, transducing it into a "wider" VPA B, which accepts all the words obtainable by applying at most K edit operations on the words accepted by A. After such a preprocessing phase, the querying phase amounts to accepting or rejecting the streaming XML document considering it as a word.

5   VPTs for Edit Operations under the First Semantics

Since XML documents are nested, when we edit one call element, we also need to edit the corresponding return element. Thus, we consider an (XML) edit operation to consist of two single-symbol operations.

We want to build a visibly pushdown transducer, which given an input word u produces as output all the words v obtainable by applying not more than a certain number (say K) of edit operations on u. We define the edit operations as substitutions, deletions, and insertions of call-return matches, and computationally represent them by using VPTs.

5.1   Substitution

A *call-return match substitution* replaces in an input word a call-return match $a, \bar{a}$ by another call-return match $b, \bar{b}$.

For example, consider the XML document given in Fig. 2 [left]. By substituting `<phone>`, `</phone>` by `<tel>`, `</tel>`, we obtain the document shown in Fig. 2 [right].

```
<contact>                      <contact>
   <address>                      <address>
      <str>...</str>                 <str>...</str>
      <city>...</city>              <city>...</city>
   </address>                     </address>
   <phone>...</phone>             <tel>...</tel>
</contact>                      </contact>
```

Fig. 2. Illustration of substitution under the first semantics

In the following, given a non-negative integer K, we build a VPT which for any word u produces as output the set of all the words w obtainable from u by applying at most K substitutions. We denote this transducer by $T_\sigma^{\leq K}$ and formally define it as a VPT with

- $Q = \{q_0, q_1, q_2, \ldots, q_{2K}\}$,
- $I = O = \Sigma$, $I_c = O_c = \Sigma_c$, $I_r = O_r = \Sigma_r$ and $f = g$,
- $\Gamma = \{\gamma_a : a \in \Sigma_c\} \cup \{\sigma_{ab} : a, b \in \Sigma_c, a \neq b,\} \cup \{\bot\}$,
- $F = \{q_{2i} : 0 \leq i \leq K\}$,
- $\tau = \tau_c \cup \tau_r$, where

$$\tau_c = \{(q_i, a, a, q_i, \gamma_a) : 0 \leq i \leq 2K \text{ and } a \in \Sigma_c\} \cup$$
$$\{(q_i, a, b, q_{i+1}, \sigma_{ab}) : 0 \leq i \leq 2K - 1, a, b \in \Sigma_c \text{ and } a \neq b\},$$
$$\tau_r = \{(q_i, \bar{a}, \bar{a}, \gamma_a, q_i) : 0 \leq i \leq 2K \text{ and } \bar{a} \in \Sigma_r\} \cup$$
$$\{(q_i, \bar{a}, \bar{b}, \sigma_{ab}, q_{i+1}) : 1 \leq i \leq 2K - 1, \bar{a}, \bar{b} \in \Sigma_r \text{ and } \bar{a} \neq \bar{b}\}.$$

For illustration, in Fig. 3, we show $T_\sigma^{\leq 2}$, for alphabet $\{a, b\} \cup \{\bar{a}, \bar{b}\}$.

Intuitively, the transitions in the first set of τ_c and in the first set of τ_r leave the consumed call and return symbols unchanged.

Regarding the transitions in the second set of τ_c, they substitute a call symbol, say a, by another call symbol, say b. A substitution marking symbol σ_{ab} is pushed onto the stack. Symbol σ_{ab} in the stack is crucial in determining which occurrence of $\bar{a}$ has to be replaced by $\bar{b}$ using a transition in the second set of τ_r.

Finally, since we want to substitute 0, 1, ..., K symbols, we need $K + 1$ different final states for the $K + 1$ different cases.

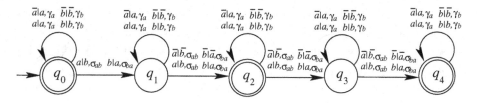

Fig. 3. VPT $T_{\bar{\sigma}}^{\leq 2}$

5.2   Deletion

A *call-return match deletion* removes in an input word a call-return match $a, \bar{a}$. Deletion is a "structure flattening" operation.

For example, consider the XML document given in Fig. 4 [left]. By deleting `<address>`, `</address>`, we (partially) flatten the document to the one shown in Fig. 4 [right].

```
<contact>                    <contact>
    <address>                    <str>...</str>
        <str>...</str>           <city>...</city>
        <city>...</city>         <phone>...</phone>
    </address>               </contact>
    <phone>...</phone>
</contact>
```

Fig. 4. Illustration of deletion under the first semantics

In the following, given a non-negative integer K, we build a VPT which for any word u produces as output the set of all the words w obtainable from u by applying at most K deletions. We denote this transducer by $T_{\delta}^{\leq K}$ and formally define it as a VPT with

- $Q = \{q_0, q_1, q_2, \ldots, q_{2K}\}$,
- $I = O = \Sigma$, $I_c = O_c = \Sigma_c$, $I_r = O_r = \Sigma_r$ and $f = g$,
- $\Gamma = \{\gamma_a : a \in \Sigma_c\} \cup \{\delta_a : a \in \Sigma_c\} \cup \{\perp\}$,
- $F = \{q_{2i} : 0 \leq i \leq K\}$,
- $\tau = \tau_c \cup \tau_r$, where

$$\tau_c = \{(q_i, a, a, q_i, \gamma_a) : 0 \leq i \leq 2K \text{ and } a \in \Sigma_c\} \cup$$
$$\{(q_i, a, \epsilon, q_{i+1}, \delta_a) : 0 \leq i \leq 2K - 1, \text{ and } a \in \Sigma_c\},$$
$$\tau_r = \{(q_i, \bar{a}, \bar{a}, \gamma_a, q_i) : 0 \leq i \leq 2K \text{ and } \bar{a} \in \Sigma_r\} \cup$$
$$\{(q_i, \bar{a}, \epsilon, \delta_a, q_{i+1}) : 1 \leq i \leq 2K - 1, \text{ and } \bar{a} \in \Sigma_r\}.$$

For illustration, in Fig. 5, we show $T_{\delta}^{\leq 2}$, for alphabet $\{a, b\} \cup \{\bar{a}, \bar{b}\}$.

Similarly with the substitution, the transitions in the first set of τ_c and in the first set of τ_r leave the consumed call and return symbols unchanged.

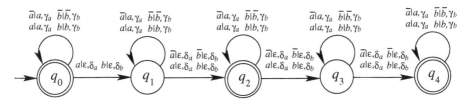

Fig. 5. VPT $T_\delta^{\leq 2}$

Regarding the transitions in the second set of τ_c, they delete a call symbol, say a. A deletion marking symbol δ_a is pushed onto the stack. Symbol δ_a in the stack is crucial in determining which occurrence of $\bar{a}$ has to be deleted by using a transition in the second set of τ_r.

Since we want to perform $0, 1, \ldots, K$ deletions, we need $K+1$ different final states for the $K+1$ different cases.

5.3   Insertion

A *call-return match insertion* inserts in an input word a call symbol a and a corresponding return symbol $\bar{a}$ while maintaining the well-formedness of the XML document. Thus, insertion is a "structure creation" operator.

For example, consider the XML document given in Fig. 6 [left]. By inserting `<address>`, `</address>`, surrounding the street and city elements, we obtain the document in Fig. 6 [right].

```
<contact>                    <contact>
  <str>...</str>               <address>
  <city>...</city>               <str>...</str>
  <phone>...</phone>             <city>...</city>
</contact>                     </address>
                               <phone>...</phone>
                             </contact>
```

Fig. 6. Illustration of insertion under the first semantics

In the following, given a non-negative integer K, we build a VPT which for any word u produces as output the set of all the words w obtainable from u by applying at most K insertions. We denote this transducer by $T_\eta^{\leq K}$ and formally define it as a VPT with

- $Q = \{q_0, q_1, q_2, \ldots, q_{2K}\}$,
- $I = O = \Sigma$, $I_c = O_c = \Sigma_c$, $I_r = O_r = \Sigma_r$ and $f = g$,
- $\Gamma = \{\gamma_a : a \in \Sigma_c\} \cup \{\eta_a : a \in \Sigma_c\} \cup \{\bot\}$,
- $F = \{q_{2i} : 0 \leq i \leq K\}$,
- $\tau = \tau_c \cup \tau_r$, where

$$\tau_c = \{(q_i, a, a, q_i, \gamma_a) : 0 \leq i \leq 2K \text{ and } a \in \Sigma_c\} \cup$$
$$\{(q_i, \epsilon, a, q_{i+1}, \eta_a) : 0 \leq i \leq 2K - 1, \text{ and } a \in \Sigma_c\},$$
$$\tau_r = \{(q_i, \bar{a}, \bar{a}, \gamma_a, q_i) : 0 \leq i \leq 2K \text{ and } \bar{a} \in \Sigma_r\} \cup$$
$$\{(q_i, \epsilon, \bar{a}, \eta_a, q_{i+1}) : 1 \leq i \leq 2K - 1, \text{ and } \bar{a} \in \Sigma_r\}.$$

For illustration, in Fig. 7, we show $T_\eta^{\leq 2}$, for alphabet $\{a, b\} \cup \{\bar{a}, \bar{b}\}$.

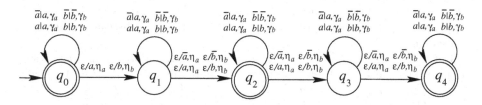

Fig. 7. VPT $T_\eta^{\leq 2}$

Again, the transitions in the first set of τ_c, and in the first set of τ_r leave the consumed call and return (respectively) symbols unchanged.

Regarding the transitions in the second set of τ_c, they insert a call symbol, say a. An insertion marking symbol η_a is inserted on the stack. Symbol η_a in the stack is crucial in determining when to insert $\bar{a}$ by using a transition in the second set of τ_r.

5.4   A VPT for All Operations

Here, for a given a non-negative integer K, we want to construct a VPT which for any word u produces as output the set of all the words w obtainable from u by applying at most K edit operations, which can be substitutions, deletions or insertions.

As can be observed above, sets Q, I, O and F are the same for all the transducers constructed so far. Notably, a VPT $T^{\leq K}$ for at most K edit operations can be simply obtained by superimposing $T_\sigma^{\leq K}$, $T_\delta^{\leq K}$ and $T_\eta^{\leq K}$.

Transducer $T^{\leq K}$ has $2K + 1$ states and $O(KR^2)$ transitions, where R is the size of the underlying alphabet.

Edit Distance. Edit operations transform a word into other words. For two given words u and w we define the *distance between u and w* as the least number of edit operations needed to transform u into w. We denote this distance by $d(u, w)$. It is easy to see that the distance defined using the edit operations under the first semantics is metric.

Given, a VPL L and a non-negative integer K, we define

$$L^{(K)} = \{u : \exists w \in L \text{ and } d(u, w) \leq K\}.$$

Now, we can show that for the above transducer $T^{\leq K}$.

Theorem 2. $T^{\leq K}(L) = L^{(K)}$.

Proof. **Basis step.** For $k = 0$, we have $L^{(k)} = L^{(0)} = L$. On the other hand, $T^{\leq 0}$ is nothing but just a single state transducer with only self-loop transitions which leave everything unchanged. Thus, $T^{\leq 0}(L) = L = L^{(0)}$.

Induction step. Suppose that $T^{\leq k}(L) = L^{(k)}$ is true for non-negative k. We want to show that $T^{\leq k+1}(L) = L^{(k+1)}$ is true as well.

Based on the construction of edit VPAs, we have that $T^{\leq k+1}$ is in fact $T^{\leq k}$ with two more additional states, q_{2k+1} and q_{2k+2}. These two states allow $T^{\leq k+1}$ to optionally perform one more operation, which can be substitution, deletion or insertion.

By the hypothesis, $T^{\leq k}(L) = L^{(k)} = \{w : \exists w' \in L \text{ and } d(w, w') \leq k\}$. Now, we have that, since $T^{\leq k+1}$ can perform one more operation, $T^{\leq k+1}(L) = \{w : \exists w' \in L \text{ and } d(w, w') \leq k \text{ or } d(w, w') = k + 1\} = \{w : \exists w' \in L \text{ and } d(w, w') \leq k + 1\}$. The latter is nothing but $L^{(k+1)}$, and this completes the proof. □

From all the above, and the construction for the language transduction of a VPT (in Section 4), we can show that

Theorem 3. *Under the first semantics, the total time for the preprocessing phase, and the space for storing the output of it, is $O(KR^2M)$.*

Proof. This claim follows from the fact that transducer $T^{\leq K}$ has $2K + 1$ states and $O(KR^2)$ transitions, and the transduction of the schema VPA with M states is done through a Cartesian product, which will have in this case $O(KM)$ states and $O(KR^2M)$ transitions. The latter is thus an upper bound for the time and space needed to compute and store the transduction of the schema VPA. This is nothing but the time and space needed for the preprocessing phase. □

6   VPTs for Edit Operations under the Second Semantics

6.1   Substitution

Under the second semantics, a substitution replaces in an input word *all* the call-return matches of a call-return pair $(a, \bar{a})$ by call-return matches of another call-return pair $(b, \bar{b})$.

Let alphabet Σ be $\{a_1, \ldots, a_R\} \cup \{\bar{a}_1, \ldots, \bar{a}_R\}$. Clearly, we can have $R(R-1)$ pairs of different call symbols (e.g. (a_1, a_2), etc). We can now construct substitution transducers which are indexed by these pairs and perform accordingly the substitution indicated by their index pair. For example, the substitution transducer indexed by (a_1, a_2), denoted in short by $T^1_{\sigma:12}$, will substitute all call-return matches of $(a_1, \bar{a}_1)$ by call-return matches of $(a_2, \bar{a}_2)$ in any word provided as input. The superscript says that this transducer is of "order one," i.e. it substitutes only the call-return matches of one call-return pair. It is not difficult to construct such transducers. Formally, $T^1_{\sigma:ij}$ (for $i \neq j$) is defined as a VPT with a single state q_0 which is both initial and final, stack alphabet

$\Gamma = \{\gamma_k : k \in \{1, \ldots, r\} \backslash \{i\}\} \cup \{\sigma_{ij}\} \cup \{\bot\}$, and transition relation

$$\tau_c = \{(q_0, a_k, a_k, q_0, \gamma_k) : k \neq i\} \cup \{(q_0, a_i, a_j, q_0, \sigma_{ij})\},$$
$$\tau_r = \{(q_0, \bar{a}_k, \bar{a}_k, \gamma_k, q_0) : k \neq i\} \cup \{(q_0, \bar{a}_i, \bar{a}_j, \sigma_{ij}, q_0)\}.$$

Similarly, we can construct transducers of "order two," which perform substitutions for the call-return matches of two call-return pairs. Such a transducer $T^2_{\sigma:ij,kl}$ (for $i \neq j, k$ and $k \neq l$) is defined again as a VPT with a single state q_0 which is both initial and final, stack alphabet $\Gamma = \{\gamma_m : m \in \{1, \ldots, r\} \backslash \{i, k\}\} \cup \{\sigma_{ij}, \sigma_{kl}\} \cup \{\bot\}$, and transition relation

$$\tau_c = \{(q_0, a_m, a_m, q_0, \gamma_m) : m \neq i, k\} \cup \{(q_0, a_i, a_j, q_0, \sigma_{ij}), (q_0, a_k, a_l, q_0, \sigma_{kl})\},$$
$$\tau_r = \{(q_0, \bar{a}_m, \bar{a}_m, \gamma_m, q_0) : m \neq i, k\} \cup \{(q_0, \bar{a}_i, \bar{a}_j, \sigma_{ij}, q_0), (q_0, \bar{a}_k, \bar{a}_l, \sigma_{kl})\}.$$

We can observe that, regardless of H, the number of transitions in these one-state transducers is $2R$. The transitions in each of the τ_c and τ_r sets are divided into: "leave unchanged" transitions and "modify symbol" transitions.

In general, we can construct substitution transducers of any order up to the size R of the alphabet. Let T^H_σ be the union of all substitution transducers of order H. Then, we construct transducer $T^{\leq K}_\sigma = \bigcup_{H=0}^{K} T^H_\sigma$ (considering also T^0_σ which leaves everything unchanged). Given a word u as input, $T^{\leq K}_\sigma$ produces as output the set of all words w obtainable from u by applying at most K substitutions under the second semantics.

Now the question is: For a given H, how many transducers of order H can be created? We show that

Theorem 4. *Given a non-negative integer $H \leq R$, the number of substitution transducers of order H is $C^H_R \cdot (R-1)^H$.*

Proof. For this, recall that a transducer of order H substitutes call-return matches of H call-return pairs. We have C^H_R choices for these pairs. In each choice, any chosen pair, say $(a_i, \bar{a}_i)$, can be substituted by any of the $(R-1)$ remaining pairs $(a_1, \bar{a}_1), \ldots, (a_{i-1}, \bar{a}_{i-1}), (a_{i+1}, \bar{a}_{i+1}), \ldots, (a_R, \bar{a}_R)$. $\square$

6.2   Deletion

A deletion removes from an input word *all* the call-return matches of a call-return pair $(a, \bar{a})$. A deletion transducer of order H deletes all the call-return matches of H call-return pairs, say $(a_{i_1}, \bar{a}_{i_1}), \ldots, (a_{i_H}, \bar{a}_{i_H})$, in an input word. This transducer, denoted by $T^H_{\delta:i_1,\ldots,i_H}$, has a single state which is both initial and final, stack alphabet $\{\gamma_j : j \in \{1, \ldots, R\} \backslash \{i_1, \ldots, i_H\}\} \cup \{\delta_{i_1}, \ldots, \delta_{i_H}\} \cup \{\bot\}$ and transition relation

$$\tau_c = \{(q_0, a_j, a_j, q_0, \gamma_j) : j \neq i_1, \ldots, i_H\} \cup \{(q_0, a_{i_1}, \epsilon, q_0, \delta_{i_1}), \ldots, (q_0, a_{i_H}, \epsilon, q_0, \delta_{i_H})\},$$
$$\tau_r = \{(q_0, \bar{a}_j, \bar{a}_j, \gamma_j, q_0) : j \neq i_1, \ldots, i_H\} \cup \{(q_0, \bar{a}_{i_1}, \epsilon, \delta_{i_1}, q_0), \ldots, (q_0, \bar{a}_{i_H}, \epsilon, \delta_{i_H}, q_0)\}.$$

We can observe that, regardless of H, the number of transitions in these one-state transducers is $2R$. Also, reasoning similarly as for the substitution, we can show that

Theorem 5. *Given a non-negative integer $H \leq R$, the number of deletion transducers of order H is C_R^H.*

Proof. This follows from the fact that a deletion transducer of order H deletes call-return matches of H call-return pairs, and we have C_R^H choices for these pairs.   □

In general, we can construct deletion transducers of any order up to the size R of the alphabet.

6.3   Insertion

An insertion operation under the second semantics non-deterministically inserts in an input word *any* number of a single call symbol a balancing those insertions by inserting in the right places the corresponding return symbol $\bar{a}$.

An insertion transducer of order H inserts call-return matches for H call-return pairs, say $(a_{i_1}, \bar{a}_{i_1}), \ldots, (a_{i_H}, \bar{a}_{i_H})$, in an input word. This transducer, denoted by $T_{\eta : i_1, \ldots, i_H}^H$, has a single state which is both initial and final, stack alphabet $\{\gamma_j : j \in \{1, \ldots, R\}\} \cup \{\eta_{i_1}, \ldots, \eta_{i_H}\} \cup \{\perp\}$ and transition relation

$$\tau_c = \{(q_0, a_j, a_j, q_0, \gamma_j) : j \in \{1, \ldots, R\}\} \cup \{(q_0, \epsilon, a_{i_1}, q_0, \eta_{i_1}), \ldots, (q_0, \epsilon, a_{i_H}, q_0, \eta_{i_H})\},$$
$$\tau_r = \{(q_0, \bar{a}_j, \bar{a}_j, \gamma_j, q_0) : j \in \{1, \ldots, R\}\} \cup \{(q_0, \epsilon, \bar{a}_{i_1}, \eta_{i_1}, q_0), \ldots, (q_0, \epsilon, \bar{a}_{i_H}, \eta_{i_H}, q_0)\}.$$

We can observe that, the number of transitions in these one-state transducers is $2(R + H)$. Also, as for the deletion, we have that

Theorem 6. *Given a non-negative integer $H \leq R$, the number of insertion transducers of order H is C_R^H.*

Proof. This follows from the fact that an insertion transducer of order H inserts call-return matches of H call-return pairs, and we have C_R^H choices for these pairs.   □

In general, we can construct insertion transducers of any order up to the size R of the alphabet.

6.4   A VPT for All Operations

We now can easily create edit transducers of order H, by superimposing substitution, deletion and insertion transducers of orders H_1, H_2 and H_3, such that $H_1 + H_2 + H_3 = H$.

Formally, let $T_{\sigma : i_1 j_1, \ldots, i_{H_1} j_{H_1}}^{H_1}$, $T_{\delta : k_1, \ldots, k_{H_2}}^{H_2}$ and $T_{\eta : l_1, \ldots, l_{H_3}}^{H_3}$ be substitution, deletion and insertion transducers. If $\{i_1, \ldots, i_{H_1}\} \cap \{k_1, \ldots, k_{H_2}\} = \emptyset$, then we superimpose these three transducers to obtain a transducer of order $H = H_1 + H_2 + H_3$.

The condition $\{i_1, \ldots, i_{H_1}\} \cap \{k_1, \ldots, k_{H_2}\} = \emptyset$ says that the call-return pairs we substitute must be different from those we delete. This is because under the second semantics of edit operations, all call-return matches of a call-return pair have "the same fate." If the transducer both substitutes and deletes the call-return matches of a call-return pair, then we will have a situation where some of these call-return matches have been substituted and some other ones have been deleted.

Of course, in a superimposition, there does not need to be a transducer for each kind of edit operation. For example, we can create a transducer of order H by superimposing a substitution transducer and a deletion transducer of orders H_1 and H_2 respectively, such that $H = H_1 + H_2$.

Now, based on the above as well as theorems 4, 5 and 6, we can state that

Theorem 7. *Given a non-negative integer $H \leq R$, the number of edit transducers of order H under the second semantics is $O\left(R^{2H} \cdot H^2\right)$.*

Proof. We can create an edit transducer of order $H = H_1 + H_2 + H_3$ by selecting for the superimposition one of $C_R^{H_1} \cdot (R-1)^{H_1}$, $C_R^{H_2}$ and $C_R^{H_3}$ substitution, deletion and insertion transducers respectively. Of course, one or two of H_1, H_2 and H_3 might be zero.

Clearly, $C_R^{H_1} \cdot (R-1)^{H_1}, C_R^{H_2}$ and $C_R^{H_3}$ are bounded by $R^{H_1} \cdot (R-1)^{H_1}, R^{H_2}$ and R^{H_3} respectively. Thus, given H_1, H_2 and H_3 we have $R^{H_1} \cdot (R-1)^{H_1} \cdot R^{H_2} \cdot R^{H_3} = R^H \cdot (R-1)^{H_1}$, or $O(R^{2H})$ edit transducers.

Now, the claimed upper bound follows from the above and the fact that we have $O(H^2)$ possibilities of choosing H_1, H_2 and H_3 such that $H = H_1 + H_2 + H_3$. $\square$

Let T^H be the union of all edit transducers of order H. Then, we construct transducer $T^{\leq K} = \bigcup_{H=0}^{K} T^H$. Given a word u as input, $T^{\leq K}$ produces as output the set of all words w obtainable from u by applying at most K edit operations under the second semantics.

Based on Theorem 7 and the construction given for the union of transducers in Section 4, we can state that

Theorem 8. *Transducer $T^{\leq K}$ has $O\left(R^{2K} \cdot K^3\right)$ states.*

Proof. By the construction for the union of transducers, $T^{\leq K} = \bigcup_{H=0}^{K} T^H$ has $O\left(\Sigma_{H=0}^{K} R^{2H} \cdot H^2\right)$ states. This is subsumed by $O\left(K \cdot R^{2K} \cdot K^2\right)$, which is $O\left(R^{2K} \cdot K^3\right)$, i.e. the upper bound in our claim. $\square$

Finally, by the fact that, in a superimposition, each of the three transducers has $O(R)$ transitions, we can state that

Theorem 9. *Transducer $T^{\leq K}$ has $O\left(R^{2K+1} \cdot K^3\right)$ transitions.*

Proof. Direct from the constructions for the superimposition and union of VPTs. $\square$

Edit Distance. Similarly as for the first semantics, given words u and w, we define the *distance between words u and w* as the least number of edit operations (under the second semantics) needed to transform u into w. Here as well, it is easy to see that this distance is metric.

Given, a VPL L and a non-negative integer K, we define $L^{(K)}$ as in Subsection 5.4, but considering instead the distance under the second semantics. Here we can show that, similarly with Theorem 2, $T^{\leq K}(L) = L^{(K)}$.

From the above and theorems 8 and 9, we can state that

Theorem 10. *Under the second semantics, the total time for the preprocessing phase, and the space for storing the output of it, is $O(R^{2K+1}K^3M)$.*

Proof. This claim follows from the fact that transducer $T^{\leq K}$ has $O\left(R^{2K} \cdot K^3\right)$ states, and the transduction of the schema VPA with M states is done through a Cartesian product, which will have in this case $O(R^{2K} \cdot K^3 \cdot M)$ states and $O(R \cdot R^{2K} \cdot K^3 \cdot M)$ transitions. The latter is thus an upper bound for the time and space needed to compute and store the transduction of the schema VPA. This is nothing but the time and space needed for the preprocessing phase under the second semantics. □

We believe that this exponential penalty in K is an artifact of the requirement that the queries be answered using only one pass through the XML document, while using auxilliary storage space only bounded by the depth of the document. If we were allowed to use auxilliary space polynomial in the size N of the document, we believe that a *cubic* in N algorithm similar in spirit to [1] could possibly be devised to use storage space only polynomial in K and M. However, such an algorithm is useful only in a non-streaming context and when the document size is not large. This is a topic for our future investigation.

We can also observe that the edit distance between any XML document and a VPA in the second semantics is at most $2R$, i.e. twice the size of the underlying alphabet. This is because we can first delete all the call-return matches of the document using at most R delete operations and then create a string in the language of the VPA using at most R insert operations.

7   Repairs

Now, suppose that we are also interested in obtaining the set $L_{w,K}$ of words accepted by the original schema VPA A that can be transformed to match an XML document w by applying on them at most K edit operations based on either semantics. The words in $L_{w,K}$ are the possible *repairs* of XML document w.

For computing $L_{w,K}$, we need to "enrich" the construction of the transduction to remember the lineage of its words. For this, instead of VPA B, we can construct a VPT T_B, which coincides with B when considering only the input of its transitions. On the other hand, the output of its transitions "remembers" the input of T's transitions that matched the A's transitions. Formally, the transition relation of T_B is $\tau^B = \tau_c^B \cup \tau_r^B$, where

$$\tau_c^B = \{(q,p), b, a, (q',p'), (\gamma^A, \gamma^T)) : (q,a,q',\gamma^A) \in \tau_c^A \text{ and } (p,a,b,p',\gamma^T) \in \tau_c^T\} \cup$$
$$\{((q,p),\epsilon,a,(q',p')) : (q,a,q',\gamma^A) \in \tau_c^A, a \neq \epsilon \text{ and } (p,a,\epsilon,p',\gamma^T) \in \tau_c^T\} \cup$$
$$\{((q,p),b,\epsilon,(q,p'),(\dagger,\gamma^T)) : q \in Q \text{ and } (p,\epsilon,b,p',\gamma^T) \in \tau_c^T\}$$
$$\tau_r^B = \{((q,p),\bar{b},\bar{a},(\gamma^A,\gamma^T),(q',p')) : (q,\bar{a},\gamma^A,q') \in \tau_r^A \text{ and } (p,\bar{a},\bar{b},\gamma^T,p') \in \tau_r^T\} \cup$$
$$\{((q,p),\epsilon,\bar{a},(q',p')) : (q,\bar{a},\gamma^A,q') \in \tau_r^A, a \neq \epsilon \text{ and } (p,\bar{a},\epsilon,\gamma^T,p') \in \tau_r^T\} \cup$$
$$\{((q,p),\bar{b},\epsilon,(\dagger,\gamma^T),(q,p')) : q \in Q \text{ and } (p,\epsilon,\bar{b},\gamma^T,p') \in \tau_r^T\}.$$

Now, it can be easily seen that for a given document w, we have $L_{w,K} = T_B(w)$.

8 Concluding Remarks

In this work, we have investigated the problem of approximate XML validation, an important problem in XML processing. Useful contributions of this paper include the introduction of VPTs and their application to building two algorithms for checking approximate XML validity in the streaming model. We also believe that the new semantics introduced in this paper is interesting in the context of XML and merits further investigation.

References

1. Aho, V.A., Peterson, G.T.: A Minimum Distance Error-Correcting Parser for Context-Free Languages. SIAM J. Comput. 1(4), 305–312 (1972)
2. Alur, R., Kumar, V., Madhusudan, P., Viswanathan, M.: Congruences for Visibly Pushdown Languages. In: Caires, L., Italiano, G.F., Monteiro, L., Palamidessi, C., Yung, M. (eds.) ICALP 2005. LNCS, vol. 3580, pp. 1102–1114. Springer, Heidelberg (2005)
3. Alur, R., Madhusudan, P.: Visibly Pushdown Languages. In: Proc. 36th ACM Symp. on Theory of Computing, Chicago, Illinois, June 13–15, 2004, pp. 202–211 (2004)
4. Balmin, A., Papakonstantinou, Y., Vianu, V.: Incremental Validation of XML Documents. ACM Trans. Database Syst. 29(4), 710–754 (2004)
5. Barbosa, D., Leighton, G., Smith, A.: Efficient Incremental Validation of XML Documents After Composite Updates. In: Proc. of 2nd Int. XML Database Symp., Seoul Korea, September 10–11, 2006, pp. 107–121 (2006)
6. Barbosa, D., Mendelzon, A.O., Libkin, L., Mignet, L., Arenas, M.: Efficient Incremental Validation of XML Documents. In: Proc. of 20th Int. Conf. on Data Engineering, Boston, USA, March 30–April 2, 2004, pp. 671–682 (2004)
7. Boobna, U., de Rougemont, M.: Correctors for XML Data. In: Proc. 2nd International XML Database Symposium, Toronto, Canada, August 29–30, 2004, pp. 97–111 (2004)
8. Clark, J., Murata, M.: RELAX NG Specification. OASIS (December 2001)
9. Flesca, S., Furfaro, F., Greco, S., Zumpano, E.: Querying and Repairing Inconsistent XML Data. In: Proc. 6th International Conference on Web Information Systems Engineering, New York, USA, November 20–22, 2005, pp. 175–188 (2005)

10. Grahne, G., Thomo, A.: Approximate Reasoning in Semistructured Data. In: Proc. of the 8th International Workshop on Knowledge Representation meets Databases, Rome, Italy (September 15, 2001)

11. Grahne, G., Thomo, A.: Query Answering and Containment for Regular Path Queries under Distortions. In: Proc. of 3rd International Symposium on Foundations of Information and Knowledge Systems, Wilhelmminenburg Castle, Austria, February 17–20, 2004, pp. 98–115 (2004)

12. Grahne, G., Thomo, A.: Regular Path Queries under Approximate Semantics. Ann. Math. Artif. Intell. 46(1–2), 165–190 (2006)

13. Green, T.J., Gupta, A., Miklau, G., Onizuka, M., Suciu, A.: Processing XML Streams with Deterministic Automata and Stream Indexes. ACM Trans. Database Syst. 29(4), 752–788 (2004)

14. Kumar, V., Madhusudan, P., Viswanathan, M.: Visibly Pushdown Automata for Streaming XML. In: Proc. of Int. Conf. on World Wide Web, Alberta, Canada, May 8–12, 2007, pp. 1053–1062 (2007)

15. Neven, F.: Automata Theory for XML Researchers. SIGMOD Record 31(3), 39–46 (2002)

16. Martens, W., Neven, F., Schwentick, T., Bex, G.J.: Expressiveness and complexity of XML Schema. ACM Trans. Database Syst. 31(3), 770–813 (2006)

17. Papakonstantinou, Y., Vianu, V.: DTD Inference for Views of XML Data. In: Proc. 19th ACM Symp. on Principles of Database Systems, Dallas, Texas, May 15–17, 2000, pp. 35–46 (2000)

18. Segoufin, L., Vianu, V.: Validating Streaming XML Documents. In: Proc. 21st ACM Symp. on Principles of Database Systems, Madison, Wisconsin, June 3–5, 2002, pp. 53–64 (2002)

19. Sperberg-McQueen, C.M., Thomson, H.: XML Schema 1.0. (2005), http://www.w3.org/XML/Schema

20. Staworko, S., Chomicki, J.: Validity-Sensitive Querying of XML Data-bases. In: Proc. of 2nd International Workshop on Database Technologies for Handling XML Information on the Web, EDBT Workshops, Munich, Germany, March 26–31, 2006, pp. 164–177 (2006)

21. Segoufin, L., Sirangelo, C.: Constant-Memory Validation of Streaming XML Documents Against DTDs. In: Proc. 11th International Conference on Database Theory, Barcelona, Spain, January 10–12, 2007, pp. 299–313 (2007)

22. Suciu, D.: The XML Typechecking Problem. SIGMOD record 31(1), 89–96 (2002)

A Probabilistic Logic with Polynomial Weight Formulas

Aleksandar Perović[1], Zoran Ognjanović[2], Miodrag Rašković[2],
and Zoran Marković[2]

[1] Saobraćajni fakultet, Vojvode Stepe 305, 11000 Beograd, Srbija
pera@sf.bg.ac.yu
[2] Matematički institut SANU, Kneza Mihaila 35, 11000 Beograd, Sbija
{zorano, miodragr, zoranm}@mi.sanu.ac.yu

Abstract. The paper presents a sound and strongly complete axiomatization of reasoning about polynomial weight formulas. In addition, the PSPACE decision procedure for polynomial weight formulas developed by Fagin, Halpern and Megiddo works for our logic as well. The introduced formalism allows the expression of qualitative probability statements, conditional probability and Bayesian inference.

1 Introduction

The present paper continues our previous work, see [11,12,13,15,16,17,18,19], on the formal development of probabilistic logics, where probability statements are expressed by probabilistic operators expressing bounds on the probability of a propositional formula. Developing logics for probabilistic beliefs is a well-worked area. We have listed some of the references (see [2,3,4,5,6,7,8,9,10]) that are relevant to this paper. For more references see [14].

Renewed interest in this type of logic was initiated by [10], which introduced a mechanism for drawing conclusions from probable knowledge. This paper stimulated extensive investigations of formal systems for reasoning in the presence of uncertainty. The first attempt at developing a logic along these lines was given in [2]. In this paper a notion of weight formula was introduced, which makes it possible to express statements such as "the probability of E_1 is at least the product of probabilities of E_2 and E_3", where E_1, E_2 and E_3 are arbitrary events. In the case of linear weight formulas, a simply-complete ("every consistent formula has a model", in contrast to the strong completeness: "every consistent set of formulas has a model") axiomatization was given, together with a companion NP-complete decision procedure. Fagin, Halpern and Megiddo also discussed the case of polynomial weight formulas for which they provided a PSPACE decision procedure. However, they have failed to obtain a corresponding complete axiomatization. Instead, they have completely axiomatized the reasoning about polynomial weight formulas within the first order logic in the following way[1]: the

---

[1] The approach given in [2] is slightly different, though equivalent to the one given here.

S. Hartmann and G. Kern-Isberner (Eds.): FoIKS 2008, LNCS 4932, pp. 239–252, 2008.

language of ordered fields is extended by countably many new constant symbols $w(\alpha)$, and the theory of real closed fields is extended by the following axioms:

- $w(\alpha) \geqslant 0$.
- $w(\alpha) = w(\beta)$, whenever $\alpha \leftrightarrow \beta$ is a tautology.
- $w(\bot) = 0$.
- $w(\top) = 1$.
- $w(\alpha \vee \beta) = w(\alpha) + w(\beta) - w(\alpha \wedge \beta)$.

As a consequence, the companion decision procedure is EXPSPACE hard.

Uncertain reasoning arises also in the area of game theory with applications in economics. A finitary axiomatization for type spaces within the framework of probabilistic logic was given in [3]. The proposed axiomatization is simply-complete with respect to the introduced semantics. In the single-agent case, that logic can be easily embedded into our system. A strongly-complete infinitary axiomatization for type spaces was given in [8]. The main difference between that system and our approach is that infinitary formulas are allowed in [8]. As a consequence, that logic is undecidable, due to cardinality argument. Though they are not explicitly mentioned, qualitative statements are possible in [8]. For instance, "the ith player thinks that B is at least probable as A" can be formally written as

$$\bigwedge_{s \in [0,1] \cap \mathbb{Q}} p_i^s(A) \rightarrow p_i^s(B).$$

A discussion of possibility of axiomatizing qualitative probability (a Boolean algebra with an ordering relation instead of a probability measure) is given in [9].

A complete[2] and decidable logic that handles conditional probabilities (following de Finetti) by means of fuzzy connectives[3] and probabilistic operators was given in [7]. As a consequence, comparative statements are possible in that system. For instance, "the conditional event $\varphi|\chi$ is at least probable as the conditional event $\psi|\delta$" can be formally expressed by

$$P(\psi|\delta) \rightarrow_\Pi P(\varphi|\chi).$$

Conditional probability formulas with semantics similar to ours, but without any axiomatization, are used in the field of nonmonotonic reasoning in [5,6].

The main technical novelty of this paper is a sound and strongly complete axiomatization of the reasoning about polynomial weight formulas introduced in Section 5 of [2], where the PSPACE containment of the logic was proved. The rich language allows purely comparative statements, expression of Kolmogorovian conditional probability, Bayesian inference etc.

The rest of the paper is organized as follows. In Section 2 syntax of the logic is given and the class of measurable probabilistic models is described. Section 3

---

[2] With respect to finite theories.

[3] Łukasiewicz implication $\rightarrow_L$, product conjunction $\odot$, product implication $\rightarrow_\Pi$ and the truth constants $\underline{0}$ and $\underline{\frac{1}{2}}$.

contains the corresponding axiomatization and introduces the notion of deduction. A proof of the completeness theorem is presented in Section 4. Decidability of the logic is analyzed in Section 5. We conclude in Section 6 and describe possible applications of the logic.

2   Syntax and Semantics

Let $Var = \{p_n \mid n = 0, 1, 2, \ldots\}$. The elements of Var will be called propositional variables.

Definition 1. *We define the set For_C of all propositional formulas over the set Var recursively as follows:*

- $For_C(0) = Var$.
- $For_C(n+1) = For_C(n) \cup \{\neg\alpha \mid \alpha \in For_C(n)\} \cup \{(\alpha \wedge \beta) \mid \alpha, \beta \in For_C(n)\}$.
- $For_C = \bigcup\limits_{n=0}^{\infty} For_C(n)$.   □

Propositional formulas will be denoted by α, β and γ, possibly with indices. The rest of connectives ($\vee, \rightarrow$ and $\leftrightarrow$) are introduced in the usual way. $\top$ denotes an arbitrary tautology, while $\perp$ denotes an arbitrary contradiction.

Definition 2. *The set $Term$ of all probabilistic terms is recursively defined as follows:*

- $Term(0) = \{\underline{s} \mid s \in \mathbb{Q}\} \cup \{P(\alpha) \mid \alpha \in For_C\}$.
- $Term(n+1) = Term(n) \cup \{(\mathbf{f} + \mathbf{g}), (\mathbf{f} \cdot \mathbf{g}), (-\mathbf{f}) \mid \mathbf{f}, \mathbf{g} \in Term(n)\}$
- $Term = \bigcup\limits_{n=0}^{\infty} Term(n)$.   □

Probabilistic terms will be denoted by $\mathbf{f}, \mathbf{g}$ and $\mathbf{h}$, possibly with indices. To simplify notation, we introduce the following convention: $\mathbf{f} + \mathbf{g}$ is $(\mathbf{f} + \mathbf{g})$, $\mathbf{f} + \mathbf{g} + \mathbf{h}$ is $((\mathbf{f} + \mathbf{g}) + \mathbf{h})$. For $n > 3$, $\sum\limits_{i=1}^{n} \mathbf{f}_i$ is $((\cdots((\mathbf{f}_1 + \mathbf{f}_2) + \mathbf{f}_3) + \cdots) + \mathbf{f}_n)$. Similarly, $\mathbf{f} \cdot \mathbf{g}$ is $(\mathbf{f} \cdot \mathbf{g})$ etc. Finally, $-\mathbf{f}$ is $(-\mathbf{f})$ and $\mathbf{f} - \mathbf{g}$ is $(\mathbf{f} + (-\mathbf{g}))$.

Definition 3. *A basic probabilistic formula is any formula of the form*

$$\mathbf{f} \geqslant \underline{0}.$$

A probabilistic formula is a Boolean combination of basic probabilistic formulas.

□

As in the propositional case, $\neg$ and $\wedge$ are the primitive connectives, while all other connectives are introduced in the usual way. Furthermore, we define the following abbreviations:

- $\mathbf{f} \leqslant \underline{0}$ is $-\mathbf{f} \geqslant \underline{0}$.
- $\mathbf{f} > \underline{0}$ is $\neg(\mathbf{f} \leqslant \underline{0})$.

- $\mathtt{f} < \underline{0}$ is $\neg(\mathtt{f} \geqslant \underline{0})$.
- $\mathtt{f} = \underline{0}$ is $\mathtt{f} \leqslant \underline{0} \wedge \mathtt{f} \geqslant \underline{0}$.
- $\mathtt{f} \neq \underline{0}$ is $\neg(\mathtt{f} = \underline{0})$.
- $\mathtt{f} \geqslant \mathtt{g}$ is $\mathtt{f} - \mathtt{g} \geqslant \underline{0}$. Similarly are defined $\mathtt{f} \leqslant \mathtt{g}$, $\mathtt{f} > \mathtt{g}$, $\mathtt{f} < \mathtt{g}$, $\mathtt{f} = \mathtt{g}$ and $\mathtt{f} \neq \mathtt{g}$.

Probabilistic formulas will be denoted by ϕ, ψ and θ, possibly with indices. The set of all probabilistic formulas will be denoted by For_P.

By "formula" we mean either a classical formula or a probabilistic formula. We do not allow mixing of those types of formulas, nor the nesting of the probability operator P. For instance, $p_i \rightarrow P(p_j) \geqslant 0,3$ and $P(P(p_i) \geqslant 0,1) \geqslant 0,5$ are syntactically incorrect. Formulas will be denoted by Φ, Ψ and Θ, possibly with indices. The set of all formulas will be denoted by For.

We define the notion of a model as a special kind of Kripke model. Namely, a *model* M is any tuple $\langle W, H, \mu, v \rangle$ such that:

- W is a nonempty set. As usual, its elements will be called worlds.
- H is an algebra of sets over W.
- $\mu : H \longrightarrow [0,1]$ is a finitely additive probability measure.
- $v : For_C \times W \longrightarrow \{0,1\}$ is a truth assignment compatible with $\neg$ and $\wedge$. That is, $v(\neg\alpha, w) = 1 - v(\alpha, w)$ and $v(\alpha \wedge \beta, w) = v(\alpha, w) \cdot v(\beta, w)$.

For the given model M, let $[\alpha]_M$ be the set of all $w \in W$ such that $v(\alpha, w) = 1$. If the context is clear, we will write $[\alpha]$ instead of $[\alpha]_M$. We say that M is *measurable* if $[\alpha] \in H$ for all $\alpha \in For_C$.

Definition 4. *Let* $M = \langle W, H, \mu, v \rangle$ *be any measurable model. We define the satisfiability relation* $\models$ *recursively as follows:*

- $M \models \alpha$ *if* $v(\alpha, w) = 1$ *for all* $w \in W$.
- $M \models \mathtt{f} \geqslant \underline{0}$ *iff* $\mathtt{f}^M \geqslant 0$, *where* $\mathtt{f}^M$ *is recursively defined in the following way:*
  - $\underline{s}^M = s$.
  - $P(\alpha)^M = \mu([\alpha])$.
  - $(\mathtt{f} + \mathtt{g})^M = \mathtt{f}^M + \mathtt{g}^M$.
  - $(\mathtt{f} \cdot \mathtt{g})^M = \mathtt{f}^M \cdot \mathtt{g}^M$.
  - $(-\mathtt{f})^M = -(\mathtt{f}^M)$.
- $M \models \neg\phi$ *if* $M \not\models \phi$.
- $M \models \phi \wedge \psi$ *if* $M \models \phi$ *and* $M \models \psi$. □

A formula Φ is *satisfiable* if there is a measurable model M such that $M \models \Phi$; Φ is *valid* if it is satisfied in every measurable model. We say that the set T of formulas is *satisfiable* if there is a measurable model M such that $M \models \Phi$ for all $\Phi \in T$.

Notice that the last two clauses of Definition 4 provide validity of each tautology instance.

3   Axiomatization

In this section we will introduce the axioms and inference rules and prove that the proposed axiomatization is sound and strongly complete with respect to the class of all measurable models. The set of axioms for our axiomatic system, which we denote AX_{LPWF}, is divided into four parts:

- axioms for propositional reasoning,
- axioms for probabilistic reasoning,
- axioms about rational numbers and
- axioms about commutative ordered rings.

Axioms for propositional reasoning

  A1. $\tau(\Phi_1, \ldots, \Phi_n)$, where $\tau(p_1, \ldots, p_n) \in For_C$ is any tautology and all Φ_i are either propositional or probabilistic.

Axioms for probabilistic reasoning

  A2. $P(\alpha) \geqslant \underline{0}$.
  A3. $P(\top) = \underline{1}$.
  A4. $P(\bot) = \underline{0}$.
  A5. $P(\alpha \leftrightarrow \beta) = \underline{1} \ \rightarrow \ P(\alpha) = P(\beta)$.
  A6. $P(\alpha \vee \beta) = P(\alpha) + P(\beta) - P(\alpha \wedge \beta)$.

Axioms about rational numbers

  A7. $\underline{r} \geqslant \underline{s}$, iff $r \geqslant s$.

Axioms about commutative ordered rings

  A8. $\mathbf{f} + \mathbf{g} = \mathbf{g} + \mathbf{f}$.
  A9. $(\mathbf{f} + \mathbf{g}) + \mathbf{h} = \mathbf{f} + (\mathbf{g} + \mathbf{h})$.
 A10. $\mathbf{f} + \underline{0} = \mathbf{f}$.
 A11. $\mathbf{f} - \mathbf{f} = \underline{0}$.
 A12. $\mathbf{f} \cdot \mathbf{g} = \mathbf{g} \cdot \mathbf{f}$.
 A13. $\mathbf{f} \cdot (\mathbf{g} \cdot \mathbf{h}) = (\mathbf{f} \cdot \mathbf{g}) \cdot \mathbf{h}$.
 A14. $\mathbf{f} \cdot \underline{1} = \mathbf{f}$.
 A15. $\mathbf{f} \cdot (\mathbf{g} + \mathbf{h}) = (\mathbf{f} \cdot \mathbf{g}) + (\mathbf{f} \cdot \mathbf{h})$.
 A16. $\mathbf{f} \geqslant \mathbf{g} \ \vee \ \mathbf{g} \geqslant \mathbf{f}$.
 A17. $(\mathbf{f} \geqslant \mathbf{g} \ \wedge \ \mathbf{g} \geqslant \mathbf{h}) \rightarrow \mathbf{f} \geqslant \mathbf{h}$.
 A18. $\mathbf{f} \geqslant \mathbf{g} \ \rightarrow \ \mathbf{f} + \mathbf{h} \geqslant \mathbf{g} + \mathbf{h}$.
 A19. $(\mathbf{f} \geqslant \mathbf{g} \ \wedge \ \mathbf{h} > 0) \rightarrow \mathbf{f} \cdot \mathbf{h} \geqslant \mathbf{g} \cdot \mathbf{h}$.

Inference rules

  R1. From Φ and $\Phi \rightarrow \Psi$ infer Ψ.
  R2. From α infer $P(\alpha) = \underline{1}$.

R3. From the set of premises

$$\{\phi \ \rightarrow \ \mathtt{f} \geqslant \underline{-n^{-1}} \mid n = 1, 2, 3, \ldots\}$$

infer $\phi \rightarrow \mathtt{f} \geqslant \underline{0}$.

Let us briefly comment axioms and inference rules. A1, together with Definition 4, provides completeness for tautology instances; A2 provides nonnegativity of probability measures; A3, A4 and A6 provide non-triviality and finite additivity of probability measures; A5 provides that equivalent formulas have the same probability; A7 provides formal unification of terms like $\underline{1} + \underline{1}$ and $\underline{2}$, $\underline{0.2} + \underline{0.3}$ and $\underline{0.5}$ etc. The rest of the axioms provide computational properties of a commutative ordered ring. Rule R1 is modus ponens, while Rule R2 resembles necessitation. Finaly, Rule R3 tames the well known non-compactness phenomena. Namely, consider the theory

$$T = \{P(\alpha) > 0\} \cup \{P(\alpha) \leqslant \underline{n^{-1}} \mid n = 1, 2, 3, \ldots\}.$$

Clearly, it is unsatisfiable (any measure of α must be a proper infinitesimal, while the semantics for polynomial weight formulas allows only the real-valued probability functions), but finitely satisfiable (every finite subset of T is satisfiable). In the presence of R3, T becomes inconsistent.

Definition 5. *A formula Φ is a theorem $(\vdash \Phi)$ if there is an at most countable sequence of formulas $\Phi_0, \Phi_1, \ldots, \Phi$, such that every Φ_i is an axiom or it is derived from the preceding formulas of the sequence by an inference rule. In this paper we will also use the notion of deducibility. A formula Φ is deducible from a set T of sentences $(T \vdash \Phi)$ if there is an at most countable sequence of formulas $\Phi_0, \Phi_1, \ldots, \Phi$, such that every Φ_i is an axiom or a formula from the set T, or it is derived from the preceding formulas by an inference rule. A formula Φ is a theorem $(\vdash \Phi)$ if it is deducible from the empty set. A set T of sentences is consistent if there is at least one formula from For_C, and at least one formula from For_P that are not deducible from T, otherwise T is inconsistent. A consistent set T of sentences is said to be maximal consistent if the following holds:*

- *for every $\alpha \in For_C$, if $T \vdash \alpha$, then $\alpha \in T$ and $P(\alpha) \geqslant \underline{1} \in T$, and*
- *for every $\phi \in For_P$, either $\phi \in T$ or $\neg\phi \in T$.*

A set T is deductively closed if for every $\Phi \in For$, if $T \vdash \Phi$, then $\Phi \in T$. $\square$

Observe that the length of inference may be any successor ordinal lesser than the first uncountable ordinal ω_1. Using a straightforward induction on the length of the inference, one can easily show that the above axiomatization is sound with respect to the class of all measurable models.

4   Completeness

Theorem 1 (Deduction theorem). *Suppose that T is an arbitrary set of formulas and that $\Phi, \Psi \in For$. Then, $T \vdash \Phi \rightarrow \Psi$ iff $T \cup \{\Phi\} \vdash \Psi$.*

Proof. If $T \vdash \Phi \to \Psi$, then clearly $T \cup \{\Phi\} \vdash \Phi \to \Psi$, so, by modus ponens (R1), $T \cup \{\Phi\} \vdash \Psi$. Conversely, let $T \cup \{\Phi\} \vdash \Psi$. As in the classical case, we will use the induction on the length of inference to prove that $T \vdash \Phi \to \Psi$. The proof differs from the classical only in the cases when we apply the infinitary inference rule R3.

Suppose that Ψ is the formula $\phi \to \mathbf{f} \geqslant \underline{0}$ and that

$$T \vdash \Phi \to (\phi \to \mathbf{f} \geqslant \underline{-n^{-1}})$$

for all n. Since the formula

$$(p_0 \to (p_1 \to p_2)) \leftrightarrow ((p_0 \wedge p_1) \to p_2),$$

is tautology, we obtain

$$T \vdash (\Phi \wedge \phi) \to \mathbf{f} \geqslant \underline{-n^{-1}}$$

for all n (A1). Now, by R3,

$$T \vdash (\Phi \wedge \phi) \to \mathbf{f} \geqslant \underline{0}.$$

Hence, by the same tautology, $T \vdash \Phi \to \Psi$. $\square$

Example 1. Let us prove that the formula

$$P(\alpha \to \beta) = \underline{1} \;\to\; P(\alpha) \leqslant P(\beta)$$

is a theorem. This formula is a reminiscence of the well known modal-axiom K. We will need the following claim:

$$\vdash P(\neg\alpha) = \underline{1} - P(\alpha). \tag{1}$$

Indeed,

$$P(\alpha \vee \neg\alpha) \stackrel{A6}{=} P(\alpha) + P(\neg\alpha) - P(\alpha \wedge \neg\alpha)$$
$$\stackrel{A4}{=} P(\alpha) + P(\neg\alpha) - \underline{0}$$
$$\stackrel{A11}{=} P(\alpha) + P(\neg\alpha).$$

By A3, $P(\alpha \vee \neg\alpha) = \underline{1}$, so

$$\underline{1} = P(\alpha) + P(\neg\alpha).$$

Hence,

$$\underline{1} - P(\neg\alpha) = (P(\alpha) + P(\neg\alpha)) - P(\neg\alpha),$$

so

$$\underline{1} - P(\neg\alpha) = (P(\alpha) + P(\neg\alpha)) - P(\neg\alpha)$$
$$\stackrel{A9}{=} P(\alpha) + (P(\neg\alpha) - P(\neg\alpha))$$
$$\stackrel{A11}{=} P(\alpha) + \underline{0}$$
$$\stackrel{A10}{=} P(\alpha).$$

Thus, if we take $\neg\alpha$ instead of α, we directly obtain (1) by A5.

$$P(\alpha \to \beta) \overset{A5}{=} P(\neg\alpha \vee \beta)$$
$$\overset{A6}{=} P(\neg\alpha) + P(\beta) - P(\neg\alpha \wedge \beta)$$
$$\overset{(1)}{=} \underline{1} - P(\alpha) + P(\beta) - P(\neg\alpha \wedge \beta).$$

Hence, $P(\alpha \to \beta) - \underline{1} + P(\alpha) = P(\beta) - P(\neg\alpha \wedge \beta)$. Since

$$P(\alpha \to \beta) = \underline{1},$$

we obtain $P(\alpha) = P(\beta) - P(\neg\alpha \wedge \beta)$. It remains to prove that

$$P(\beta) \geqslant P(\beta) - P(\neg\alpha \wedge \beta).$$

The above is equivalent to $\underline{0} \geqslant -P(\neg\alpha \wedge \beta)$, which is equivalent to

$$P(\neg\alpha \wedge \beta) \geqslant \underline{0}.$$

Finally, the last formula is an instance of A2.   □

The next technical lemma will be used in the construction of a maximal consistent extension of some consistent set of formulas.

Lemma 1. *Suppose that T is a consistent set of formulas. If $T \cup \{\phi \to \mathbf{f} \geqslant \underline{0}\}$ is inconsistent, then there is a positive integer n such that $T \cup \{\phi \to \mathbf{f} < \underline{-n^{-1}}\}$ is consistent.*

Proof. The proof is based on the reductio ad absurdum argument. Thus, let us suppose that $T \cup \{\phi \to \mathbf{f} < \underline{-n^{-1}}\}$ is inconsistent for all n. Due to Deduction theorem, we can conclude that

$$T \vdash \phi \to \mathbf{f} \geqslant \underline{-n^{-1}}$$

for all n. By R3, $T \vdash \phi \to \mathbf{f} \geqslant \underline{0}$, so T is inconsistent; a contradiction.   □

Definition 6. *Suppose that T is a consistent set of formulas and that $For_P = \{\phi_i \mid i = 0, 1, 2, 3, \ldots\}$. We define a completion T^* of T recursively as follows:*

1. $T_0 = T \cup \{\alpha \in For_C \mid T \vdash \alpha\} \cup \{P(\alpha) = 1 \mid T \vdash \alpha\}$.
2. *If $T_i \cup \{\phi_i\}$ is consistent, then $T_{i+1} = T_i \cup \{\phi_i\}$.*
3. *If $T_i \cup \{\phi_i\}$ is inconsistent, then:*
   (a) If ϕ_i has the form $\psi \to \mathbf{f} \geqslant \underline{0}$, then

   $$T_{i+1} = T_i \cup \{\psi \to \mathbf{f} < \underline{-n^{-1}}\},$$

   where n is a positive integer such that T_{i+1} is consistent. The existence of such n is provided by Lemma 1.
   (b) Otherwise, $T_{i+1} = T_i$.   □

Obviously, each T_i is consistent. In the next theorem we will prove that T^* is deductively closed, consistent and maximal with respect to For_P.

Theorem 2. *Suppose that T is a consistent set of formulas and that T^* is constructed as above. Then:*

1. *T^* is deductively closed, id est, $T^* \vdash \Phi$ implies $\Phi \in T^*$.*
2. *There is $\phi \in For_P$ such that $\phi \notin T^*$.*
3. *For each $\phi \in For_P$, either $\phi \in T^*$, or $\neg\phi \in T^*$.*

Proof. We will prove only the first clause, since the remaining clauses can be proved in the same way as in the classical case. In order to do so, it is sufficient to prove the following four claims:

(i) Each instance of any axiom is in T^*.
(ii) If $\Phi \in T^*$ and $\Phi \to \Psi \in T^*$, then $\Psi \in T^*$.
(iii) If $\alpha \in T^*$, then $P(\alpha) = 1 \in T^*$.
(iv) If $\{\phi \to f \geqslant \underline{-n^{-1}} \mid n = 1, 2, 3, \ldots\}$ is a subset of T^*, then $\phi \to f \geqslant \underline{0} \in T^*$.

(i): If $\Phi \in For_C$, then $\Phi \in T_0$. Otherwise, there is a nonnegative integer i such that $\Phi = \phi_i$. Since $\vdash \phi_i$, $T_i \vdash \phi_i$ as well, so $\phi_i \in T_{i+1}$.

(ii): If $\Phi, \Phi \to \Psi \in For_C$, then $\Psi \in T_0$. Otherwise, let $\Phi = \phi_i$, $\Psi = \phi_j$, and $\Phi \to \Psi = \phi_k$. Then, Ψ is a deductive consequence of each T_l, where $l \geqslant \max(i, k) + 1$. Let $\neg\Psi = \phi_m$. If $\phi_m \in T_{m+1}$, then $\neg\Psi$ is a deductive consequence of each T_n, where $n \geqslant m+1$. So, for every $n \geqslant \max(i, k, m)+1$, $T_n \vdash \Psi \wedge \neg\Psi$, a contradiction. Thus, $\neg\Psi \notin T^*$. On the other hand, if also $\Psi \notin T^*$, we have that $T_n \cup \{\Psi\} \vdash \bot$, and $T_n \cup \{\neg\Psi\} \vdash \bot$, for $n \geqslant \max(j, m) + 1$, a contradiction with the consistency of T_n. Thus, $\Psi \in T^*$.

(iii): If $\alpha \in T^*$, then $\alpha \in T_0$, so $P(\alpha) = \underline{1} \in T_0$.

(iv): Suppose that $\{\phi \to P(\alpha) \geqslant \underline{-n^{-1}} \mid n = 0, 1, 2, \ldots\}$ is a subset of T^*. We want to prove that $\phi \to P(\alpha) \geqslant \underline{0} \in T^*$. The proof uses reductio ad absurdum argument. So, let $\phi \to P(\alpha) \geqslant \underline{0} = \phi_i$ and let us suppose that $T_i \cup \{\phi_i\}$ is inconsistent. By 3.(a) of Definition 6, there is a positive integer n such that

$$T_{i+1} = T_i \cup \{\phi \to P(\alpha) < \underline{-n^{-1}}\}$$

and T_{i+1} is consistent. Then, for all sufficiently large k, $T_k \vdash \phi \to P(\alpha) < \underline{-n^{-1}}$ and $T_k \vdash \phi \to P(\alpha) \geqslant \underline{-n^{-1}}$, so $T_k \vdash \phi \to \psi$ for all $\psi \in For_P$. In particular, $T_k \vdash \phi \to P(\alpha) \geqslant \underline{0}$, i.e., $T_k \vdash \phi_i$ for all sufficiently large k. But, $\phi_i \notin T^*$, so ϕ_i is inconsistent with all T_k, $k \geqslant i$. It follows that each T_k is inconsistent for sufficiently large k, a contradiction.

Thus, $T_i \cup \{\phi_i\}$ is consistent, so $\phi \to P(\alpha) \geqslant \underline{0} \in T_{i+1}$. $\qquad\square$

For the given completion T^*, we define a *canonical model* M^* as follows:

- W is the set of all truth assignments $w : For_C \longrightarrow \{0,1\}$ that satisfy all propositional formulas from T^*.
- $v : For_C \times W \longrightarrow \{0,1\}$ is defined by $v(\alpha, w) = 1$ iff $w(\alpha) = 1$.
- $H = \{[\alpha] \mid \alpha \in For_C\}$.
- $\mu : H \longrightarrow [0,1]$ is defined by

$$\mu([\alpha]) = \sup\{s \in [0,1] \cap \mathbb{Q} \mid T^* \vdash P(\alpha) \geqslant \underline{s}\}.$$

Lemma 2. M^* *is a measurable model.*

Proof. We need to prove that H is an algebra of sets and that μ is a finitely additive probability measure. It is easy to see that H is an algebra of sets, since $[\alpha] \cap [\beta] = [\alpha \wedge \beta]$, $[\alpha] \cup [\beta] = [\alpha \vee \beta]$ and $H \setminus [\alpha] = [\neg\alpha]$.

Concerning μ, first note that it is well defined, i.e., $\mu([\alpha]) = \mu([\beta])$ whenever $\alpha \leftrightarrow \beta$ is a theorem of T^*. Indeed, if $T^* \vdash \alpha \leftrightarrow \beta$, then by R2 and A5 we have that $T^* \vdash P(\alpha) = P(\beta)$. Next, we want to prove that μ is a probability measure, i.e., that:

1. $\mu([\alpha]) \geqslant 0$.
2. $\mu([\alpha]) = 1 - \mu([\neg\alpha])$.
3. $\mu([\alpha \vee \beta]) = \mu([\alpha]) + \mu([\beta]) - \mu([\alpha \wedge \beta])$.

Formally, instead of $\mu([\alpha \vee \beta])$ and $\mu([\alpha \wedge \beta])$, there should be $\mu([\alpha] \cup [\beta])$ and $\mu([\alpha] \cap [\beta])$, respectively. However, $[\alpha] \cup [\beta] = [\alpha \vee \beta]$ and $[\alpha] \cap [\beta] = [\alpha \wedge \beta]$ by the definition of $[\ ]$. The only item that is not immediate is the last one, so we will prove it. Thus, let $a = \mu([\alpha])$, $b = \mu([\beta])$ and $c = \mu([\alpha \wedge \beta])$. Since $\mathbb{Q}$ is dense in $\mathbb{R}$, we can chose sequences $r_0 < r_1 < r_2 < \cdots$, $s_0 < s_1 < s_2 < \cdots$, $k_0 < k_1 < k_2 < \cdots$ and $l_0 > l_1 > l_2 > \cdots$ in $\mathbb{Q} \cap [0,1]$ so that $\lim r_n = a$, $\lim s_n = b$ and $\lim k_n = \lim l_n = c$. By the definition of μ,

$$T^* \vdash P(\alpha) \geqslant \underline{r_n} \wedge P(\beta) \geqslant \underline{s_n} \wedge \underline{k_n} \leqslant P(\alpha \wedge \beta) \leqslant \underline{l_n}$$

for all n. Applying axioms about commutative ordered rings, we obtain that

$$T^* \vdash \underline{r_n} + \underline{s_n} - \underline{l_n} \leqslant P(\alpha) + P(\beta) - P(\alpha \wedge \beta) \leqslant \underline{r_n} + \underline{s_n} - \underline{k_n}$$

for all n. Since $T^* \vdash \underline{r} + \underline{s} = \underline{r+s}$ for all $r, s \in \mathbb{Q}$, we have that

$$T^* \vdash \underline{r_n + s_n - l_n} \leqslant P(\alpha) + P(\beta) - P(\alpha \wedge \beta) \leqslant \underline{r_n + s_n - k_n}$$

for all n. As a consequence,

$$\begin{aligned}
\mu([\alpha \vee \beta]) &= \sup\{s \mid T^* \vdash P(\alpha \vee \beta) \geqslant \underline{s}\} \\
&= \sup\{s \mid T^* \vdash P(\alpha) + P(\beta) - P(\alpha \wedge \beta) \geqslant \underline{s}\} \\
&= \lim_{n \to \infty} (r_n + s_n - l_n) = \lim_{n \to \infty} (r_n + s_n - k_n) \\
&= a + b - c \\
&= \mu([\alpha]) + \mu([\beta]) - \mu([\alpha \wedge \beta]). \qquad \square
\end{aligned}$$

Theorem 3 (Strong completeness theorem). *Every consistent set of formulas has a measurable model.*

Proof. Let T be a consistent set of formulas. We can extend it to a maximal consistent set T^* and define a canonical model M^*, as above. By the induction on the complexity of formulas we can prove that $M^* \models \Phi$ iff $\Phi \in T^*$.

To begin the induction, let $\Phi = \alpha \in For_C$. If $\alpha \in T^*$, i.e., $T^* \vdash \alpha$, then by the definition of M^*, $M^* \models \alpha$. Conversely, if $M^* \models \alpha$, then $w(\alpha) = 1$ for all $w \in W$. Since W is the set of all classical models of T_0, we have that $T_0 \models \alpha$. By the completeness of classical propositional logic, $T_0 \vdash \alpha$, and $\alpha \in T^*$.

Let us suppose that $\mathbf{f} \geqslant \underline{0} \in T^*$. Then, using the axioms for ordered commutative rings, we can prove that

$$T^* \vdash \mathbf{f} = \sum_{i=1}^{m} \underline{s_i} \cdot P(\alpha_1)^{n_{i1}} \cdots P(\alpha_k)^{n_{ik}}$$

and

$$T^* \vdash \sum_{i=1}^{m} \underline{s_i} \cdot P(\alpha_1)^{n_{i1}} \cdots P(\alpha_k)^{n_{ik}} \geqslant \underline{0},$$

for some $s_i \in \mathbb{Q}$ and some $\alpha_j \in For_C$ such that $T^* \vdash P(\alpha_j) > \underline{0}$. Let $a_j = \mu([\alpha_j])$. It remains to prove that

$$\sum_{i=1}^{m} s_i \cdot a_1^{n_{i1}} \cdots a_k^{n_{ik}} \geqslant 0. \tag{2}$$

Notice that (2) is an immediate consequence of the following facts:

- $\mu([\gamma]) = \sup\{s \in \mathbb{Q} \mid T^* \vdash P(\gamma) \geqslant \underline{s}\}$, $\gamma \in For_C$.
- The real function $F(x_1, \ldots, x_k) = \sum_{i=1}^{n} s_i \cdot x_1^{n_{i1}} \cdots x_k^{n_{ik}}$ is continuous.
- For each $r, s \in \mathbb{Q}$, $T^* \vdash \underline{r} \geqslant \underline{s}$ iff $r \geqslant s$.
- $\mathbb{Q}^k$ is dense in $\mathbb{R}^k$.

For the other direction, let $M^* \models \mathbf{f} \geqslant \underline{0}$. If $\mathbf{f} \geqslant \underline{0} \notin T^*$, by the construction of T^*, there is a positive integer n such that

$$\mathbf{f} < \underline{-n^{-1}} \in T^*.$$

Reasoning as above, we have that $\mathbf{f}^{M^*} < 0$, a contradiction. So, $\mathbf{f} \geqslant \underline{0} \in T^*$.

Let $\Phi = \neg\phi \in For_P$. Then $M^* \models \neg\phi$ iff $M^* \not\models \phi$ iff $\phi \notin T^*$ iff (by Theorem 2) $\neg\phi \in T^*$.

Finally, let $\Phi = \phi \wedge \psi \in For_P$. $M^* \models \phi \wedge \psi$ iff $M^* \models \phi$ and $M^* \models \psi$ iff ϕ, $\psi \in T^*$ iff (by Theorem 2) $\phi \wedge \psi \in T^*$. $\square$

5   Decidability and Complexity

Concerning decidability, we notice that the PSPACE decision procedure described in [2] works for our logic as well. It is based on the small model theorems (the theorems 5.1 and 5.2 from [2]) and on the Canny's PSPACE decision procedure for the existential theory of the reals [1].

6   Conclusion

In this paper we introduced a sound and strongly-complete infinitary axiomatic system for the probabilistic logic with polynomial weight formulas. Polynomial weight formulas were introduced in [2]. As it was noticed in [20,13], it is not possible to give any finitary strongly complete axiomatization for that logic. The reason is that the compactness theorem follows easily from the extended completeness theorem (if we have a finitary axiomatization), while, as we noted above, compactness does not hold for the considered logic.

In our case the strong completeness was made possible by adding an infinitary rule of inference. However, since our formulas are finite polynomial weight formulas, it is still possible to prove the decidability using the method from [2]. It should be noted that in [2] the first order theory of polynomial weight formulas has an EXPSPACE decision procedure, while for our system the procedure is in PSPACE.

The obtained formalism is quite expressive and allows representation of uncertain knowledge, where uncertainty is modelled by probability formulas. For instance, conditional statement of the form "the probability of α given β is at least 0.95" can be written as

$$P(\alpha \wedge \beta) \geqslant \underline{0.95} \cdot P(\beta).$$

Due to the possibility of making products of terms, more complicated formulas can be expressed. An example of this kind would be Bayes' formula which connects the posterior and the prior probabilities ($\{\alpha_i\}$ forms a partition of the event space):

$$CP(\alpha_i|\beta) = \frac{CP(\beta|\alpha_i)P(\alpha_i)}{\sum\limits_{j} CP(\beta|\alpha_j)P(\alpha_j)}$$

can be written as

$$P(\alpha_i \wedge \beta) \cdot \sum_{j} P(\alpha_j \wedge \beta) = P(\beta) \cdot P(\alpha_i \wedge \beta).$$

As a consequence, Bayesian inference can be represented in our system.

Although numerical coefficients may appear in formulas, pure qualitative statements can also be made. For example, the statement "the probability of α_1 is at least the product of probabilities of α_2 and α_3" can be written as

$$P(\alpha_1) \geqslant P(\alpha_2) \cdot P(\alpha_3).$$

As we noted in Section 2, the considered language does not allow mixing of propositional and probabilistic formulas, nor nesting of the probability operator

P. However, using ideas from [13], we can overcome those restrictions. Also, note that although we cannot express statements like "q is true and its probability is at least s", we can consider a theory T which contains the formulas q and $P(q) \geqslant \underline{s}$.

An approach similar to the one presented here can be applied to de Finetti style conditional probabilities. Future research could also consider a possibility of dealing with probabilistic first-order formulas.

References

1. Canny, J.: Some algebraic and geometric computations in PSPACE. In: Proc. of XX ACM Symposium on theory of computing, pp. 460–467 (1978)
2. Fagin, R., Halpern, J., Megiddo, N.: A logic for reasoning about probabilities. Information and Computation 87(1–2), 78–128 (1990)
3. Heifetz, A., Mongin, P.: Probability logic for type spaces. Games and economic behavior 35, 31–53 (2001)
4. Lehmann, D.: Generalized qualitative probability: Savage revisited. In: Horvitz, E., Jensen, F. (eds.) UAI 1996. Procs. of 12th Conference on Uncertainty in Artificial Intelligence, pp. 381–388 (1996)
5. Lukasiewicz, T.: Probabilistic Default Reasoning with Conditional Constraints. Annals of Mathematics and Artificial Intelligence 34, 35–88 (2002)
6. Lukasiewicz, T.: Nonmonotonic probabilistic logics under variable-strength inheritance with overriding: Complexity, algorithms, and implementation. International Journal of Approximate Reasoning 44(3), 301–321 (2007)
7. Marchioni, E., Godo, L.: A Logic for Reasoning about Coherent Conditional Probability: A Modal Fuzzy Logic Approach. In: Alferes, J.J., Leite, J.A. (eds.) JELIA 2004. LNCS (LNAI), vol. 3229, pp. 213–225. Springer, Heidelberg (2004)
8. Meier, M.: An infinitary probability logic for type spaces. Israel J. of Mathematics, ∞
9. Narens, L.: On qualitative axiomatizations for probability theory. Journal of Philosophical Logic 9(2), 143–151 (1980)
10. Nilsson, N.: Probabilistic logic. Artificial intelligence 28, 71–87 (1986)
11. Ognjanović, Z., Rašković, M.: A logic with higher order probabilities. Publications de l'institut mathematique, Nouvelle série, tome 60(74), 1–4 (1996)
12. Ognjanović, Z., Rašković, M.: Some probability logics with new types of probability operators. J. Logic Computat. 9(2), 181–195 (1999)
13. Ognjanović, Z., Rašković, M.: Some first-order probability logics. Theoretical Computer Science 247(1–2), 191–212 (2000)
14. Ognjanović, Z., Timotijević, T., Stanojević, A.: Database of papers about probability logics. Mathematical institute Belgrade (2005),
http://problog.mi.sanu.ac.yu/
15. Ognjanović, Z., Marković, Z., Rašković, M.: Completeness Theorem for a Logic with imprecise and conditional probabilities. Publications de L'Institute Matematique (Beograd) 78(92), 35–49 (2005)
16. Ognjanović, Z., Perović, A., Rašković, M.: Logic with the qualitative probability operator. Logic journal of IGPL, doi:10.1093/jigpal/jzm031

17. Rašković, M.: Classical logic with some probability operators. Publications de l'institut mathematique, Nouvelle série, tome 53(67), 1–3 (1993)
18. Rašković, M., Ognjanović, Z.: A first order probability logic LP_Q. Publications de l'institut mathematique, Nouvelle série, tome 65(79), 1–7 (1999)
19. Rašković, M., Ognjanović, Z., Marković, Z.: A logic with Conditional Probabilities. In: Alferes, J.J., Leite, J.A. (eds.) JELIA 2004. LNCS (LNAI), vol. 3229, pp. 226–238. Springer, Heidelberg (2004)
20. van der Hoek, W.: Some considerations on the logic $P_F D$: a logic combining modality and probability. Journal of Applied Non-Classical Logics 7(3), 287–307 (1997)

A Transformation-Based Approach to View Updating in Stratifiable Deductive Databases

Andreas Behrend and Rainer Manthey

University of Bonn, Institute of Computer Science III
Roemerstr. 164, D-53117 Bonn, Germany
{behrend,manthey}@cs.uni-bonn.de

Abstract. In this paper we present a new rule-based approach for consistency preserving view updating in deductive databases. Based on rule transformations performed during schema design, fixpoint evaluations of these rules at run time compute consistent realizations of view update requests. Alternative realizations are expressed using disjunctive Datalog internally. The approach extends and integrates standard techniques for efficient query answering and integrity checking (based on transformation techniques and fixpoint computation, too). Views may be stratifiably recursive. The set-orientedness of the approach makes it potentially useful in the context of (commercial) SQL systems, too.

1 Introduction

In deductive databases, views are usually updated indirectly, via changes of the underlying base relations. If update requests for derived relations (view updates) are accepted at all, they cannot be directly executed. Instead, possible realizations of the view update (i.e. base relation updates inducing the requested change of the view) have to be determined and one of them chosen for execution, if alternatives exist. Realizations are (at least) expected to be consistency preserving. In "real life" SQL databases, view updating (VU) is heavily restricted in order to guarantee existence of a unique realization. In the context of Datalog and logic programming, however, the approach followed by most researchers is to systematically analyze all "reasonable" realizations and to leave the final choice to the user who has issued the request. We systematically compute alter-native consistent realizations, too, restricting computation to those realizations which are (in a particular sense) "small enough" for being meaningful.

Our approach is transformation-based, i.e., we compute a set of VU rules from the rules of the given deductive database. VU requests as well as their realizations are represented by VU facts, either explicitly introduced (the request to be satisfied) or derived via the VU rules (intermediate requests or realizations). VU rules are expressed in disjunctive Datalog, disjunctive facts representing choices between alternative realizations. Computation of VU facts is performed by means of fixpoint computation (which can be viewed as a set-oriented version of model generation). We rely on fixpoint computation as the "engine" for view updating

S. Hartmann and G. Kern-Isberner (Eds.): FoIKS 2008, LNCS 4932, pp. 253–271, 2008.

as this will be the method of choice once the limited, but quite powerful forms of recursion defined in the SQL standard will at last be implemented by the commercial DBMS vendors. Furthermore, this choice enables us to employ, e.g., the Magic Sets approach (transformation-based, too) for optimizing evaluation of recursive queries and rules.

A related standard mode of inference driven by updates is update propagation (UP), i.e. the incremental computation of all consequences of a given base data change on derived data (induced updates). Whereas VU is inherently a top-down problem (from the request down to the realizations), UP is bottom-up (from the base updates up to the induced updates). There are various transformation- and fixpoint-based approaches to UP around, e.g [3,6,11,13,17,19,22]. UP has been introduced for checking integrity over views as well as for adapting materialized views efficiently. In a VU context, UP is ideal for controlling integrity of the proposed realizations, too, and may in addition be used for computing any kind of further side effect caused by the chosen realization, as such side effects will certainly correspond to updates induced by the respective realization. We make use of "Magic Updates", our own variant of rule-based UP [6], which makes use of Magic Sets for ensuring goal-directedness of UP.

After an initial VU phase determining a first set of candidate realizations, and a subsequent first UP phase eliminating inconsistent branches of the resulting tree of alternatives, we try to identify cases where integrity violations caused by the resp. realization can be repaired by generating additional VU requests. Similar extensions to the initial VU request are used for compensating side effects sometimes contradicting the effects of the chosen realization steps, e.g. by introducing new derivations for a fact intended to disappear). The resulting overall process thus alternates between clearly distinguishable phases of "local" fixpoint computations using either VU rules or UP rules only. Phases are linked by additional transition rules triggering UP after a VU subprocess has stopped, or triggering further VU inferences (for compensation and repair) once UP has terminated. Termination is guaranteed unless no finite consistent realization exists.

An important advantage of our approach is that it uses other transformation- and fixpoint-based inference modes, present in a set-oriented (e.g. an SQL) context anyway (Magic Sets for efficient query evaluation, Magic Updates for efficient integrity checking and materialized view maintenance) rather than introducing a completely new style of computation, which would be hard to integrate with existing view-related DBMS services. From a theoretical perspective, covering all aspects of VU in a coherent formal manner (transformed internal rules as specifications, fixpoint operators as interpreters) appears a promising basis for investigating general problems and properties of view updating such as, e.g., termination, completeness, minimality and the like.

There are various well-known approaches to VU based on disjunctive fixpoints/model generation, e.g. [7,10,12]. Our approach is probably closest to that of Bry [7], combining the given application rules with an application-independent set of meta rules expressing the logical relation between consecutive database states. Our (application-specific) VU rules can be seen as optimized versions of partial interpretations of Bry's meta rules over the respective application rules, UP rules are incremental (and thus more efficient) specifications of differences between old and new state. Furthermore, set-oriented fixpoint computation replaces Bry's instance-oriented model generator.

Another closely related rule-based approach to both, VU and UP, is the internal events method [18,22,23]. In this approach, the same transformed rules are applied for both, VU and UP purposes. SLDNF resolution is used for evaluating rules rather than fixpoint computation, thus limiting the approach in presence of recursion. In addition, the resolution engine has to maintain internal, temporary data structures for memorizing control information needed for finding realizations, thus leading to a rather "heavy" computation engine not suitable in a set-oriented (SQL) context. This is the "price to pay" for the formally elegant re-use of internal events rules (originally introduced for UP only) for finding realizations of VU requests. We prefer to keep the underlying runtime engine simple (and uniform) rather than the transformation process (applied only once during schema design). There are many other approaches to VU exhibiting similarities and differences to our approach if analyzed more deeply - such a discussion is beyond the scope of this paper.

In Section 2, we briefly summarize necessary basic notions and notations for stratifiable deductive databases and their extension to disjunctive databases. In Section 3 we recall our approach to update propagation ([6,11,17]), as UP is needed as a subprocess in our VU method and because the style of rule transformation used in the UP case serves as a model for the newly introduced VU rules. In Section 4, we first present these new transformations for computing the VU analysis rules, then we address the role of UP in view updating. Afterwards, we introduce the fixpoint algorithm "driving" both, VU and UP rules according to an alternating control strategy. In Section 5, we briefly discuss pros and cons of the new approach.

2   Basic Concepts

In this section, we recall basic concepts for stratifiable definite and indefinite (disjunctive) databases [5,16]. The fixpoint semantics of indefinite databases induces a model generation procedure that can handle our transformed VU rules in which alternative view update realizations are represented in form of disjunctions.

2.1   Deductive Databases

A Datalog *rule* is a function-free clause of the form $H_1 \leftarrow L_1 \wedge \ldots \wedge L_m$ with $m \geq 1$ where H_1 is an atom denoting the rule's head, and $L_1, \ldots, L_m$ are literals, i.e. positive or negative atoms, representing its body. We assume all deductive rules to be *safe*, i.e., all variables occurring in the head or in any negated literal of a rule must be also present in a positive literal in its body. If $A \equiv p(t_1, \ldots, t_n)$

with $n \geq 0$ is a literal, we use $\mathtt{vars}(A)$ to denote the set of variables occurring in A and $\mathtt{pred}(A)$ to refer to the predicate symbol p of A. If A is the head of a given rule R, we use $\mathtt{pred}(R)$ to refer to the predicate symbol of A. For a set of rules $\mathcal{R}$, $\mathtt{pred}(\mathcal{R})$ is defined as $\cup_{r \in \mathcal{R}}\{\mathtt{pred}(r)\}$. A *fact* is a ground atom in which every t_i is a constant.

A *deductive database* $\mathcal{D}$ is a triple $\langle \mathcal{F}, \mathcal{R}, \mathcal{I} \rangle$ where $\mathcal{F}$ is a finite set of facts (called *base facts*), $\mathcal{I}$ is a finite set of integrity constraints and $\mathcal{R}$ a finite set of rules such that $\mathtt{pred}(\mathcal{F}) \cap \mathtt{pred}(\mathcal{R}) = \emptyset$ and $\mathtt{pred}(\mathcal{I}) \subseteq \mathtt{pred}(\mathcal{F} \cup \mathcal{R})$. Within a deductive database $\mathcal{D}$, a predicate symbol p is called derived (view predicate), if $p \in \mathtt{pred}(\mathcal{R})$. The predicate p is called extensional (or base predicate), if $p \in \mathtt{pred}(\mathcal{F})$. The *state* of a database $\mathcal{D}$ is defined as the set of all facts that can be derived by the rules $\mathcal{R}$ including the facts in $\mathcal{F}$. An *integrity constraint* is represented by a ground atom which is required to be derivable in every state of the database. For the sake of simplicity of exposition, and without loss of generality, we assume that a predicate is either base or derived, but not both, which can be easily achieved by rewriting a given database. Additionally, we solely consider *stratifiable rules* [19] which do not allow recursion through negative predicate occurrences. Given a stratifiable deductive database $\mathcal{D} = \langle \mathcal{F}, \mathcal{R}, \mathcal{I} \rangle$, its semantics is defined by the perfect model $PM_\mathcal{D}$ [19] of $\mathcal{F} \cup \mathcal{R}$ which satisfies all integrity constraints $\mathcal{I}$, i.e., $\mathcal{I} \subseteq PM_\mathcal{D}$. Otherwise, the semantics of $\mathcal{D}$ is undefined.

For determining the semantics of a stratifiable database, the iterated fixpoint computation can be used which constructs bottom-up a sequence of least Herbrand models according to a given stratification of the rules. As an example, consider the following consistent deductive database $\mathcal{D} = \langle \mathcal{F}, \mathcal{R}, \mathcal{I} \rangle$:

$\underline{\mathcal{R}}$		$\underline{\mathcal{I}}$	$\underline{\mathcal{F}}$
$\mathtt{h}(X, Y) \leftarrow \mathtt{p}(X, Y) \wedge \neg \mathtt{e}(X, Y)$	$\mathtt{ic}_1 \leftarrow \mathtt{p}(X, Y)$	$\mathtt{ic}_1$	$\mathtt{e}(1, 2)$
$\mathtt{p}(X, Y) \leftarrow \mathtt{e}(X, Y)$	$\mathtt{ic}_2 \leftarrow \neg \mathtt{aux}$	$\mathtt{ic}_2$	$\mathtt{e}(1, 4)$
$\mathtt{p}(X, Y) \leftarrow \mathtt{e}(X, Z) \wedge \mathtt{p}(Z, Y)$	$\mathtt{aux} \leftarrow \mathtt{p}(X, X)$		$\mathtt{e}(2, 3)$

Relation p is the transitive closure of e. Relation h is the set of derivable paths within p. Constraint ic_1 ensures that in every consistent database state at least one p-tuple exists. Constraint ic_2 is used to prevent cycles in p. Note that it is necessary to determine relation p before checking ic_2 due to the negative dependency. The last model computed by iterated fixpoint computation is represented by the set $\mathcal{F} \cup \{p(1,2), p(1,4), p(2,3), p(1,3)\} \cup \{ic_1, ic_2\}$ and coincides with the positive portion of the perfect model $PM_\mathcal{D}$ of $\mathcal{D}$.

2.2   Normalized Definite Rules

In this paper, we will recall a transformation-based approach to update propagation and introduce a related transformation-based approach to view updating, both leading to the generation of rather complex propagation rules if applied to an arbitrary set of stratifiable definite rules. In order to reduce syntactic complexity, our transformations will be expressed for *normalized rules*. Except for the occurrence of 0-ary predicates, each normalized rule directly corresponds to one operation in relational algebra. Note that every given set of deductive rules can be systematically transformed into an equivalent set of normalized ones by unfolding [8]. A normalized definition of relation p from Section 2.1 is

$$p(X, Y) \leftarrow e(X, Y) \qquad\qquad aux_1(X, Y) \leftarrow aux_2(X, Y, Z)$$
$$p(X, Y) \leftarrow aux_1(X, Y) \qquad\quad aux_2(X, Y, Z) \leftarrow e(X, Z) \wedge p(Z, Y)$$

where the recursive rule for p is replaced by three new rules to separate the implicit union, projection and join operator, respectively.

2.3 Disjunctive Deductive Databases

A disjunctive Datalog rule is a function-free clause of the form $H_1 \vee \ldots \vee H_m \leftarrow L_1 \wedge \ldots \wedge L_n$ with $m, n \geq 1$ where the rule's head $H_1 \vee \ldots \vee H_m$ is a disjunction of positive atoms, and the rule's body $L_1 \wedge \ldots \wedge L_n$ consists of literals, i.e. positive or negative atoms. If $H \equiv H_1 \vee \ldots \vee H_m$ is the head of a given rule R, we use $\mathtt{pred}(R)$ to refer to the set of predicate symbols of H, i.e. $\mathtt{pred}(R) = \{\mathtt{pred}(H_1), \ldots, \mathtt{pred}(H_m)\}$. For a set of rules $\mathcal{R}$, $\mathtt{pred}(\mathcal{R})$ is defined again as $\cup_{r \in \mathcal{R}} \mathtt{pred}(r)$. A *disjunctive fact* $f \equiv f_1 \vee \ldots \vee f_k$ is a disjunction of ground atoms f_i with $i \geq 1$. f is called *definite* if $i = 1$. In the following, we identify a disjunctive fact with a set of atoms such that the occurrence of a ground atom A within a fact f can also be written as $A \in f$. The set difference operator can then be used to exclude certain atoms from a disjunction while the empty set is interpreted as the boolean constant *false*.

A *disjunctive deductive database* $\mathcal{DD}$ is a pair $\langle \mathcal{F}, \mathcal{R} \rangle$ where $\mathcal{F}$ is a finite set of disjunctive facts and $\mathcal{R}$ a finite set of disjunctive rules such that $\mathtt{pred}(\mathcal{F}) \cap \mathtt{pred}(\mathcal{R}) = \emptyset$. Again, stratifiable rules are considered only, that is, recursion through negative predicate occurrences is not permitted. In addition to the usual stratification concept for definite rules it is required that all predicates within a rule's head are assigned to the same stratum. Note that integrity constraints are omitted this time because our approach transforms a set of definite rules together with the given integrity constraints into a single set of disjunctive rules. This integration allows for providing a uniform approach to handling constraint violations as well as detecting erroneous derivation paths.

In contrast to a stratifiable definite database, there is no unique perfect model for a stratifiable disjunctive database. Instead, Minker et al. have shown that there is generally a set of perfect models which captures the semantics of a disjunctive database [14]. For their computation, an extended model generation approach based on a DNF representation of conclusions has been proposed in [21]. For the purpose of computing alternative VU realizations, however, the corresponding perfect model state based on CNF representation seems to be more appropriate. This is especially the case if an application scenario deals with substantially more definite than indefinite facts, this more compact representation allows for reducing the total number of derivations [5]. In following the positive portion of a perfect model state for a disjunctive database $\mathcal{DD}$ is denoted as

$\mathcal{MS_{DD}}$. As an example, consider the following stratifiable disjunctive database $\mathcal{DD} = \langle \mathcal{F}, \mathcal{R} \rangle$:

$\mathcal{R}$	$\mathcal{F}$
$s(X) \vee t(X) \leftarrow r(X) \wedge \neg p(X)$	$q(2)$
$p(X) \leftarrow b(X,Y) \wedge p(Y)$	$b(1,2)$
$p(X) \leftarrow q(X)$	$r(a) \vee r(b) \vee r(c) \vee r(1)$

Similar to the stratification concept in definite databases, it is necessary to postpone the determination of relations s and t because of the negative dependency to p. $\mathcal{MS_{DD}}$ is then given by the set $\mathcal{F} \cup \{p(1), p(2)\} \cup \{s(a) \vee t(a) \vee r(b) \vee r(c) \vee r(1), s(b) \vee t(b) \vee r(a) \vee r(c) \vee r(1), s(c) \vee t(c) \vee r(a) \vee r(b) \vee r(1)\}$. Note that a model generation approach based on DNF would have to deal with 26 minimal models in order to compute the set of perfect models of $\mathcal{DD}$. A constructive fixpoint-based method for computing the perfect model state has been proposed by the authors in [5].

3   Update Propagation

Determining the consequences of base relation changes is essential for maintaining materialized views as well as for efficiently checking integrity. However, as in most cases an update will affect only a small portion of the database, it is rarely reasonable to compute the induced changes by comparing the entire old and new database states. Instead, update propagation methods have been proposed aiming at the efficient computation of implicit changes of derived relations resulting from explicitly performed updates of extensional facts [3,6,11,13,17,19,22]. In the context of deductive databases transformation-based approaches to UP have been studied which allow for an incremental computation of induced changes and utilize a uniform fixpoint computation mechanism. Bearing in mind the manifold benefits of these well-established approaches, it seems worthwhile to develop a similar, i.e., incremental and transformation-based, approach to the dual problem of view updating. Before doing so, we need to recall basic principles of UP in deductive databases.

In the following a specific method for UP is considered in more detail which fits well to the semantics of deductive databases introduced above. We will use the notion *update* to denote the 'true' changes caused by a transaction only; that is, we solely consider sets of updates where compensation effects (i.e., given by an insertion and deletion of the same fact or the insertion of facts which already existed, for example) have already been taken into account [11].

Definition 1. *Let $\mathcal{D} = \langle \mathcal{F}, \mathcal{R}, \mathcal{I} \rangle$ be a stratifiable deductive database. An update u_D is a pair $\langle u_D^+, u_D^- \rangle$ where u_D^+ and u_D^- are sets of base facts with $\mathrm{pred}(u_D^+ \cup u_D^-) \subseteq \mathrm{pred}(\mathcal{F})$, $u_D^+ \cap u_D^- = \emptyset$, $u_D^+ \nsubseteq \mathcal{F}$ and $u_D^- \subseteq \mathcal{F}$. The atoms u_D^+ represent insertions into $\mathcal{D}$, whereas u_D^- contains the facts to be deleted from $\mathcal{D}$.*

We consider true updates only, i.e., ground atoms which are presently not derivable for atoms to be inserted, or are derivable for atoms to be deleted, respectively [11]. The notion *induced update* is used to refer to the entire set of facts in which the new state of the database (including derived facts) differs from the former after an update of base tables has been applied.

Definition 2. *Let $\mathcal{D}$ be a stratifiable database, $PM_{\mathcal{D}}$ the semantics of $\mathcal{D}$ and $u_{\mathcal{D}}$ an update. Then $u_{\mathcal{D}}$ leads to an induced update $u_{D \to D'}$ from D to the updated database D' which is a pair $\langle u_{D \to D'}^+, u_{D \to D'}^- \rangle$ of sets of ground atoms such that $u_{D \to D'}^+ = PM_{D'} \backslash PM_D$ and $u_{D \to D'}^- = PM_D \backslash PM_{D'}$. The atoms $u_{D \to D'}^+$ represent the induced insertions, whereas $u_{D \to D'}^-$ consists of the induced deletions.*

The task of update propagation is to systematically compute the set of all induced modifications in $u_{D \to D'}$ starting from the physical changes of base data. Technically, this is a set of delta facts for any affected relation which may be stored in corresponding delta relations. For each predicate symbol $p \in \mathtt{pred}(\mathcal{D})$, we will use a pair of delta relations $\langle \Delta_p^+, \Delta_p^- \rangle$ representing the insertions and deletions induced on p by an update $u_{\mathcal{D}}$. The initial set of delta facts directly results from the given update $u_{\mathcal{D}}$ and represents the so-called *UP seeds*.

Definition 3. *Let $\mathcal{D}$ be a stratifiable deductive database and $u_D = \langle u_D^+, u_D^- \rangle$ a base update. The set of UP seeds $\mathtt{prop\_seeds}(u_D)$ with respect to u_D is defined as follows:*

$$\mathtt{prop\_seeds}(u_D) := \{ \ \Delta_p^\pi(c_1, \ldots, c_n) \mid p(c_1, \ldots, c_n) \in u_D^\pi \text{ and } \pi \in \{+, -\} \}.$$

In the following we will recall a transformation-based approach to UP which uses the stratifiable rules given in a database schema and the UP seeds to derive *deductive propagation rules* for computing delta relations [6,11]. A propagation rule refers to at least one delta relation in its body in order to provide a focus on the underlying changes when computing induced updates. For showing the effectiveness of an induced update, however, references to the state of a relation before and after the base update has been performed are necessary. We call these states the old and the new state, respectively. The state relations are never completely computed but are queried with bindings from the delta relation in the propagation rule body and thus act as a test of effectiveness. For evaluating rule bodies, fixpoint computation combined with a Magic Sets transformation is used, as proposed by the authors in [6].

Definition 4 (Update Propagation Rules). *Let $\mathcal{R}$ be a normalized stratifiable deductive rule set. The set of UP rules for true updates with respect to $\mathcal{R}$ is denoted by $\mathcal{R}^\Delta$ and is defined as the smallest set satisfying the following conditions:*

1. *For each rule of the form $p(\boldsymbol{x}) \leftarrow q(\boldsymbol{y}) \wedge r(\boldsymbol{z}) \in \mathcal{R}$ with $vars(p(\boldsymbol{x})) = (vars(q(\boldsymbol{y})) \cup vars(r(\boldsymbol{z})))$ four UP rules of the form*

$$\Delta_p^+(\boldsymbol{x}) \leftarrow \Delta_q^+(\boldsymbol{y}) \wedge r^{new}(\boldsymbol{z}) \qquad \Delta_p^-(\boldsymbol{x}) \leftarrow \Delta_q^-(\boldsymbol{y}) \wedge r(\boldsymbol{z})$$
$$\Delta_p^+(\boldsymbol{x}) \leftarrow \Delta_r^+(\boldsymbol{z}) \wedge q^{new}(\boldsymbol{y}) \qquad \Delta_p^-(\boldsymbol{x}) \leftarrow \Delta_r^-(\boldsymbol{z}) \wedge q(\boldsymbol{y})$$

are in $\mathcal{R}^{\Delta}$. For $\boldsymbol{y} = \boldsymbol{z}$ these rules correspond to an intersection, and for $\boldsymbol{y} \neq \boldsymbol{z}$ to a join in relational algebra (RA).

2. For each rule of the form $p(\boldsymbol{x}) \leftarrow q(\boldsymbol{x}) \wedge \neg r(\boldsymbol{x}) \in \mathcal{R}$ four UP rules of the form

$$\Delta_p^+(\boldsymbol{x}) \leftarrow \Delta_q^+(\boldsymbol{x}) \wedge \neg r^{new}(\boldsymbol{x}) \qquad \Delta_p^-(\boldsymbol{x}) \leftarrow \Delta_q^-(\boldsymbol{x}) \wedge \neg r(\boldsymbol{x})$$
$$\Delta_p^+(\boldsymbol{x}) \leftarrow \Delta_r^-(\boldsymbol{x}) \wedge q^{new}(\boldsymbol{x}) \qquad \Delta_p^-(\boldsymbol{x}) \leftarrow \Delta_r^+(\boldsymbol{x}) \wedge q(\boldsymbol{x})$$

are in $\mathcal{R}^{\Delta}$. This kind of rules corresponds to the difference operator in RA.

3. For each two rules of the form $p(\boldsymbol{x}) \leftarrow q(\boldsymbol{x})$ and $p(\boldsymbol{x}) \leftarrow r(\boldsymbol{x})$ four UP rules of the form

$$\Delta_p^+(\boldsymbol{x}) \leftarrow \Delta_q^+(\boldsymbol{x}) \wedge \neg p(\boldsymbol{x}) \qquad \Delta_p^-(\boldsymbol{x}) \leftarrow \Delta_q^-(\boldsymbol{x}) \wedge \neg p^{new}(\boldsymbol{x})$$
$$\Delta_p^+(\boldsymbol{x}) \leftarrow \Delta_r^+(\boldsymbol{x}) \wedge \neg p(\boldsymbol{x}) \qquad \Delta_p^-(\boldsymbol{x}) \leftarrow \Delta_r^-(\boldsymbol{x}) \wedge \neg p^{new}(\boldsymbol{x})$$

are in $\mathcal{R}^{\Delta}$. This kind of rules corresponds to the union operator in RA. Note that the additional references to p and p^{new} are necessary for excluding alternative derivations.

4. a) For each relation p defined by a single rule $p(\boldsymbol{x}) \leftarrow q(\boldsymbol{y}) \in \mathcal{R}$ with $vars(p(\boldsymbol{x})) = vars(q(\boldsymbol{y}))$ two UP rules of the form

$$\Delta_p^+(\boldsymbol{x}) \leftarrow \Delta_q^+(\boldsymbol{y}) \qquad\qquad \Delta_p^-(\boldsymbol{x}) \leftarrow \Delta_q^-(\boldsymbol{y})$$

are in $\mathcal{R}^{\Delta}$. These rules correspond to the special case where projections turns into permutation of variables.

b) For each relation p defined by a single rule $p \leftarrow \neg q \in \mathcal{R}$ two UP rules of the form

$$\Delta_p^+ \leftarrow \Delta_q^- \qquad\qquad \Delta_p^- \leftarrow \Delta_q^+$$

are in $\mathcal{R}^{\Delta}$. These rules cover the special case of negated 0-ary predicates.

5. For each relation p defined by a single projection rule $p(\boldsymbol{x}) \leftarrow q(\boldsymbol{y}) \in \mathcal{R}$ with $vars(p(\boldsymbol{x})) \supset vars(q(\boldsymbol{y}))$ two UP rules of the form

$$\Delta_p^+(\boldsymbol{x}) \leftarrow \Delta_q^+(\boldsymbol{y}) \wedge \neg p(\boldsymbol{x}) \qquad \Delta_p^-(\boldsymbol{x}) \leftarrow \Delta_q^-(\boldsymbol{y}) \wedge \neg p^{new}(\boldsymbol{x})$$

are in $\mathcal{R}^{\Delta}$. Again, additional references to p and p^{new} are necessary for detecting alternative derivations.

As the selection operator of relational algebra is not fully expressible in Datalog, a corresponding transformation rule is missing in the above definition. However, Datalog can be easily extended by built-in terms which would make it relationally complete. In this paper we will not consider Datalog extensions for simplicity reasons although corresponding view update rules can be easily defined.

The UP rules defined above reference both, the old and new database state. We assume the old state to be present, thus only references to the new database are indicated by the adornment *new* while unadorned predicate symbols refer to the old state. For simulating the new database state from a given update and the old state, so called *transition rules* [11,18] are used:

Definition 5 (Update Propagation Transition Rules). *Let $\mathcal{R}$ be a stratifiable deductive rule set. The set of UP transition rules for new state simulation with respect to $\mathcal{R}$ denoted $\mathcal{R}_\tau^\Delta$ is defined as the smallest set satisfying the following conditions:*

1. *For each n-ary extensional predicate symbol $p \in \mathbf{pred}(\mathcal{F})$, the direct transition rules*

$$p^{new}(x_1, \ldots, x_n) \leftarrow p(x_1, \ldots, x_n) \wedge \neg\Delta_p^-(x_1, \ldots, x_n)$$
$$p^{new}(x_1, \ldots, x_n) \leftarrow \Delta_p^+(x_1, \ldots, x_n)$$

   are in $\mathcal{R}_\tau^\Delta$ where the x_i are distinct variables.
2. *For each rule $H \leftarrow L_1 \wedge \ldots \wedge L_n \in \mathcal{R}$, an indirect transition rule of the form*

$$\mathbf{new}(H) \leftarrow \mathbf{new}(L_1) \wedge \ldots \wedge \mathbf{new}(L_n)$$

   is in $\mathcal{R}_\tau^\Delta$ where the mapping $\mathbf{new}$ for a literal $A \equiv r(t_1, \ldots, t_n)$ is defined as $\mathbf{new}(A) = r^{new}(t_1, \ldots, t_n)$ and $\mathbf{new}(\neg A) = \neg\mathbf{new}(A)$.

Note that all adorned predicates are assumed to introduce new names not found in the database yet, e.g., $\forall p \in \mathbf{pred}(\mathcal{R} \cup \mathcal{F}) : \{p^{new}, \Delta_p^+\} \cap \mathbf{pred}(\mathcal{R} \cup \mathcal{F}) = \emptyset$. Neither the references to the new nor to the old database state have to be completely evaluated. Instead they are queried (e.g. by using Magic Sets) with respect to the bindings provided by the delta relation within the body of a propagation rule and thus, play the role of test predicates. As an example of this propagation approach, consider again the normalized rules for relation p from the example in Section 2.1. The UP rules $\mathcal{R}^\Delta$ with respect to insertions into e are as follows:

$$\Delta_p^+(X, Y) \leftarrow \Delta_e^+(X, Y) \wedge \neg p(X, Y) \qquad \Delta_{aux_2}^+(X, Y, Z) \leftarrow \Delta_e^+(X, Z) \wedge p^{new}(Z, Y)$$
$$\Delta_p^+(X, Y) \leftarrow \Delta_{aux_1}^+(X, Y) \wedge \neg p(X, Y) \qquad \Delta_{aux_2}^+(X, Y, Z) \leftarrow \Delta_p^+(Z, Y) \wedge e^{new}(X, Z)$$
$$\Delta_{aux_1}^+(X, Y) \leftarrow \Delta_{aux_2}^+(X, Y, Z) \wedge \neg aux_1(X, Y).$$

The transition rules $\mathcal{R}_\tau^\Delta$ with respect to p are given by:

$$e^{new}(X, Y) \leftarrow e(X, Y) \wedge \neg\Delta_e^-(X, Y) \qquad p^{new}(X, Y) \leftarrow e^{new}(X, Y)$$
$$e^{new}(X, Y) \leftarrow \Delta_e^+(X, Y) \qquad p^{new}(X, Y) \leftarrow aux_1^{new}(X, Z)$$
$$aux_1^{new}(X, Y) \leftarrow aux_2^{new}(X, Y, Z)$$
$$aux_2^{new}(X, Y, Z) \leftarrow e^{new}(X, Z) \wedge p^{new}(Z, Y)$$

Note that these rules can be determined at schema definition time and don't have to be recompiled each time a new transaction is applied. It is obvious that if $\mathcal{R}$ is stratifiable, the rule set $\mathcal{R} \cup \mathcal{R}^\Delta \cup \mathcal{R}_\tau^\Delta$ will be stratifiable, too [11]. The following proposition states the correctness of this approach:

Proposition 1. *Let $\mathcal{D} = \langle \mathcal{F}, \mathcal{R}, \mathcal{I} \rangle$ be a stratifiable database, $u_\mathcal{D}$ an update and $u_{D \to D'} = \langle u_{D \to D'}^+, u_{D \to D'}^- \rangle$ the corresponding induced update from $\mathcal{D}$ to $\mathcal{D}'$. Let $\mathcal{D}^\Delta = \langle \mathcal{F} \cup \mathbf{prop\_seeds}(u_\mathcal{D}), \mathcal{R} \cup \mathcal{R}^\Delta \cup \mathcal{R}_\tau^\Delta \rangle$ be the transformed deductive database of $\mathcal{D}$. Then the delta relations defined by the UP rules $\mathcal{R}^\Delta$ correctly* *represent the induced update* $u_{D \to D'}$. *Hence, for each relation* $p \in \texttt{pred}(\mathcal{D})$ *the following conditions hold:*

$$\Delta_p^+(\boldsymbol{t}\,) \in PM_{\mathcal{D}^\Delta} \iff p(\boldsymbol{t}\,) \in u_{D \to D'}^+$$
$$\Delta_p^-(\boldsymbol{t}\,) \in PM_{\mathcal{D}^\Delta} \iff p(\boldsymbol{t}\,) \in u_{D \to D'}^- .$$

Proof. cf. [11, p. 161-163].

4 View Updating

In contrast to update propagation, view updating aims at determining one or more base relation updates such that all given update requests with respect to derived relations are satisfied after the base updates have been successfully applied. In the following, we develop a transformation-based approach to incrementally compute such base updates. After introducing some basic definitions, in Sections 4.1 and 4.2 two rule forms are presented playing a similar role as the UP rules and UP transition rules from Section 3. As the application of these disjunctive rules is followed by an integrity check the overall process incorporates the UP rules from Section 3 and is described in Section 4.3.

Definition 6. *Let* $\mathcal{D} = \langle \mathcal{F}, \mathcal{R}, \mathcal{I} \rangle$ *be a stratifiable deductive database. A VU request* $\nu_{\mathcal{D}}$ *is a pair* $\langle \nu_{\mathcal{D}}^+, \nu_{\mathcal{D}}^- \rangle$ *where* $\nu_{\mathcal{D}}^+$ *and* $\nu_{\mathcal{D}}^-$ *are sets of ground atoms representing the facts to be inserted into* $\mathcal{D}$ *or deleted from* $\mathcal{D}$, *resp., such that* $\texttt{pred}(\nu_{\mathcal{D}}^+ \cup \nu_{\mathcal{D}}^-) \subseteq \texttt{pred}(\mathcal{R})$, $\nu_{\mathcal{D}}^+ \cap \nu_{\mathcal{D}}^- = \emptyset$, $\nu_{\mathcal{D}}^+ \cap PM_{\mathcal{D}} = \emptyset$ *and* $\nu_{\mathcal{D}}^- \subseteq PM_{\mathcal{D}}$.

Note that we consider again true view updates only, i.e., ground atoms which are presently not derivable for atoms to be inserted, or are derivable for atoms to be deleted, respectively. A method for view updating determines sets of alternative updates satisfying a given request. A set of updates leaving the given database consistent after its execution is called *VU realization*.

Definition 7. *Let* $\mathcal{D} = \langle \mathcal{F}, \mathcal{R}, \mathcal{I} \rangle$ *be a stratifiable deductive database and* $\nu_{\mathcal{D}}$ *a VU request. A VU realization is a base update* $u_{\mathcal{D}}$ *which leads to an induced update* $u_{\mathcal{D} \to \mathcal{D}'}$ *from* $\mathcal{D}$ *to* $\mathcal{D}'$ *such that* $\nu_{\mathcal{D}}^+ \subseteq PM_{\mathcal{D}'}$ *and* $\nu_{\mathcal{D}}^- \cap PM_{\mathcal{D}'} = \emptyset$.

There may be infinitely many realizations and even realizations of infinite size which satisfy a given VU request. In our approach, a breadth-first search (BFS) is employed for determining a set of minimal realizations $\tau_{\mathcal{D}} = \{u_{\mathcal{D}}^1, \ldots, u_{\mathcal{D}}^i\}$. Any $u_{\mathcal{D}}^i$ is minimal in the sense that none of its updates can be removed without losing the property of being a realization for $\nu_{\mathcal{D}}$. As each level of the search tree is completely explored, the result usually consists of more than one realization. If only VU realizations of infinite size exist, our method will not terminate.

4.1 Top-Down Computation of Realizations

Given a VU request $\nu_{\mathcal{D}}$, view updating methods usually determine further VU requests in order to find relevant base updates. Similar to delta relations for

UP we will use the notion *VU relation* to access individual view updates with respect to the relations of our system. For each relation $p \in \mathbf{pred}(\mathcal{R} \cup \mathcal{F})$ we use the VU relation $\nabla_p^+(\boldsymbol{x})$ for tuples to be inserted into $\mathcal{D}$ and $\nabla_p^-(\boldsymbol{x})$ for tuples to be deleted from $\mathcal{D}$. The initial set of delta facts resulting from a given VU request is again represented by so-called *VU seeds* similar to the UP seeds.

Definition 8. *Let $\mathcal{D}$ be a deductive database and $\nu_D = \langle \nu_D^+, \nu_D^- \rangle$ a VU request. The set of VU seeds* $\mathbf{vu\_seeds}(\nu_D)$ *with respect to ν_D is defined as follows:*

$$\mathbf{vu\_seeds}(\nu_D) := \{ \ \nabla_p^\pi(c_1, \ldots, c_n) \mid p(c_1, \ldots, c_n) \in \nu_D^\pi \text{ and } \pi \in \{+, -\}\}.$$

Starting from the seeds, view updating methods as well as UP methods analyze the deductive rules of $\mathcal{D}$ in order to find subsequent VU requests systematically. For finding the direct consequences of a given view update request, we employ so-called VU rules. These rules are used to perform a top-down analysis in a similar way as the bottom-up analysis implemented by the UP rules from Section 3. In order to enhance readability and to syntactically distinguish VU rules from UP rules, the notation $Head \leftarrow Body$ is changed to $Body \rightarrow Head$.

Definition 9 (View Update Rules). *Let $\mathcal{R}$ be a normalized stratifiable deductive rule set. The set of VU rules for true view updates is denoted $\mathcal{R}^\nabla$ and is defined as the smallest set satisfying the following conditions:*

1. *For each rule of the form $p(\boldsymbol{x}) \leftarrow q(\boldsymbol{y}) \wedge r(\boldsymbol{z}) \in \mathcal{R}$ with $vars(p(\boldsymbol{x})) = (vars(q(\boldsymbol{y})) \cup vars(r(\boldsymbol{z})))$ the following three VU rules are in $\mathcal{R}^\nabla$:*

$$\nabla_p^+(\boldsymbol{x}) \wedge \neg q(\boldsymbol{y}) \rightarrow \nabla_q^+(\boldsymbol{y}) \qquad \nabla_p^-(\boldsymbol{x}) \rightarrow \nabla_q^-(\boldsymbol{y}) \vee \nabla_r^-(\boldsymbol{z})$$
$$\nabla_p^+(\boldsymbol{x}) \wedge \neg r(\boldsymbol{z}) \rightarrow \nabla_r^+(\boldsymbol{z})$$

2. *For each rule of the form $p(\boldsymbol{x}) \leftarrow q(\boldsymbol{x}) \wedge \neg r(\boldsymbol{x}) \in \mathcal{R}$ the following three VU rules are in $\mathcal{R}^\nabla$:*

$$\nabla_p^+(\boldsymbol{x}) \wedge \neg q(\boldsymbol{x}) \rightarrow \nabla_q^+(\boldsymbol{x}) \qquad \nabla_p^-(\boldsymbol{x}) \rightarrow \nabla_q^-(\boldsymbol{x}) \vee \nabla_r^+(\boldsymbol{x})$$
$$\nabla_p^+(\boldsymbol{x}) \wedge r(\boldsymbol{x}) \rightarrow \nabla_r^-(\boldsymbol{x})$$

3. *For each two rules of the form $p(\boldsymbol{x}) \leftarrow q(\boldsymbol{x})$ and $p(\boldsymbol{x}) \leftarrow r(\boldsymbol{x})$ the following three VU rules are in $\mathcal{R}^\nabla$:*

$$\nabla_p^-(\boldsymbol{x}) \wedge q(\boldsymbol{x}) \rightarrow \nabla_q^-(\boldsymbol{x}) \qquad \nabla_p^+(\boldsymbol{x}) \rightarrow \nabla_q^+(\boldsymbol{x}) \vee \nabla_r^+(\boldsymbol{x})$$
$$\nabla_p^-(\boldsymbol{x}) \wedge r(\boldsymbol{x}) \rightarrow \nabla_r^-(\boldsymbol{x})$$

4. *a) For each relation p defined by a single rule $p(\boldsymbol{x}) \leftarrow q(\boldsymbol{y}) \in \mathcal{R}$ with $vars(p(\boldsymbol{x})) = vars(q(\boldsymbol{y}))$ the following two VU rules are in $\mathcal{R}^\nabla$:*

$$\nabla_p^+(\boldsymbol{x}) \rightarrow \nabla_q^+(\boldsymbol{y}) \qquad\qquad \nabla_p^-(\boldsymbol{x}) \rightarrow \nabla_q^-(\boldsymbol{y})$$

b) For each relation p defined by a single rule $p \leftarrow \neg q \in \mathcal{R}$ the following two VU rules are in $\mathcal{R}^\nabla$:

$$\nabla_p^+ \rightarrow \nabla_q^- \qquad\qquad \nabla_p^- \rightarrow \nabla_q^+$$

5. *Assume without loss of generality that each projection rule in $\mathcal{R}$ is of the form $p(\boldsymbol{x}) \leftarrow q(\boldsymbol{x}, Y) \in \mathcal{R}$ with $Y \notin vars(p(\boldsymbol{x}))$. Then the following two VU rules*

$$\nabla_p^-(\boldsymbol{x}) \wedge q(\boldsymbol{x}, Y) \rightarrow \nabla_q^-(\boldsymbol{x}, Y)$$
$$\nabla_p^+(\boldsymbol{x}) \rightarrow \nabla_q^+(\boldsymbol{x}, c_1) \vee \ldots \vee \nabla_q^+(\boldsymbol{x}, c_n) \vee \nabla_q^+(\boldsymbol{x}, c^{new})$$

are in $\mathcal{R}^\nabla$ where all c_i are constants from the Herbrand universe $\mathcal{U}_\mathcal{D}$ of $\mathcal{D}$ and c^{new} is a new constant, i.e., $c^{new} \notin \mathcal{U}_\mathcal{D}$.

In contrast to the UP rules from Definition 4, no explicit references to the new database state are included in the above VU rules. The reason is that these rules are applied iteratively over several intermediate database states before the minimal set of realizations has been found. Hence, the apparent references to the old state really refer to the current state which is continuously modified while computing VU realizations. The need for a multi-level compensation of side effects, however, will be discussed in more detail in Section 4.2. Nevertheless, these predicates solely act as tests again queried with respect to bindings from VU relations and thus will never be completely evaluated. The following theorem states that for every realization for a given VU request a corresponding solution path will be started by the application of $\mathcal{R} \cup \mathcal{R}^\nabla$. That is, for all realizations at least one base update is found implying the completeness of our VU rules.

Theorem 1. *Let $\mathcal{D} = \langle \mathcal{F}, \mathcal{R}, \mathcal{I} \rangle$ be a stratifiable database, $\nu_\mathcal{D}$ a view update request and $\tau_\mathcal{D} = \{u_\mathcal{D}^1, \ldots, u_\mathcal{D}^n\}$ the corresponding set of minimal realizations. Let $\mathcal{D}^\nabla = \langle \mathcal{F} \cup \texttt{vu\_seeds}(\nu_\mathcal{D}), \mathcal{R} \cup \mathcal{R}^\nabla \rangle$ be the transformed deductive database of $\mathcal{D}$. Then the VU relations in $\mathcal{MS}_{\mathcal{D}^\nabla}$ with respect to base relations of $\mathcal{D}$ correctly represent all direct consequences of $\nu_\mathcal{D}$. That is, for each realization $u_\mathcal{D}^i = \langle u_\mathcal{D}^{i^+}, u_\mathcal{D}^{i^-} \rangle \in \tau_\mathcal{D}$ the following condition holds:*

$$\exists p(\boldsymbol{t}) \in u_\mathcal{D}^{i^+} : \nabla_p^+(\boldsymbol{t}) \in \mathcal{MS}_{\mathcal{D}^\nabla} \vee \exists p(\boldsymbol{t}) \in u_\mathcal{D}^{i^-} : \nabla_p^-(\boldsymbol{t}) \in \mathcal{MS}_{\mathcal{D}^\nabla}.$$

Proof. (Sketch) Every realization $u_\mathcal{D}^i \in \tau_\mathcal{D}$ leads to an induced update $u_{\mathcal{D} \rightarrow \mathcal{D}^i}^+$ from $\mathcal{D}$ to $\mathcal{D}^i$ which is correctly represented by delta relations defined by UP rules $\mathcal{R}^\Delta$ according to Proposition 1. One can show that there is at least one derivation path from $u_\mathcal{D}^i$ to $\nu_\mathcal{D}$ using the rules $\mathcal{R} \cup \mathcal{R}^\Delta$ which is conversely present in $\mathcal{R} \cup \mathcal{R}^\nabla$, too. From this, the condition above directly follows. $\qquad\square$

Note that only VU rules resulting from projection rules may introduce new constants in order to find a realization. These constants are accessible subsequently when further view updates are to be determined. Recursion combined with projection may lead to an infinite solution path by introducing an infinite number of new constants. However, during the top-down analysis phase in which $\mathcal{R}^\nabla$ is applied to the facts of $\mathcal{D}$ only one constant is introduced for all projection rules. Further constants are introduced in subsequent iteration rounds for repairing undesired side effects. For example, the VU rules above do not restrict sequences of ∇-relations to those leading to a consistent database state only. For repairing violated constraints corresponding new update requests are generated and a re-application of $\mathcal{R}^\nabla$ is necessary. The handling of erroneous 'solution' paths, however, will be discussed in more detail in Section 4.2.

Although only one new constant is introduced for projection rules per iteration round, the considered number of constants within the corresponding disjunctive facts is still impractically high. However, it is usually not necessary to consider all constants from $\mathcal{U}_\mathcal{D}$ as proposed in Definition 9. For example, only those constants which are also provided by the codomain of q have to be considered. In fact, the set of constants can be even further reduced without losing the completeness of our approach. Additionally, the handling of the existentially quantified new constants c^{new} can be optimized as, e.g., proposed in [2]. We will refrain from a discussion of implementation issues, however, as the focus of this paper is to discover general properties of view updating methods rather than efficiency considerations.

We illustrate via an example the basic mechanism behind the application of $\mathcal{R}^\nabla$. Consider the following deductive database $\mathcal{D} = \langle \mathcal{F}, \mathcal{R}, \mathcal{I} \rangle$ with $\mathcal{F} = \{r_2(2), s(2)\}$, $\mathcal{I} = \{ic(2)\}$ and the normalized deductive rules $\mathcal{R}$:

$$p(X) \leftarrow q_1(X) \qquad\qquad q_1(X) \leftarrow r_1(X) \wedge s(X)$$
$$p(X) \leftarrow q_2(X) \qquad\qquad q_2(X) \leftarrow r_2(X) \wedge \neg s(X)$$
$$ic(2) \leftarrow \neg au(2) \qquad\qquad au(X) \leftarrow q_2(X) \wedge \neg q_1(X)$$

The corresponding set of VU rules with respect to the VU request $\nu_\mathcal{D}^+ = \{p(2)\}$ is then given by:

$$\nabla_p^+(X) \rightarrow \nabla_{q_1}^+(X) \vee \nabla_{q_1}^+(X)$$
$$\nabla_{q_1}^+(X) \wedge \neg r_1(X) \rightarrow \nabla_{r_1}^+(X) \qquad\qquad \nabla_{q_2}^+(X) \wedge \neg r_2(X) \rightarrow \nabla_{r_2}^+(X)$$
$$\nabla_{q_1}^+(X) \wedge \neg s(X) \rightarrow \nabla_s^+(X) \qquad\qquad \nabla_{q_2}^+(X) \wedge s(X) \rightarrow \nabla_s^-(X)$$

Applying these rules using disjunctive fixpoint computation leads to two alternative updates $u_{\mathcal{D} \rightarrow \mathcal{D}^1}^+ = \{r_1(2)\}$ and $u_{\mathcal{D} \rightarrow \mathcal{D}^2}^- = \{s(2)\}$ induced by the derived disjunction $\nabla_{r_1}^+(2) \vee \nabla_s^-(2)$. Obviously, the second update represented by $\nabla_s^-(2)$ would lead to an undesired side effect. In order to provide a complete method, however, such erroneous/incomplete paths must be also explored and side effects repaired if possible.

4.2   Bottom-Up Assessment of Realizations

Determining whether a computed update will lead to a consistent database state or not can be done by applying a bottom-up UP process at the end of the top-down phase leading to an irreparable constraint violation with respect to $\nabla_s^-(2)$:

$$\nabla_s^-(2) \Rightarrow \Delta_{q_2}^+(2) \Rightarrow \Delta_p^+(2), \Delta_{au}^+(2) \Rightarrow \Delta_{ic}^-(2) \rightsquigarrow false$$

In order to see whether the violated constraint can be repaired, the subsequent view update request $\nu_{\mathcal{D}^2}^+ = \{ic(2)\}$ with respect to $\mathcal{D}^2$ ought to be answered. The application of $\mathcal{R}^\nabla$ yields

$$\Rightarrow \nabla_{q_2}^-(2), \nabla_{q_2}^+(2) \rightsquigarrow false$$
$$\nabla_{ic}^+(2) \Rightarrow \nabla_{aux}^-(2) \updownarrow$$
$$\Rightarrow \nabla_{q_1}^+(2) \Rightarrow \nabla_s^+(2), \nabla_s^-(2) \rightsquigarrow false$$

showing that this request cannot be satisfied as inconsistent subsequent view update requests are generated on this path. In the following, erroneous derivation paths will be indicated by the keyword *false*. To eliminate those paths, we will employ the operation **reduce** which extracts the occurrences of *false* from given disjunctions. The reduced set of updates - each of them leading to a consistent database state only - represents the set of realizations $\tau_\mathcal{D} = \{\langle u^+_{\mathcal{D} \to \mathcal{D}^1}, \emptyset \rangle\}$ with $u^+_{\mathcal{D} \to \mathcal{D}^1} = \{r_1(2)\}$.

An induced deletion of an integrity constraint predicate can be seen as a side effect of an 'erroneous' VU. Similar side effects, however, can be also found when induced changes to the database caused by a VU request may include derived facts which had been actually used for deriving this view update. This effect is shown in the following example for a non-normalized deductive database $\mathcal{D} = \langle \mathcal{R}, \mathcal{F}, \mathcal{I} \rangle$ with $\mathcal{R} = \{h(X) \leftarrow p(X) \wedge q(X) \wedge i, i \leftarrow p(X) \wedge \neg q(X)\}$, $\mathcal{F} = \{p(1)\}$, and $\mathcal{I} = \emptyset$. Given the VU request $\nu^+_\mathcal{D} = \{h(1)\}$, the overall evaluation scheme for determining the only realization $\tau_\mathcal{D} = \{\langle u^+_{\mathcal{D} \to \mathcal{D}^1}, \emptyset \rangle\}$ with $u^+_{\mathcal{D} \to \mathcal{D}^1} = \{q(1), p(c^{new_1})\}$ would be as follows:

$$\nabla^+_h(1) \Rightarrow \nabla^+_q(1) \Rightarrow \Delta^+_q(1) \Rightarrow \Delta^-_i \Rightarrow \nabla^+_i \updownarrow \begin{matrix} \Rightarrow \nabla^+_p(c^{new_1}) \\ \Rightarrow \nabla^-_q(1), \nabla^+_q(1) \rightsquigarrow false \end{matrix}$$

The example shows the necessity of compensating side effects, i.e., the compensation of the 'deletion' Δ^-_i (that prevents the 'insertion' $\Delta^+_h(1)$) caused by the tuple $\nabla^+_q(1)$. In general the compensation of side effects, however, may in turn cause additional side effects which have to be 'repaired'. Thus, the view updating method must alternate between top-down and bottom-up phases until all possibilities for compensating side effects (including integrity constraint violations) have been considered, or a solution has been found. Therefore, we introduce so-called *VU transition rules* which are used to restart the VU analysis:

Definition 10 (View Update Transition Rules). *Let $\mathcal{D} = \langle \mathcal{F}, \mathcal{R}, \mathcal{I} \rangle$ be a stratifiable database, $\mathcal{R}^\nabla$ the set of corresponding VU rules, $Rel_\mathcal{D}$ the set of all predicate symbols occurring in $\langle \mathcal{F}, \mathcal{R}, \mathcal{I} \rangle$. The set of VU transition rules is defined as the smallest rule set satisfying the following conditions:*

1. *For each n-ary predicate symbol $p \in Rel_\mathcal{D}$ the following rule is in $\mathcal{R}^\nabla_\tau$*

   $$\nabla^+_p(x_1, \ldots, x_n) \wedge \nabla^-_p(x_1, \ldots, x_n) \to false$$

   where x_i $(i = 1, \ldots, n)$ are distinct variables. This kind of rule is employed for detecting and deleting erroneous derivation paths.
2. *For each ground literal $ic(\boldsymbol{c}) \in \mathcal{I}$ the following rule is in $\mathcal{R}^\nabla_\tau$:*

   $$\Delta^-_{ic}(\boldsymbol{c}) \to \nabla^+_{ic}(\boldsymbol{c})$$

   These rules are used to repair violated integrity constraints.
3. *For each rule $\nabla^\pi_p(\boldsymbol{x}) \wedge q(\boldsymbol{y}) \to \nabla^-_q(\boldsymbol{y}) \in \mathcal{R}^\nabla$ with $\pi \in \{+, -\}$ the following rule is in $\mathcal{R}^\nabla_\tau$*

$$\nabla_p^{\pi}(\boldsymbol{x}) \wedge \neg q(\boldsymbol{y}) \wedge \Delta_q^{+}(\boldsymbol{y}) \rightarrow \nabla_q^{-}(\boldsymbol{y})$$

whereas for each rule $\nabla_p^{\pi}(\boldsymbol{x}) \wedge \neg q(\boldsymbol{y}) \rightarrow \nabla_q^{+}(\boldsymbol{y}) \in \mathcal{R}^{\nabla}$ *a rule of the form*

$$\nabla_p^{\pi}(\boldsymbol{x}) \wedge q(\boldsymbol{y}) \wedge \Delta_q^{-}(\boldsymbol{y}) \rightarrow \nabla_q^{+}(\boldsymbol{y})$$

is in $\mathcal{R}_{\tau}^{\nabla}$. *These rules identify side effects and are employed for initiating a compensation attempt.*

4. *For each n-ary predicate symbol* $p \in \mathcal{F}$ *the following rule is in* $\mathcal{R}_{\tau}^{\nabla}$

$$\nabla_p^{+}(x_1, \ldots, x_n) \wedge \neg p(x_1, \ldots, x_n) \rightarrow \Delta_p^{+}(x_1, \ldots, x_n)$$
$$\nabla_p^{-}(x_1, \ldots, x_n) \wedge p(x_1, \ldots, x_n) \rightarrow \Delta_p^{-}(x_1, \ldots, x_n)$$

where x_i $(i = 1, \ldots, n)$ *are distinct variables. These rules initiate the bottom-up consequence analysis after the top-down analysis phase has been finished. Note that an effectiveness test is needed as the accumulated VU request will not represent true updates anymore after the first iteration round [11].*

The rules in $\mathcal{R}_{\tau}^{\nabla}$ make sure that erroneous solutions are evaluated to *false* and side effects are repaired. The following theorem establishes the correctness of the view update rules $\mathcal{R}^{\nabla}$ and $\mathcal{R}_{\tau}^{\nabla}$.

Theorem 2. *Let* $\mathcal{D} = \langle \mathcal{F}, \mathcal{R}, \mathcal{I} \rangle$ *be a stratifiable database,* $\nu_{\mathcal{D}}$ *a view update request and* $\tau_{\mathcal{D}} = \{u_{\mathcal{D}}^1, \ldots, u_{\mathcal{D}}^n\}$ *the corresponding set of minimal realizations. Let* $\mathcal{D}^{\nabla} = \langle \mathcal{F} \cup \text{vu\_seeds}(\nu_{\mathcal{D}}), \mathcal{R} \cup \mathcal{R}^{\nabla} \rangle$ *be the transformed deductive database of* $\mathcal{D}$ *for determining the direct consequences of* $\nu_{\mathcal{D}}$ *and* $\mathcal{D}^{\nabla \Delta} = \langle \mathcal{MS}_{\mathcal{D}^{\nabla}}, \mathcal{R} \cup \mathcal{R}^{\Delta} \cup \mathcal{R}_{\tau}^{\Delta} \cup \mathcal{R}_{\tau}^{\nabla} \rangle$ *be the database for determining compensating VU requests. Then the VU facts* ∇_{new}^{π} *in* $\mathcal{MS}_{\mathcal{D}^{\nabla \Delta}}$ *which are not contained in* $\mathcal{MS}_{\mathcal{D}^{\nabla}}$ *correctly represent all indirect consequences of* $\nu_{\mathcal{D}}$. *That is, for every realization* $u_{\mathcal{D}}^i \in \tau_{\mathcal{D}}$ *which has not been completely determined by* $\mathcal{MS}_{\mathcal{D}^{\nabla}}$ *exists a fact f in* ∇_{new}^{π} *with* $\pi \in \{+, -\}$ *such that* $u_{\mathcal{D}}^i$ *is also a realization for* $\text{vu\_seeds}(\nu_{\mathcal{D}}) \cup f$.

Proof. (Sketch) Since $u_{\mathcal{D}}^i$ is a realization which has not yet completely determined by $\mathcal{MS}_{\mathcal{D}^{\nabla}}$, there must be side effects that inhibit the derivation of one or more delta facts in $\text{vu\_seeds}(\nu_{\mathcal{D}})$ or lead to the violation of an integrity constraint. Compensation paths for violated constraints are correctly started by the corresponding rules in $\mathcal{R}_{\tau}^{\nabla}$. The other side effects must be handled by introducing new VU request which lead to the generation of at least one of the missing base updates in $u_{\mathcal{D}}^i$. From Theorem 1 it follows that the rules $\mathcal{R}^{\nabla}$ are correct and complete. Thus, the missing VU request result from VU rules where the body VU literal was satisfied but not the side literal with reference to the current database state. As for every rule in $\mathcal{R}^{\nabla}$ a corresponding rule is present in $\mathcal{R}_{\tau}^{\nabla}$ which react to the changed database state, all possible new VU request are determined. Correctness follows from the correctness of $\mathcal{R}^{\nabla}$ and $\mathcal{R} \cup \mathcal{R}^{\Delta} \cup \mathcal{R}_{\tau}^{\Delta}$. Completeness follows from the fact, that all possible new VU request from $\mathcal{R}^{\nabla}$ are considered in $\mathcal{R}_{\tau}^{\nabla}$, too. □

Algorithm 1. BFS determination of view update realizations

Input: normalized stratifiable deductive database $\mathcal{D} = \langle \mathcal{F}, \mathcal{R}, \mathcal{I} \rangle$,
          transformed rule sets $\mathcal{R}^\nabla, \mathcal{R}_\tau^\nabla, \mathcal{R}^\Delta, \mathcal{R}_\tau^\Delta$ w.r.t. $\mathcal{R}$ and $\mathcal{I}$,
          view update request $\nu_\mathcal{D} = \langle \nu_\mathcal{D}^+, \nu_\mathcal{D}^- \rangle$
Output: set of realizations $\tau_\mathcal{D} = \{u_{\mathcal{D} \to \mathcal{D}^1}, \ldots, u_{\mathcal{D} \to \mathcal{D}^n}\}$ for $\nu_\mathcal{D}$

$\quad i \quad := 0;$
$\quad F_0^\nabla \quad := \mathtt{vu\_seeds}(\nu_\mathcal{D});$
$\quad F_0 \quad := \mathcal{F};$
$\quad \tau_\mathcal{D} \quad := \emptyset;$
$\quad \mathbf{repeat}$
$\quad\quad i \quad := i + 1;$
$\quad\quad MS_{D_i^\nabla} \quad := \mathcal{MS}_{\langle F_{i-1} \cup F_{i-1}^\nabla, \mathcal{R} \cup \mathcal{R}^\nabla \rangle_n};$
$\quad\quad MS_{D_i^{\nabla\Delta}} \quad := \mathcal{MS}_{\langle MS_{D_i^\nabla}, \mathcal{R} \cup \mathcal{R}^\Delta \cup \mathcal{R}_\tau^\Delta \cup \mathcal{R}_\tau^\nabla \rangle_n};$
$\quad\quad MS_i \quad := \mathtt{reduce}(MS_{D_i^{\nabla\Delta}});$
$\quad\quad F_i^\nabla \quad := \mathtt{get\_\nabla}(MS_i);$
$\quad\quad F_i^\Delta \quad := \mathtt{get\_\Delta}(MS_i);$
$\quad\quad F_i \quad := \mathtt{update}(\mathcal{F}, F_i^\Delta);$
$\quad\quad \tau_\mathcal{D} \quad := \mathtt{get\_realizations}(MS_i, \nu_\mathcal{D});$
$\quad \mathbf{until} \ (\tau_\mathcal{D} \neq \emptyset) \vee (false \in MS_{D_i^{\nabla\Delta}} \vee (F_{i-1}^\nabla = F_i^\nabla))$
$\quad \mathbf{return} \quad \tau_\mathcal{D};$

Having the rules for the direct and indirect consequences of a given VU request, we can now define a general application scheme for systematically determining VU realizations.

4.3   Overall Organization of View Updating

The top-down analysis rules $\mathcal{R} \cup \mathcal{R}^\nabla$ and the bottom-up consequence analysis rules $\mathcal{R} \cup \mathcal{R}^\Delta \cup \mathcal{R}_\tau^\Delta \cup \mathcal{R}_\tau^\nabla$ are iteratively applied using the general evaluation scheme depicted in Algorithm 1. Note that these disjunctive rules are stratifiable such that the computation of the perfect model state can be applied. The algorithm continuously extends the sets of alternative base updates by new ones until a realization for the given VU request has been found.

During the initialization phase, the operator $\mathtt{vu\_seeds}$ is applied to the given VU request $\nu_\mathcal{D}$ for determining the initial ∇-seed facts similar to the propagation seeds from Definition 3. In the following iteration phase the VU rules and UP rules are consecutively employed and their iterated fixpoint model state $MS_{D_i^\nabla}$ respectively $MS_{D_i^{\nabla\Delta}}$ is determined. The $\mathtt{reduce}$ operation eliminates all occurrences of $false$ from disjunctive facts in $MS_{D_i^{\nabla\Delta}}$ in order to discard erroneous derivation paths. If the fact $false$ has been derived, no solution for $\nu_\mathcal{D}$ could be found and the algorithm terminates. Operation $\mathtt{get\_\nabla}$ accumulates previous as well as new compensating VU requests. If no new compensation requests are generated by $\mathcal{R}_\tau^\nabla$, no more alternative solution paths must be explored and the iteration stops. Operation get_Δ accumulates all determined base relation updates over all iteration round. These updates are then applied to $\mathcal{F}$ such that the new compensation requests are processed with respect to the new (possibly disjunctive) database state. Based on the minimal models $\mathcal{D}^1, \ldots, \mathcal{D}^n$ of MS_i satisfying the VU request $\nu_\mathcal{D}$, the operator get_realizations constructs the updates $u_{\mathcal{D} \to \mathcal{D}^j}$ which cause the transition from $\mathcal{D}$ to $\mathcal{D}^j$. The correctness of Algorithm 1 directly follows from Theorem 1 and Theorem 2.

5 Discussion

In the previous chapter we introduced a new method for consistency preserving view updating. In our systematic approach, the top-down analysis phase and the bottom-up consequence analysis are separated showing the duality of update propagation and view updating analysis. Our method is suited for being implemented in a database context because of the transformation-based approach and the close relationship of normalized rules to relational algebra operators. In order to deal with alternative realizations, however, an extension to disjunctive databases becomes necessary. Further implementation details remain undiscussed, too, even though various possibilities for enhancing efficiency exist. For example, the employed rules can be further improved by incorporating transformation-based query optimization methods such as Magic Sets [4] or by a delayed evaluation of existentially quantified variables in projections [2].

Our method illustrates further interesting properties relevant for view updating in general. It is well-known that the view update problem: "Does there exist a VU realization for $\mathcal{D}$ satisfying $\nu_\mathcal{D}$?" is undecidable given an arbitrary stratifiable deductive database and an arbitrary VU request. This can be shown, e.g., by reducing the query containment problem for positive deductive rules to a VU problem with respect to a stratifiable database. Since the query containment problem is undecidable for positive Datalog with at least one derived relation having more than one attribute, the VU problem must be undecidable, too [9,15]. Note that undecidability is not caused by the combination of deductive rules and integrity constraints but rather by the combination of rules with a view update which corresponds to an existential condition. The main reason for undecidability of the VU problem is that there are cases where only infinite fact bases are able to satisfy a given VU request.

The existence of infinite solutions, of course, is not a proof of undecidability of the VU problem but already indicates a fundamental problem of every view updating method. The infinite number of possible solutions usually can be handled by using a breadth-first search and cycle tests within the derivation tree. However, if the problem has an infinite solution only, these approaches need infinite time to compute this solution, too. Usually recursive rule sets have been identified as a source for infinite solutions. But our algorithm indicates that even a non-recursive rule set can lead to an infinite realization set in case that the compensation of side effects lead to a continuous introduction of new constants.

As an example, consider the following non-recursive example having an infinite model only that satisfies the given request $\nu_{\mathcal{D}}^+ = \{h\}$:

$$h \leftarrow e(X, Y) \wedge \neg i$$

$$
\begin{array}{ll}
i \leftarrow e(Y, X) \wedge e(Z, X) \wedge \neg eq(Y, Z) & j(X) \leftarrow e(X, Y) \wedge e(X, Z) \wedge \neg eq(Y, Z) \\
i \leftarrow e(X, Y) \wedge \neg j(X) & eq(X, X) \leftarrow e(X, Y) \\
i \leftarrow e(X, Y) \wedge \neg j(Y) & eq(Y, Y) \leftarrow e(X, Y)
\end{array}
$$

Relation e must contain an infinite number of facts in order to satisfy the VU request $\nu_{\mathcal{D}}^+ = \{h\}$. In this example, relation e corresponds to a mapping from the second attribute Y to the first attribute X such that each value of X is assigned to at least two values of Y and each value of Y is also value of X.

But from Algorithm 1 it can be directly followed that if no projection is employed within compensation paths of $\mathcal{R}^\nabla \cup \mathcal{R}^\Delta$, termination can be assured. In the example above, the path $\nabla_h^+(\boldsymbol{x}) \to \nabla_e^+(\boldsymbol{y}) \to \Delta_e^+(\boldsymbol{y}) \to \Delta_i^+ \to \Delta_h^-(\boldsymbol{x})$ violates this condition as it contains a (π sign changing) negation and a projection rule which together represents a possibly infinite generator of new constants. Although this - quite restrictive - condition uses the UP rules introduced above, it can be also formulated by means of the original deductive rules. The class of deductive databases for which the VU problem is decidable, however, may still include recursion, projections and arbitrary n-ary derived relations. This condition offers a general approach to finding better static criteria as it depends closely on the chosen propagation method and the way of optimizing them. This includes dynamic criteria taking into account the specific instance $\nu_{\mathcal{D}}$ as well as the current database state for optimizing the propagation rules.

Although our set-oriented approach fits well into a database context, it cannot be generally preferred over instance-oriented methods. In fact, approaches based on SLDNF [18,22] or tableaux calculus [1] may perform better in cases where only a small number of alternative solution paths have to be exploited. In general, however, none of the approaches can be preferred over the other as similar optimization effects can be incorporated in all those methods. As our search tree is completely materialized till a solution has been found, the space and time complexity for computing a finite model growths linear with the size of this solution. Because Ackermann's function can be modelled by stratifiable Datalog rules without function terms, however, the growth of constants in such a model can reach the same order of magnitude as the computation of Ackermann terms [8].

References

1. Aravindan, C., Baumgartner, P.: Theorem Proving Techniques for View Deletion in Databases. Journal of Symbolic Computation 29(2), 119–147 (2000)
2. Abdennadher, S., Schütz, H.: Model Generation with Existentially Quantified Variables and Constraints. ALP/HOA, 256–272 (1997)

3. Bry, F., Decker, H., Manthey, R.: A Uniform Approach to Constraint Satisfaction and Constraint Satisfiability in Deductive Databases. In: Schmidt, J.W., Missikoff, M., Ceri, S. (eds.) EDBT 1988. LNCS, vol. 303, pp. 488–505. Springer, Heidelberg (1988)

4. Behrend, A.: Soft stratification for magic set based query evaluation in deductive databases. In: PODS 2003, New York, pp. 102–110 (June 9–12, 2003)

5. Behrend, A.: A Fixpoint Approach to State Generation for Stratifiable Disjunctive Deductive Databases. In: Ioannidis, Y., Novikov, B., Rachev, B. (eds.) ADBIS 2007. LNCS, vol. 4690, Springer, Heidelberg (2007)

6. Behrend, A., Manthey, R.: Update Propagation in Deductive Databases Using Soft Stratification. In: Benczúr, A.A., Demetrovics, J., Gottlob, G. (eds.) ADBIS 2004. LNCS, vol. 3255, pp. 22–36. Springer, Heidelberg (2004)

7. Bry, F.: Intensional Updates: Abduction via Deduction. In: ICLP 1990, pp. 561–575 (1990)

8. Eckert, H.: Ein regelbasierter Ansatz zur Analyse von Sichtenänderungswünschen in stratifizierbaren deduktiven Datenbanken. Master Thesis, University of Bonn (2004)

9. Farré, C., Teniente, E., Urpí, T.: Query Containment Checking as a View Updating Problem. In: Quirchmayr, G., Bench-Capon, T.J.M., Schweighofer, E. (eds.) DEXA 1998. LNCS, vol. 1460, pp. 310–321. Springer, Heidelberg (1998)

10. Grant, J., Horty, J., Lobo, J., Minker, J.: View Updates in Stratified Disjunctive Databases. JAR 11(2), 249–267 (1993)

11. Griefahn, U.: Reactive Model Computation - A Uniform Approach to the Implementation of Deductive Databases. Dissertation, University of Bonn (1997), http://www.cs.uni-bonn.de/~idb/publications/diss_griefahn.ps.gz

12. Inoue, K., Sakama, C.: A Fixpoint Characterization of Abductive Logic Programs. JLP 27(2), 107–136 (1996)

13. Küchenhoff, V.: On the Efficient Computation of the Difference Between Consecutive Database States. In: Delobel, C., Masunaga, Y., Kifer, M. (eds.) DOOD 1991. LNCS, vol. 566, pp. 478–502. Springer, Heidelberg (1991)

14. Lobo, J., Minker, J., Rajasekar, A.: Foundations of Disjunctive Logic Programming. MIT Press, Cambridge (1992)

15. Levy, A., Mumick, I., Sagiv, Y., Shmueli, O.: Equivalence, Query-Reachability, and Satisfiability in Datalog Extensions. In: PODS 1993, pp. 109–122 (1993)

16. Lloyd87, J.W.: Foundations of Logic Programming, 2nd edn. Springer, Berlin (1987)

17. Manthey, R.: Reflections on some fundamental issues of rule-based incremental update propagation. In: DAISD 1994, pp. 255–276 (1994)

18. Olivé, A.: Integrity Constraints Checking In Deductive Databases. In: VLDB 1991, pp. 513–523 (1991)

19. Przymusinski, T.: Every Logic Program Has a Natural Stratification And an Iterated Least Fixed Point Model. In: PODS 1989, pp. 11–21 (1989)

20. Sagiv, Y.: Optimizing Datalog Programs. Foundations of Deductive Databases and Logic Programming, 659–698 (1988)

21. Seipel, D., Minker, J., Ruiz, C.: Model Generation and State Generation for Disjunctive Logic Programs. Journal of Logic Programming 32(1), 49–69 (1997)

22. Teniente, E., Olivé, A.: The Events Method for View Updating in Deductive Databases. In: Pirotte, A., Delobel, C., Gottlob, G. (eds.) EDBT 1992. LNCS, vol. 580, pp. 245–260. Springer, Heidelberg (1992)

23. Teniente, E., Urpi, T.: On the abductive or deductive nature of database schema validation and update processing problems. TPLP 3(3), 287–327 (2003)

Algorithms for Effective Argumentation in Classical Propositional Logic: A Connection Graph Approach

Vasiliki Efstathiou and Anthony Hunter

Department of Computer Science
University College London
Gower Street, London WC1E 6BT, UK
{v.efstathiou,a.hunter}@cs.ucl.ac.uk

Abstract. There are a number of frameworks for modelling argumentation in logic. They incorporate a formal representation of individual arguments and techniques for comparing conflicting arguments. A common assumption for logic-based argumentation is that an argument is a pair $\langle \Phi, \alpha \rangle$ where Φ is minimal subset of the knowledgebase such that Φ is consistent and Φ entails the claim α. Different logics provide different definitions for consistency and entailment and hence give us different options for argumentation. Classical propositional logic is an appealing option for argumentation but the computational viability of generating an argument is an issue. Here we propose ameliorating this problem by using connection graphs to give information on the ways that formulae of the knowledgebase can be used to minimally and consistently entail a claim. Using a connection graph allows for a substantially reduced search space to be used when seeking all the arguments for a claim from a knowledgebase. We provide a theoretical framework and algorithms for this proposal, together with some theoretical results and some preliminary experimental results to indicate the potential of the approach.

1 Introduction

Argumentation is a vital aspect of intelligent behaviour by humans. Consider diverse professionals such as politicians, journalists, clinicians, scientists, and administrators, who all need to collate and analyse information looking for pros and cons for consequences of importance when attempting to understand problems and make decisions.

There are a number of proposals for logic-based formalisations of argumentation (for reviews see [9,21,5]). These proposals allow for the representation of arguments for and against some claim, and for counterargument relationships between arguments. In a number of key examples of argumentation systems, an argument is a pair where the first item in the pair is a minimal consistent set of formulae that proves the second item which is a formula (see for example [2,14,3,1,15,4]). Furthermore, in these approaches, a key form of counterargument is an undercut: One argument undercuts another argument when the claim of the first argument negates the premises of the second argument. Proof procedures and algorithms have been developed for finding preferred arguments

S. Hartmann and G. Kern-Isberner (Eds.): FoIKS 2008, LNCS 4932, pp. 272–290, 2008.

from a knowledgebase using defeasible logic and following for example Dung's preferred semantics (see for example [7,23,20,17,8,11,12]). However, these techniques and analyses do not offer any ways of ameliorating the computational complexity inherent in finding arguments and counterarguments for classical logic. Furthermore, we wish to find all arguments for a particular claim, and this means a pruning strategy, such as incorporated into defeasible logic programming [15,10], would not meet our requirements since some undercuts would not be obtained.

In this paper we restrict the language used to a language of (disjunctive) clauses and for this language we propose algorithms for finding arguments using search tree structures that correspond to the steps of a systematic application of the connection graph proof procedure [18,19]. We describe how this method can be efficient regarding the computational cost of finding arguments.

2   Logical Argumentation for a Language of Clauses

In this section, we adapt an existing proposal for logic-based argumentation [3] by restricting the language to being disjunctive clauses so that the premises of an argument is a set of clauses and the claim of an argument is a literal.

Definition 1. *A language of clauses C is composed from a set of atoms A as follows: If α is an atom, then α is a* **positive literal***, and $\neg\alpha$ is a* **negative literal***. If β is a positive literal, or β is a negative literal, then β is a* **literal***. If $\beta_1, .., \beta_n$ are literals, then $\beta_1 \vee ... \vee \beta_n$ is a* **clause***. A* **clause knowledgebase** *is a set of clauses.*

We use $\phi, \psi, \ldots$ to denote disjunctive clauses and $\Delta, \Phi, \Psi, \ldots$ to denote sets of clauses. For the following definitions, we first assume a clause knowledgebase Δ (a finite set of clauses) and use this Δ throughout. The paradigm for the approach is that there is a large repository of information, represented by Δ, from which arguments can be constructed for and against arbitrary claims. Apart from information being understood as declarative statements, there is no *a priori* restriction on the contents, and the pieces of information in the repository can be arbitrarily complex. Therefore, Δ is not expected to be consistent.

The framework adopts a very common intuitive notion of an argument. Essentially, an argument is a set of clauses that can be used to prove some claim, together with that claim. In this paper, we assume each claim is represented by a literal.

Definition 2. *A* **literal argument** *is a pair $\langle \Phi, \alpha \rangle$ such that: (1) α is a literal (2) $\Phi \subseteq \Delta$; (3) $\Phi \not\vdash \bot$; (4) $\Phi \vdash \alpha$; and (5) there is no $\Phi' \subset \Phi$ such that $\Phi' \vdash \alpha$. We say that $\langle \Phi, \alpha \rangle$ is a literal argument for α. We call α the* **claim** *of the argument and Φ the* **support** *of the argument (we also say that Φ is a support for α).*

Example 1. Let $\Delta = \{a, \neg a \vee b, \neg b \vee c, b \vee \neg d, \neg a, a \vee b, \neg c, \neg b \vee \neg c, c \vee a\}$. Some literal arguments are:

$$\langle \{a, \neg a \vee b\}, b \rangle$$
$$\langle \{\neg a\}, \neg a \rangle$$
$$\langle \{a, \neg a \vee b, \neg b \vee c\}, c \rangle$$
$$\langle \{a \vee b, \neg b \vee \neg c, c \vee a\}, a \rangle$$

Some arguments oppose the claim or the support of other arguments. This leads to the notion of a counterargument as follows.

Definition 3. *Let $\langle \Phi, \alpha \rangle$ and $\langle \Psi, \beta \rangle$ be literal arguments*

- $\langle \Psi, \beta \rangle$ *is a* **rebut** *of* $\langle \Phi, \alpha \rangle$ *iff* $\{\beta, \alpha\} \vdash \perp$.
- $\langle \Psi, \beta \rangle$ *is an* **undercut** *of* $\langle \Phi, \alpha \rangle$ *iff* $\Phi \vdash \neg\beta$.
- $\langle \Psi, \beta \rangle$ *is a* **counteragument** *of* $\langle \Phi, \alpha \rangle$ *iff* $\langle \Psi, \beta \rangle$ *is a rebut or an undecut of* $\langle \Phi, \alpha \rangle$.

Example 2. $\langle \{\neg c \vee b, c\}, b \rangle$ is a rebut of $\langle \{\neg a, a \vee \neg d, d \vee \neg b \vee c, \neg c\}, \neg b \rangle$.
$\langle \{\neg c \vee b, c\}, b \rangle$ is an undercut of $\langle \{a, d, \neg a \vee \neg d \vee \neg b, b \vee e\}, e \rangle$.

Following a number of proposals for argumentation (e.g. [3,1,15,20,12]), logical arguments and counterarguments can be presented in a graph: Each node denotes an argument and each arc (A_1, A_2) denotes that argument A_2 is a counterargument to argument A_1. Various constraints have been imposed on the nature of such graphs leading to a range of options for evaluating whether a particular argument in the graph is "defeated" or "undefeated". We will not consider this aspect of logic-based argumentation further in this paper. We are only concerned in this paper with how we can construct the arguments from the knowledgebase, and not how to compare them.

3   Towards Effective Algorithms for Generating Arguments

We now turn to automating the construction of arguments and counterarguments. Unfortunately automated theorem proving technology cannot do this directly for us. For each argument, we need a minimal and consistent set of formulae that proves the claim. An automated theorem prover (an ATP) may use a "goal-directed" approach, bringing in extra premises when required, but they are not guaranteed to be minimal and consistent. For example, supposing we have a clause knowledgebase $\{\neg\alpha \vee \beta, \beta\}$, for proving β, the ATP may start with the premise $\neg\alpha \vee \beta$, then to prove β, a second premise is required, which would be β, and so the net result is $\{\neg\alpha \vee \beta, \beta\}$, which does not involve a minimal set of premises. In addition, an ATP is not guaranteed to use a consistent set of premises since by classical logic it is valid to prove anything from an inconsistency.

So if we seek arguments for a particular claim δ, we need to post queries to an ATP to ensure that a particular set of premises entails δ, that the set of premises is minimal for this, and that it is consistent. So finding arguments for a claim α involves considering subsets Φ of Δ and testing them with the ATP to ascertain whether $\Phi \vdash \alpha$ and $\Phi \nvdash \perp$ hold. For $\Phi \subseteq \Delta$, and a formula α, let $\Phi?\alpha$ denote a call (a query) to an ATP. If Φ classically entails α, then we get the answer $\Phi \vdash \alpha$, otherwise we get the answer $\Phi \nvdash \alpha$. In this way, we do not give the whole of Δ to the ATP. Rather we call it with particular subsets of Δ. So for example, if we want to know if $\langle \Phi, \alpha \rangle$ is an argument, then we have a series of calls $\Phi?\alpha$, $\Phi?\perp$, $\Phi \setminus \{\phi_1\}?\alpha, ..., \Phi \setminus \{\phi_k\}?\alpha$, where $\Phi = \{\phi_1, .., \phi_k\}$. So the first call is to ensure that $\Phi \vdash \alpha$, the second call is to ensure that $\Phi \nvdash \perp$, the remaining calls are to ensure that there is no subset Φ' of Φ such that $\Phi' \vdash \alpha$. This then raises the question of which subsets Φ of Δ to investigate to determine whether $\langle \Phi, \alpha \rangle$ holds when we are seeking for an argument for α.

A further problem we need to consider is that if we want to generate all arguments for a particular claim in the worst case we may have to send each subset Φ of Δ to the ATP to determine whether $\Phi \vdash \alpha$ and $\Phi \not\vdash \perp$. So in the worst case, if $|\Delta| = n$, then we may need to make 2^{n+1} calls to the ATP. Even for a small knowledgebase of say 20 or 30 formulae, this can become prohibitively expensive.

It is with these issues in mind that we explore an alternative way of finding all the arguments from a knowledgebase Δ for a claim α. Our approach is to adapt the idea of connection graphs to enable us to find arguments.

4   Connection Graphs

Connection graphs were initially proposed by Kowalski (see [18,19]) for reducing the search space for applying resolution to clauses in logic programming. They have also been developed more generally for classical logic [6]. In this section we will adapt the definition of a connection graph to give us the notion of a focal graph which for a knowledgebase Δ and a claim α essentially delineates the subset of the knowledgebase that may have a role in an argument for the claim α. For example, for $\Delta = \{a, \neg a \vee f, \neg a \vee b, \neg b \vee c, \neg n \vee \neg m, b \vee d, b \vee e, \neg e \vee a, \neg d \vee a \vee \neg c, \neg g \vee m, \neg q \vee r \vee p, \neg p\}$ and the claim a we require that the delineated subset is $\{a, \neg a \vee b, \neg b \vee c, b \vee d, b \vee e, \neg e \vee a, \neg d \vee a \vee \neg c\}$. In this way, formulae that cannot possibly be a premise in an argument will be excluded. This provides the potential for substantially reducing the set of formulae to be considered for constructing arguments.

So in this section we will formalize the notion of a focal graph, then in the next section we consider how we can search the focal graph, and in the subsequent section we provide algorithms for efficiently searching the focal graph so as to return all the arguments for the claim of interest.

We start by introducing some relations on the elements of $\mathcal{C}$, that will be used to determine the links of the connection graphs and how these can be used by the search algorithms.

Definition 4. *The* Disjuncts *function takes a clause and returns the set of disjuncts in the clause*

$$\mathsf{Disjuncts}(\beta_1 \vee .. \vee \beta_n) = \{\beta_1, .., \beta_n\}.$$

Definition 5. *Let ϕ and ψ be clauses. Then,* Preattacks$(\phi, \psi) = \{\beta \mid \beta \in$ Disjuncts(ϕ) and $\neg \beta \in$ Disjuncts$(\psi)\}$.

Example 3. Preattacks$(a \vee \neg b \vee \neg c \vee d, a \vee b \vee \neg d \vee e) = \{\neg b, d\}$, Preattacks$(a \vee b \vee \neg d \vee e, a \vee \neg b \vee \neg c \vee d) = \{b, \neg d\}$, Preattacks$(a \vee b \vee \neg d, a \vee b \vee c) = \emptyset$.

Definition 6. *Let ϕ and ψ be clauses. If* Preattacks$(\phi, \psi) = \{\beta\}$ *for some β, then* Attacks$(\phi, \psi) = \beta$ *otherwise* Attacks$(\phi, \psi) = null$.

Example 4. Attacks$(a \vee \neg b \vee \neg c \vee d, a \vee b \vee \neg d \vee e) = null$, Attacks$(a \vee b \vee \neg d, a \vee b \vee c) = null$, Attacks$(a \vee b \vee \neg d, a \vee b \vee d) = \neg d$, Attacks$(a \vee b \vee \neg d, e \vee c \vee d) = \neg d$.

Hence, the Preattacks relation is defined for any pair of clauses ϕ, ψ while the Attacks relation is defined for a pair of clauses ϕ, ψ such that $|$Preattacks$(\phi, \psi)| = 1$.

Lemma 1. *We can see that for two clauses ϕ and ψ if* $\mathsf{Preattacks}(\phi, \psi) \neq \emptyset$ *then from the resolution proof rule it follows that* $\forall \beta \in \mathsf{Preattacks}(\phi, \psi)$,

$$\{\phi, \psi\} \vdash \bigvee((\mathsf{Disjuncts}(\phi) \setminus \{\beta\}) \cup (\mathsf{Disjuncts}(\psi) \setminus \{\neg\beta\}))$$

Example 5. For $\phi = \neg a \vee b \vee c \vee d$ and $\psi = a \vee b \vee e \vee \neg c$, $\mathsf{Preattacks}(\phi, \psi) = \{\neg a, c\}$ hence

$$\{\phi, \psi\} \vdash \bigvee((\mathsf{Disjuncts}(\phi) \setminus \{\neg a\}) \cup (\mathsf{Disjuncts}(\psi) \setminus \{a\}))$$
$$= \bigvee((\{\neg a, b, c, d\} \setminus \{\neg a\}) \cup (\{a, b, e, \neg c\} \setminus \{a\}))$$
$$= \bigvee(\{b, c, d, e, \neg c\}) = b \vee c \vee d \vee e \vee \neg c.$$

$$\{\phi, \psi\} \vdash \bigvee((\mathsf{Disjuncts}(\phi) \setminus \{c\}) \cup (\mathsf{Disjuncts}(\psi) \setminus \{\neg c\}))$$
$$= \bigvee((\{\neg a, b, c, d\} \setminus \{c\}) \cup (\{a, b, e, \neg c\} \setminus \{\neg c\}))$$
$$= \bigvee(\{\neg a, b, d, e, a\}) = \neg a \vee b \vee d \vee e \vee a.$$

From Lemma 1 we can see that $\mathsf{Preattacks}(\phi, \psi) \neq \mathsf{Attacks}(\phi, \psi)$ iff ϕ with ψ resolve to a tautology.

We now introduce some types of graphs whose nodes correspond to a set of clauses and the links between each pair of clauses are determined according to the attack relations defined above. In the following examples of graphs we use the $|$, $\diagup$, $\diagdown$ and — symbols to denote arcs in the pictorial representation of a graph.

Definition 7. *Let Δ be a clause knowledgebase. The* **connection graph** *for Δ, denoted* $\mathsf{Connect}(\Delta)$, *is a graph* (N, A) *where* $N = \Delta$ *and* $A = \{(\phi, \psi) \mid$ *there is a* $\beta \in$ $\mathsf{Disjuncts}(\phi)$ *such that* $\beta \in \mathsf{Preattacks}(\phi, \psi)\}$.

Example 6. The following is the connection graph for $\Delta = \{k, \neg k \vee l, \neg l, \neg k \vee \neg m, k \vee m \vee d, \neg d, \neg e \vee c \vee \neg d, \neg c \vee d, e \vee \neg c, c \vee f, \neg f \vee g, \neg e, \neg f, f \vee \neg g, a \vee q \vee n, \neg n \vee \neg q, \neg n, a \vee r, a \vee t, \neg r \vee \neg t\}$.

$$
\begin{array}{llll}
k \;—\; \neg k \vee \neg m \;—\; k \vee m \vee d \;—\; \neg d & \qquad & a \vee q \vee n \;—\; \neg n \vee \neg q \\
| & & | \\
\neg k \vee l & \neg e \vee c \vee \neg d \;—\; \neg c \vee d & \neg n \\
| & | \qquad\qquad | \\
\neg l & e \vee \neg c \;—\; c \vee f \;—\; \neg f \vee g & a \vee r \qquad a \vee t \\
& | \qquad\quad | \qquad\quad | \\
& \neg e \qquad \neg f \qquad f \vee \neg g & \neg r \vee \neg t
\end{array}
$$

We now need to go beyond Kowalski's idea of a connection graph and introduce the following types of graph. The attack graph defined below is a subgraph of the connection graph identified using the Attacks function.

Definition 8. *Let Δ be a clause knowledgebase. The* **attack graph** *for Δ, denoted* $\mathsf{AttackGraph}(\Delta)$, *is a graph* (N, A) *where* $N = \Delta$ *and* $A = \{(\phi, \psi) \mid$ *there is a* $\beta \in \mathsf{Disjuncts}(\phi)$ *such that* $\mathsf{Attacks}(\phi, \psi) = \beta\}$.

Example 7. Continuing Example 6, the following is the attack graph for Δ.

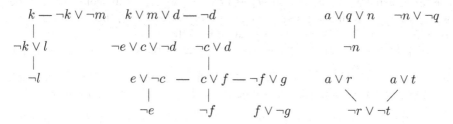

The following definition of closed graph gives a kind of connected subgraph of the attack graph where connectivity is determined in terms of the attack relation among its nodes.

Definition 9. *Let Δ be a clause knowledgebase. The **closed graph** for Δ, denoted* $\mathsf{Closed}(\Delta)$, *is the largest subgraph (N, A) of* $\mathsf{AttackGraph}(\Delta)$, *such that for each $\phi \in N$, for each $\beta \in \mathsf{Disjuncts}(\phi)$ there is a $\psi \in N$ with* $\mathsf{Attacks}(\phi, \psi) = \beta$.

The above definition assumes that there is a unique largest subgraph of the attack graph that meets the conditions presented. This is justified because having a node from the attack graph in the closed graph does not exclude any other node from the attack graph also being in the closed graph. Any subset of nodes is included when each of the disjuncts is negated by disjuncts in the other nodes. Moreover, we can consider the closed graph being composed of components where for each component Y, and for each node ϕ in Y, and for each disjunct β in ϕ, there is another node ψ in Y such that there is a disjunct $\neg\beta$ in ψ. So the nodes in each component work together to ensure each disjunct is negated by a disjunct in another node in the component, and the largest subgraph of the attack graph is obtained by just taking the union of these components.

Example 8. Continuing Example 7, the following is the closed graph for Δ.

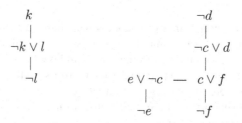

The focal graph (defined next) is a subgraph of the closed graph for Δ which is delineated by a clause ϕ from Δ and corresponds to the part of the closed graph that contains ϕ. In the following, we assume a component of a graph means that each node in the component is connected to any other node in the component by a path.

Definition 10. *Let Δ be a clause knowledgebase and ϕ be a clause in Δ. The **focal graph** of ϕ in Δ denoted* $\mathsf{Focal}(\Delta, \phi)$ *is defined as follows: If there is a component X in* $\mathsf{Closed}(\Delta)$ *containing the node ϕ, then* $\mathsf{Focal}(\Delta, \phi) = X$, *otherwise* $\mathsf{Focal}(\Delta, \phi)$ *is the empty graph.*

Example 9. Continuing Example 8, if $C_2 = (N_1, A_1)$ is the component of the closed graph for Δ with $N_1 = \{k, \neg k \vee l, \neg l\}$ and $C_2 = (N_2, A_2)$ is the component of the closed graph for Δ with $N_2 = \{\neg d, \neg c \vee d, e \vee \neg c, c \vee f, \neg e, \neg f\}$ then the focal graph of ϕ in Δ is C_1 for $\phi \in \{k, \neg k \vee l, \neg l\}$, and it is C_2 for $\phi \in \{\neg d, \neg c \vee d, e \vee \neg c, c \vee f, \neg e, \neg f\}$. For any other ϕ, the focal graph of ϕ in Δ corresponds to the empty graph.

The query graph of a literal α in a clause knowledgebase Δ defined below is the graph whose elements, as we will see, determine all the literal arguments for α, if there are any.

Definition 11. *Let Δ be a clause knowledgebase and α be a literal. The* **query graph** *of α in Δ, denoted $\mathsf{Query}(\Delta, \alpha)$, is the focal graph of $\neg\alpha$ in $\Delta \cup \{\neg\alpha\}$. Hence,* $\mathsf{Query}(\Delta, \alpha) = \mathsf{Focal}(\Delta \cup \{\neg\alpha\}, \neg\alpha)$.

Example 10. For knowledgebase Δ given in Example 6, the following is the query graph of $\neg m$ in Δ,

$$
\begin{array}{ccc}
k & \!\!\!\!—\!\!\!\! & \neg k \vee \neg m \\
| & & | \\
\neg k \vee l & & m \\
| & & \\
\neg l & &
\end{array}
$$

and the following is the query graph of $\neg c$ in Δ.

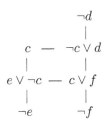

The query graph of a literal α in a clause knowledgebase Δ delineates the subset of the Δ that contains formulae that may be a premise in a literal argument for α. Furthermore, the query graph contains information about how the formulae relate to each other in the sense of how they can potentially form proofs for the claim. Now, in order to determine whether there are any arguments that can be obtained from these formulae and to determine the support for these arguments, we need to search the query graph. This is the subject of the next section.

5   Searching Query Graphs

The set of nodes of the query graph of α in Δ contains all the subsets of the knowledgebase that can be used as supports for literal arguments for α. The appropriate subsets can be obtained by selecting the nodes of the query graph of α in Δ that obey certain conditions. For this we will use the notion of a support tree which represents the support set for a literal argument for α together with the negation of α in a tree structure where

$\neg\alpha$ is the root. Essentially a support tree is constructed from a subgraph of the query graph of α in Δ.

In order to define the notion of a support tree we will first introduce the notion of the presupport tree which is a tree with $\neg\alpha$ as the root and some clauses from Query(Δ, α) as nodes on its branches. Then we will introduce some additional constraints that define a support tree as a special kind of a presupport tree.

Definition 12. *Let Δ be a clause knowledgebase and let α be a literal. A* **presupport tree** *for Δ and α is tuple (N, A, f) where (N, A) is a tree, and f is a mapping from N to Δ such that*

(1) *if x is the root of the tree, then $f(x) = \neg\alpha$ and there is exactly one child y of x s.t.*
    Attacks$(f(y), f(x)) = \alpha$,
(2) *for any nodes x, y in the same branch, if $x \neq y$, then $f(x) \neq f(y)$,*
(3) *for any node x in the tree, if y is a child of x,*
    then there is a $\neg\beta_i \in$ Disjuncts$(f(x))$ s.t. Attacks$(f(y), f(x)) = \beta_i$
    and for each $\beta_j \in$ Disjuncts$(f(y)) \setminus \{\beta_i\}$,
        i) either there is exactly one child z of y s.t. Attacks$(f(z), f(y)) = \neg\beta_j$,
        ii) or there is an arc (w, w') in the branch containing y s.t.
        Attacks$(f(w), f(w')) = \beta_j$ *and w' is the parent of w.*

The first condition of the definition initialises the tree structure of the presupport tree by setting the negated claim as the clause identifying the root and ensures that it will be attacked by some other clause from the presupport tree otherwise the tree will be empty. The fact that the root can only have one child guarantees that the width of the first level of the tree will be minimized. The second condition of the definition ensures that for a finite Δ there can only be presupport trees of finite depth. A clause from Δ can have its value assigned to exactly one node in a branch ensuring that no repetitions of the same clause will be allowed in this branch. The third condition of the definition is the equivalent of condition 1 for the general case of non-root nodes. It ensures that all the disjuncts of the clause identifying a node are attacked by a node of the same branch. Each node has as many children as the number of its disjuncts that do not appear as attack values on the branch earlier, ensuring that only the necessary number of children will be in the tree at each level.

Example 11. Going back to Example 10, for $\Delta = \{k, \neg k \vee l, \neg l, \neg k \vee \neg m, k \vee m \vee d, \neg d, \neg e \vee c \vee \neg d, \neg c \vee d, e \vee \neg c, c \vee f, \neg f \vee g, \neg e, \neg f, f \vee \neg g, a \vee q \vee n, \neg n \vee \neg q, \neg n, a \vee r, a \vee t, \neg r \vee \neg t\}$ and $\alpha = \neg c$ there are two presupport trees for Δ and α.

$$
\begin{array}{cc}
c & c \\
\uparrow & \uparrow \\
e \vee \neg c & \neg c \vee d \\
\uparrow & \uparrow \\
\neg e & \neg d
\end{array}
$$

Example 12. The following is a presupport tree for $\Delta = \{\neg d, \neg a \lor b \lor c, \neg b \lor \neg e, a \lor \neg e, \neg e, e, e \lor d, e \lor \neg a, a\}$ and $\alpha = c$.

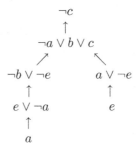

Example 13. The following is a presupport tree for $\Delta = \{a, \neg c, \neg b \lor c \lor \neg a, \neg d, b, \neg e, d \lor b \lor \neg f, \neg b, f\}$ and $\alpha = c$.

Example 14. The following is a presupport tree for $\Delta = \{\neg e \lor d, e \lor a, \neg a \lor b \lor c, \neg c \lor f, \neg b \lor e \lor d \lor \neg a, \neg f \lor e, \neg a \lor g, \neg g \lor h\}$ and $\alpha = d$.

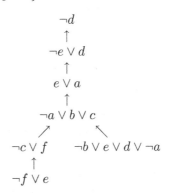

Proposition 1. *If (N, A, f) is a presupport tree for a finite clause knowledgebase Δ and α then (N, A) is a finite tree.*

We will now introduce two special cases of presupport trees each of which amounts to the notions of minimal entailment and consistent entailment.

Definition 13. *Let Δ be a clause knowledgebase and let α be a literal. A **consistent presupport tree** for Δ and α is a presupport tree (N, A, f) for Δ and α such that for any nodes x and y where x' is the parent of x and y' is the parent of y, $\mathsf{Attacks}(f(x), f(x')) \neq \neg\mathsf{Attacks}(f(y), f(y'))$.*

So, a presupport tree is consistent if does not contain any pair of arcs $(x, x'), (y, y')$ such that $\mathsf{Attacks}(f(x), f(x')) = \neg\beta$ and $\mathsf{Attacks}(f(y), f(y')) = \beta$ for some $\neg\beta \in \mathsf{Disjuncts}(f(x))$ and $\beta \in \mathsf{Disjuncts}(f(y))$.

Example 15. The following is a consistent presupport tree for $\Delta = \{\neg d, \neg a \vee b \vee c, \neg b \vee \neg e, a \vee \neg e, \neg e, e, e \vee d, \neg d\}$ and $\alpha = c$.

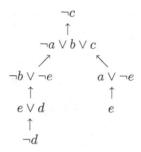

Example 16. The following is not a consistent presupport tree for $\Delta = \{\neg d, \neg a \vee b \vee c, \neg b \vee e, a \vee \neg e, \neg e, e\}$ and $\alpha = c$: for $f(x') = \neg b \vee e, f(x) = \neg e, f(y') = a \vee \neg e, f(y) = e$ we get $\mathsf{Attacks}(f(x), f(x')) = \neg e$ and $\mathsf{Attacks}(f(y), f(y')) = e$.

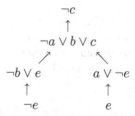

Definition 14. *Let Δ be a clause knowledgebase and let α be a literal. A* **minimal pre-support tree** *for Δ and α is a presupport tree (N, A, f) for Δ and α such that:*

> (1) *for any nodes x, y in the same branch where*
>   *x' is the parent of x and y' is the parent of y*
>   *$\mathsf{Attacks}(f(x), f(x')) \neq \mathsf{Attacks}(f(y), f(y'))$*
> (2) *if for two nodes x and y, where x' is the parent*
>   *of x and y' is the parent of y,*
>   *$\mathsf{Attacks}(f(x), f(x')) = \mathsf{Attacks}(f(y), f(y'))$*
>   *then $\mathsf{Subtree}(x) \subseteq \mathsf{Subtree}(y')$ or $\mathsf{Subtree}(y) \subseteq \mathsf{Subtree}(x')$*

Where $\mathsf{Subtree}(x)$ is the set of formulae in the subtree rooted at x.

The first condition of this definition ensures that nodes that are not necessary for the entailment of the given claim cannot be added on the branches of a minimal presupport tree. The second condition ensures that if two nodes x and y need to be attacked on the same disjunct then common nodes will be used to attack both, ensuring that there will be no more than one set of nodes contributing to the entailment of the claim in the same way.

Example 17. The presupport tree of Example 16 is a minimal presupport tree for Δ and α. The presupport tree of Example 15 is not a minimal presupport tree for Δ and α because it violates the second condition of Definition 14. If in the presupport tree of Example 15 we replace Subtree(x_2) by a copy of Subtree(x_1) for x_1, x_2 such that $f(x_1) = e \vee d$ and $f(x_2) = e$ then both conditions of the definition will be satisfied and we will obtain the following minimal presupport tree for Δ and α:

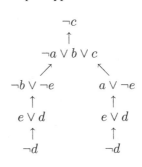

Definition 15. *A presupport tree* (N, A, f) *is a* **support tree** *iff it is a minimal and consistent presupport tree.*

Example 18. Each of the presupport trees of Examples 11, 13 and 14 is a support tree.

We will now introduce some theoretical results illustrating why support trees can be useful for our purposes in seeking arguments for a claim from a knowledgebase. First we give the definition of a minimal inconsistent subset of a knowledgebase Δ and then we give a proposition illustrating how these sets can be used in argumentation and how they relate to support trees.

Definition 16. *For a set of formulae* Δ, *a* **minimal inconsistent subset** Φ *of* Δ *is such that:*

(1) $\Phi \vdash \perp$
(2) *For all* $\Psi \subseteq \Delta$, *if* $\Psi \subset \Phi$, *then* $\Psi \not\vdash \perp$.

Proposition 2. *For a literal* α, $\langle \Phi, \alpha \rangle$ *is a literal argument iff* $\Phi \cup \{\neg\alpha\}$ *is a minimal inconsistent subset of* $\Delta \cup \{\neg\alpha\}$.

Proposition 3. *If* (N, A, f) *is a support tree for* Δ *and* α, *and* $\Gamma = \{f(x) \mid x \in N\}$, *then* $\Gamma \vdash \perp$ *and for any* $\Gamma' \subset \Gamma, \Gamma' \not\vdash \perp$.

From proposition 3 we get that the clause knowledgebase that corresponds to a support tree for Δ and α is a minimal inconsistent set and hence the following proposition holds.

Proposition 4. *If* (N, A, f) *is a support tree for* Δ *and* α, *then* $\{f(x) \mid x \in N\} \setminus \{\neg\alpha\} \vdash \alpha$.

According to the following proposition, the clause knowledgebase that corresponds to the nodes of a support tree for Δ and α cannot contain another knowledgebase that can be arranged is a support tree structure for Δ and α.

Proposition 5. *If* (N, A, f) *is a support tree for* Δ *and* α *and* (N', A', f') *is a support tree for* Δ *and* α, *then* $\{f(x) \mid x \in N\} \not\subseteq \{f'(x') \mid x' \in N'\}$.

From the last four propositions it follows that for any minimal inconsistent set of clauses that contains a literal $\neg\alpha$ there is a support tree for Δ and α.

Proposition 6. *Let* Δ *be a clause knowledgebase and let* $\Phi \subseteq \Delta$. $\langle\Phi, \alpha\rangle$ *is a literal argument iff there is a support tree* (N, A, f) *for* Δ *and* α *such that* $\Phi = \{f(x) \mid x \in N\}$.

Therefore, given a clause knowledgebase Δ and a literal α, we can find all the arguments for α by finding all the subgraphs of the query graph of α in Δ whose clauses can be arranged in a support tree for Δ and α. This helps reduce the computational cost of the process in two ways. First, the search space used when searching for arguments is reduced: instead of an algorithm searching through the whole knowledgebase it can search through the part of the knowledgebase that corresponds to the query graph of α in Δ. Potentially this offers very substantial savings since the query graph may involve a relatively small subset of the formulae in the knowledgebase. Second, the query graph also provides useful information on the attack relation among the clauses its nodes contain. The existence of links among the clauses of the knowledgebase motivates the use of algorithms that follow the paths in the query graph rather than searching through arbitrary subsets of the graph.

The algorithms for searching a query graph are introduced below. The links of the query graph are used to trace paths when searching for arguments and the attack values to which they correspond are used to identify the arcs on the branches of a presupport tree or a support tree.

6   Algorithms

In this section we present the algorithms that can be used to construct and search a query graph in order to find all the literal arguments for α from Δ.

6.1   Algorithm for Building the Query Graph

First we will give a brief description of the GetFocal(Δ, ϕ) algorithm which retrieves the focal graph of a clause ϕ in a clause knowledgebase Δ, and therefore can be used to retrieve the query graph of α in Δ.

The GetFocal(Δ, ϕ) algorithm finds the focal graph of ϕ in Δ by doing a depth-first search which follows the links of the component of the attack graph for Δ that is linked to ϕ. Initially all the clauses from Δ are considered as candidates for being clauses in the focal graph of ϕ in Δ and then during the search they can be rejected if they do not satisfy the conditions of the definition of the focal graph. The algorithm chooses the appropriate nodes by using the boolean method isConnected(C, ψ) which tests whether a clause ψ of the attack graph C is such that each literal $\beta \in$ Disjuncts(ψ) corresponds to at least one arc to a clause from Δ that has not been rejected . Given the adjacency matrix for the attack graph for Δ, the algorithm locates which clauses of the attack graph need to be visited. Only those that are linked to ϕ either directly or indirectly with a sequence of arcs from the attack graph for Δ will be visited. From the set of the visited clauses only the ones that satisfy the condition of being connected according to the isConnected function will be clauses in the focal graph.

The algorithm starts by locating clause ϕ in the attack graph. If $\phi \notin \Delta$ or the function isConnected(C, ϕ) returns false, the algorithm returns the empty graph. Otherwise the algorithm, starting from ϕ, follows in a depth-first way all the possible paths through clauses from Δ, indicated by the links of the attack graph and tests whether the isConnected function returns true for the visited nodes. If the function returns false for some clause, then this clause is marked as rejected and the algorithm backtracks to retest whether the rest of the clauses in this path remain connected after this clause has been rejected.

6.2   Algorithm for Finding the Formulae for Each Presupport Tree (Algorithm 1)

The GetPresupportsTree algorithm constructs a search tree representing an exhaustive search of the query graph of α in Δ in order to find all the different subsets of Δ that can be arranged in a presupport tree structure. Each branch of the search tree is a linked list of nodes which can be accepted or rejected according to the conditions of definition 12. Each of the accepted branches identifies a unique subset of Δ that can be arranged in a presupport tree for Δ and α.

Each node of the search tree denoted $Node$ contains a set of candidates for a presupport tree where each candidate is identified by a clause from the query graph of α in Δ. The set of candidates in a node, denoted $Candidates$, corresponds to a level of a presupport tree for Δ and α. Apart from the value $Candidates$ each $Node$ contains the value $Parent$ as a pointer to its previous $Node$ on the branch, and the value $Ancestors$ which is the set of all the nodes that appear on the same branch above this node.

Each element in $Candidates$ is of the form $Candidate_\phi = (\phi, Attacked_\phi)$ s.t. $\phi \in \Delta$ and $Attacked_\phi \subseteq \mathsf{Disjuncts}(\phi)$ where ϕ represents a potential node of a presupport tree for Δ and α. So each such candidate contains a clause ϕ from the query graph of α in Δ and a subset of $\mathsf{Disjuncts}(\phi)$ denoted $Attacked_\phi$ which keeps track of the disjuncts on which ϕ is attacked by clauses of other candidates from the same branch of the search tree. Each $Candidate_\phi$ is in the $Candidates$ of a $Node$ if there is at least one $Candidate_\psi$ in the $Candidates$ of the parent of the given $Node$ such that ϕ attacks ψ on a disjunct that has not been already attacked by clauses of candidates in the preceding levels (i.e. ancestor nodes).

The root of the search tree, which also represents the root of each of the presupport trees generated by the algorithm, containsas its value for $Candidates$ the set $\{Candidate_{\neg\alpha}\}$. The algorithm then proceeds in a depth-first way in order to construct each branch. Each step of this search corresponds to retrieving all the possible different ϕ s.t. $Candidate_\phi$ can be in the $Candidates$ of a node. A stack is used to store temporarily each $Node$ which will then be replaced by all the possible children nodes for that $Node$. The branch continues being expanded until there is no possible new level, which is either the case when the formulae in the nodes in the branch satisfy all the conditions of being a presupport tree for Δ and α or the case when this set of formulae violates some of the conditions of definition 12. In the first case, when the formulae on

---

Algorithm 1. GetPresupportsTree(Δ, α)

Let S be an empty Stack
Let $QueryGraph = $ GetFocal($\Delta \cup \{\neg\alpha\}, \neg\alpha$)
Let AcceptedBranches $= \emptyset$
Let $rootNode = Node(\{\neg\alpha\}, null)$
push $rootNode$ onto S
while S is not empty **do**
  Let $topNode$ be the top of S
  Let $newNodes = $ getNewNodes($QueryGraph$, $topNode$)
  if $newNodes = \emptyset$ **then**
    if there is $branch \in$ AcceptedBranches $s.t.\ branch = $ getFormulae($Node$) **then**
      pop S
    else
      AcceptedBranches $= $ AcceptedBranches $\bigcup \{$getFormulae($topNode$)$\}$
      pop S
    end if
  else
    pop S
    for all $Node \in newNodes$ **do**
      UpdateAttackValues($Node$)
      push $Node$ onto S
    end for
  end if
end while
return AcceptedBranches

---

the current branch can be arranged in a presupport tree for Δ and α, the formulae of this set excluding $\neg\alpha$ is stored, as long as the same set of formulae has not been stored previously. It is in the last two cases when the algorithm reaches the end of a branch that it moves to the next branch.

In order to control the number of nodes that need to be created, the algorithm is using the subsidiary function UpdateAttackValues($Node$) which updates the value $Attacked_\phi$ of each $Candidate_\phi$ of a newly created $Node$ by testing the attack relation of ϕ with the clauses contained in the rest of the $Candidates$ of its $Ancestors$ nodes.

In order to facilitate the search of the query graph of α in Δ denoted $QueryGraph$, the algorithm is using the function getNewNodes($QueryGraph$, $Node$) which retrieves from the $QueryGraph$ the clauses that attack each candidate of the given node on a disjunct that is not already attacked by candidates of previous nodes and combines them to get all the possible sets of candidates in a way that there is a 1-1 correspondence between the elements of each given candidate and these non-attacked disjuncts. If the $Candidates$ of the updated node contains at least one node $Candidate_\phi$ with Disjuncts(ϕ) $\neq Attacked_\phi$ then the getNewNodes function returns a non-empty set of all the possible next levels of the given $Node$, otherwise it returns the empty set.

Finally, the getFormulae($Node$) function is used when a leaf node is found and returns the set of formulae on the branch where the given leaf node belongs.

Hence, each of the accepted branches of the algorithm introduced above gives us the set of formulae that can be arranged as a presupport tree for Δ and α. Each node of an accepted branch is selected so as to represent a level of a presupport tree. Each $Candidate_\psi$ in a node's $Candidates$ is such that there is a $Candidate_{\psi'}$ in the $Candidates$ of its parent with $\mathsf{Attacks}(\psi, \psi') \neq null$ and (ψ, ψ') defines an arc of the presupport tree.

The $\mathsf{UpdateAttackValues}(Node)$ algorithm updates each of the candidates of $Node$ according to their attack relation with the candidates of the previous nodes on the same branch in order to ensure that the $\mathsf{getNewNodes}$ algorithm will return a set of children nodes $Children = \{Node_1, \ldots, Node_n\}$ such that for each $Node_i$ from the set $Children$ if $Candidates_i$ is the set of candidates in $Node_i$ and $Candidate_\psi \in Candidates_i$ and $\beta \in \mathsf{Disjuncts}(\psi)$, then there is a no ancestor $Node_a$ of $Node_i$ s.t. $Candidate_{\psi'}$ is in the candidates of $Node_a$ and $\mathsf{Attacks}(\psi, \psi') = \beta$. This ensures that conditions 1) and 3) of the definition of the presupport tree are satisfied. The fact that the candidates of a newly created node cannot be from the candidates that appear on the branch before ensures that condition 2) of the definition of the presupport tree will be satisfied. As a result, all the conditions for an accepted branch to be a presupport tree are met by the algorithm.

Example 19. For $\Delta = \{a \vee b, \neg b, a \vee c \vee d, \neg c \vee f, \neg d \vee e, \neg f, \neg e, \neg d \vee g, \neg g \vee h, c \vee j, \neg k \vee m \vee n, \neg n \vee \neg j, \neg g\}$ and $\alpha = a$, following is the query graph of a in Δ:

$$\neg d \vee g \;\text{---}\; \neg g$$
$$|$$
$$\neg a \;\text{---}\; a \vee c \vee d \;\text{---}\; \neg d \vee e$$
$$| \qquad\qquad | \qquad\qquad |$$
$$a \vee b \qquad \neg c \vee f \qquad \neg e$$
$$| \qquad\qquad |$$
$$\neg b \qquad\qquad \neg f$$

The $\mathsf{GetPresupportsTree}(\Delta, \alpha)$ algorithm generates the following search tree from the above query graph.

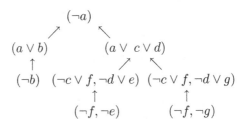

Hence, for the branches (numbered from left to right) we have the following sets of formulae:

From branch 1, $\{a \vee b, \neg b\}$
From branch 2, $\{a \vee c \vee d, \neg c \vee f, \neg d \vee e, \neg f, \neg e\}$
From branch 3, $\{a \vee c \vee d, \neg c \vee f, \neg d \vee g, \neg f, \neg g\}$

So each of these sets of formulae can be arranged as a presupport tree.

Since we require arguments for α from Δ, we need to take the output of the algorithm GetPresupportsTree(Δ, α) and determine whether each set of formulae corresponding to a presupport tree can be arranged as a support tree. That is the role of our next algorithm in the next section.

6.3   Algorithm for Checking Support Tree Conditions

We now describe the GetSupports algorithm which, using the output of the algorithm presented in section 6.2 (i.e. GetPresupportsTree(Δ, α)), tests whether the set of clauses from each of the accepted branches of the search tree can be arranged as a support tree for Δ and α.

Let *Branches* denote the output of the GetPresupportsTree(Δ, α). So *Branches* is a set of sets of formulae. The GetSupports(*Branches*) algorithm uses the function hasSupport(Γ, α) to test each set $\Gamma \in$ *Branches* individually. Given a set of clauses Γ, the hasSupport(Γ, α) function generates the presupport trees (N, A, f) where $N = \bigcup \{x \mid f(x) \in \Gamma\}$ and $\neg \alpha$ is the root. The algorithm keeps track of the attack values among the clauses in each presupport tree it generates and these are then used to test whether the additional conditions that differentiate a support tree from a presupport tree are satisfied. When the first such presupport tree that satisfies the conditions of being minimal and consistent is found, Γ is stored with the set of the accepted supports for literal arguments for α and the next set from *Branches* is tested. If no such presupport tree exists, then Γ is rejected for being a support for a literal argument for α and the algorithm proceeds by testing with the hasSupport function the next set from *Branches*.

Example 20. Given the results of example 19, if *Branches* is the output given by the GetPresupportsTree(Δ, α) algorithm, then for each of the sets $\Gamma_1, \Gamma_2, \Gamma_3 \in$ *Branches* that correspond to to branches 1,2 and 3 respectively, the hasSupport(Γ_i, α), $i = 1 \ldots 3$ function returns true and therefore the output of the GetSupports(*Branches*) algorithm is the set $\Gamma_1, \Gamma_2, \Gamma_3$. Hence, there are three literal arguments for α: $\langle \Gamma_1, \alpha \rangle$, $\langle \Gamma_2, \alpha \rangle$, and $\langle \Gamma_3, \alpha \rangle$.

7   Experimental Results

This section covers a preliminary experimental evaluation of the algorithms presented in Section 6 using a prototype implementation programmed in java running on a modest PC (Core2 Duo 1.8GHz).

The experimental data were obtained using randomly generated clause knowledge-bases according to the fixed clause length model K-SAT ([22,16]) where the chosen length (i.e. K) for each clause was 3 literals. The 3 disjuncts of each clause were chosen out of a set of N distinct variables (i.e. atoms). Each variable was randomly chosen out of the N available and negated with probability 0.5. For a fixed number of clauses, the number of distinct variables that occur in the disjuncts of all the clauses determines the size of the query graph which in turn determines the size of the search space and hence influences the perfomance of the system. For this reason, 10 different clauses-to-variables ratios were used for each of the different cardinalities tested (where this ratio varied from 1 to 10). For the definition of the ratio we take the integer part of the division of the number of clauses in Δ by the number of variables N (i.e. $\lfloor |\Delta|/|N| \rfloor$).

The evaluation was based on the time consumed by the system when searching for all the literal arguments for a given literal and the randomly generated knowledgebases of 15 to 30 clauses. Hence, for the results presented the smallest number of variables used was 1 and so for the case of a 15 clause knowledgebase, the clauses-to-variables ratio is 10. The largest number of variables used was 30 and so for the case of a 30 clause knowledgebase, and clauses-to-variables ratio is 1.

The preliminary results are presented in Table 1 which contains the median time consumed in milliseconds for 100 repetitions of running the system for each different cardinality and each ratio from 1 to 10. In other words, each field of the table is the median time obtained from finding all the arguments in 100 different knowledgebases of fixed cardinality where the cardinality is determined by the column of the table and the different clauses-to-variables ratios is determined by the row.

Table 1. Experimental data

clauses-to-variables ratio	$\vert\Delta\vert = 15$	$\vert\Delta\vert = 20$	$\vert\Delta\vert = 25$	$\vert\Delta\vert = 30$
1	3.000	6.000	9.000	13.00
2	3.000	6.000	11.00	17.00
3	2.000	6.000	12.50	238.0
4	2.000	5.000	14.00	466.5
5	2.000	4.000	8.000	178.0
6	1.000	3.000	6.500	71.00
7	1.000	5.000	4.000	9.000
8	0.000	1.000	4.000	6.000
9	1.000	1.000	2.000	6.000
10	1.000	2.000	2.000	7.000

From the preliminary results in Table 1, we see that for a low clauses-to-variables ratio (≤ 2) the number of variables is large enough to allow a distribution of the variables amongst the clauses such that it is likely for a literal to occur in a clause without its opposite occurring in another clause from the set. As a result, the query graph tends to contain a small subset of the knowledgebase and the system perfoms relatively quickly. The query graph tends also to be small in the case when a relatively small number of variables is distributed amongst the clauses of the knowledgebase (i.e. when the ratio is high) and this makes the occurrence of a variable and its negation in different clauses more frequent. As a result, it is likely for a pair of clauses ϕ, ψ from Δ to be such that $|\mathsf{Preattacks}(\phi, \psi)| > 1$ which will then allow the Attacks relation to be defined among a small number of clauses and therefore the attack graph will involve only a small subset of the knowledgebase. Hence, a large clauses-to-variables ratio also makes the system perform quickly. From these preliminary results the worst case occurs for ratio 4, and this appears to be because the size of the query graph tends to be maximized. This indicates that the clauses-to-variables ratio, rather than the cardinality of the knowledgebase is the dominant factor determining the time perfomance for the system. In future experiments we want to further characterize this worst case perfomance. In particular, we want to better understand the effect of increasing the value for K and so consider clauses with more literals, and we want to better understand the relationship between the number of arguments for a claim that can be obtained from a knowledgebase and the time taken.

8 Discussion

Classical logic has many advantages over defeasible logic for representing and reasoning with knowledge including syntax, proof theory and semantics for the intuitive language incorporating negation, conjunction, disjunction and implication. However, for argumentation, it is computationally challenging to generate arguments from a knowledgebase using classical logic. If we consider the problem as an abduction problem, where we seek the existence of a minimal subset of a set of formulae that implies the consequent, then the problem is in the second level of the polynomial hierarchy [13].

In this paper, we have proposed the use of a connection graph approach as a way of ameliorating the computation cost. The framework we have presented focuses the search for arguments in way that ensures that formulae that have no role as a premise in an argument will not be considered. We have provided theoretical results to ensure the correctness of the proposal, and we have provided provisional empirical results to indicate the potential advantages of the approach. In furture work, we will extend the empirical evaluation, and extend the theory and algorithms for dealing with arbitrary formulae as claims of arguments.

References

1. Amgoud, L., Cayrol, C.: A model of reasoning based on the production of acceptable arguments. Annals of Mathematics and Artificial Intelligence 34, 197–216 (2002)
2. Benferhat, S., Dubois, D., Prade, H.: Argumentative inference in uncertain and inconsistent knowledge bases. In: UAI 1993. Proceedings of the 9th Annual Conference on Uncertainty in Artificial Intelligence, pp. 1445–1449. Morgan Kaufmann, San Francisco (1993)
3. Besnard, Ph., Hunter, A.: A logic-based theory of deductive arguments. Artificial Intelligence 128, 203–235 (2001)
4. Besnard, Ph., Hunter, A.: Practical first-order argumentation. In: AAAI 2005. Proc. of the 20th National Conference on Artificial Intelligence, pp. 590–595. MIT Press, Cambridge (2005)
5. Besnard, Ph., Hunter, A.: Elements of Argumentation. MIT Press, Cambridge (2008)
6. Bibel, W.: Deduction: Automated Logic. Academic Press, London (1993)
7. Bryant, D., Krause, P., Vreeswijk, G.: Argue tuProlog: A lightweight argumentation engine for agent applications. In: Comma 2006. Computational Models of Argument, pp. 27–32. IOS Press, Amsterdam (2006)
8. Cayrol, C., Doutre, S., Mengin, J.: Dialectical proof theories for the credulous preferred semantics of argumentation frameworks. In: Benferhat, S., Besnard, P. (eds.) ECSQARU 2001. LNCS (LNAI), vol. 2143, pp. 668–679. Springer, Heidelberg (2001)
9. Chesñevar, C., Maguitman, A., Loui, R.: Logical models of argument. ACM Computing Surveys 32, 337–383 (2000)

10. Chesñevar, C., Simari, G., Godo, L.: Computing dialectical trees efficiently in possibilistic defeasible logic programming. In: Baral, C., et al. (eds.) LPNMR 2005. LNCS (LNAI), vol. 3662, Springer, Heidelberg (2005)
11. Dimopoulos, Y., Nebel, B., Toni, F.: On the computational complexity of assumption-based argumentation for default reasoning. Artificial Intelligence 141, 57–78 (2002)
12. Dung, P., Kowalski, R., Toni, F.: Dialectical proof procedures for assumption-based admissible argumentation. Artificial Intelligence 170, 114–159 (2006)
13. Eiter, T., Gottlob, G.: The complexity of logic-based abduction. Journal of the ACM 42, 3–42 (1995)
14. Elvang-Gøransson, M., Krause, P., Fox, J.: Dialectic reasoning with classically inconsistent information. In: UAI 1993. Proceedings of the 9th Conference on Uncertainty in Artificial Intelligence, pp. 114–121. Morgan Kaufmann, San Francisco (1993)
15. García, A., Simari, G.: Defeasible logic programming: An argumentative approach. Theory and Practice of Logic Programming 4(1), 95–138 (2004)
16. Gent, I.P., Walsh, T.: Easy problems are sometimes hard. Artificial Intelligence 70(1–2), 335–345 (1994)
17. Kakas, A., Toni, F.: Computing argumentation in logic programming. Journal of Logic and Computation 9, 515–562 (1999)
18. Kowalski, R.: A proof procedure using connection graphs. Journal of the ACM 22, 572–595 (1975)
19. Kowalski, R.: Logic for problem solving. North-Holland, Amsterdam (1979)
20. Prakken, H., Sartor, G.: Argument-based extended logic programming with defeasible priorities. Journal of Applied Non-Classical Logics 7, 25–75 (1997)
21. Prakken, H., Vreeswijk, G.: Logical systems for defeasible argumentation. In: Gabbay, D. (ed.) Handbook of Philosophical Logic. Kluwer, Dordrecht (2000)
22. Selman, B., Mitchell, D.G., Levesque, H.J.: Generating hard satisfiability problems. Artificial Intelligence 81(1–2), 17–29 (1996)
23. Vreeswijk, G.: An algorithm to compute minimally grounded and admissible defence sets in argument systems. In: Comma 2006. Computational Models of Argument, pp. 109–120. IOS Press, Amsterdam (2006)

Database Preferences Queries – A Possibilistic Logic Approach with Symbolic Priorities

Allel Hadjali[1], Souhila Kaci[2], and Henri Prade[3]

[1] IRISA/ENSSAT, Université Rennes I, 6 rue de Kérampont
22305 Lannion Cedex, France
hadjali@enssat.fr
[2] CRIL, IUT de Lens, Rue de l'Université SP 16
62300 Lens, France
kaci@cril.univ-artois.fr
[3] IRIT, Université Paul Sabatier, 118 route de Narbonne
31062 Toulouse Cedex 9, France
prade@irit.fr

Abstract. The paper presents a new approach to database preferences queries, where preferences are represented in a possibilistic logic manner, using symbolic weights. The symbolic weights may be processed without assessing their precise value, which leaves the freedom for the user to not specify any priority among the preferences. The user may also enforce a (partial) ordering between them, if necessary. The approach can be related to the processing of fuzzy queries whose components are conditionally weighted in terms of importance. Here, importance levels are symbolically processed, and refinements of both Pareto ordering and minimum ordering are used. The representational power of the proposed setting is stressed, while the approach is compared with database Best operator-like methods and with the CP-net approach developed in artificial intelligence. The paper also provides a structured and rather broad overview of the different lines of research in the literature dealing with the handling of preferences in database queries.

Keywords: Preferences queries, possibilistic logic.

1 Introduction

For more than two decades now, there has been an increasing interest in expressing preferences inside database queries. Motivations for such a concern are manifold. First, it has appeared to be desirable to offer more expressive query languages that can be more faithful to what a user intends to say. Second, the introduction of preferences in queries provides a basis for rank-ordering the retrieved items, which is especially valuable in case of large sets of items satisfying a query. Third, on the contrary, a classical query may also have an empty set of answers, while a relaxed (and thus less restrictive) version of the query might be matched by items in the database. This research trend has motivated several

S. Hartmann and G. Kern-Isberner (Eds.): FoIKS 2008, LNCS 4932, pp. 291–310, 2008.

distinct lines of research, in particular in the mainstream database literature and in fuzzy set applications to information systems.

Early mainstream database proposals either explicitly distinguish between mandatory conditions and secondary conditions, or use similarity relations. Lacroix and Lavency [22] use Boolean expressions for the secondary conditions that refine conditions that are higher in the hierarchy of priorities. Flexibility may be explicitly stated in the queries, or may be implicit, as it is often the case when using similarity relations. Thus, Motro [24] extends usual equality by means of a similarity relation relying on a notion of distance between attribute values of the same domain. Queries are transformed into Boolean conditions using thresholds, and then an ordering process takes place based on the distances.

Fuzzy set-based approaches use fuzzy set membership functions for describing the preference profiles of the user on each attribute domain involved in the query. This is especially convenient and suitable when dealing with numerical domains, where a continuum of values is to be interfaced for each domain with satisfaction degrees in the unit interval scale. Then individual satisfaction degrees associated with elementary conditions are combined using a panoply of fuzzy set connectives, which may go beyond conjunctive and disjunctive aggregations. Generally speaking, it should be emphasized that fuzzy set-based approaches rely on a commensurability hypothesis between the satisfaction degrees pertaining to the different attributes taking part to a query. Bosc and Pivert [6] provide a comparative discussion of the mainstream database approaches to preference handling and of the fuzzy set-based methods developed in the 80's, pointing out how the former could be encoded in the latter.

The present decade has seen a revival of interest in preference queries with the publication of many algorithms aiming at the efficient computation of non Pareto-dominated answers (viewed as points in a multi-dimensional space, their set constitutes a so-called skyline), in case of queries referring to totally ordered attribute domains (such as numerical domains), starting with the pioneering works of Börzsönyi et al. [5]. Clearly, the skyline computation approach does not require any commensurability hypothesis between satisfaction degrees pertaining to elementary requirements that refer to different attribute domains, as needed in the fuzzy set-based approach. Thus, some skyline points may represent very poor answers with respect to some elementary requirements (while they are excellent w.r.t. others, and Pareto ordering yields a strict partial order only, while fuzzy set-based approaches lead to complete pre-orders). Kiessling [21] has provided foundations for a Pareto-based preference model for database systems. A preference algebra has also been proposed by Chomicki [10] for an embedding of preference formulas into a relational setting (and SQL). See also Torlone and Ciaccia [25], who have focused on the so-called Best operator aiming at iteratively returning the non-dominated tuples of a relation, after excluding the tuples retrieved in previous steps.

Meanwhile, Brafman and Domshlak [8] have advocated the use, in database preference queries, of an approach developed in artificial intelligence in the last decade, which is based on a graphical representation, called CP-nets [7], of conditional ceteris paribus preference statements. The underlying idea is that users' preferences generally express that, in a given context, a partially described state of affairs is strictly preferred to another mutually exclusive partially described state of affairs, in a ceteris paribus way, i.e. everything else being equal in the description of the two compared states of affairs.

The paper provides a new approach to the handling of database preference queries, which takes its roots in the fuzzy set-based approaches, but remains as symbolic as possible, while remaining compatible with the Best operator-like approach, and improving the CP-net based method from a preference-based ranking point of view. It is organized as follows. Section 2 provides the necessary background on the three main types of methods considered here: i) the Pareto ordering-based approaches, ii) the fuzzy set-based approaches, including the encoding of Lacroix and Lavency [22]' hierarchical expression of preferences by means of importance levels, and the refinement of minimum (or maximum) operation in agreement with Pareto ordering, iii) the CP-net approach. Section 3 presents and discusses the possibilistic approach that handles symbolic priorities, both in the cases of binary and non-binary attributes.

2   Background

As said in the introduction, the idea of preferences and flexibility have raised interest in different sub-communities, which have identified and emphasized different points related to these issues, in the last decades. Preferences may pertain to binary or non-binary, to numerical or non-numerical attributes. Attributes with totally ordered domains induce strict Pareto partial pre-orders. Attributes may be a matter of importance and weighting. Flexibility does not only mean that values that are somewhat less preferred may be still acceptable, but may also refer to the use of similarity relations associated with attribute domains for expressing that if a value is acceptable to some extent, the values that are (sufficiently) close to this value should be also considered as being acceptable, maybe to a less extent. The number of elementary requirements in a query that are sufficiently satisfied may be also taken into account, when rank-ordering the retrieved items. These different points are considered in the different approaches that are now recalled, namely databases works that develop and refine the skyline point of view, fuzzy logic-based methods for evaluating flexible queries, and the ceteris paribus principle-based method for preference processing. However, it is worth noticing that the mainstream database approaches rather focus on the retrieval of a classical set of "best" possible items, while the fuzzy set approaches rank-order the items according to their level of acceptability, and yield a layered set of items.

2.1   Databases Approaches to Preference Handling in Querying

Approaches to database preference queries can be classified into qualitative and quantitative ones. In the latter (e.g., [1]), preferences are expressed quantitatively by a monotone scoring function (the overall score is positively correlated with partial scores), often taken as a weighted linear combination of attributes' values (which have therefore to be numerical). Since the scoring function associates each tuple with a numerical score, tuple t_1 is preferred to tuple t_2 if the score of t_1 is higher than the score of t_2. It is well known that scoring functions cannot represent all preferences that are strict partial orders [18], not even those that occur in database applications in a natural way [10]. For example, scoring functions cannot capture skylines (discussed below). Another issue is that devising the scoring function may not be simple.

In the qualitative approach [22,5,21,10], preferences are defined through binary preference relations. Since binary preference relations can be defined in terms of scoring functions, the qualitative approach is more general than the quantitative one. Note that it is reasonable to require that qualitative preferences, even if not based on the comparison of overall scores, still have a monotonic behavior with respect to partial scores. Hereafter, we review the main qualitative approaches proposed in the literature. In the system Preferences [22], queries of the form "find the tuples which satisfy necessarily S with a preference for those which satisfy also P", involve a main selection condition S and a component P devoted to preferences, both relying on Boolean expressions. The system returns the tuples satisfying S and P if they exist, or by default those satisfying only S. This system combines preferences by nesting (hierarchy of preferences) and juxtaposition (preferences having the same importance). It frees the user from a set of successive questions/answers, which is often necessary to reach a desired number of responses with classical queries.

Börzsönyi et al. [5] have introduced skyline queries that aim at finding the set of all the tuples in a relation that are not dominated by any other tuples in the same relation in all dimensions. This returned set is called the Skyline of the query under consideration. Let us recall that a tuple t_1 dominates a tuple t_2 if t_1 is at least as good as t_2 in all dimensions and better than t_2 in at least one dimension (this is the well known principle of Pareto optimality). Skyline offers a natural way to combine multiple preference criteria in parallel. Skyline as introduced has a clear partial order semantics. However, as stressed by [19], Skyline in its original form, is limited in its expressiveness and does not capture many types of preferences and compositions people would like to support (prioritizing preferences can be accomplished to some degree by nested skyline clauses, but often not naturally).

Kiessling [21] has considered an algebraic approach to constructing a rich preference query language as an extension to SQL, called Preference SQL. It is based on the idea that people express their wishes frequently as "I like A better than B". A preference is then formulated as a strict partial order on a set of attribute values. To express preferences, a number of preference operators are introduced and the way they can be composed is defined. Preference SQL allows users to write best-match queries by composing their preference criteria via the preference operators. Let us note that all tuples returned by a Preference SQL query satisfy the Pareto principle. A compensatory strategy between different atomic conditions is not possible due to the fact that Preference SQL makes use of two different functions (level and distance) for evaluating the degree in which a tuple satisfies an atomic condition. Chomicki [10] has developed a general logic framework for preferences handling, and has proposed a relational operator winnow for composing preferences relations in relational algebra.

Definition 1. *A preference formula* $C(t_1, t_2)$ *is a first-order formula defining a preference relation* $\succ_C$ *in the standard sense, namely* $t_1 \succ_C t_2$ *if and only if* $C(t_1, t_2)$. *Tuple* t_1 *dominates (or is preferred to) a tuple* t_2 *in the sense of formula C if* $t_1 \succ_C t_2$.

To pick from a given relation the set of the most preferred tuples, according to a given preference formula, an algebraic operator, called winnow is defined.

Definition 2. *Let R be a relation schema and C a preference formula defining a preference* $\succ_C$ *over R. The winnow operator is written as* $W_C(r)$ *for any instance r of R*

$$W_C(r) = \{t \in r | \nexists t' \in r, t' \succ_C t\}.$$

This operator returns the tuples that are not dominated by any other tuple in r.

Example 1. [11] Consider the relation $Car(Make, Year)$ and the following preference relation $\succ_C$ between Car tuples: within each make, prefer a more recent car, i.e. $(m, y) \succ_C (m', y') \equiv (m = m' \land y > y')$.

The winnow operator W_C returns for every make the most recent car available. Consider the instance r_1 of Car in Table 1.a. The set $W_C(r)$ is shown in Table 1.b. Note that the preference in this example is expressed by a first-order

Table 1. (a) The Car relation, (b) winnow result

tuples	Make	Year
t_1	VW	2002
t_2	VW	1997
t_3	Kia	1997

tuples	Make	Year
t_1	VW	2002
t_2	VW	1997
t_3	Kia	1997

expression, on a non-binary attribute domain (for $Year$ attribute), which requires comparisons between tuples and makes impossible an evaluation of the tuples one by one w.r.t. individual preferences, as it is possible with a preference such as $VW \succ_C Kia$ for instance, except if one would write $2007 \succ_C 2006$, $2006 \succ_C 2005, \cdots, 1997 \succ_C 1996$, etc.

Chomicki [10] has shown that the winnow can be formulated in relational algebra, and has also studied its main properties regarding commutativity and distributivity w.r.t. relational algebra operators. In order to express a complex preference, the logical composition of preferences has been defined. Depending on the preference formulas, we can distinguish between two types of compositions: the uni-dimensional composition pertaining to the same database schema (Boolean and prioritized compositions, and a transitive closure of the preference relation) and the multi-dimensional composition (Pareto and lexicographic compositions). Composition of preference relations can raise difficulties since certain compositions can violate the partial order semantics. One of the advantages of this model is the fact that it does not impose any restriction on the preference relations. Note also that the Skyline operator is a special case of the winnow operator, also called BMO (Best Matches Only) in [21] and Best in [25].

2.2   The Fuzzy Set-Based Approach to Flexible Querying

Fuzzy set membership functions are convenient tools for modeling user's preference profiles and the large panoply of fuzzy set connectives can capture different user attitudes concerning the way the different criteria present in a query compensate or not. For instance, in case of a query such as "find an apartment which is not too expensive and not too far from downtown", there does not exist a definite threshold for which the price becomes suddenly too high, but rather we have to differentiate between prices which are perfectly acceptable for the user, and other prices, somewhat higher, which are still more or less acceptable. Obviously, the meaning of vague predicate expressions like "not too expensive" is context/user dependent, rather than universal. Observe also that here one can capture the idea of "smallest" price and "smallest" distance as well, by using decreasing functions from 1 to 0 on the respective attribute domains, and then applying a refinement of the minimum-based ordering on the pairs of scores, such as the leximin ordering that refines Pareto ordering (see the end of the section for the definition), however the commensurateness of the values of the two decreasing functions is debatable here.

Making a requirement flexible is not only a matter of gradual representation reflecting preferences. It may also refer to a possible weakening of the requirement in some way: by putting some tolerance on it (for the sake of brevity we shall not further discuss this point here), or by assessing its importance and by conditioning it [15].

Importance assignment. If it is not so important to take into account an elementary requirement P in a query, it means that to some extent, any value of the corresponding attribute may be somewhat acceptable. It amounts to modifying P into P^* such that

$$\mu_{P^*}(u) = \max(\mu_P(u), 1 - w).$$

As one can see, P^* considers any value outside the support of P as acceptable at degree $1 - w$. It means that the larger w the smaller the acceptability of values outside the support of P (i.e. the values with a zero membership grade). In case of a logical conjunctive combination of several requirements P_i performed by min-combination (min is the only idempotent and the largest associative aggregation operation on $[0, 1]$ that extends ordinary conjunction), i.e. for a tuple $d = (u_1, \cdots, u_n)$, we get

$$\min_{i=1,\cdots,n} \mu_{P_i^*}(d) = \min_{i=1,\cdots,n} \max(1 - w_i, \mu_{P_i}(d))$$

with $\mu_{P_i}(d) = \mu_{P_i}(u_i)$ where u_i is the precise value in d of the attribute i pertaining to P_i, and where the normalization condition $\max_{i=1,\cdots,n} w_i = 1$ should be satisfied by the w_i's. It expresses that the most important requirement(s), which has/have a weight w equal to 1 is/are compulsory. When $w_i = 0$, $\mu_{P_i}(d)$ is ignored in the combination.

Conditional requirement. A conditional requirement is a constraint that applies only if another one is satisfied: A requirement P_j conditioned by a non-fuzzy requirement P_i is imperative if P_i is satisfied and can be dropped otherwise. More generally, the variable level of satisfaction $\mu_{P_i}(d)$ of a fuzzy conditioning requirement P_i for an item d is viewed as the level of priority of the conditioned requirement P_j, i.e., the greater the level of satisfaction of P_i, the greater the priority of P_j is. This is represented by:

$$\mu_{P_i \to P_j}(d) = \max(\mu_{P_j}(d), 1 - \mu_{P_i}(d)).$$

It allows us to represent Lacroix and Lavency [22]' nested requirements: "P_1 should be satisfied, and among the solutions to P_1 (if any) the ones satisfying P_2 are preferred, and among those satisfying both P_1 and P_2, those satisfying P_3 are preferred, and so on", where $P_1, P_2, P_3, \cdots$ are hard constraints. There is a hierarchy between the constraints. Thus, one wants to express that P_1 should hold (with priority 1), and that if P_1 holds, P_2 holds with priority α_2, and if P_1 and P_2 hold, P_3 holds with priority α_3, with $\alpha_3 < \alpha_2 < 1$ for reflecting the hierarchy. This is represented by

$$\mu_{P^*}(d) = \min(\mu_{P_1}(d), \max[\max(\mu_{P_2}(d), 1 - \alpha_2), 1 - \mu_{P_1}(d)],$$
$$\max[\max(\mu_{P_3}(d), 1 - \alpha_3), 1 - \min(\mu_{P_1}(d), \mu_{P_2}(d))]]).$$

It can be checked that $\mu_{P^*}(d) = 1$ if P_1, P_2 and P_3 hold, $\mu_{P^*}(d) = 1 - \alpha_3$ if P_1 and P_2 hold, $\mu_{P^*}(d) = 1 - \alpha_2 < 1 - \alpha_3$ if P_2 does not hold, and $\mu_{P^*}(d) = 0$ if P_1 does not hold. Besides, $\mu_{P^*}(d)$ can be still rewritten (since $\mu_{P_i}(d) \in \{0, 1\}$ for $i = 1, 3$), as

$$\mu_{P^*}(d) = \min(\mu_{P_1}(d), \max(\mu_{P_2}(d), 1 - \alpha_2), \max(\mu_{P_3}(d), 1 - \alpha_3))$$

which is the semantic counterpart of the *possibilistic logic base* $K = \{(p_1, \alpha_1), (p_2, \alpha_2), (p_3, \alpha_3)\}$, where μ_{P_i} is the characteristic function of the set of models of proposition p_i, and the possibilistic logic formula (p_i, α_i), is understood as $N(p_i) \geq \alpha_i$, where N is a necessity measure associated with μ_{P^*}, namely $N(q) = \inf_d \max(\mu_Q(d), 1 - \mu_{P^*}(d))$ where $\mu_Q(d) = 1$ if q is true and $\mu_Q(d) = 0$ if q is false; see [3] for details. This shows that the conditional preferences here can be expressed in a non-conditional way, namely having condition P_1 satisfied (i. e. p_1 true) is imperative (priority 1), satisfying condition P_2 is less imperative (priority α_2), and satisfying condition P_3 is still less imperative (priority α_3).

Strictly speaking, Lacroix and Lavency [22] viewed the problem of finding feasible solutions (say, satisfying C) and then finding good solutions (D) among them, as satisfying the logical requirement: Find $\{d | C(d) \wedge [(\exists d', C(d') \wedge D(d')) \Rightarrow D(d)]\}$. This expresses that either C and D are consistent, and one wants to find items that both satisfy C and D, or the two requirements are inconsistent, and the items to be retrieved should just satisfy C. This is a bit different from what is done above, where the items that satisfy C and not D are ranked (as being less satisfactory) after those satisfying C and D (if any), whatever it is known or not if there exists an item in the world satisfying both C and D.

Refinements of the minimum-based ordering. In the fuzzy set approach, in case of a request made of a conjunction of several elementary requirements, if no compensation is allowed between elementary degrees of satisfaction, a pure conjunctive approach is applied, which amounts to use the minimum operation that retains the worst satisfaction degree among the elementary ones. Thus, let two items d and d' be respectively associated with tuples $d = (u_1, \cdots, u_n)$ and $d' = (u'_1, \cdots, u'_n)$, and be evaluated by means of membership functions μ_i for $i = 1, \cdots, n$, giving birth to vectors $e(d) = (v_1, \cdots, v_n)$ and $e(d') = (v'_1, \cdots, v'_n)$ with $v_i = \mu_i(u_i)$ and $v'_i = \mu'_i(u'_i)$. This leads to the following ranking: $d \succ_{\min} d'$ if and only if $\min_i v_i > min_i v'_i$. But, $d \succ_{Pareto} d'$ does not entail $d \succ_{\min} d'$ (the converse is not true either), where $\succ_{Pareto}$ denotes Pareto ordering between $e(d)$ and $e(d')$ (i.e. $d \succ_{Pareto} d'$ if and only if $\forall i, v_i \geq v'_i$ and $\exists j, v_j > v'_j$). However, there exist two noticeable orderings, called discrimin and leximin (e.g., [14]), that both refine $\succ_{Pareto}$ and $\succ_{\min}$. Discrimin is defined by ignoring the attributes that yield values regarded as equivalent w.r.t. the query. Namely, $d \succ_{discrimin} d'$ if and only if $\min_{i|v_i \neq v'_i} v_i > min_{i|v_i \neq v'_i} v'_i$. Then it can be checked that $d \succ_{Pareto} d'$ does entail $d \succ_{discrimin} d'$. Obviously, $d \succ_{\min} d'$ entails $d \succ_{discrimin} d'$. Note that $\succ_{discrimin}$ is still a partial order. Leximin ordering is a complete preorder that is obtained by first ranking non-decreasingly the values $e(d)$ and $e(d')$ of the vectors to be compared, and then applying the discrimin ordering on these rearranged vectors. Leximin refines discrimin: $d \succ_{discrimin} d'$ entails $d \succ_{leximin} d'$. Clearly, its application in the fuzzy set approach to preference queries enables us to get rid of the many ties resulting of the application of minimum aggregation. Note that the non-decreasing rearrangement of the vectors values assume that all the elementary requirements have the same importance. Discrimin partial order does not require this assumption.

2.3   The CP-Net-Based Approach to Preference Processing

This section provides a background on Conditional Preference networks (CP-nets for short), which are graphical models of conditional independence of preferences, based on the ceteris paribus semantics (which stands in Latin for "everything else being equal"), and discusses their use in database preference queries.

Basic notions on CP-nets. CP-nets [7], a popular representation format for modelling preferences, are based on the expression of conditional ceteris paribus preference statements and capture partial orders. With CP-nets, the user describes how his preferences over the values of one variable depend on the value of other variables. Assume a set of variables (or attributes) $V = \{X_1, \cdots, X_n\}$ about which the user has preferences. For each variable X_i, a set of parent variables $Pa(X_i)$ is identified. For each value assignment of $Pa(X_i)$, the user has to specify a preference order over possible values of X_i, all other things being equal. For simplicity, variables are binary in the examples, and CP-nets are acyclic for ensuring consistency.

Formally, a CP-net N over the set of variables $V = \{X_1, \cdots, X_n\}$ is a directed graph over the nodes $X_1, \cdots, X_n$, and there is a directed edge from X_i to X_j if the preference over the value X_j is conditioned on the value of X_i. Each node $X_i \in V$ is associated with a conditional preference table $CPT(X_i)$ that associates a strict (possibly empty) partial order $\succ_{CP}(u_i)$ with each possible instantiation u_i of the parents of X_i. A complete preference ordering satisfies a CP-net N iff it satisfies each conditional preference expressed in N. In this case, the preference ordering is said to be consistent with N.

Example 2. [7] Assume that black (b) jackets (resp. pants) are preferred to white (w) jackets (resp. pants), and that in case of jackets and pants of the same color red (r) shirts are preferred to white ones; otherwise, white shirts are preferred. Only the colors mentioned are supposed to be available. Thus, we have three binary variables J, P, S corresponding to the colors of the jacket, the pants and the shirt. See Figure 1 for the CP-net built on these variables.

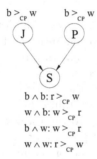

Fig. 1. CP-net for "Evening Dress: Jacket, Pants and Shirt"

Thus, there are 8 tuples corresponding to the relational schema jackets-pants-shirts: $bbr, bbw, bwr, bww, wbr, wbw, wwr, www$. The preferences associated to the induced CP-net, applying the ceteris paribus principle, are i) $bPS \succ_{CP} \neg bPS$, $\forall P \in \{b, \neg b\}$, $\forall S \in \{r, \neg r\}$; ii) $JbS \succ_{CP} J\neg bS$, $\forall J \in \{b, \neg b\}$, $\forall S \in \{r, \neg r\}$; iii) $bbr \succ_{CP} bbw$, $wwr \succ_{CP} www$, $bww \succ_{CP} bwr$, $wbw \succ_{CP} wbr$.

Then the following partial ordering holds under the ceteris paribus assumption:

$bbr \succ_{CP} bbw \succ_{CP} bww \succ_{CP} bwr \succ_{CP} wwr \succ_{CP} www$;
$bbr \succ_{CP} bbw \succ_{CP} wbw \succ_{CP} wbr \succ_{CP} wwr \succ_{CP} www$.

Using CP-nets in preference queries. Brafman and Domshlak [8] propose a qualitative approach to database queries based on CP-nets in which preferences are represented by a binary relation over a relation schema. Let R be a relation schema, a preference query Q over R consists of a set $Q = \{s_1, \cdots, s_m\}$ of preference statements (usually between sub-tuples of R, thanks to ceteris paribus semantics). These statements define preference relations $\{\succ_{CP(1)}, \cdots, \succ_{CP(m)}\}$, from which one should derive the global preference $\succ_{CP(Q)}$.

Example 3. [8] Table 2 gives a schema instance of $Car(category, make, color)$, and a query.

Table 2. An instance of the Car schema and a preference query over Car

tuples	category	make	color	for short
t_1	minivan	Chrysler	white	mcw
t_2	minivan	Chrysler	black	mcb
t_3	minivan	Ford	white	mfw
t_4	minivan	Ford	black	mfb
t_5	sedan	Chrysler	white	scw
t_6	sedan	Chrysler	black	scb
t_7	sedan	Ford	white	sfw
t_8	sedan	Ford	black	sfb

s_1 the user prefers minivan (m) cars to sedan (s) cars
s_2 for minivans, he prefers Chrysler (c) to Ford (f)
s_3 for sedans, he prefers Ford to Chrysler
s_4 in Ford cars, he prefers the black (b) ones to the white (w) ones
s_5 in Chrysler cars, he prefers the white ones to the black ones

As pointed out in [8], database preference queries can be interpreted according to two semantics: the totalitarian semantics and the ceteris paribus semantics. Authors in database community seem to implicitly favour the first semantics. Then, when evaluating a query containing s': *"the user prefers Chrysler minivans to Ford sedans"*, do we ignore the other attributes or fix their values? These two possibilities correspond to the totalitarian semantics and the ceteris paribus semantics, respectively. According to the totalitarian semantics, s' implies that any (minivan $\wedge$ Chrysler)-tuple is preferred to any (sedan $\wedge$ Ford)-tuple. In particular, in Example 3, we obtain the preference order over the database tuples $\succ_{T(s')} = \{t_1 \succ_{T(s')} t_7, t_1 \succ_{T(s')} t_8, t_2 \succ_{T(s')} t_7, t_2 \succ_{T(s')} t_8\}$. Now, according to ceteris paribus semantics, s' implies that a (minivan $\wedge$ Chrysler)-tuple is preferred to a (sedan $\wedge$ Ford)-tuple, provided that both tuples agree on the value of all other attributes. Under this semantics, we would get the preference order $\succ_{CP(s')} = \{t_1 \succ_{CP(s')} t_7, t_2 \succ_{CP(s')} t_8\}$. As can be seen, only fewer tuples are now comparable.

Since preference queries generally consist of several preference statements, the different preference orders have to be composed into a single preference order. In database community [10,21], Boolean (e.g., intersection and union) and prioritized compositions have been considered. Regarding Boolean composition, the query evaluation operator that has been proposed retrieves all non dominated database tuples. This is the operator $BEST(r, \succ_{CP(Q)}) = \{t \in r | \forall t' \in r, t' \nsucc_{CP(Q)} t\}$ with r an instance of R. Unfortunately, it has been shown that it can be computationally intractable for ceteris paribus semantics. Worst-case time complexity of $BEST$ is $O(exp(n).D^2)$, where n is the number of attributes and D the size of r.

In database community, the totalitarian interpretation with intersection composition appears to be the implicit choice. However, as shown on different examples by [8], this interpretation is semantically inappropriate and yields unintuitive or empty results even for very simple queries. So, they propose an approach based on ceteris paribus semantics and union-based composition. It looks more appealing in database preference queries for combining preference statements. Besides, Ciaccia [12] also proposes to query databases using incomplete CP-nets.

The ceteris paribus union semantics. To illustrate it, consider Example 3 again. The union of preference relations $\succ_{CP(1)}, \cdots, \succ_{CP(5)}$ under ceteris paribus semantics is depicted in Figure 2(a): The nodes stand for tuples $t_1, \cdots, t_8$, and a directed edge from t to t' means $t \succ_{CP(i)} t'$ for one of the statements s_i. The resulting preference relation is given by the transitive closure of the graph, e.g., $t_4 = mfb$ is preferred to $t_7 = sfw$.

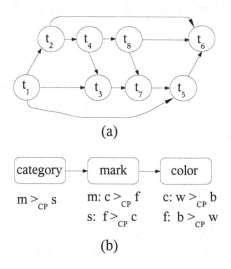

(a)

category → mark → color

$m >_{CP} s$ 　　　m: $c >_{CP} f$ 　　　c: $w >_{CP} b$

　　　　　　　　s: $f >_{CP} c$ 　　　f: $b >_{CP} w$

(b)

Fig. 2. Preferences induced by ceteris paribus semantics in Example 3 and CP-net for the query

Let us consider the computational cost for evaluating queries under ceteris paribus semantics. As already said, computing the $BEST$ operator under this semantics is prohibitive. This is because it is based on "dominance-testing" (i.e., comparing two tuples for determining the one that is better). But, as shown in [7] and [23], even simple qualitative preference representations suffer from the complexity of dominance testing. Even for queries forming acyclic CP-nets over binary-valued attributes the problem is NP-hard, and for non-binary attributes it is even not in NP. To efficiently answer ceteris paribus preference queries, Brafman and Domshlak [8] have introduced an operator called ORD, as a relaxation of $BEST$:

Definition 3. *Let R be a relation schema, and Q be a preference query inducing a strict partial order $\succ_Q$ over the Cartesian product of the attribute domains of R. Given a relation instance r of R, $ORD(r, \succ_Q)$ contains all the tuples of r, totally ordered such that for every $t, t' \in r$, if t appears before t' in $ORD(r, \succ_Q)$, then we have $t' \not\succ_Q t$.*

Clearly, ORD is based on sorting the given data set r according to relation $\succ_Q$, and providing the user with the top k tuples of r in a non-increasing order of preference. If $t \succ_Q t'$, then we are certain that ORD will show t prior t', but if t and t' are incomparable according to $\succ_Q$, then ORD will order them arbitrarily. Moreover, ORD guarantees to show any element in $BEST$ before all the elements that it dominates. Brafman and Domshlak [8] show that for a wide class of preference queries that are problematic for $BEST$, the ORD operator is computable in time $O(nDlogD)$.

3   A Symbolic Possibilistic Logic Approach

This section presents a symbolic possibilistic logic approach that sort of unifies the different approaches reviewed in Section 2. The approach is both more faithful to user's preferences than the CP-net approach as we shall see, and liable to be interfaced with the fuzzy set approach, which is important for non-binary preferences (the usual case with numerical attributes). It also refines the mainstream database approaches from Lacroix and Lavincy [22]' method to Chomicki [10] winnow operator.

3.1   Comparative Preferences

Representation. In Example 3, one prefers minivans (m) to sedans (s). For minivans, one prefers Chrysler (c) to Ford (f), and the converse for sedans. Finally, for Ford, one prefers black (b) cars to white (w) ones, while for Chrysler cars it is the converse. Such preferences can be encoded as prioritized goals in possibilistic logic.

Lemma 1. *The possibilistic encoding of the conditional preference "in context c, a is preferred to b" is a pair of possibilistic formulas: $\{(\neg c \lor a \lor b, 1), (\neg c \lor a, 1-\alpha)\}$.*

Namely if c is true, one should have a or b (the choice is only between a and b), and in context c, it is somewhat imperative to have a true. This encodes a constraint of the form $N(\neg c \lor a) \geq 1 - \alpha$, itself equivalent here to a constraint on a conditional necessity measure $N(a|c) \geq 1 - \alpha$ (see, e.g., [17]). This is still equivalent to $\Pi(\neg a|c) \leq \alpha$, where Π is the dual possibility measure associated with N. It expresses that the possibility of not having a is upper bounded by α, i. e. $\neg a$ is all the more impossible as α is small. Such a modeling has been proposed in [20] for representing preferences, and approximating CP-nets. Note that when $b \equiv \neg a$, the first clause becomes a tautology, and thus does not need to be written. Strictly speaking, the possibilistic clause $(\neg c \lor a, 1 - \alpha)$

expresses a preference for a (over $\neg a$) in context c. The clause $(\neg c \vee a \vee b, 1)$ is only needed if $a \vee b$ does not cover all the possible choices. Assume $a \vee b \equiv \neg d$ (where $\neg d$ is not a tautology), then it makes sense to understand the preference for a over b in context c, as the fact that in context c, b is a default choice if a is not available. If one wants to open the door to remaining choices, it is always possible to use $(\neg c \vee a \vee b, 1 - \alpha')$ with $1 - \alpha' > 1 - \alpha$, instead of $(\neg c \vee a \vee b, 1)$. Thus, the approach easily extends to non binary choices. For instance, "I prefer Renault (r) to Chrysler and Chrysler to Ford" is encoded as $\{(r \vee c \vee f, 1), (r \vee c, 1 - \alpha), (r, 1 - \alpha')\}$, with $1 - \alpha > 1 - \alpha'$.

It is worth noticing that the encoding of preferences in the sense of Lemma 1 also applies to Lacroix and Lavincy [22]'approach, as discussed in Section 2.2. Namely, one wants to express that "$p_1 \wedge p_2$ is preferred to $p_1 \wedge \neg p_2$" and p_1 is mandatory. It is encoded by $((p_1 \wedge p_2) \vee (p_1 \wedge \neg p_2), 1)$, equivalent to $(p_1, 1)$, and by $(p_1 \wedge p_2, 1 - \alpha)$ equivalent to $(p_1, 1 - \alpha)$ and $(p_2, 1 - \alpha)$, $(p_1, 1 - \alpha)$ being subsumed by $(p_1, 1)$. Thus, one retrieves the encoding $(p_1, 1)$ and $(p_2, 1 - \alpha)$, already encountered.

The symbolic approach. Since one does not know precisely how imperative the preferences are, the weights will be handled in a symbolic manner. However, they are assumed to belong to a linearly ordered scale (the strict order will be denoted by $\succ$ on this scale), with a top element (denoted 1) and a bottom element (denoted 0). Thus, $1 - (.)$ should be regarded here just as denoting an order-reversing map on this scale (without having a numerical flavor necessarily), with $1 - (0) = 1$, and $1 - (1) = 0$. On this scale, one has $1 \succ 1 - \alpha$, as soon as $\alpha \neq 0$. The order-reversing map exchanges two scales: the one graded in terms of necessity degrees, or if we prefer here in terms of imperativeness, and the one graded in terms of possibility degrees, i.e. here, in terms of satisfaction levels. Thus, the level of priority $1 - \alpha$ for satisfying a preference is changed by the involutive mapping $1 - (.)$ into a satisfaction level when this preference is violated.

Example 4. (Example 3 continued). Here, this leads to the following encoding of the preferences $K_1 = \{(m, 1 - \alpha), (\neg m \vee c, 1 - \beta), (\neg s \vee f, 1 - \gamma), (\neg f \vee b, 1 - \delta), (\neg c \vee w, 1 - \varepsilon)\}$[1].

Since the values of the weights $1 - \alpha, 1 - \beta, 1 - \gamma, 1 - \delta, 1 - \varepsilon$, are unknown, no particular ordering is assumed between them. Table 3 gives the satisfaction levels for the possibilistic clauses encoding the five elementary preferences, and the eight possible choices. The last column gives the global satisfaction level by minimum combination. Even if the values of the weights are unknown, a partial order between the eight choices is naturally induced by a leximin ordering of the corresponding 5-component vectors evaluating the satisfaction levels. For instance,

$$(\alpha, 1, 1, 1, 1) \succ_{leximin} (\alpha, 1, \gamma, 1, 1) \succ_{leximin} (\alpha, 1, \gamma, 1, \varepsilon)$$

---

[1] We omitted the weighted formulas $(m \vee s, 1)$, $(\neg m \vee c \vee f, 1)$, $(\neg s \vee f \vee c, 1)$, $(\neg f \vee b \vee w, 1)$ and $(\neg c \vee w \vee b, 1)$ since we are dealing with binary variables.

Table 3. Possible alternative choices in Example 3

	$(m, 1-\alpha)$	$(\neg m \vee c, 1-\beta)$	$(\neg s \vee f, 1-\gamma)$	$(\neg f \vee b, 1-\delta)$	$(\neg c \vee w, 1-\varepsilon)$	min
scw	α	1	γ	1	1	α, γ
mcw	1	1	1	1	1	1
sfw	α	1	1	δ	1	α, δ
mfw	1	β	1	δ	1	β, δ
scb	α	1	γ	1	ε	$\alpha, \gamma, \varepsilon$
mcb	1	1	1	1	ε	ε
sfb	α	1	1	1	1	α
mfb	1	β	1	1	1	β

since $(\alpha, 1, 1, 1, 1) \succ_{discrimin} (\alpha, \gamma, 1, 1, 1) \succ_{discrimin} (\alpha, \gamma, \varepsilon, 1, 1)$, whatever the values of $\alpha, \gamma, \varepsilon^2$. Thus, we get the following partial order between tuples $mcw \succ_{lex} \{mcb, sfb, scw, scb, mfb, mfw, sfw\}$; $sfb \succ_{lex} scw \succ_{lex} scb$; $mfb \succ_{lex} mfw$; $sfb \succ_{lex} sfw$; $mcb \succ_{lex} scb$ and $sfb \succ_{lex} scb$, where $\succ_{lex}$ denotes the partial order induced on tuples from the partial order $\succ_{leximin}$ between evaluation vectors with symbolic components.

It is worth noticing that this partial order amounts to rank-ordering a vector v' after a vector v, each time the set of preferences violated in v is strictly included in the set of preferences violated in v', since nothing is known on the relative values of the symbolic levels (except they are strictly smaller than 1, when different from 1). Then a vector v is greater than another v', only when the components of v are equal to 1 for those components that are different in v and v'. If constraints (e.g., $\alpha < \gamma$) are specified between symbolic levels, the $\succ_{leximin}$ order would take them into account also.

Assume we address a query represented by a set of preferences to an instance of a relational schema, and consider the tuples whose evaluation correspond to one of the vector(s) that is/are non dominated in the sense of the leximin ordering $\succ_{leximin}$ applied on symbolic values (among the vectors associated to tuples in the database). Consequently, these tuples are themselves not dominated by the other database tuples in the sense of $\succ_{lex}$, nor in the sense of Pareto (since $\succ_{leximin}$ and thus $\succ_{lex}$ refines $\succ_{Pareto}$ applied to vectors and then to tuples). Thus, the symbolic possibilistic method agrees with mainstream database approaches. More precisely, if preference orderings are associated with different attributes, for instance one looks for hotels both cheap and close to the beach, then non-dominated tuples in a database may include very different answers, such as an inexpensive hotel far from the beach, a very expensive hotel on the beach, and a reasonable price hotel rather close to the beach. If we apply the symbolic vector evaluation approach in such a case, with the constraint that the higher the price, and the larger the distance, the smaller the satisfaction levels, these three hotels will be respectively associated with vectors such as $(1, \varepsilon)$, $(\varepsilon', 1)$,

---

[2] Since the weights are symbolic, we use an arbitrary ordering of $\alpha, \gamma, \varepsilon$ for applying the leximin procedure here. But, it only leads to a partial ordering between vectors, since one can only make comparison w.r.t. 1.

(α, β), and the last hotel will be preferred as soon as $\min(\alpha, \beta) > \max(\varepsilon, \varepsilon')$. More formally,

Proposition 1. *Let a database D, the corresponding set of symbolic evaluation vectors for a set of unconditional preferences, together with constraints stating that the higher the violation of a preference the smaller the satisfaction level, and let $\succ_{lex}$ be the associated partial order. Then, $\forall t, t'$ two tuples of r (an instance of the schema of D), if $t' \succ_C t$ (t is dominated by t' in the sense of Chomicki's approach) then $t' \succ_{lex} t$.*

Proof (Sketch). It relies on the fact that $\succ_{leximin}$ refines $\succ_{Pareto}$.

Example 5. (Example 3 continued). Let us consider again the strict partial order $\succ_{lex}$ induced by leximin. Thus, if the database contains white Chrisler minivans (mcw), they will be retrieved first. If the database does not include such cars, nor black Ford sedans (sfb), the non-dominated cars will be the black Chrisler minivans (mcb), the white Chrisler sedans (scw), the black Ford minivans (mfb), and the white Ford sedans (sfw), whose respective evaluations are $(1, 1, 1, 1, \varepsilon)$, $(\alpha, 1, \gamma, 1, 1)$, $(1, \beta, 1, 1, 1)$, and $(\alpha, 1, 1, \delta, 1)$, which are indeed not comparable.

Agreement with CP-net approach. Note that CP-nets yield a partial ordering that is in general less incomplete between choices than with the above method. Namely in Example 3, we get: $mcw \succ_{CP} mcb \succ_{CP} mfb \succ_{CP} sfb \succ_{CP} sfw \succ_{CP} scw \succ_{CP} scb$; $mfb \succ_{CP} mfw \succ_{CP} sfw$. This can be observed on Example 2 as well.

Example 6. (Example 2 continued). The symbolic possibilistic logic base associated with Example 2 is $K_2 = \{i = (b_j, 1 - \alpha), ii = (b_p, 1 - \beta), iii = (\neg b_j \vee \neg b_p \vee r_s, 1 - \gamma), iv = (\neg w_j \vee \neg b_p \vee w_s, 1 - \delta), v = (\neg b_j \vee \neg w_p \vee w_s, 1 - \varepsilon), vi = (\neg w_j \vee \neg w_p \vee r_s, 1 - \eta)\}$, using the notations: b_j (black jacket), w_j (white jacket), b_p (black pants), w_p (white pants), r_s (red shirt), w_s (white shirt). Table 4 gives the satisfaction vectors for the different tuples. Note that the evaluation of the conditional preference (about shirt color) can be done using only one column, the rightest one, in Table 2, since the different contexts corresponding to the conditional part are mutually exclusive. This leads to the following ordering:

$bbr \succ_{lex} \{bbw, bww, bwr, wwr, www, wbw, wbr\}$; $bww \succ_{lex} wwr \succ_{lex} www$; $bww \succ_{lex} bwr$; $wbw \succ_{lex} wbr$ and $wbw \succ_{lex} wwr$.

This fully agrees with the partial order in the CP-net approach, but leaves more tuples incomparable.

In Examples 2 and 3, any strict ordering between two choices in the possibilistic approach, is found by the CP-net approach. This is always true, formally,

Proposition 2. *Let N be a CP-net and $\succ_{CP}$ be its associated partial order. Let K be a possibilistic logic formulas base encoding the set of preferences represented in N and let $\succ_{lex}$ be the partial order on tuples associated with K. Then, $\forall t, t'$ two tuples of an instance r of a relational schema, if $t \succ_{lex} t'$ then $t \succ_{CP} t'$.*

Table 4. Possible alternative choices in Example 2

jacket	pants	shirt	i	ii	iii	iv	v	vi	iii-iv-v-vi
b	b	r	1	1	1	1	1	1	1
b	b	w	1	1	γ	1	1	1	γ
b	w	r	1	β	1	1	ε	1	ε
b	w	w	1	β	1	1	1	1	1
w	b	r	α	1	1	δ	1	1	δ
w	b	w	α	1	1	1	1	1	1
w	w	r	α	β	1	1	1	1	1
w	w	w	α	β	1	1	1	η	η

Proof (Sketch). It relies on the fact that $\succ_{lex}$ amounts to agreeing with the strict inclusion between sets of preferences that are violated on the one hand, and that in the CP-net approach, each time a further preference is violated by a tuple w.r.t. another tuple, the former tuple is ranked strictly after the other.

Recovering more CP-net strict preferences. As it can be seen in Proposition 2 the partial order returned by leximin ordering is weaker than the one associated with the CP-net in the sense that some ceteris paribus preferences are not recovered by the leximin ordering. The authors of [20] proposed a stronger way to recover *all* ceteris paribus preferences by giving priority to parent nodes.

Example 7. (Example 2 end) If we further enforce the priority in favor of father nodes $\max(1 - \gamma, 1 - \delta, 1 - \varepsilon, 1 - \eta) < \min(1 - \alpha, 1 - \beta)$, i.e. $\max(\alpha, \beta) < \min(\gamma, \delta, \varepsilon, \eta)$, we get $bbr \succ_{lex} bbw \succ_{lex} bww \succ_{lex} bwr \succ_{lex} wwr \succ_{lex} www$; $bbr \succ_{lex} bbw \succ_{lex} wbw \succ_{lex} wbr \succ_{lex} wwr \succ_{lex} www$.

Note that we do not even need to acknowledge the fact that maybe we could enforce $\gamma = \delta = \varepsilon = \eta$, since these levels are related to the same preference statement, without any further precision about relative preferences depending on the context.

Example 8. (Example 3 end) Indeed, if we further assume the following priority ordering $1 - \alpha > \max(1 - \beta, 1 - \gamma) \geq \min(1 - \beta, 1 - \gamma) > \max(1 - \delta, 1 - \varepsilon)$, i.e., $\alpha < \min(\beta, \gamma) \leq \max(\beta, \gamma) < \min(\delta, \varepsilon)$, then it can be checked that one retrieves all the CP-net strict preferences between pairs.

Formally, we have the following result [20]:

Proposition 3. *Let N be a CP-net and $\succ_{CP}$ be its associated partial order. Let K be a possibilistic logic formulas base encoding the set of preferences represented in N and let $\succ_{lex}$ be the partial order on tuples associated with K, applied with enforcement of the priority in favor of father nodes in N. Then, $\forall t, t'$ two tuples of an instance r of a relational schema, if $t \succ_{CP} t'$ then $t \succ_{lex} t'$.*

As noticed in [20], the way preferences are expressed in CP-nets is somewhat rigid. Indeed in the example, having a black Ford (if he should have a Ford) is maybe more important for the user (who is old fashioned!), than, say, his preference for minivans over sedans. In such a case, $1 - \alpha < 1 - \delta$, i. e. $\alpha > \delta$, and this would reverse preferences w.r.t. the CP-net solution. If we introduce formal constraints between symbolic priorities for expressing partial hierarchies between preferences, this can be formally handled at a rather low computational cost in the setting of a symbolic possibilistic logic, rewritten as a two-sorted classical logic (where the formula weights become literals of a special sort, and where inequalities between symbolic weights, possibly involving max and min operations, are turned into classical logic formulas); see [4].

It is now clear that the proposed approach can handle preferences in a more flexible way than in the CP-net-based approach. Moreover, the symbolic approach can give birth to a complete preorder in a controlled manner, once it has been decided what is, according to the user, the priority ordering between his elementary preferences.

Computing an ordered set of answers to a database query in the proposed setting appears to be a two-step procedure. First, from the set of user's preferences, possibly taking into account some priorities between them, one computes the ordering, partial or complete, between the different possible alternatives. Computing the symbolic satisfaction level of an alternative can be done efficiently using results in [4]. The second step consists in browsing the database and to compute for each tuple the class to which it belongs among the alternatives. This evaluation process can be shortened using classical means if one only looks for a maximal number of answers, or only for answers whose evaluation is above some symbolic threshold.

3.2   Non-binary Attributes

Many database attributes are numerical, or at least many-valued. Preferences about the values of such attributes are conveniently represented by fuzzy sets. At the elicitation level, it usually amounts to use piecewise linear membership functions, and to identify a few thresholds separating values that are fully satisfactory from ones that are less satisfactory, and the latter from values that are rejected. Thus, it would be interesting to interface the approach proposed in the previous section, with the fuzzy set approach, or something similar. One can imagine at least two ways (apart from translating non-binary preferences into a set of binary preferences), for doing it.

The first one amounts to use a predefined satisfaction scale with a finite set of explicit levels, such as, e.g.: $1 = \alpha_1 =$"very satisfactory", $\alpha_2 =$ "satisfactory", $\alpha_3 =$"rather satisfactory", $\alpha_4 =$"half satisfactory", $\alpha_5 =$"a bit satisfactory", $\alpha_6 =$"not at all satisfactory". It would be associated with a priority scale such as $1 - \alpha_6 =$"mandatory", $1 - \alpha_5 =$"very important", $1 - \alpha_4 =$"important", $1 - \alpha_3 =$ "rather important", $1 - \alpha_2 =$"has a low priority", $1 - \alpha_1 = 0 =$"indifferent". Assuming that the levels in the scales are equally distant, $[0, 1]$-valued fuzzy set membership degrees can then be approximated on such a discrete scale. Then, preferences pertaining both to numerical attributes and to binary ones can be handled in the way outlined now.

Example 9. Considering the query asking for "a sufficiently large (L) house, with a reasonable price (R)". The house may be an apartment or a villa, but the user prefers a villa to an apartment, and accepts to pay a bit more (M) if it is a villa. An instance of a relational schema (identifier,surface,price,house_type) of the database is a tuple (n, s, p, t) with $t = a$ (apartment), or $t = v$ (villa). Its evaluation is of the form:

$$(ap(\mu_L(s)), \max(\mu_v(t), \min(\mu_a(t), \alpha)), \min(\max(1 - \mu_v(t), ap(\mu_M(p))),$$
$$\max(1 - \mu_a(t), ap(\mu_R(p))))),$$

where ap denotes the approximation operator, $\mu_v(t)$ (resp. $(\mu_a(t)) = 1$) if $t = v$ (resp. $t = a$), and $v(t) = 0$ (resp. $(\mu_a(t)) = 0$) otherwise. μ_L, μ_R, μ_M are fuzzy set membership functions for 'sufficiently large', 'reasonably-priced', and 'a bit more than reasonably- priced', whose definitions clearly depend on the user. For getting a complete preorder, a commensurability hypothesis between the priorities of binary preferences and fuzzy set satisfaction degrees is necessary, stating for instance that not having a villa is only α_4 ="half satisfactory"=α, which corresponds to a priority $1 - \alpha_4$ ="important" for having a villa. Then leximin ordering can be applied on evaluations.

A second approach, which would not require a commensurability hypothesis nor an approximation of the membership degrees, would consist in applying the approach of Section 3.1 on the propositions corresponding to the supports of the fuzzy requirements, and then refining the obtained result by applying in a second step a winnow operator-based approach. This means, in the above example, that the above evaluation will be replaced in the first step by

$$(\mu_{sL}(s), \max(\mu_v(t), \min(\mu_a(t), \alpha)), \min(\max(1 - \mu_v(t), \mu_{sM}(p)),$$
$$\max(1 - \mu_a(t), \mu_{sR}(p)))),$$

where sF is the support of fuzzy set $F(\mu_{sF}(x)) = 1$ if $\mu_F(x) > 0$, and $\mu_{sF}(x) = 0$ if $\mu_F(x) = 0$).

4   Concluding Remarks

The contributions of the paper are twofold. First, it provides a comparative view of the main currents of research that have aimed in the last two decades at handling flexible queries to a database. Still some artificial intelligence approaches to the modeling of preferences have not been covered because of the lack of space, and also because their application to data base querying have not been specially investigated; some are anyway close to CP-nets [13,26].

The second main contribution of the paper is an approach based on possibilistic logic with symbolic weights that can handle conditional preferences in a more cautious way than CP-nets (i.e. without introducing any default priority ordering induced by the structure of the preference graph). The representation of the preferences may be completed by explicit partial order among elementary preferences, if available. It is compatible both with the fuzzy set approach and the Best operator of database approaches. Let us also briefly mention the relation between our possibilistic logic representation setting and qualitative choice logic [9]. Indeed if, for instance one considers preferences such as "I prefer Renault (r) to Chrysler (c) and Chrysler to Ford (f)", it can be conveniently expressed as a "weighted" disjunction of the three choices r, c and f, stating that r is fully satisfactory, c is less satisfactory, and that f is still less satisfactory. This would lead to a fully equivalent treatment of the preferences here. Indeed, the representational equivalence between qualitative choice logic and guaranteed possibility logic, which can be viewed itself as a DNF-like counterpart of possibilistic logic at the representation level, has been established [2].

It is also worth pointing out that the possibilistic representation setting allows for different representation formats that are all equivalent to the possibilistic logic format, but which may be of interest depending on the way people express their preferences [3]. One may also distinguish in a bipolar setting between preferences that should be more or less imperatively taken into account, and simple wishes that are not compulsory at all but provide a bonus when they are satisfied [16]. Moreover, the capabilities of possibilistic logic [17] for handling nonmonotonic reasoning and belief revision processes could be exploited for preference revision and query modification in the spirit of [11].

References

1. Agrawal, R., Wimmers, E.L.: A framework for expressing and combining preferences. In: Proc. ACM SIGMOD, pp. 297–306 (2000)
2. Benferhat, S., Brewka, G., Le Berre, D.: On the relation between qualitative choice logic and possibilistic logic. In: Proc. 10th International Conference IPMU, pp. 951–957 (2004)
3. Benferhat, S., Dubois, D., Prade, H.: Towards a possibilistic logic handling of preferences. Applied Intelligence 14, 303–317 (2001)
4. Benferhat, S., Prade, H.: Encoding formulas with partially constrained weights in a possibilistic-like many-sorted propositional logic. In: IJCAI 2005. Proc. of the 9th Inter. Joint Conference on Artificiel Intelligence, pp. 1281–1286 (2005)
5. Börzsönyi, S., Kossmann, D., Stocker, K.: The skyline operator. In: Proc. 17th IEEE International Conference on Data Engineering, pp. 421–430 (2001)
6. Bosc, P., Pivert, O.: Some approaches for relational databases flexible querying. Journal of Intelligent Information Systems 1, 323–354 (1992)
7. Boutilier, C., Brafman, R.I., Domshlak, C., Hoos, H., Poole, D.: CP-nets: A tool for representing and reasoning with conditional ceteris paribus preference statements. J. Artificial Intelligence Reasoning (JAIR) 21, 135–191 (2004)
8. Brafman, R.I., Domshlak, C.: Database preference queries revisited. Technical Report TR2004-1934, Cornell University, Computing and Information Science (2004)
9. Brewka, G., Benferhat, S., Le Berre, D.: Qualitative choice logic. Artificial Intelligence 157, 203–237 (2004)
10. Chomicki, J.: Preference formulas in relational queries. ACM Transactions on Database Systems 28, 1–40 (2003)
11. Chomicki, J.: Database querying under changing preferences. Annals of Mathematics and Artificial Intelligence 50, 79–109 (2007)
12. Ciaccia, P.: Querying databases with incomplete cp-nets. In: M-PREF 2007. Multidisciplinary Workshop on Advances in Preference Handling (2007)

13. Domshlak, C., Venable, B., Rossi, F.: Reasoning about soft constraints and conditional preferences. In: IJCAI 2003. Proc. 18th International Joint Conference on Artificial Intelligence, pp. 215–220 (2003)
14. Dubois, D., Fargier, H., Prade, H.: Beyond min aggregation in multicriteria decision (ordered) weighted min, discri-min, leximin. In: Yager, R.R., Kacprzyk, J. (eds.) The Ordered Weighted Averaging Operators - Theory and Applications, pp. 181–192. Kluwer Acad. Publ., Dordrecht (1997)
15. Dubois, D., Prade, H.: Using fuzzy sets in flexible querying: Why and how? In: Andreasen, T., Christiansen, H., Larsen, H. (eds.) Flexible Query Answering Systems, pp. 45–60 (1997)
16. Dubois, D., Prade, H.: Bipolarity in flexible querying. In: Andreasen, T., Motro, A., Christiansen, H., Larsen, H.L. (eds.) FQAS 2002. LNCS (LNAI), vol. 2522, pp. 174–182. Springer, Heidelberg (2002)
17. Dubois, D., Prade, H.: Possibilistic logic: A retrospective and prospective view. Fuzzy Sets and Systems 144, 3–23 (2004)
18. Fishburn, P.C.: Preferences structures and their numerical representation. Theoretical Computer Science 217, 359–383 (1999)
19. Godfrey, P., Ning, W.: Relational preferences queries via stable skyline. Technical Report CS-2004-03, York University, pp. 1–14 (2004)
20. Kaci, S., Prade, H.: Relaxing ceteris paribus preferences with partially ordered priorities. In: Mellouli, K. (ed.) ECSQARU 2007. LNCS (LNAI), vol. 4724, pp. 660–671. Springer, Heidelberg (2007)
21. Kiessling, W.: Foundations of preferences in database systems. In: Bressan, S., Chaudhri, A.B., Lee, M.L., Yu, J.X., Lacroix, Z. (eds.) VLDB 2002. LNCS, vol. 2590, pp. 311–322. Springer, Heidelberg (2003)
22. Lacroix, M., Lavency, P.: Preferences: Putting more knowledge into queries. In: VLDB 1987. Proc. of the 13th Inter. Conference on Very Large Databases, pp. 217–225 (1987)
23. Lang, J.: From preferences to combinatorial vote. In: Proc. of KR Conferences, pp. 277–288 (2002)
24. Motro, A.: A user interface to relational databases that permits vague queries. ACM Transactions on Information Systems 6, 187–214 (1988)
25. Torlone, R., Ciaccia, P.: Finding the best when it's a matter of preference. In: SEBD 2002. Proc. 10th Italian National Conference on Advanced Data Base Systems, pp. 347–360 (2002)
26. Wilson, N.: Extending CP-nets with stronger conditional preference statements. In: AAAI 2004. Proc. 19th National Conference on Artificial Intelligence, pp. 735–741 (2004)

Defeasible Reasoning and Partial Order Planning

Diego R. García, Alejandro J. García, and Guillermo R. Simari

Artificial Intelligence Research and Development Laboratory
Department of Computer Science and Engineering
Universidad Nacional del Sur – Av. Alem 1253, (8000) Bahía Blanca
Consejo Nacional de Investigaciones Científicas y Técnicas (CONICET)[*]

Abstract. Argumentation-based formalisms provide a way of considering the defeasible nature of reasoning with partial and often erroneous knowledge in a given environment. This problem affects every aspect of a planning process. We will present an argumentation-based formalism that an agent could use for constructing plans starting from a previously introduced formalism. In such a formalism, agents represent their knowledge about their environment in Defeasible Logic Programming, and have a set of actions they can execute to affect their environment. These actions are defined in combination with a defeasible argumentation formalism. We will analyze the interplay of arguments and actions when constructing plans using Partial Order Planning techniques.

1 Introduction

In this paper, we introduce an argumentation-based formalism an agent could use for constructing plans using partial order planning techniques. In our proposed approach, actions and arguments will be combined by the agent to construct plans. As we will explain next, actions' preconditions can be satisfied by other actions' effects (as usual) or by conclusions supported by arguments that are based on inference rules and other actions effects. We will also show that besides those effects declared in the definition of actions, there could be more effects that the agent will be able to deduce using the argumentation-based reasoning formalism.

Defeasible argumentation is a powerful formalism suitable for reasoning with potentially contradictory information and in dynamic environments (see [11, 3, 2, 10, 8]). The formalism presented here is based on Defeasible Logic Programming (DELP) [3], a defeasible argumentation formalism grounded in Logic Programming. For dealing with contradictory and dynamic information in DELP, arguments for conflicting pieces of information are build and then compared to decide which one prevails. The argument that prevails provides a warrant for the information that it supports.

---

[*] Partially supported by SGCyT Universidad Nacional del Sur, CONICET (PIP 5050) and Agencia Nacional de Promoción Cientfica y Tecnológica (PICT 2002 Nro 13096).

S. Hartmann and G. Kern-Isberner (Eds.): FoIKS 2008, LNCS 4932, pp. 311–328, 2008.

In [13] a formalism that combines actions and defeasible argumentation was introduced. There, they show the problems related to the combination of their formalism with simple planning algorithms.

Extending the mentioned work, in this paper we will analyze the interaction of arguments and actions when they are combined to construct plans using Partial Order Planning techniques. When actions and arguments are combined in a partial order plan, new types of interferences appear (called threats in [7]). These interferences need to be identified and resolved to obtain valid plans.

The main contribution of this paper will be to show meaningful examples and the description of the proposed solution for the combination of Partial Order Planning and Defeasible Argumentation. Thus, our work focuses on improving the capabilities and scope of current planning technology and not in improving the efficiency of current planning implementations.

2   Motivation

To solve a planning problem, an agent should be provided with an appropriate set of actions to perform. The representation of these actions must consider all the preconditions and effects that are relevant to solve the problem. Consider for example the consequences of the action of striking a match. A relevant effect can be to produce fire. However, there are many other consequences that can be entailed and could be considered irrelevant and not be included in the representation of the action (e. g., to produce light, to raise the temperature of the room, to make smoke, etc).

We propose that, instead of considering all the possible effects in the representation of the actions, agents should be provided with a defeasible reasoning formalism for obtaining those consequences that are entailed from action's effects. For example, an action for turning the switch on to light a room (called "*turn_switch_on*" from now on) should have as a precondition that the *switch is set to off* and the effect should be that the *switch is set to on*. The effect "*there is light in the room*" should be entailed as a consequence of the effect the *switch is on*. Thus, besides the action *turn_switch_on*, we propose to include the (defeasible) rule: "*if the switch is set to on then there is a reason to believe that there is light in the room*". It is important to note that if "*there is light in the room*" is considered as an effect of the action *turn_switch_on*, then it will be difficult to consider exceptions like "*there is no electricity*". However, this kind of problems are easily handled by the argumentation formalism. For example, this situation could be represented by the defeasible rule "*if the switch is set to on but there is no electricity then there is a reason to believe that there is no light in the room*".

3   Defeasible Argumentation and Actions

In this section, we introduce a formalism that combines actions and defeasible argumentation based on a previous work reported in [13, 12, 4]. Our formalism follows the logic programming paradigm for knowledge representation called Defeasible Logic Programming (DELP) [3]. Thus, the agent's knowledge will be represented by a DELP program and the agent will be able to perform defeasible reasoning over this knowledge.

The agent's knowledge base will be a defeasible logic program $\mathcal{K} = (\Psi, \Delta)$, where Ψ should be a consistent set of *facts*, and Δ a set of *defeasible rules*. Defeasible Rules are denoted $L_0 \prec L_1, \ldots, L_n$, where L_0 is a ground literal and $\{L_i\}_{i>0}$ is a set of ground literals. A defeasible rule "*Head* $\prec$ *Body*" is the key element for introducing *defeasibility* [8] and is understood as expressing that "*reasons to believe in the antecedent Body of a rule provide reasons to believe in its consequent, Head*" [14]. Following Lifschitz [6], DELP rules could be represented as *schematic rules* with variables, making abstraction of the object constants. The resulting DELP programs are therefore *schematic programs*.

Strong negation "$\sim$" can appear in the head of defeasible rules, and it could be used to represent conflicting information. In DELP arguments for conflicting pieces of information are built and then compared to decide which one prevails. Since the notion of argument will be extensively used in this paper its definition adapted from [3] is included below:

Definition 1 [Argument]
Let L be a literal, and $\mathcal{K} = (\Psi, \Delta)$ a defeasible logic program. We say that $\langle \mathcal{A}, L \rangle$ is an argument for L (or L is supported by $\mathcal{A}$) if $\mathcal{A}$ is a set of defeasible rules of Δ, such that:

1. there exists a derivation for L from $\Psi \cup \mathcal{A}$,
2. the set $\Psi \cup \mathcal{A}$ is non-contradictory, and
3. $\mathcal{A}$ is minimal: there is no proper subset $\mathcal{A}'$ of $\mathcal{A}$ such that $\mathcal{A}'$ satisfies conditions (1) and (2).

Example 1. Let (Ψ, Δ) be a knowledge base, where $\Psi = \{a, b, c, d\}$ and $\Delta = \{(p \prec b), (q \prec r), (r \prec d), (\sim r \prec s), (s \prec b), (\sim s \prec a, b), (w \prec b), (\sim w \prec b, c)\}$. From the defeasible logic program (Ψ, Δ) the literal p is supported by the argument $\mathcal{A} = \{p \prec b\}$, the literal q by $\mathcal{A}_1 = \{(q \prec r), (r \prec d)\}$, the literal $\sim r$ by $\mathcal{A}_2 = \{(\sim r \prec s), (s \prec b)\}$, the argument $\mathcal{A}_3 = \{(\sim s \prec a, b)\}$ supports $\sim s$, and $\mathcal{A}_4 = \{s \prec b\}$ supports the liteal s. Observe that $\mathcal{A}_4$ is a subargument of $\mathcal{A}_2$, *i.e.*, $\mathcal{A}_4$ is a subset of $\mathcal{A}_2$ that supports an inner conclusion in $\mathcal{A}_2$.

Given a defeasible logic program it is possible to generate arguments that are in conflict. For instance, in Example 1, $\mathcal{A}_3$ and $\mathcal{A}_4$ are in conflict because both support contradictory conclusions. Thus, $\mathcal{A}_3$ is a *counterargument* for $\mathcal{A}_4$ (and viceversa). Observe that a counterargument can also be in conflict with an inner part of other argument. For instance, $\mathcal{A}_2$ is a counterargument for $\mathcal{A}_1$, because $\mathcal{A}_2$ is in conflict with the subargument $\{r \prec d\}$ of $\mathcal{A}_1$. In this case, we also say that $\mathcal{A}_2$ attacks $\mathcal{A}_1$ at the point r.

Since conflicting arguments can be generated, DELP provides a mechanism for deciding which argument prevails and therefore, which literals are warranted. Next, we will describe briefly this mechanism (see [3] for further details).

In DeLP, a literal L is *warranted* from (Ψ, Δ) if there exists a non-defeated *argument* $\mathcal{A}$ supporting L. To establish whether $\langle \mathcal{A}, L \rangle$ is a non-defeated argument, *counter-arguments* that could be *defeaters* for $\langle \mathcal{A}, L \rangle$ are considered. An argument $\mathcal{B}$ is a defeater for $\mathcal{A}$, if $\mathcal{B}$ is counter-argument for $\mathcal{A}$ and by some comparison criterion is preferred to $\langle \mathcal{A}, L \rangle$. In the examples in this paper we will use *generalized specificity* [15], a criterion that favors two aspects in an argument: it prefers (1) a *more precise* argument (*i.e.*, with greater information content) or (2) a *more concise* argument (*i.e.*, with less use of rules). A defeater $\mathcal{D}$ for an argument $\mathcal{A}$ can be *proper* ($\mathcal{D}$ is preferred to $\mathcal{A}$) or *blocking* (same strength). In Example 1, $\mathcal{A}_3$ is preferred to $\mathcal{A}_2$ (more precise) hence $\mathcal{A}_3$ is a proper defeater for $\mathcal{A}_2$. As stated above, the argument $\mathcal{A}_1$ is a counterargument for $\mathcal{A}_2$. Since the subargument $\{r \prec d\}$ of $\mathcal{A}_1$ and $\mathcal{A}_2$ have the same strength, then $\mathcal{A}_2$ is a blocking defeater for $\mathcal{A}_1$. It is important to note that in DeLP the argument comparison criterion is modular and can be replaced. Thus, the most appropriate criterion for the domain that is being represented can be selected.

Since defeaters are arguments, there may exist defeaters for them, and defeaters for these defeaters, and so on. Thus, a sequence of arguments called *argumentation line* appears, where each argument defeats its predecessor in the line (see Example 2).

To avoid undesirable sequences, that may represent circular or fallacious argumentation lines, in DeLP an argumentation line is *acceptable* if it satisfies certain constraints. That is, the argumentation line has to be finite, an argument can not appear twice, and supporting arguments, *i.e.*, arguments in odd positions, (resp. interfering arguments) have to be not contradictory (see [3] for details). Given an acceptable argumentation line $[\mathcal{A}_1, \ldots, \mathcal{A}_n]$ we will say that C *is acceptable wrt* $[\mathcal{A}_1, \ldots, \mathcal{A}_n]$ if $[\mathcal{A}_1, \ldots, \mathcal{A}_n, \mathcal{C}]$ is an acceptable argumentation line.

Clearly, there can be more than one defeater for a particular argument $\mathcal{A}$. Therefore, many acceptable argumentation lines could arise from $\mathcal{A}$, leading to a tree structure. Given an argument $\langle \mathcal{A}, h \rangle$, a *dialectical tree* [3] for $\langle \mathcal{A}, h \rangle$, denoted $\mathcal{T}(\langle \mathcal{A}, h \rangle)$, is a tree where every node is an argument. The root of $\mathcal{T}(\langle \mathcal{A}, h \rangle)$ is $\langle \mathcal{A}, h \rangle$, and every inner node is a defeater (proper or blocking) of its parent. Leaves correspond to non-defeated arguments. In a dialectical tree every path from the root to a leaf corresponds to a different acceptable argumentation line. Thus, a dialectical tree provides a structure for considering all the possible acceptable argumentation lines that can be generated for deciding whether an argument is defeated. We call this tree *dialectical* because it represents an exhaustive dialectical analysis for the argument in its root.

Given a literal h and an argument $\langle \mathcal{A}, h \rangle$, to decide whether a literal h is warranted, every node in the dialectical tree $\mathcal{T}(\langle \mathcal{A}, h \rangle)$ is recursively marked as "D" (*defeated*) or "U" (*undefeated*), obtaining a marked dialectical tree $\mathcal{T}^*(\langle \mathcal{A}, h \rangle)$. Nodes are marked by a bottom-up procedure that starts marking all leaves in $\mathcal{T}^*(\langle \mathcal{A}, h \rangle)$ as "U"s. Then, for each inner node $\langle \mathcal{B}, q \rangle$ of $\mathcal{T}^*(\langle \mathcal{A}, h \rangle)$, $\langle \mathcal{B}, q \rangle$ will be marked as "U" iff every child of $\langle \mathcal{B}, q \rangle$ is marked as "D", or $\langle \mathcal{B}, q \rangle$ will be marked as "D" iff it has at least a child marked as "U".

Given an argument $\langle \mathcal{A}, h \rangle$ obtained from (Ψ, Δ), if the root of $T^*(\langle \mathcal{A}, h \rangle)$ is marked as "U", then we will say that $T^*(\langle \mathcal{A}, h \rangle)$ *warrants* h and that h is *warranted* from (Ψ, Δ).

Example 2 (Extends Example 1). Argument $\mathcal{A}$ for p is undefeated because there is no counter-argument for it. Hence, p is warranted.

The literal q has the argument $\mathcal{A}_1$ that is defeated by $\mathcal{A}_2$ that attacks r, an inner point in $\mathcal{A}_1$. The argument $\mathcal{A}_2$ is in turn defeated by $\mathcal{A}_3 = \{(\sim s \prec a, b)\}$. Thus, the argumentation line $[\mathcal{A}_1, \mathcal{A}_2, \mathcal{A}_3]$ is obtained. The literal q is warranted because its supporting argument $\mathcal{A}_1$ has only one defeater $\mathcal{A}_2$ that is defeated by $\mathcal{A}_3$, and $\mathcal{A}_3$ has no defeaters.

Observe that there is no warrant for $\sim r$ because $\mathcal{A}_2$ is defeated by $\mathcal{A}_3$. The literals t and $\sim t$ have no argument, so neither of them is warranted. Finally note that every fact of Ψ is warranted, because no counter-argument can defeat a fact. Thus, the set of warranted literals from (Ψ, Δ) is $\{a, b, c, d, p, q, r, \sim s, \sim w\}$.

Besides its knowledge base $\mathcal{K}$, an agent will have a set of actions Γ that it may use to change its world. The formal definitions that were introduced in [12, 13] are recalled below.

Definition 2 [Action]. An action A is an ordered triple $A = \langle \mathsf{X}, \mathsf{P}, \mathsf{C} \rangle$, where A is a ground atom representing the name of the action, X is a consistent set of ground literals representing consequences of executing A, P is a set of ground literals representing preconditions for A, and C is a set of constraints of the form *not* L, where L is a ground literal. We will denote actions as follows:

$$\{X_1, \ldots, X_n\} \xleftarrow{A} \{P_1, \ldots, P_m\}, not \ \{C_1, \ldots, C_k\}$$

where *not* $\{C_1, \ldots, C_k\}$ represents $\{not \ C_1, \ldots, not \ C_k\}$.

Example 3. Let Γ be the set of available actions of for agent:

$$\Gamma = \left\{ \begin{array}{c} \{\sim a, d, x\} \xleftarrow{Ac_1} \{a, p, q\}, not \ \{t, \sim t, w\} \\ \{e\} \xleftarrow{Ac_2} \{p\}, not \ \{\} \\ \{e\} \xleftarrow{Ac_3} \{t\}, not \ \{w\} \\ \{\sim p\} \xleftarrow{Ac_4} \{b\}, not \ \{q\} \end{array} \right\}$$

Note that all the atoms and literals considered in an action definition are ground. However, *schematic actions* (operators) can be defined using non-ground atoms and literals. An *schematic action* stands for the set of all possible ground action instances generated using the object constants.

The condition that must be satisfied before an action $A = \langle \mathsf{X}, \mathsf{P}, \mathsf{C} \rangle$ can be executed contains two parts: P, which mentions the literals that *must* be warranted, and C, which mentions the literals that *must not* be warranted. In this way, the conditions that must be satisfied to execute an action could also depend on the fact that some information is unknown (*un-warranted*).

Definition 3 [Applicable Action]. Let $\mathcal{K} = (\Psi, \Delta)$ be an agent's knowledge base. Let Γ be the set of actions available to this agent. An action A in Γ, is applicable if every precondition P_i in P has a warrant built from (Ψ, Δ) and every constraint C_i in C fails to be warranted.

Example 4. Let $\mathcal{K} = (\Psi, \Delta)$ be the agent's knowledge base as defined in Example 1, and the set of actions Γ from Example 3. As shown in Example 2, the set of warranted literals from (Ψ, Δ) is $\{a, b, c, d, p, q, r, \sim s, \sim w\}$. Then, from (Ψ, Δ) the action Ac_1 is applicable because every literal in its precondition set ($\{a, p, q\}$) is warranted, and no constraints in $\{t, \sim t, w\}$ are warranted. The action Ac_2 is also applicable because it has no constraint and its precondition p is warranted. Finally, action Ac_3 is not applicable because its precondition t is not warranted, and action Ac_4 is not applicable because its constraint q it is warranted.

It is clear that only applicable actions can be executed, and the agent has to plan which one to execute. Once the agent selects an action to apply, its execution will affect directly the agent environment. That is, once an action has been applied, the effect of the action will change both the environment and the set $\mathcal{K}$. The effect of the execution of an applicable action in our formalism is defined below:

Definition 4 [Action Effect]. Let $\mathcal{K} = (\Psi, \Delta)$ be an agent's knowledge base. Let Γ be the set of actions available to this agent. Let A be an applicable action in Γ defined by:

$$\{X_1, \ldots, X_n\} \xleftarrow{A} \{P_1, \ldots, P_m\}, not \ \{C_1, \ldots, C_k\}$$

The effect of executing A is the revision of Ψ by X, i.e. $\Psi^{*X} = \Psi^{*\{X_1,\ldots,X_n\}}$. Revision will consist of removing any literal in Ψ that is complementary of any literal in X and then adding X to the resulting set. Formally:

$$\Psi^{*X} = \Psi^{*\{X_1,\ldots,X_n\}} = (\Psi \setminus \overline{X}) \cup X$$

where $\overline{X}$ is the set of complements of members of X.

Example 5 (Extends Example 4). The action Ac_1 was shown to be applicable from (Ψ, Δ). If Ac_1 is executed, Ψ then becomes $\Psi_1 = \{b, c, \sim a, d, x\}$. Observe that the precondition a was "consumed" by the action, and the literals $\sim a$ and x were added. It is important to note that from (Ψ_1, Δ) the set of warranted literals changes to $\{\sim a, x, b, c, d, p, s, \sim w\}$. Therefore, the action Ac_1 is now not applicable again because from (Ψ_1, Δ) there is no warrant for a and q. However, the action Ac_4 that was not applicable from (Ψ, Δ) is now applicable from (Ψ_1, Δ) because its constraint q is not warranted. Observe that action Ac_2 remains applicable.

The argumentation formalism described above allows an agent to represent knowledge about the environment and to define the actions it can perform. It also defines when an action is applicable and how to compute its effects. However, it does not describe how to construct a plan to achieve the agent's goals.

In our approach a *planning problem* is defined by the tuple $(\Psi, \Delta, Goal, \Gamma)$ where Ψ is a set of literals that represents the *initial state*, Δ is the set of defeasible rules that agent can use for reasoning, *Goal* is a set of literals representing the agent's *goals* and Γ is a set of *actions* that the agent can perform. The agent will satisfy its goals when, through the execution of a sequence of actions, it reaches some state Ψ' where each literal of *Goal* is warranted from (Ψ', Δ).

Next, we will describe how partial order planning techniques can be combined with the formalism described above to provide the agent with the ability to build plans.

4   Argumentation in Partial Order Planning

The basic idea behind a regression Partial Order Planning (POP) algorithm [7] is to search through the plan space. The planner starts with an initial plan consisting solely of a *start* step (whose effects encode the initial state conditions) and a *finish* step (whose preconditions encode the goals) (see Figure 1(a)). Then it attempts to complete this initial plan by adding new steps (actions) and constraints until all step's preconditions are guaranteed to be satisfied. The main loop in a traditional POP algorithm makes two types of choices:

— *Supporting unsatisfied preconditions*: all steps that could possibly achieve a selected unsatisfied precondition are considered. It chooses one step nondeterministically and then adds a causal link to the plan to record that the selected precondition is achieved by the chosen step.
— *Resolve threats*: If a step might possibly interfere with the precondition being supported by a casual link, it nondeterministically chooses a method to resolve this *threat*: either by reordering steps in the plan (adding *ordering constraints*) or posting additional subgoals.

Recall that in the argumentation formalism described in the previous section, an action is applicable if every precondition of the action has a warrant built from the agent's current knowledge base, and every constraint fails to be warranted. To combine this formalism with POP, we must consider the use of arguments for supporting unsatisfied preconditions, besides actions.

In this section we will illustrate with an example how to build a plan using actions and arguments. When actions and arguments are combined to construct plans, new types of interferences (threats) appear that need to be resolved to obtain a valid plan. In Section 5 we will identify these new types of threats and methods to resolve each of them will be proposed. Finally, in Section 6 we will propose an extension to the traditional POP algorithm that use actions and arguments to built plans and resolve the new types of threats.

The following definitions are introduced for identifying different sets of literals present in an argument, that will be considered when a plan is constructed.

Definition 5 [Heads-Bodies-Literals]. Given an argument $\langle B, h \rangle$, *heads*($\mathcal{B}$) is the set of all literals that appear as heads of rules in $\mathcal{B}$. Similarly, *bodies*($\mathcal{B}$)

is the set of all literals that appear in the bodies of rules in $\mathcal{B}$. The set of all literals appearing in $\mathcal{B}$, denoted $literals(\mathcal{B})$ is the set $heads(\mathcal{B}) \cup bodies(\mathcal{B})$.

Definition 6 [Argument base]. Given an argument $\langle B, h \rangle$ we will say that the base of $\mathcal{B}$ is the set $base(\mathcal{B}) = bodies(\mathcal{B}) - heads(\mathcal{B})$.

Definition 7 [Conclusion]. Given an argument $\langle B, h \rangle$ we will say that the conclusion of $\mathcal{B}$ is the literal $conclusion(\mathcal{B}) = heads(\mathcal{B}) - bodies(\mathcal{B})$.

Example 6. Given the argument $\langle B, b \rangle$ where $\mathcal{B}= \{(b \prec c, d), (c \prec e)\}$, the corresponding sets are:

$$
\begin{array}{ll}
heads(\mathcal{B}) = \{b, c\} & base(\mathcal{B}) = \{d, e\} \\
bodies(\mathcal{B}) = \{c, d, e\} & conclusion(\mathcal{B}) = \{b\} \\
literals(\mathcal{B}) = \{b, c, d, e\} &
\end{array}
$$

The combined use of argumentation and actions to build plans introduces new issues not present in the traditional POP algorithm that need to be addressed. Following, we will present an example to illustrate how traditional POP algorithm can be extended to consider arguments as planning steps. We will also introduce the basic terminology and graphical representation that will be used in the rest of the paper. For simplicity, we present a propositional planning problem that defines actions without constraints.

Example 7. Consider an agent that works at night and its job is cleaning rooms in a building. The agent arrives to a room where the light switch is set to off and has to build a plan for having that room cleaned. The agent has the following knowledge base: $\Psi = \{switch\_off\}$ and

$$
\Delta = \left\{ \begin{array}{c}
light\_in\_room \prec switch\_on \\
\sim light\_in\_room \prec switch\_on, \sim electricity
\end{array} \right\}
$$

The agent's goal is $G = \{room\_clean\}$, and the available actions are:

$$
\{room\_clean\} \overset{clean\_room}{\longleftarrow} \{light\_in\_room\}, not \; \{\}
$$
$$
\{switch\_on, \sim switch\_off\} \overset{turn\_switch\_on}{\longleftarrow} \{switch\_off\}, not \; \{\}.
$$

Figure 1(a) shows the initial plan for example 7 and Figure 1(b) depicts an incomplete plan where only actions (not arguments) were considered to achieve the unsatisfied preconditions. Finally, Figure 1(c) shows a complete plan obtained using actions and arguments.

In our approach, we will distinguish between two types of steps: *action steps* (*i.e.*, steps that represent the execution of an action) and *argument steps* (*i.e.*, arguments used in the plan to support the precondition of some action step). *Action steps* are depicted by square nodes labeled with the action name. The squares labeled START and FINISH represent the *start* and *finish* steps respectively. The literals that appear below an action step represent the preconditions

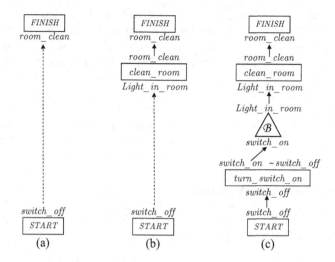

Fig. 1. Different partial plans for Example 7

of the action step, and the literals that appear above represent its effects. Triangles represent *argument steps* and are labeled with the argument name. The literal at the top of the triangle is the conclusion of the argument (Definition 7), and the literals at the base of the triangle represent the base of the argument (Definition 6).

The solid arrows that link an effect of an action step with a precondition of another action step, or with a literal in the base of an argument step, represent *causal links*. The solid arrows between the conclusion of an argument step and a precondition of an action step represent *support links*. *Causal* and *support links* are used to explicitly record the source for each literal during planning.

Dashed arrows represent *ordering constraints* and are used to explicitly establish an order between two steps. By definition the *start* step comes before the *finish* step and the rest of the steps are constrained to come after the *start* step and before the *finish* step. All causes are constrained to come before their effects, so a causal link also represents an ordering constraint.

In Figure 1(a) there is only one unsatisfied precondition *room_clean*, and the only possible way to satisfy it is by the action *clean_room*. A new step *clean_room* is added (Figure 1(b)) to the plan and it's precondition *light_in_room* becomes a new unsatisfied subgoal. Observe that none of the actions available achieve *light_in_room*, then it is not possible to obtain a plan if only actions are considered.

However, from the rules Δ of the agent's knowledge base it is possible to construct the (potential) argument $\mathcal{B}=\{\ (light\_in\_room \prec switch\_on)\ \}$ that supports *light_in_room*. Therefore, an alternative way to achieve *light_in_room* would be to use $\mathcal{B}$ for supporting *light_in_room*, and then to find a plan for satisfying all the literals in the base of $\mathcal{B}$ ($base(\mathcal{B}) = \{switch\_on\}$). Figure 1(c) shows this situation. The argument $\mathcal{B}$ is chosen to support *light_in_room* and the literal *switch_on* becomes a new subgoal of the plan. Then, the action *turn_switch_on* is selected to satisfy *switch_on* and the corresponding step is added to the plan. The precondition *switch_off* of the step *turn_switch_on* is achieved by the *start* step, so the corresponding causal link is added and a plan is obtained.

Note that $\mathcal{B}=\{$ (*light_in_room* $\prec$ *switch_on*) $\}$ is a "*potential argument*" because it is conditioned to the existence of a plan that satisfies its base. This argument can not be constructed from a set of facts, as usual in DeLP. The reason is that at the moment of the argument construction it is impossible to know which literals are true, because they depend on steps that will be chosen later in the planning process. A formal definition of potential argument follows:

Definition 8 [Potential Argument]
Let h be a literal, and Δ a set of defeasible rules. We say that $\langle\langle\mathcal{A}, h\rangle\rangle$ is a potential argument for h (or that $\mathcal{A}$ is a potential argument supporting h), if $\mathcal{A}$ is a set of defeasible rules of Δ, such that:

1. there exists a defeasible derivation for h from $base(\mathcal{A}) \cup \mathcal{A}$
2. the set $base(\mathcal{A}) \cup \mathcal{A}$ is non-contradictory, and
3. $\mathcal{A}$ is minimal: there is no proper subset $\mathcal{A}'$ of $\mathcal{A}$ such that $\mathcal{A}'$ satisfies conditions 1. and 2.

Another thing to consider is that the existence of the argument $\mathcal{B}$ in the plan shown in Figure 1(c) is not enough to have a warrant for *light_in_room*, because it could exist a defeater for $\mathcal{B}$ (for example, when there is no electricity in the building). Recall that to be able to apply an action, all its preconditions have to be warranted. The existence of a defeater for $\mathcal{B}$ will depend on the decisions made later in the planning process, that is, a defeater for $\mathcal{B}$ could appear as new action steps are added to the plan.

5   Interferences among Actions and Arguments

When only actions are considered, there is only one type of destructive interference that can arise in a plan. In the traditional POP algorithm, this interference is captured by the notion of *threat* (see Figure 2). When actions and arguments are combined to construct plans, new types of interferences appear that need to be identified and resolved to obtain a valid plan. We will extend the notion of *threat* to identify all the different types of interferences that could arise in a plan and propose methods to resolve each of them.

Figure 2(a) shows the kind of threat that appears in the traditional POP algorithm: the precondition p of A_1, supported by A_2, is threatened by the action step A_3 because it negates p. Note that $\overline{p}$ is an effect of A_3, where $\overline{p}$ stands for the complement of p with respect to strong negation, *i.e.*, $\overline{p}$ is $\sim p$ and $\overline{\sim p}$ is p. The way to resolve this threat is to add an ordering constraint to make sure that A_3 is not executed between A_2 and A_1. There are two alternatives: A_3 is forced to come before A_2 (called *demotion*, see Figure 2(b)) or A_3 is forced to come after A_1 (called *promotion*, see Figure 2(c)).

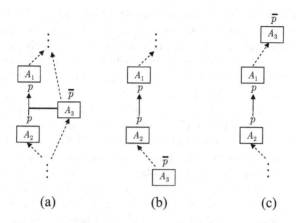

Fig. 2. An action step threatens the precondition supported by another action step

This type of threat involves only action steps and will be called *action-action threat*. However, as we consider also arguments to construct a plan, a different kind of threat could arise involving action steps and argument steps. Consider the situation shown in Figure 3. In this case, the action step A_3 threatens the argument step $\mathcal{B}$ because it negates a literal present in the argument $\mathcal{B}$. Note that $\overline{n}$ is an effect of A_3 and $n \in literals(\mathcal{B})$ (see definition 5). The argument step $\mathcal{B}$ was added to the plan to support the precondition b of the action step A_1. If A_3 makes $\overline{n}$ true before A_1 is executed, the argument $\mathcal{B}$ will not exist at the moment a warrant for b is needed to execute A_1. This type of threat will be called *action-argument threat*.

This *action-argument threat* can be resolved ensuring that A_3 does not make $\overline{n}$ true just before the execution A_1. In general, this can be accomplished adding

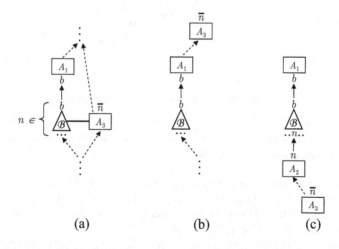

Fig. 3. An action step threaten and argument step

an ordering constraint to force A_3 to come after A_1. This method will be called *promotion** (see figure 3(b)). However, this is not the only way to resolve this threat. In the particular case that the literal n is present in the base of $\mathcal{B}$, there will be an action step A_2 in the plan that achieves it. If $A_2 \neq start$, then the threat can be resolved adding an ordering constraint to force A_3 to come before A_2. This method will de called *demotion** (see figure 3(c)).

Finally, there is another type of threat to consider involving only arguments. Arguments are introduced in the plan to have a warrant for the precondition of some action step. Since the arguments could be defeated by a counter-argument, the precondition would not be warranted. This situation is shown in figure 4(a). The argument step $\mathcal{B}$ was added to the plan to support the precondition b of the action step A_1. However, at the moment a warrant for b is needed to execute A_1, there exists a defeater $\mathcal{C}$ for $\mathcal{B}$. We will refer to this type of threat as *argument-argument threat*. Observe that the defeater $\mathcal{C}$ is not necessarily an argument step of the plan. The argument $\mathcal{C}$ could be any defeater that can be built from the effects of the actions steps that could be ordered to come before A_1 in the plan.

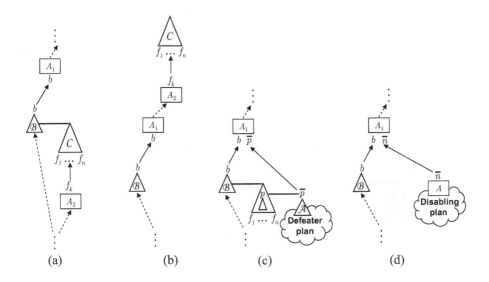

(a)   (b)   (c)   (d)

Fig. 4. An argument step is attacked by a counter-argument

This *argument-argument threat* is resolved ensuring that $\mathcal{B}$ is not defeated by $\mathcal{C}$. There are three alternative methods:

- *delaying the defeater*. Add orderings constraint to the plan to force *every* action step A_2 in the plan, that achieve a literal $f_k \in base(\mathcal{C})$, to come after A_1 (see figure 4(b)). This forces $\mathcal{C}$ to exist after A_1, therefore b is warranted at the moment is needed to execute A_1.
- *defeating the defeater*: add steps to the plan to force the defeater $\mathcal{C}$ to be defeated (see figure 4(c)). To accomplish this, the precondition $\bar{p}$ is added to

the step A_1, where $p \in heads(\mathcal{C})$, and the plan is completed to achieve this precondition. Since the intention is to defeat $\mathcal{C}$, the step chosen to support $\overline{p}$ must be an argument step.

– *disabling the defeater*: add steps to the plan to prohibit the existence of the defeater $\mathcal{C}$ (see figure 4(d)). To accomplish this, the precondition $\overline{n}$ is added to the step A_1, where $n \in literals(\mathcal{C})$, and the plan is completed to achieve this precondition. The step chosen to support $\overline{n}$ must be an action step, because the intention is to prohibit the existence of $\mathcal{C}$.

The proposed solutions for solving the new types of threats will be used in an algorithm presented in following section (See Figure 6).

6   Proposed Algorithm

In this section we will present an extension of the traditional POP algorithm that we will call APOP. Figure 5, 6 and 7 show an outline of the APOP algorithm. The ↓ identify input parameters and the ↑ identify output parameters. The statements **choose** and **fail** are used to describe nondeterminism. The primitive **choose** allow the algorithm to make a choice between different alternatives and keeps track of the pending choices. If the algorithm encounter a **fail** statement the control is resumed at the point in the algorithm where the choice was made and the pending choices are considered.

```
function APOP(Ψ↓, Δ↓, Goal↓, Γ↓): Plan;
begin
   Plan:= Make_Initial_Plan(Ψ, Goal);
      loop do
         if Plan.Subgoals = ∅ then return Plan;
         Let (SubGoal, Step, SubGoalType, ArgLine) ∈ Plan.Subgoals;
         Plan.Subgoals:= Plan.Subgoals − {(SubGoal, Step, SubGoalType, ArgLine)};
         Choose_Step(Plan, Δ, Γ, Step, SubGoal, SubGoalType, ArgLine);
         Resolve_Threats(Plan, Δ, Γ);
      end
end

function Make_Initial_Plan(Ψ↓, Goal↓): Plan;
begin
   Plan.Action_Steps:= {(START,{},Ψ), (FINISH, Goal,{})};

   Plan.Argument_Steps:= ∅;
   Plan.Orderings:= {START ≺ FINISH};
   Plan.Subgoals:= {(g,FINISH, any,[])| g ∈ Goal};

   Plan.Causal_Links:= ∅;
   Plan.Support_Links:= ∅;
   Plan.Defeated_Args:= ∅;
end
```

Fig. 5. Outline of the APOP algorithm part 1

The function APOP (Figure 5) proceeds as the traditional pop algorithm: starts with an initial plan and attempts to complete it by adding new steps, resolving the threats that may appear. However, it will consider arguments and actions as planning steps and resolve the new types of threats introduced in Section 5. Besides the initial state Ψ and the *Goal*, function APOP takes Δ and Γ as input parameters. The set Δ contains defeasible rules that can be used for building arguments and Γ is the set of available actions. In APOP a plan is represented by seven sets (see Make_Initial_Plan in Figure 5).

In contrast with the traditional POP algorithm the procedure Choose_Step (Figure 6) will consider arguments, besides actions, to support unsatisfied sub-goals (Plan.Subgoals). Note that if no arguments can be constructed to support a subgoal (Arg_Steps=$\emptyset$ in statement (2)) then only actions steps will be considered and the algorithm will proceed as POP (statement (6)). However, if Arg_Steps$\neq \emptyset$, the algorithm will consider the inclusion of an argument step to support a subgoal (statement (7)). In each case the plan will updated acordingly.

```
procedure Choose_Step(Plan, Δ, Γ, S_need, p, SubGoalType, ArgLine)
begin

(1)Act_Steps:= {S | S ∈ Plan.Action_Steps that possibly S ≺ S_need
                    or S is created using an action A ∈ Γ and  p ∈ Effects(S) };
(2)Arg_Steps:= {S | S is created using a potential argument ⟨⟨B,p⟩⟩ from Δ
                    acceptable wrt ArgLine};
(3) case SubGoalType of
        action:   Steps:= Act_Steps;
        argument: Steps:= Arg_Steps;
        any:      Steps:= Act_Steps ∪ Arg_Steps;
     endcase
(4) if Steps= ∅  then fail;
(5) choose S_add  from Steps;
(6) if S_add ∈ Act_Steps  then
        Plan.Orderings := Plan.Orderings ∪ {S_add≺S_need};
        Plan.Causal_Links:= Plan.Causal_Links ∪ {S_add ──P──> S_need};
        if S_add ∉ Plan.Action_Steps  then   // S_add is a newly added step
            Plan.Action_Steps:= Plan.Action_Steps ∪ {S_add};
            Plan.Orderings:= Plan.Orderings ∪ {START ≺ S_add, S_add ≺ FINISH};
            Plan.Subgoals:= Plan.Subgoals ∪ {(g, S_add, any,[])| g ∈ Preconditions(S_add)};
        endif
     endif
(7) if S_add ∈ Arg_Steps  then
        Plan.Argument_Steps:= Plan.Argument_Steps ∪ {S_add};
        Plan.Orderings:= Plan.Orderings ∪ {S_add≺S_need };
        Plan.Support_Links:= Plan.Support_Links ∪ {S_add ──P──< S_need};
        Plan.Subgoals:= Plan.Subgoals ∪ {(g, S_add, action,[])|g ∈ Argument_Base(S_add)};
     endif
end
```

Fig. 6. Outline of the APOP algorithm. Part 2: choosing and adding steps.

As new steps are added to the plan, new threats could appear. The function Resolve_threats (Figure 7) detects these threats and tries to resolve them using the methods proposed in section 5.

The statement (1) considers all *action-action threats* present in the plan and tries to resolve each of them choosing either *promotion* or *demotion* (see Figure 2). Following [7], we use "*possibly* $A_1 \prec A_3$" in the algorithm to express that $A_1 \prec A_3$ is consistent with the ordering constraint of the plan (Plan.Orderings).

procedure Resolve_Threats($\overset{\downarrow\uparrow}{Plan}$, $\overset{\downarrow}{\Delta}$, $\overset{\downarrow}{\Gamma}$)
begin

(1)**for each** $A_2 \overset{p}{\longrightarrow} A_1 \in$ **Plan.Causal_Links** and $A_1 \in$ **Plan.Action_Steps do**
    for each $A_3 \in \{A|\ A \in$ Plan.Action_Steps $\land\ \overline{p} \in effects(A) \land possibly\ A_2 \prec A \prec A_1\}$ **do**
        choose either
           Promotion:
             if $possibly\ A_1 \prec A_3$ **then**
                Plan.Orderings:= Plan.Orderings $\cup \{A_1 \prec A_3\}$;
             else fail;
           Demotion:
             if $possibly\ A_3 \prec A_2$ **then**
                Plan.Orderings:= Plan.Orderings $\cup \{A_3 \prec A_2\}$;
             else fail;
    end

(2)**for each** $\mathcal{B} \overset{b}{\succ\!\!\!-} A_1 \in$ **Plan.Support_Links do**
    begin
        $S_{A_1}:=\{(Add,l)|\ \exists\ Add\ \in$ Plan.Action_Steps $\land\ l\ \in$ Effects(Add) $\land\ possibly\ Add \prec A_1 \land$
                  $\not\exists\ Dell\ \in$ Plan.Action_Steps $\land\ \overline{l} \in$ Effects($Dell$) $\land\ Add \prec Dell \prec A_1$ };
        $\Psi_{A_1}:=\{l\ |\ (Add,l)\ \in\ S_{A_1}\}$;

(2.a) **for each** $(A_3,\overline{n})\ \in\ \{(Add,\overline{n})|\ (Add,\overline{n})\ \in\ S_{A_1} \land\ n\ \in literals(\mathcal{B})\}$ **do**
    choose either
        Promotion*:
           if $possibly\ A_1 \prec A_3$ **then**
              Plan.Orderings:= Plan.Orderings $\cup \{A_1 \prec A_3\}$;
           else fail;
        Demotion*:
           if n $\in Base(\mathcal{B})$ **and** $A_2 \overset{n}{\longrightarrow} \mathcal{B} \in$ Plan.Causal_Links **and** $possibly\ A_3 \prec A_2$ **then**
             Plan.Orderings:= Plan.Orderings $\cup \{A_3 \prec A_2\}$;
           else fail;
    end
(2.b) **for each** $\mathcal{C} \in \{\mathcal{C}|\ \langle\langle\mathcal{C},q\rangle\rangle$ is a *potential argument* from Δ that is a defeater for $\mathcal{B}$,
    base($\mathcal{C}$)$\subseteq\ \Psi_{A_1}$, $(\mathcal{C},\mathcal{B})\not\in$ Plan.Defeated_Args and $\mathcal{C}$ is acceptable wrt $Arg\_Line(\mathcal{B})$} **do**
    choose either
        Delaying the defeater:
           choose f **from** base($\mathcal{C}$)
           for each $Add\ \in \{Add\ |(Add,f)\ \in\ S_{A_1}\}$ **do**
             if $possibly\ A_1 \prec Add$ **then**
               Plan.Orderings:= Plan.Orderings $\cup \{A_1 \prec Add\}$;
             else fail;
        Defeating the defeater:
           choose p **from** heads($\mathcal{C}$);
           Plan.Subgoals:= Plan.Subgoals $\cup (\overline{p},A_1,argument,Arg\_Line(\mathcal{B}) + [\langle\langle\mathcal{C},q\rangle\rangle])$;
           Plan.Defeated_Args:= Plan.Defeated_Args $\cup \{(\mathcal{C},\mathcal{B})\}$

        Desabling the defeater:
           choose n **from** literals($\mathcal{C}$);
           Plan.Subgoals:= Plan.Subgoals $\cup (\overline{n},A_1,\ action,[\,])$;

    end
  end
end

Fig. 7. Outline of the APOP algorithm. Part 3: Threats Resolution.

It is important to note that if the plan contains no argument steps then Plan.Support_Links= $\emptyset$, hence, statement (2) will not be executed and Resolve_threats will proceed as in traditional POP algorithm.

If the plan contains arguments, statement (2.a) considers all *action-argument threats* present in the plan and tries to resolve each of them choosing either *promotion** or *demotion** (see Figure 3). Finally, statement (2.b) considers all *argument-argument threats* and tries to resolve each of them choosing either *delaying the defeater, defeating the defeater* or *disabling the defeater* (see Figure 4).

7 Related Work

The combination of defeasible reasoning and planning is not new [9,11]. However, in these works, the whole plan is viewed as an argument and then, defeasible reasoning is performed about complete plans. In contrast, our approach uses arguments for warranting subgoals, and hence, defeasible reasoning is used in a single step of the plan.

In [9], a planner is proposed that performs essentially the same search as POP, but by reasoning defeasibly about plans. That is, it reasons backwards from goals to subgoals, planning for conjunctive goals separately and then merging the plans for the individual goals into a combined plan for the conjunctive goal. Pollock argues that planning must be done defeasibly, making the default assumption that there are no threats and then modifying the plan as threats are discovered. Therefore, a planning agent will *infer defeasibly* that the merged plan is a solution to the planning problem. A *defeater* for this defeasible inference consists of discovering that the plan contains destructive interference. This interference refers to the traditional notion of *threat* that involves only actions. Although this approach combines defeasible reasoning and partial order planning, defeasible reasoning is not used in the same way as we propose in this work. He uses defeasible reasoning to reason about the plan as a whole, while we use defeasible reasoning to warrant subgoals *during* the planning process.

In [11], an argumentation-based approach for agents following the BDI model is introduced. They introduce different instantiations of Dung's abstract argumentation framework for generating consistent desires and consistent plans for achieving those desires. Although their work relates argumentation and plans, their approach differs considerably from ours. For them, a complete plan for a desire *d* is an "instrumental argument", *i.e.*, a set of planning rules that support *d*. Since complete plans are arguments, they introduce the notion of conflict among plans (arguments) and their approach defines which plan prevails after an argumentative analysis. Therefore, plans are more related to our notion of argument, than to our definition of plan. Finally, they use plans to justify the selected intentions of the agent.

8 Conclusions

In this paper we have introduced an argumentation-based formalism an agent could use for constructing plans using partial order planning technics. We have described how the traditional POP algorithm can be extended to consider arguments as planning steps.

When actions and arguments are combined to construct plans, new types of interferences appear. Therefore, we have extended the notion of threat to consider: *action-action, action-argument,* and *argument-argument threats.* Methods to resolve each type of threat have been proposed.

We have presented an algorithm called APOP, that extends the traditional POP algorithm to consider actions and arguments as planning steps and resolve the new types of threats using the proposed methods. A prototype implementation of this algorithm was implemented in Prolog.

This work was focused on improving the capabilities and scope of current planning technology and not in improving the efficiency of current planning implementations. Therefore, we have not made any comparison of the efficiency with other existing planners.

Future work includes the extension of our formalism to consider other features (e. g., conditional effects) that are present in other action representation languages like AL [1] and PDDL [5].

References

1. Baral, C., Gelfond, M.: Reasonig agents in dynamic domains. In: In Minker, J. (ed.) Logic-Based artificial intelligence, pp. 257–259. Kluwer Academic Publishers, Dordrecht (2000)
2. Chesñevar, C.I., Maguitman, A.G., Loui, R.P.: Logical Models of Argument. ACM Comp. Surveys 32(4) (December 2000)
3. García, A.J., Simari, G.R.: Defeasible logic programming: An argumentative approach. Theory and Practice of Logic Programming 4(1), 95–138 (2004)
4. Garcia, D.R., Simari, G.R., Garcia, A.J.: Planning and Defeasible Reasoning. In: AAMAS 2007. Proceedings of the Sixth Intl. Joint Conf. on Autonomous Agents and Multi-Agent Systems, pp. 856–858 (2007)
5. Ghallab, M., Howe, A., Knoblock, C., McDermott, D., Ram, A., Veloso, M., Weld, D., Wilkins, D.: Pddl—the planning domain definition language (1998)
6. Lifschitz, V.: Foundations of logic programs. In: Brewka, G. (ed.) Principles of Knowledge Representation, pp. 69–128. CSLI Pub., Stanford (1996)
7. Penberthy, J., Weld, D.S.: UCPOP: A Sound, Complete, Partial Order Planner for ADL. In: Proc. of the 3rd. Int. Conf. on Principles of Knowledge Representation and Resoning, pp. 113–124 (1992)
8. Pollock, J.: Cognitive Carpentry: A Blueprint for How to Build a Person. MIT Press, Cambridge (1995)
9. Pollock, J.: Defeasible Planning. In: Bergmann, R., Kott, A. (Cochairs) Integrating Planning, Scheduling, and Execution in Dynamic and Uncertain Environments AIPS Workshop (1998)
10. Prakken, H., Vreeswijk, G.: Logical systems for defeasible argumentation. In: Gabbay, D. (ed.) Handbook of Philosophical Logic, 2nd edn. Kluwer Academic Pub., Dordrecht (2000)
11. Rahwan, I., Amgoud, L.: An argumentation-based approach for practical reasoning. In: Proc. AAMAS, pp. 347–354 (2006)

12. Simari, G.R., García, A.J.: Actions and arguments: Preliminaries and examples. In: Proc. VII Congreso Argentino en Ciencias de la Computación, Argentina, pp. 273–283 (October 2001)
13. Simari, G.R., García, A.J., Capobianco, M.: Actions, Planning and Defeasible Reasoning. In: NMR 2004. Proceedings of the 10th International Workshop on Non-Monotonic Reasoning, pp. 377–384 (2004), ISBN: 92-990021-0-X
14. Simari, G.R., Loui, R.P.: A Mathematical Treatment of Defeasible Reasoning and its Implementation. Artificial Intelligence 53, 125–157 (1992)
15. Stolzenburg, F., García, A., Chesñevar, C.I., Simari, G.R.: Computing Generalized Specificity. Journal of Non-Classical Logics 13(1), 87–113 (2003)

Lossless Decompositions in Complex-Valued Databases

Henning Koehler and Sebastian Link*

Massey University, Palmerston North, New Zealand
{h.koehler,s.link}@massey.ac.nz

Abstract. When decomposing database schemas, it is desirable that a decomposition is lossless and dependency preserving. A well-known and frequently used result for the relational model states that a functional dependency preserving decomposition is lossless if and only if it contains a key. We will show that this result does not always hold when domains are allowed to be finite, but provide conditions under which it can be preserved. We then extend our work to a complex-valued data model based on record, list, set and multiset constructor, where finite domains occur naturally for subattributes, even if the domains of flat attributes are infinite.

1 Introduction

In order to obtain well-designed databases, it is often necessary to decompose a schema into smaller subschemas. When doing so, we must take care not to lose any information - decompositions which preserve information are called *lossless*. In addition, it is important to preserve dependencies specified over the schema, which restrict what data can be stored in the database, thereby avoiding inconsistencies. Thus one typically wants decompositions that are not only lossless, but also *dependency preserving* [2,3,8,9,11,12,16,19].

For the relational data model, lossless and dependency preserving decompositions have often been studied under the (sometimes implicit) assumption that attribute domains are infinite. A well-known and frequently used result (Theorem 1) states that a dependency preserving decomposition of a schema R with a set Σ of functional dependencies (FDs) as constraints is lossless if and only if it contains a key [3]. It is the basis for the well-known synthesis approach introduced by Biskup, Dayal and Bernstein in [3] for finding dependency preserving decompositions into 3NF. Their work has since been extended to find different lossless and dependency preserving decompositions, e.g. into EKNF [19] and BCNF [4,10,18]. In order to extend the synthesis approach to other, e.g. complex-valued data models, it is thus vital to investigate whether Theorem 1 still holds in these models.

---

* This research is supported by the Marsden fund council from Government funding, administered by the Royal Society of New Zealand.

S. Hartmann and G. Kern-Isberner (Eds.): FoIKS 2008, LNCS 4932, pp. 329–347, 2008.

We will demonstrate in this paper that the characterization of lossless and dependency preserving decompositions is only valid when domains of attributes are assumed to be infinite. This fact is particularly relevant in practice since many domains naturally only carry a finite number of elements. Moreover, when studying lossless and dependency preserving decompositions in complex-valued data models, then finite domains can often not be avoided as well.

More precisely, we will show that Theorem 1 is still valid if LHS-attributes, i.e., attributes which appear in the left hand side X of FDs $X \to Y \in \Sigma$, have infinite domains while domains of other attributes may be finite. This result is then extended to complex-valued databases that can be generated by a finite number of recursive applications of record, list, set and multiset constructor.

The rest of the paper is organized as follows. We first give a brief introduction to the relational model in Section 2, then examine the relational case in the presence of finite domains in Section 3. In Section 4 we introduce the complex-valued model which we shall use for our investigation. We then show how Theorem 1 can be extended to this complex model in Section 5.

2   The Relational Model of Data

We begin by introducing some basic terms for the relational model. More details can be found e.g. in [13,15,17].

A *relational database schema* consists of a set of relation schemas. A *relation schema* $R = \{A_1, A_2, \ldots, A_n\}$ is a finite set of *attributes*. Each attribute A_i has a *domain* $dom(A_i)$ associated with it. Domains are arbitrary sets, but unless explicitly stated otherwise, we will assume that domains are countably infinite.

A *relation* r over a relation schema R is a finite set of tuples, while a tuple t on R is a mapping $t : R \to \bigcup_{A \in R} dom(A)$ with $t(A) \in dom(A)$. Relations over a schema are also commonly referred to as schema instances or tables. Sets of schema instances, one for each relation schema in a database schema, are called database instances. Note that one could also consider relations which are infinite. Since we will only consider functional dependencies (see below), this would not affect our results.

With each relation schema we associate a set Σ of *functional dependencies* (FD). A functional dependency on R is an expression of the form $X \to Y$ (read "*X determines Y*") where X and Y are subsets of R. For attribute sets X, Y and attribute A we will write XY short for $X \cup Y$ and A short for $\{A\}$. We say that an FD $X \to Y$ *holds* on a relation r over R if every pair of tuples in r that coincides on all attributes in X also coincides on all attributes in Y.

A set Σ of FDs over R *implies* an FD (or set of FDs) Σ', written $\Sigma \models \Sigma'$, if Σ' holds on every relation r over R for which all FDs in Σ hold. If two sets of FDs Σ and Σ' imply each other, we call Σ a *cover* of Σ' (and vice versa) and write $\Sigma \equiv \Sigma'$. We write Σ^* for the set of all FDs on R implied by Σ.

The *closure* X^* of a set $X \subseteq R$ w.r.t. a set Σ of FDs is the set of all attributes determined by X:

$$X^* := \{A \in R \mid \Sigma \models X \to A\}$$

We call X *closed* under Σ if $X = X^*$.

A set $X \subseteq R$ is a *key* of R w.r.t. a set Σ of FDs on R, if Σ implies $X \to R$. Note that some authors use the term 'key' only for minimal keys, and call keys which may not be minimal 'superkeys'. For us, keys need not be minimal.

The *projection* of a set Σ of FDs onto a subschema $R_i \subseteq R$ is

$$\Sigma[R_i] := \{X \to Y \in \Sigma \mid XY \subseteq R_i\}$$

For a set $X \subseteq R$ we denote the *projection* of r onto the attributes in X by $r[X]$. The *join* of two relations $r[X]$ and $r[Y]$ is a relation on $X \cup Y$:

$$r[X] \bowtie r[Y] := \left\{ t \,\middle|\, \begin{array}{l} \exists t_1 \in r[X], t_2 \in r[Y]. \\ t[X] = t_1 \wedge t[Y] = t_2 \end{array} \right\}$$

A decomposition $\mathcal{D} = \{R_1, \ldots, R_n\}$ of R is a set of subschemas $R_i \subseteq R$ of R such that

$$\bigcup \mathcal{D} = R$$

To ensure that the schemas $R_1, \ldots, R_n$ of a decomposition $\mathcal{D}$ of R can hold the same data as the original schema R, we must ask for a decomposition that is *lossless*, i.e., that for all relations r on R for which Σ holds we have

$$r[R_1] \bowtie \ldots \bowtie r[R_n] = r$$

Furthermore, the decomposition should be dependency preserving, i.e., the dependencies on the schemas R_i which are implied by Σ should form a cover of the original functional dependencies Σ:

$$\Sigma \equiv \bigcup \Sigma^*[R_i]$$

This allows a database management system to check constraints for individual relations only, without having to compute their join. In practice FDs *are* only checked for individual relations, so for a given schema (R, Σ), losslessness and dependency preservation of a decomposition ensure that the semantics of (R, Σ) are adequately preserved.

In the relational model there is a nice characterization when a dependency-preserving decomposition is lossless. Note again that we are currently assuming all domains to be infinite.

Theorem 1. *[3] A dependency-preserving decomposition of R is lossless if and only if it contains a subschema which forms a key of R.*

As a result, the task of finding a dependency preserving and lossless decomposition becomes easier: We can first concentrate on finding a dependency preserving decomposition. If this decomposition is not lossless, Theorem 1 shows that we only need to add a minimal key to the decomposition to make it lossless.

Example 1. Let $R = \{ABC\}$ with FDs $\Sigma = \{A \to C, B \to C\}$. Then the decomposition $\mathcal{D} = \{AC, BC\}$ is dependency preserving, but not lossless: for the relation r on R

$$r = \begin{array}{|c|c|c|} \hline A & B & C \\ \hline 0 & 1 & 0 \\ \hline 1 & 0 & 0 \\ \hline \end{array}$$

the FDs in Σ hold, but we get

$$r[AC] \bowtie r[BC] = \begin{array}{|c|c|} \hline A & C \\ \hline 0 & 0 \\ \hline 1 & 0 \\ \hline \end{array} \bowtie \begin{array}{|c|c|} \hline B & C \\ \hline 1 & 0 \\ \hline 0 & 0 \\ \hline \end{array} = \begin{array}{|c|c|c|} \hline A & B & C \\ \hline 0 & 1 & 0 \\ \hline 1 & 0 & 0 \\ \hline 0 & 0 & 0 \\ \hline 1 & 1 & 0 \\ \hline \end{array} \neq r$$

We can turn $\mathcal{D}$ into a lossless decomposition $\mathcal{D}'$ by adding the key schema AB. Hence,

$$\mathcal{D}' := \mathcal{D} \cup \{AB\} = \{AB, AC, BC\}$$

is both lossless and dependency-preserving.

In order to extend the synthesis approach to other, complex-valued data models, it is thus vital to investigate whether Theorem 1 still holds in these models.

3 Lossless Decompositions - The Relational Case

As shown in [3], Theorem 1 holds when all attribute domains are infinite. If domains are allowed to be finite, the "if" direction still holds - this becomes obvious when we consider that instances over finite domains are also instances over infinite supersets of these domains.

To see where the "only if" direction fails, we first look into the corresponding proof [3], adapting the correctness proof for the chase procedure [1].

Lemma 1. *Let R be a relational schema and Σ a set of FDs on R. Then every lossless decomposition $\mathcal{D}$ of R contains a key of R.*

Proof. Let $\mathcal{D} = \{R_1, \ldots, R_n\}$ not contain a key of R. We show that $\mathcal{D}$ is not lossless by constructing a sample relation r on R as follows. Let t be an arbitrary tuple on R. Then for every schema $R_i \in \mathcal{D}$ add a tuple t_i to r which is identical to t on the closure R_i^* of R_i w.r.t. Σ, and contains unique values for attributes outside R_i^*.

Then for every FD $X \to Y \in \Sigma$, two tuples t_i, t_j are identical on X if and only if $X \subseteq R_i^* \cap R_j^*$. If $X \subseteq R_i^* \cap R_j^*$ holds then Y is also included in $R_i^* \cap R_j^*$, since $R_i^* \cap R_j^*$ is closed under Σ. Thus $X \to Y$ holds on r.

Since $\mathcal{D}$ contains no key, the tuple t does not lie in r. It does however lie in

$$\bowtie r[\mathcal{D}] := r[R_1] \bowtie \ldots \bowtie r[R_n]$$

which makes $\mathcal{D}$ not lossless.

We first note that the proof requires a sufficient number of distinct attribute values for the construction. In this, we implicitly assume that attribute domains are infinite. For the remainder of this paper we will allow attribute domains to be finite, although we still require them to contain at least two elements.

Example 2. Let $R = ABCD$ with FDs $\Sigma = \{A \rightarrow BC, BC \rightarrow A\}$ and decomposition $\mathcal{D} = \{ABC, BD, CD\}$. Let furthermore the domain of A contain only two values, say $dom(A) = \{0, 1\}$. Then $\mathcal{D}$ contains no key of R, but is lossless.

Proof (of Example 2). Assume $\mathcal{D}$ was not lossless, i.e., for some relation r on R there exists a tuple t with

$$t \in (\bowtie r[\mathcal{D}]) \setminus r$$

Then for every $R_i \in \mathcal{D}$ there must be a tuple $t_i \in r$ with $t_i[R_i] = t[R_i]$. We may assume w.l.o.g. that $t = (0, 0, 0, 0)$. This gives us the following subset of r:

$$r' = \begin{array}{|c|c|c|c|} \hline A & B & C & D \\ \hline 0 & 0 & 0 & d \\ \hline a_1 & 0 & c & 0 \\ \hline a_2 & b & 0 & 0 \\ \hline \end{array}$$

for some values a_1, a_2, b, c, d with $a_1, a_2 \in dom(A) = \{0, 1\}$. If $a_1 = 0$ or $a_2 = 0$ then the FD $A \rightarrow BC$ implies $c = 0$ or $b = 0$, respectively, and thus $t \in r$ which contradicts the assumption. If $a_1 = a_2 = 1$ then $A \rightarrow BC$ again implies $b = 0$ and $c = 0$, which gives us the following subset of r:

$$r' = \begin{array}{|c|c|c|c|} \hline A & B & C & D \\ \hline 0 & 0 & 0 & d \\ \hline 1 & 0 & 0 & 0 \\ \hline \end{array}$$

But then the FD $BC \rightarrow A$ would be violated on r' and thus on r.

Note that this example can easily be generalized to work for arbitrary finite domains. For $dom(A) = \{1, \ldots, k\}$ chose

$$R = AB_1 \ldots B_k C$$
$$\Sigma = \{A \rightarrow B_1 \ldots B_k, B_1 \ldots B_k \rightarrow A\}$$
$$\mathcal{D} = \{AB_1 \ldots B_k, R \setminus AB_1, \ldots, R \setminus AB_k\}$$

Obviously we need some restrictions on the domains for Lemma 1 to work. However, requiring all domains of attributes occurring in R to be infinite is too strong. While finite domains could perhaps be ignored in relational databases, we will see that they arise naturally in complex-valued databases for subattributes such as $\{\lambda\}$, for which the domain contains only the two values $\emptyset$ and $\{ok\}$.

In the example above, the attribute A occurred in the LHS of an FD in Σ. It turns out that requiring that these attributes have infinite domains is sufficient. Recall that even if we allow the domain of an attribute to be finite, we still require it to contain at least two different values.

Lemma 2. *Let R be a relational schema and Σ a set of FDs on R. If all attributes which occur in the LHS of an FD in Σ have infinite domains, then every lossless decomposition $\mathcal{D}$ of R contains a key of R.*

Proof. As for Lemma 1, except that for attributes not occurring in the LHS of an FD in Σ we do not require the t_i to have unique values on these attributes, but only values different from t.

Note that instead of working with Σ, we could use a cover of Σ instead. This can affect which attributes appear in the LHS of FDs, and thus give us different requirements for the domains in Lemma 2. However, it can easily be shown (although we will not do that here) that all *reduced* covers [15] - these are covers in which no attributes can be removed from FDs while maintaining a cover - share the same set of LHS-attributes. Clearly these form a subset of the LHS-attributes for any cover Σ' of Σ, since we can transform Σ' into a reduced cover by removing attributes.

4   The Complex-Valued Data Model

A number of complex-valued data models have been suggested. We will follow the approach of Hartmann, Link and Schewe [5,6,7], since it provides a general and unifying framework for the study of many existing data models. In particular, one can focus on the main data structures under consideration and, thereby, appreciate the direct impact of the type constructors on the results.

Instead of dealing with relation schemas we utilize nested attributes. These can be generated from flat attributes (which are the same as attributes in the relational model) by an arbitrary finite number of recursive applications of record, list, set and multiset constructors.

Definition 1. *[7] A universe is a finite set $\mathcal{U}$ together with domains (i.e., sets of values) $dom(A)$ for all $A \in \mathcal{U}$. The elements of $\mathcal{U}$ are called flat attributes.*

For the relational data model a universe was sufficient. That is, a relation schema is defined as a finite and non-empty subset $R \subseteq \mathcal{U}$. For data models supporting complex objects, however, nested attributes are needed. In the following definition we use a set $\mathcal{L}$ of labels, and assume that the symbol λ is neither a flat attribute nor a label, i.e., $\lambda \notin \mathcal{U} \cup \mathcal{L}$. Moreover, flat attributes are not labels and vice versa, i.e., $\mathcal{U} \cap \mathcal{L} = \emptyset$.

Definition 2. *[7] Let $\mathcal{U}$ be a universe and $\mathcal{L}$ a set of labels. The set $\mathcal{N}A(\mathcal{U}, \mathcal{L})$ of nested attributes over $\mathcal{U}$ and $\mathcal{L}$ is the smallest set satisfying the following conditions:*

1. *$\lambda \in \mathcal{N}A(\mathcal{U}, \mathcal{L})$,*
2. *$\mathcal{U} \subseteq \mathcal{N}A(\mathcal{U}, \mathcal{L})$,*
3. *for $L \in \mathcal{L}$ and $N_1, \ldots, N_k \in \mathcal{N}A(\mathcal{U}, \mathcal{L})$ with $k \geq 1$ we have $L(N_1, \ldots, N_k) \in \mathcal{N}A(\mathcal{U}, \mathcal{L})$,*

4. *for $L \in \mathcal{L}$ and $N \in \mathcal{N}A(\mathcal{U}, \mathcal{L})$ we have $L[N] \in \mathcal{N}A(\mathcal{U}, \mathcal{L})$,*
5. *for $L \in \mathcal{L}$ and $N \in \mathcal{N}A(\mathcal{U}, \mathcal{L})$ we have $L\{N\} \in \mathcal{N}A(\mathcal{U}, \mathcal{L})$,*
6. *for $L \in \mathcal{L}$ and $N \in \mathcal{N}A(\mathcal{U}, \mathcal{L})$ we have $L\langle N\rangle \in \mathcal{N}A(\mathcal{U}, \mathcal{L})$.*

We call λ null attribute, $L(N_1, \ldots, N_k)$ record-valued attribute, $L[N]$ list-valued attribute, $L\{N\}$ set-valued attribute, and $L\langle N\rangle$ multiset-valued attribute.

From now on we will assume that a universe $\mathcal{U}$ and a set $\mathcal{L}$ of labels are fixed. Instead of writing $\mathcal{N}A(\mathcal{U}, \mathcal{L})$ we simply write $\mathcal{N}A$.

A relation schema $R = \{A_1, \ldots, A_n\}$ can be viewed as the record-valued attribute $R(A_1, \ldots, A_n)$ using the name R as a label. The null attribute λ must not be confused with a null value, which is a distinguished element of a certain domain. The null attribute rather indicates that some information of the underlying nested attribute, i.e. some information on the schema level, has been left out. Further explanations follow.

The mapping *dom* can be extended from flat to nested attributes, i.e., we define a set $dom(N)$ of values for every nested attribute $N \in \mathcal{N}A$. We denote empty set, empty multiset, and empty list by $\emptyset, \langle\,\rangle, [\,]$, respectively.

Definition 3. *[7] For a nested attribute $N \in \mathcal{N}A$ we define the domain $dom(N)$ as follows:*

1. $dom(\lambda) = \{ok\}$,
2. $dom(A)$ *as above (Definition 1) for all $A \in \mathcal{U}$,*
3. $dom(L(N_1, \ldots, N_k)) = \{(v_1, \ldots, v_k) \mid v_i \in dom(N_i)$ for $i = 1, \ldots, k\}$, i.e., the set of all k-tuples $(v_1, \ldots, v_k)$ with $v_i \in dom(N_i)$ for all $i = 1, \ldots, k$,
4. $dom(L[N]) = \{[v_1, \ldots, v_n] \mid v_i \in dom(N)$ for $i = 1, \ldots, n\} \cup \{[\,]\}$, i.e., $dom(L[N])$ is the set of all finite lists with elements in $dom(N)$,
5. $dom(L\{N\}) = \{\{v_1, \ldots, v_n\} \mid v_i \in dom(N)$ for $i = 1, \ldots, n\} \cup \{\emptyset\}$, i.e., $dom(L\{N\})$ is the set of all finite subsets of $dom(N)$,
6. $dom(L\langle N\rangle) = \{\langle v_1, \ldots, v_n\rangle \mid v_i \in dom(N)$ for $i = 1, \ldots, n\} \cup \{\langle\,\rangle\}$, i.e., $dom(L\langle N\rangle)$ is the set of all finite multisets with elements in $dom(N)$.

The domain of the record-valued attribute $R(A_1, \ldots, A_n)$ is a set of n-tuples, i.e., an n-ary relation. The value *ok* can be interpreted as the null value "some information exists, but is currently omitted".

The replacement of flat attribute names by the null attribute λ within a nested attribute decreases the amount of information that is modelled by the corresponding attributes. This fact allows us to introduce an order between nested attributes.

Definition 4. *[7] The subattribute relation $\leq$ on the set of nested attributes $\mathcal{N}A$ over $\mathcal{U}$ and $\mathcal{L}$ is defined by the following rules, and the following rules only:*

1. $N \leq N$ *for all nested attributes $N \in \mathcal{N}A$,*
2. $\lambda \leq A$ *for all flat attributes $A \in \mathcal{U}$,*
3. $\lambda \leq N$ *for all set-valued, multiset-valued and list-valued attributes $N \in \mathcal{N}A$,*
4. $L(N_1, \ldots, N_k) \leq L(M_1, \ldots, M_k)$ *whenever $N_i \leq M_i$ for all $i = 1, \ldots, k$,*

5. $L[N] \leq L[M]$ whenever $N \leq M$,
6. $L\{N\} \leq L\{M\}$ whenever $N \leq M$,
7. $L\langle N \rangle \leq L\langle M \rangle$ whenever $N \leq M$.

For $N, M \in \mathcal{N}A$ we say that M is a subattribute of N if and only if $M \leq N$ holds. We write $M \not\leq N$ if and only if M is not a subattribute of N.

The finite set of all subattributes of N is denoted by $Sub(N)$. It has a a smallest element with respect to the subattribute relation, called the bottom element.

Lemma 3. *[5, Lemma 38] The bottom element* λ_N *of* $Sub(N)$ *is given by* $\lambda_N = L(\lambda_{N_1}, \ldots, \lambda_{N_k})$ *whenever* $N = L(N_1, \ldots, N_k)$, *and* $\lambda_N = \lambda$ *whenever* N *is not a record-valued attribute.*

The binary operators *join* $\sqcup$ and *meet* $\sqcap$ are the equivalent to union $\cup$ and intersection $\cap$ in the relational case. For $X, Y \leq N$ we get:

- $X \sqcup Y$ is the minimal subattribute of N with $X, Y \leq X \sqcup Y$
- $X \sqcap Y$ is the maximal subattribute of N with $X, Y \geq X \sqcap Y$

It has been shown in [7] that join and meet always exist.

Given the relation schema $R = \{A, B, C\}$, the attribute set $\{A, C\}$ can be viewed as the subattribute $R(A, \lambda, C)$ of the record-valued attribute $R(A, B, C)$. The occurrence of the null attribute λ in $R(A, \lambda, C)$ indicates that the information about the attribute B has been masked out. The inclusion order $\subseteq$ on attribute sets in the relational data model is now generalized to the subattribute relation $\leq$ in complex-valued data models.

Lemma 4. *The subattribute relation is a partial order on nested attributes.* $\square$

Informally, $M \leq N$ for $N, M \in \mathcal{N}A$ if and only if M comprises at most as much information as N does. The informal description of the subattribute relation is formally documented by the existence of a projection function $\pi_M^N : dom(N) \to dom(M)$ in case $M \leq N$ holds.

Definition 5. *[7] Let* $N, M \in \mathcal{N}A$ *with* $M \leq N$. *The projection function* $\pi_M^N : dom(N) \to dom(M)$ *is defined as follows:*

1. *if* $N = M$, *then* $\pi_M^N = id_{dom(N)}$ *is the identity on* $dom(N)$,
2. *if* $M = \lambda$, *then* $\pi_\lambda^N : dom(N) \to \{ok\}$ *is the constant function that maps every* $v \in dom(N)$ *to* ok,
3. *if* $N = L(N_1, \ldots, N_k)$ *and* $M = L(M_1, \ldots, M_k)$, *then* $\pi_M^N = \pi_{M_1}^{N_1} \times \cdots \times \pi_{M_k}^{N_k}$ *which maps every tuple* $(v_1, \ldots, v_k) \in dom(N)$ *to* $(\pi_{M_1}^{N_1}(v_1), \ldots, \pi_{M_k}^{N_k}(v_k)) \in dom(M)$,
4. *if* $N = L[N']$ *and* $M = L[M']$, *then* $\pi_M^N : dom(N) \to dom(M)$ *maps every list* $[v_1, \ldots, v_n] \in dom(N)$ *to the list* $[\pi_{M'}^{N'}(v_1), \ldots, \pi_{M'}^{N'}(v_n)] \in dom(M)$,
5. *if* $N = L\{N'\}$ *and* $M = L\{M'\}$, *then* $\pi_M^N : dom(N) \to dom(M)$ *maps every set* $S \in dom(N)$ *to the set* $\{\pi_{M'}^{N'}(s) : s \in S\} \in dom(M)$, *and*

6. *if* $N = L\langle N'\rangle$ *and* $M = L\langle M'\rangle$, *then* $\pi_M^N : dom(N) \to dom(M)$ *maps every multiset* $S \in dom(N)$ *to the multiset* $\langle \pi_{M'}^{N'}(s) \ : \ s \in S\rangle \in dom(M)$.

It follows, in particular, that $\emptyset, \langle \, \rangle, [\,]$ are always mapped to themselves, except when projected on the null attribute λ in which each of them is mapped to *ok*. Note that for $Y \leq X$ we have $\pi_Y^N = \pi_Y^X \circ \pi_X^N$ where $\circ$ denotes the composition of functions.

We are now ready to repeat the syntax and semantics of functional dependencies in complex-valued databases.

Definition 6. *[7] Let $N \in \mathcal{N}A$ be a nested attribute. A functional dependency on N is an expression of the form $\mathcal{X} \to \mathcal{Y}$ where $\mathcal{X}, \mathcal{Y} \subseteq Sub(N)$ are non-empty. A set $r \subseteq dom(N)$ satisfies the functional dependency $\mathcal{X} \to \mathcal{Y}$ on N, denoted by $\models_r \mathcal{X} \to \mathcal{Y}$, if and only if $\pi_Y^N(t_1) = \pi_Y^N(t_2)$ holds for all $Y \in \mathcal{Y}$ whenever $\pi_X^N(t_1) = \pi_X^N(t_2)$ holds for all $X \in \mathcal{X}$ and any $t_1, t_2 \in r$.*

The requirement that $\mathcal{X}, \mathcal{Y}$ are sets of subattributes cannot be weakened without losing expressiveness. In fact, if a set of subattributes in an FD is replaced by its join, then this may result in an FD with a different semantics. We illustrate this fact by the following example.

Example 3. [14] Suppose we store sets of tennis matches using the nested attribute

$$\text{Tennis}\{\text{Match(Winner, Loser)}\}.$$

Consider the following instance r over Tennis{Match(Winner, Loser)}:

$$\{ \ \{(\text{Becker, Agassi}), (\text{Stich, McEnroe})\},$$
$$\{(\text{Becker, McEnroe}), (\text{Stich, Agassi})\} \ \}.$$

The second element of this set results from the first by simply switching opponents. We can see that $\models_r$ Tennis{Match(Winner, λ)} $\to$ Tennis{Match(λ, Loser)} holds. In fact, the set of winners {Becker, Stich} is the same for both elements and so is the set of losers {Agassi, McEnroe}.

However, $\not\models_r$ Tennis{Match(Winner, λ)} $\to$ Tennis{Match(Winner, Loser)} since the matches stored in both elements are different from one another. The instance r is therefore a prime example for the failure of the extension rule

$$\frac{X \to Y}{X \to X \sqcup Y}$$

in the presence of sets. The same is true for multisets as a set is just a multiset in which every element occurs exactly once.

Sufficient and necessary conditions when projections on subattributes X and Y do determine the projection on $X \sqcup Y$ have been identified.

Definition 7. *[7] Let $N \in \mathcal{N}A$. The subattributes $X, Y \in Sub(N)$ are reconcilable if and only if one of the following conditions is satisfied*

- $Y \leq X$ or $X \leq Y$,
- $N = L(N_1, \ldots, N_k), X = L(X_1, \ldots, X_k), Y = L(Y_1, \ldots, Y_k)$ where X_i and Y_i are reconcilable for all $i = 1, \ldots, k$,
- $N = L[N'], X = L[X'], Y = L[Y']$ where X' and Y' are reconcilable.

In Example 3 the subattributes Tennis{Match(Winner, λ)}, Tennis{Match(λ, Loser)} are not reconcilable. However, if we use Tennis[Match(Winner, Loser)] to store (sets of) *lists* of tennis matches, rather than (sets of) sets, then the subattributes Tennis[Match(Winner, λ)] and Tennis[Match(λ, Loser)] are reconcilable.

In fact, projections of complex data tuples to reconcilable subattributes X, Y uniquely determine the projection on the join $X \sqcup Y$.

Lemma 5. *[7, Lemmas 16 and 21] Let $N \in \mathcal{N}A$, $X, Y \in Sub(N)$. Then the following are equivalent:*

(i) X and Y are reconcilable
(ii) for all $t_1, t_2 \in dom(N)$ with $\pi_X^N(t_1) = \pi_X^N(t_2)$ and $\pi_Y^N(t_1) = \pi_Y^N(t_2)$ we have $\pi_{X \sqcup Y}^N(t_1) = \pi_{X \sqcup Y}^N(t_2)$

Lemma 5 can be illustrated using Example 3: Tennis{Match(Winner, λ)} and Tennis{Match(λ, Loser)} are not reconcilable, and the two different elements in

$$\{ \{(\text{Becker, Agassi}), (\text{Stich, McEnroe})\},$$
$$\{(\text{Becker, McEnroe}), (\text{Stich, Agassi})\} \}.$$

are identical on Tennis{Match(Winner, λ)} and Tennis{Match(λ, Loser)}.

In order to simplify the implication problem for FDs, attributes are split into maximal reconcilable subattributes.

Definition 8. *[5] Let $N \in \mathcal{N}A$. A nested attribute $N_i \in \mathcal{N}A$ is a unit of N if and only if*

1. $N_i \leq N$,
2. $\forall X, Y \leq N_i$, *if X and Y are reconcilable, then $X \leq Y$ or $Y \leq X$,*
3. N_i *is $\leq$-maximal with properties 1. and 2.*

The set of all units of N is denoted by $\mathcal{U}(N)$.

We will be interested in all subattributes with properties 1. and 2. of Definition 8, not just maximal ones.

Definition 9. *We call a nested attribute N unitary if it is a unit of itself. We write $N^{\downarrow}$ for the set of all unitary subattributes of N.*

Note that $N^{\downarrow}$ can be shown to be exactly the *extended subattribute basis* $\mathcal{E}(N)$ of N as defined in [5], though we will not need that here.

Example 4. In Example 3 where $N = \text{Tennis}\{\text{Match}(\text{Winner}, \text{Loser})\}$, the sub-attribute $\text{Tennis}\{\text{Match}(\text{Winner}, \lambda)\}$ meets conditions 1. and 2. of Definition 8, but not 3. Thus it is unitary but not a unit of N. The only unit of N is N itself.

For the nested attribute $N' = \text{Tennis}[\text{Match}(\text{Winner}, \text{Loser})]$ we get the units $\text{Tennis}[\text{Match}(\text{Winner}, \lambda)]$ and $\text{Tennis}[\text{Match}(\lambda, \text{Loser})]$. Further unitary subattributes of N' are $\text{Tennis}[\text{Match}(\lambda,\lambda)]$ and λ.

When representing FDs, we will want to replace non-unitary attributes in FDs with their units. As a result, we will only need to deal with sets of unitary attributes, which makes it easier to formulate and prove some lemmas.

In [5] a characterization for the units of a nested attribute is given.

Lemma 6. *[5, Lemma 26] Let $N \in \mathcal{N}A$. Then*

$$\mathcal{U}(N) = \bigcup_{i=1}^{k} \{L(\lambda_{N_1}, \ldots, M, \ldots, \lambda_{N_k}) : M \in \mathcal{U}(N_i) \text{ and } N_i \neq \lambda_{N_i}\}$$

if $N = L(N_1, \ldots, N_k)$ and $N \neq \lambda_N$,

$$\mathcal{U}(N) = \{L[M'] : M' \in \mathcal{U}(M)\}$$

if $N = L[M]$ holds and $\mathcal{U}(N) = \{N\}$ in any other case.

Furthermore, it is clear from the results in [5], that replacing attributes with their units does not change the semantics of FDs. This means that the FDs $\mathcal{X} \to \mathcal{Y}$ and $\mathcal{X}' \to \mathcal{Y}'$ imply each other, where

$$\mathcal{X}' := \bigcup_{X \in \mathcal{X}} \mathcal{U}(X), \quad \mathcal{Y}' := \bigcup_{Y \in \mathcal{Y}} \mathcal{U}(Y)$$

For ease of reading, we will not always distinguish between an attribute and the singleton set containing that attribute, and leave out brackets, commas, labels and 'λ's where this does not cause ambiguities. Thus we would write e.g. the nested attribute $\text{Tennis}\{\text{Match}(\text{Winner}, \text{Loser})\}$ short as $\{\text{Winner}, \text{Loser}\}$, and the subattribute $\text{Tennis}\{\text{Match}(\text{Winner}, \lambda)\}$ as $\{\text{Winner}\}$.

For a more thorough discussion of the complex-valued model introduced here see e.g. [5,7].

5 Lossless Decomposition

When decomposing a nested attribute we need to decide what we decompose it into. Here we will be more general than in [6] where decompositions contain only subattributes.

Definition 10. *We call a set S of unitary subattributes of N an extended sub-attribute of N if no attribute in S is a subattribute of any other attribute in S (i.e., S is a $\leq$-antichain of unitary subattributes). We say that an extended subattribute S' of N is an extended subattribute of S if every attribute in S' is a subattribute of some attribute in S.*

From now on, when talking about decompositions of nested attributes, we will mean decompositions into extended subattributes. Note that, unlike substituting attributes by their units, using extended subattributes rather than just subattributes for decomposition changes the semantics. An extended subattribute corresponds to a set of (not necessarily unitary) subattributes rather than just a single subattribute. This is different to the relational case where all "subattributes" (subsets of attributes) are reconcilable.

Example 5. Let $N = \{AB\}C$ with FDs $\Sigma = \{\{A\}\{B\} \rightarrow C\}$. Then we can decompose N into $N_1 = \{AB\}$ and $N_2 = \{A\}\{B\}C$. This decomposition is dependency preserving and lossless (by Lemma 7). If we were to restrict ourselves to decompositions into subattributes, we could not decompose N any further without loss of information and dependencies.

The requirement that subattributes in S be unitary identifies semantically equivalent sets, such as the following sets of subattributes of $N = \{AB\}CD$:

$$\{\{A\}, \{B\}, C\}, \{\{A\}C, \{B\}\}, \{\{A\}, \{B\}C\} \text{ and } \{\{A\}C, \{B\}C\}$$

The first set $\{\{A\}, \{B\}, C\}$ is an extended subattribute of N, but the others are not. E.g. the second set contains the subattribute $\{A\}C$ which is not unitary as its units are $\{A\}$ and C.

Prohibiting subattributes of another attribute in a tuple prevents obvious redundancy.

Definition 11. *For two extended subattributes $\mathcal{X}, \mathcal{Y} \subseteq N^{\downarrow}$ we call $\mathcal{X}$ an extended subattribute of $\mathcal{Y}$, written $\mathcal{X} \leq_{ext} \mathcal{Y}$, if for every $X \in \mathcal{X}$ there exists a $Y \in \mathcal{Y}$ with $X \leq Y$. Join $\sqcap$ and meet $\sqcup$ between extended subattributes are then defined w.r.t. $\leq_{ext}$. In particular we get*

$$\mathcal{X} \sqcup \mathcal{Y} = max(\mathcal{X} \cup \mathcal{Y})$$

i.e., we take the union but then remove (redundant) non-maximal elements.

Definition 12. *The domain of an extended subattribute $S \subseteq N^{\downarrow}$ consists of all functions which map each subattribute $s \in S$ to an element in $dom(s)$, in a consistent manner:*

$$dom(S) := \left\{ f : S \rightarrow \bigcup_{s \in S} dom(s) \,\middle|\, \begin{array}{l} f(s) \in dom(s) \wedge \forall X, Y \in S. \\ \pi_{X \sqcap Y}^{X}(f(X)) = \pi_{X \sqcap Y}^{Y}(f(Y)) \end{array} \right\}$$

We can now define projection, join and losslessness as one would expect.

Definition 13. *Let N be a nested attribute, $S \subseteq N^{\downarrow}$ an extended subattribute of N. Then the projection $\pi_S^N : dom(N) \rightarrow dom(S)$ is defined as follows:*

$$\pi_S^N(t) := \bigcup_{s \in S} \{s \mapsto \pi_s^N(t)\}$$

Definition 14. *The* join *of two relations* r_1, r_2 *on the extended subattributes* $S_1, S_2 \subseteq N^{\downarrow}$ *is a relation on* $S_1 \sqcup S_2$:

$$r_1 \bowtie r_2 := \{t \in dom(S_1 \sqcup S_2) \mid \exists t_i \in r_i.\pi_{S_i}^{S_1 \sqcup S_2}(t) = t_i, i = 1, 2\}$$

As in the relational case, we call a decomposition $\mathcal{D} = \{N_1, \ldots, N_n\}$ of (N, Σ) into extended subattributes N_i lossless, if for every relation r on N for which Σ holds we have[1]

$$r[N_1] \bowtie \ldots \bowtie r[N_n] = r$$

We are now ready to investigate how Theorem 1 behaves in the complex-valued model. One direction (the "if" part) is straight forward.

Lemma 7. *Every dependency preserving decomposition of N containing an extended subattribute which forms a key of N is lossless.*

Proof. As in the relational case [3].

Extending Lemma 2 to the complex-valued case turns out to be more challenging. We will do so next.

Definition 15. *Let N be a nested attribute and S a unitary subattribute of N. We call S restricted if $dom(S)$ is finite. A FD $X \to Y$ is LHS-restricted if some element in X is restricted, and a set Σ of FDs over N is LHS-restricted if any of its FDs are.*

To prove Lemma 2 for the complex-valued data model, we need to be able to construct tuples t_i which are identical to some tuple t on an extended subattribute (corresponding to R_i^* in the relational case), but unique (or at least different when domains are finite) for subattributes "outside" these extended subattributes. For this we use and adapt some lemmas from [7,14].

Note that in [7] a set $\mathcal{X} \subseteq Sub(N)$ of subattributes is an *ideal* w.r.t. $\leq$ if and only if the following holds for all $Y \in Sub(N)$:

$$X \in \mathcal{X} \text{ and } Y \leq X \quad \Rightarrow \quad Y \in \mathcal{X}$$

We will adopt this definition of ideal as well.

Lemma 8. *[7, Lemma 21] Let $N \in \mathcal{N}A$, and $\emptyset \neq \mathcal{X} \subseteq Sub(N)$ an ideal with respect to $\leq$ with the property that for reconcilable $X, Y \in \mathcal{X}$ also $X \sqcup Y \in \mathcal{X}$ holds. Then there are $t_N, t_N' \in dom(N)$ with $\pi_W^N(t_N) = \pi_W^N(t_N')$ if and only if $W \in \mathcal{X}$.*

Note that in the last lemma we could also consider only unitary subattributes of N, and thus no longer need to worry about reconcilable subattributes. We will take that view when formulating our main lemma in the following.

Intuitively, Lemma 9 ensures the existence of tuples similar to those used in the proof of Lemmas 1 and 2 for constructing a counter example in the relational case. Recall that $N^{\downarrow}$ denotes the set of all unitary subattributes of N, and that we allow the domains of flat attributes to be finite.

---

[1] We identify N with the extended subattribute $\mathcal{U}(N)$.

Lemma 9. *Let $N \in \mathcal{N}A$, and $\emptyset \neq \mathcal{X}_i \subseteq N^{\downarrow}$ for $i = 1, \ldots, n$ be ideals with respect to $\leq$. Then there are $t, t_1, \ldots, t_n \in dom(N)$ with the following properties (for all $W \in N^{\downarrow}$):*

(i) $\pi_W^N(t_i) = \pi_W^N(t)$ if $W \in \mathcal{X}_i$
(ii) $\pi_W^N(t_i) \neq \pi_W^N(t)$ if $W \notin \mathcal{X}_i$ and W unrestricted or a unit of N
(iii) $\pi_W^N(t_i) \neq \pi_W^N(t_j)$ if $W \notin \mathcal{X}_i \cap \mathcal{X}_j$ and W unrestricted

Proof. We distinguish several cases, depending on the form of N. The numbering of these cases follows the one given in Definition 2. The case (1) where $N = \lambda$ is trivial.

(2) For $N = A$ chose $t = a$ and

$$t_i = \begin{cases} a & \text{if } \mathcal{X}_i = \{\lambda, A\} \\ a_i \neq a & \text{if } \mathcal{X}_i = \{\lambda\} \end{cases}$$

and the a_i pairwise different if $dom(A)$ is infinite. This obviously meets the conditions (i)-(iii).

(3) For $N = (M_1, \ldots, M_k)$ we construct $t, t_1, \ldots, t_n$ inductively on the structure of N. Let $t^{M_j}, t_i^{M_j}$ denote the tuples constructed on the nested attribute M_j w.r.t. $\pi_{M_j}^N(\mathcal{X}_i)^2$. We chose $t = (t^{M_1}, \ldots, t^{M_k})$ and $t_i = (t_i^{M_1}, \ldots, t_i^{M_k})$. It is easy to see that the tuples t, t_i meet conditions (i)-(iii) if the tuples $t^{M_j}, t_i^{M_j}$ do.

(4) For $N = [M]$, we have $\mathcal{X}_i = \{[X] \mid X \in \mathcal{Y}_i\} \cup \{\lambda\}$ for some ideal $\mathcal{Y}_i \subseteq M^{\downarrow}$. If $\mathcal{Y}_i \neq \emptyset$ then by Lemma 8 there exist tuples $t_{M,i}, t'_{M,i} \in dom(M)$ which are identical exactly on $\mathcal{Y}_i$. Otherwise let $t_{M,i}$ be arbitrary. We then construct $t, t_1, \ldots, t_n$ as follows:

$$t = [t_{M,1}, \ldots, t_{M,n}]$$
$$t_i = \begin{cases} [t_{M,1}, \ldots, t'_{M,i}, \ldots, t_{M,n}] & \text{if } \mathcal{Y}_i \neq \emptyset \\ \text{arbitrary of length } n + i & \text{otherwise, i.e., if } \mathcal{X}_i = \{\lambda\} \end{cases}$$

With this definition, tuples t_i with $\mathcal{X}_i = \{\lambda\}$ are of unique length, and thus differ from tuples t, t_j with $j \neq i$ on all subattributes except λ. For other tuples t_i, t_j we have

$$\pi_W^N(t_i) = \pi_W^N(t) \quad \text{if and only if} \quad \pi_W^N(t'_{M,i}) = \pi_W^N(t_{M,i})$$

which shows conditions (i) and (ii), as well as

$$\pi_W^N(t_i) = \pi_W^N(t_j) \quad \text{if and only if} \quad \pi_W^N(t'_{M,i}) = \pi_W^N(t_{M,i}) \wedge \pi_W^N(t'_{M,j}) = \pi_W^N(t_{M,j})$$

from which (iii) follows.

---

[2] Strictly speaking we would have to distinguish between M_j and the subattribute $(\lambda, \ldots, M_j, \ldots, \lambda) \leq N$. We will neglect this subtlety for ease of readability.

(6) For $N = \langle M \rangle$ let $t_{N,i}, t'_{N,i} \in dom(N)$ be tuple pairs with $\pi_W^N(t_{N,i}) = \pi_W^N(t'_{N,i})$ if and only if $W \in \mathcal{X}_i$, which exist by Lemma 8. We then define

$$t = c \cdot t_{N,1} \cup \ldots \cup c^n \cdot t_{N,n}$$
$$t_i = c \cdot t_{N,1} \cup \ldots \cup c^i \cdot t'_{N,i} \ldots \cup c^n \cdot t_{N,n}$$

with $c \in \mathbf{N}$ sufficiently large, i.e., larger than the multiplicity of any element in the multisets $t_{N,1}, \ldots, t_{N,n}, t'_{N,1}, \ldots, t'_{N,n}$. Here $c \cdot t_{N,1}$ denotes the multiset obtained by multiplying the multiplicity of all elements of $t_{N,1}$ with c. This allows us to determine the "origin" of an element e of t or t_i from its multiplicity, i.e. the number of times the element e occurs in the multiset t or t_i. We shall denote this number by $mult_e(t)$ or $mult_e(t_i)$, respectively. For every tuple $t, t_1, \ldots, t_n$ we have

$$mult_e(t/t_i) = mult_e(c \cdot M_1 \cup \ldots \cup c^n \cdot M_n)$$
$$= c \cdot mult_e(M_1) + \ldots + c^n \cdot mult_e(M_n) \qquad (1)$$
$$= c \cdot a_1 + \ldots + c^n \cdot a_n$$

for some multisets $M_1, \ldots, M_n$ and some values $a_1, \ldots, a_n \in \{0, \ldots, c-1\}$. Given a multiplicity $mult_e(t/t_i)$ there exists only one set of values $a_1, \ldots, a_n \in \{0, \ldots, c-1\}$ such that (1) holds. Consequently, we can uniquely determine the multisets $M_1, \ldots, M_n$ given t or some t_i. Conditions (i)-(iii) can now be shown as in the case of lists.

(5) What remains is the case of sets. For $N = \{M\}$ with N restricted, we use that by Lemma 8 there exist two tuples $t_N \neq t'_N$ which are identical on $N^\downarrow \setminus N$. We then set $t = t_N$ and

$$t_i = \begin{cases} t_N & \text{if } \mathcal{X}_i = N^\downarrow \\ t'_N & \text{otherwise} \end{cases}$$

which clearly meet condition (i)-(iii).

For $N = \{M\}$ with N unrestricted, we adapt the proof of Lemma 25 in [7], which shows Lemma 8 for the case of sets. For every index $i = 1, \ldots, n$ we define different identifying terms:

- $\tau_\lambda^i(\lambda) = ok$,
- $\tau_A^i(\lambda) = a, \tau_A^i(A) = a_i$ with $a, a_i \in dom(A), a \neq a_i$ and $a_1, \ldots, a_n$ pairwise different if $dom(A)$ is infinite
- $\tau_{L(N_1, \ldots, N_n)}^i(L(M_1, \ldots, M_n)) = (\tau_{N_1}^i(M_1), \ldots, \tau_{N_n}^i(M_n))$,
- $\tau_{L\{N\}}^i(L\{M\}) = \{\tau_N^i(M)\}$ and $\tau_{L\{N\}}^i(\lambda) = \emptyset$,
- $\tau_{L\langle N \rangle}^i(L\langle M \rangle) = \langle \tau_N^i(M), \ldots, \tau_N^i(M) \rangle$ of cardinality i and $\tau_{L\langle N \rangle}^i(\lambda) = \langle \rangle$,
- $\tau_{L[N]}^i(L[M]) = [\tau_N^i(M), \ldots, \tau_N^i(M)]$ of length i and $\tau_{L[N]}^i(\lambda) = [\,]$.

Note that with this definition the identifying terms $\tau_N^i(M)$ of a subattribute M with infinite domain are pairwise different, i.e., $\tau_N^i(M) \neq \tau_N^j(M)$ for $i \neq j$.

We then create pairs of tuples $t_{N,i}, t'_{N,i} \in dom(N)$ with $\pi_W^N(t_{N,i}) = \pi_W^N(t'_{N,i})$ if and only if $W \in \mathcal{X}_i$ as in [7, Lemma 25], but using the corresponding identifying terms $\tau_N^i(\ldots)$. That is, we have $\mathcal{X} = \{L\{X\} : X \in \mathcal{Y}\} \cup \{\lambda\}$ for some $\mathcal{Y} \subseteq Sub(M)$, and define

$$t_{N,i} = \{\tau_M^i(X) : X \leq M\}$$
$$t'_{N,i} = \{\tau_M^i(X) : X \in \mathcal{Y}\}$$

The tuple pairs $t_{N,i}, t'_{N,i}$ are used to construct the tuples $t, t_1, \ldots, t_k$ as follows:

$$t = t_{N,1} \cup \ldots \cup t_{N,n}$$
$$t_i = t_{N,1} \cup \ldots \cup t_{N,i-1} \cup t'_{N,i} \cup t_{N,i+1} \cup \ldots \cup t_{N,n}$$

Condition (i) clearly holds, since equivalence of $t_{N,i}$ and $t'_{N,i}$ on W implies equivalence of t and t_i on W. Now let $W = \{V\}$ meet the 'if' condition of (ii). By construction we have

$$\tau_M^i(V) \in \pi_W^N(t_{N,i}) \setminus \pi_W^N(t'_{N,i})$$

Furthermore, W must be unrestricted, since N is unrestricted and the only unit of N. As the identifying terms of unrestricted subattributes are all different, $\tau_M^i(V)$ does not lie in any set $\pi_W^N(t_{N,j})$ for $j \neq i$ either. Thus $\pi_W^N(t_i) \neq \pi_W^N(t)$, which shows (ii). Condition (iii) is proven analogous to (ii).

The following example illustrates the construction for multisets and sets.

Example 6. [**Multiset**] Let $N = \langle\{\{A\}\}\rangle$ and

$$\mathcal{X}_1 = \{\lambda, \langle\lambda\rangle, \langle\{\lambda\}\rangle, \langle\{\{\lambda\}\}\rangle\}$$
$$\mathcal{X}_2 = \{\lambda, \langle\lambda\rangle, \langle\{\lambda\}\rangle\}$$
$$\mathcal{X}_3 = \{\lambda, \langle\lambda\rangle\}$$
$$\mathcal{X}_4 = \{\lambda\}$$

Lemma 8 might give us the following tuples:

$$t_{N,1} = \langle\{\{a_1\}\}\rangle, t'_{N,1} = \langle\{\{a_2\}\}\rangle$$
$$t_{N,2} = \langle\{\{a_1\}\}\rangle, t'_{N,2} = \langle\{\emptyset\}\rangle$$
$$t_{N,3} = \langle\{\emptyset\}\rangle, \quad t'_{N,3} = \langle\emptyset\rangle$$
$$t_{N,4} = \langle\rangle, \quad\quad t'_{N,4} = \langle\emptyset\rangle$$

Since all multisets above contain only a single element (or less), it suffices to choose $c = 2$. Using our construction we obtain

$$t = 2 \cdot t_{N,1} \cup 4 \cdot t_{N,2} \cup 8 \cdot t_{N,3} \cup 16 \cdot t_{N,4}$$
$$= 2 \cdot \langle\{\{a_1\}\}\rangle \cup 4 \cdot \langle\{\{a_1\}\}\rangle \cup 8 \cdot \langle\{\emptyset\}\rangle \cup 16 \cdot \langle\rangle$$
$$= \langle\{\{a_1\}\}^6, \{\emptyset\}^8\rangle$$
$$t_1 = \langle\{\{a_2\}\}^2, \{\{a_1\}\}^4, \{\emptyset\}^8\rangle$$
$$t_2 = \langle\{\{a_1\}\}^2, \{\emptyset\}^{12}\rangle$$
$$t_3 = \langle\{\{a_1\}\}^6, \emptyset^8\rangle$$
$$t_4 = \langle\{\{a_1\}\}^6, \{\emptyset\}^8, \emptyset^{16}\rangle$$

where $\langle E^n\rangle$ means that element E occurs n times in the multiset.

[**Set**] Now let $N = \{AB\}$ with $dom(A), dom(B)$ infinite, and

$$\mathcal{X}_1 = \{\lambda, \{\lambda\}, \{A\}, \{B\}\}$$
$$\mathcal{X}_2 = \{\lambda, \{\lambda\}, \{A\}\}$$
$$\mathcal{X}_3 = \{\lambda, \{\lambda\}, \{B\}\}$$

Using our construction for sets we get

$$t_{N,1} = \{(a,b), (a_1,b), (a,b_1), (a_1,b_1)\}, t'_{N,1} = \{(a,b), (a_1,b), (a,b_1)\}$$
$$t_{N,2} = \{(a,b), (a_2,b), (a,b_2), (a_2,b_2)\}, t'_{N,2} = \{(a,b), (a_2,b)\}$$
$$t_{N,3} = \{(a,b), (a_3,b), (a,b_3), (a_3,b_3)\}, t'_{N,3} = \{(a,b), (a,b_3)\}$$

and from this

$$
\begin{aligned}
t\ &= t_{N,1} \cup t_{N,2} \cup t_{N,3}\\
&= \{(a,b), (a_1,b), (a,b_1), (a_1,b_1), (a_2,b), (a,b_2), (a_2,b_2), (a_3,b), (a,b_3), (a_3,b_3)\}\\
t_1 &= \{(a,b), (a_1,b), (a,b_1), \qquad\qquad (a_2,b), (a,b_2), (a_2,b_2), (a_3,b), (a,b_3), (a_3,b_3)\}\\
t_2 &= \{(a,b), (a_1,b), (a,b_1), (a_1,b_1), (a_2,b), \qquad\qquad\qquad (a_3,b), (a,b_3), (a_3,b_3)\}\\
t_3 &= \{(a,b), (a_1,b), (a,b_1), (a_1,b_1), (a_2,b), (a,b_2), (a_2,b_2), \qquad\quad (a,b_3) \qquad\qquad\ \}
\end{aligned}
$$

In both cases (multiset and set) it is easy to check that the t, t_i constructed meet the conditions (i)-(iii) of Lemma 9.

Comparing the last lemma to Lemma 8, one may wonder why we require in condition (ii) that W is unrestricted or a unit, and whether this requirement can be omitted. The following example shows that it is really needed.

Example 7. Let $N = \{AB\}$ and $\mathcal{X}_1 = \mathcal{X}_2 = \{\lambda\}, \mathcal{X}_3 = \{\lambda, \{\lambda\}, \{B\}\}$. Let further t, t_1, t_2 be tuples on N which meet the conditions of Lemma 9. If $t = \emptyset$ then by (i) it follows that $t_3 = \emptyset$, which violates condition (ii) for $W = \{A\}$. So $t \neq \emptyset$. If $t_1 = t_2 = \emptyset$ then condition (iii) is violated for $W = \{A\}$. So $t_1 \neq \emptyset$ or $t_2 \neq \emptyset$, say $t_1 \neq \emptyset$. But then we have

$$\pi^N_{\{\lambda\}}(t_1) = \{ok\} = \pi^N_{\{\lambda\}}(t)$$

which shows that the restriction on W in condition (ii) is necessary.

We are now ready to prove Lemma 2 in the complex-valued model.

Lemma 10. *Let N be a nested attribute with FDs Σ. If Σ is not LHS-restricted, then every lossless decomposition of N contains an extended subattribute which forms a key of N.*

Proof. We will proceed as in the proof of Lemma 1. Let $\mathcal{D} = \{N_1, \ldots, N_n\}$ be a decomposition of N not containing a key schema, i.e., an extended subattribute which forms a key of N. For $N_i \in \mathcal{D}$ define $\mathcal{X}_i := (N_i^*)^{\downarrow}$, where N_i^* is the closure of N_i w.r.t. Σ, and $\mathcal{X}_i$ is the ideal generated by it. Then there exist tuples $t, t_1, \ldots, t_n$ on N which meet the conditions (i)-(iii) of Lemma 9. We define $r = \{t_1, \ldots, t_n\}$, and start by showing that Σ holds on r.

So let $X \to Y \in \Sigma$ and $t_i, t_j \in r$ be two tuples with $\pi_X^N(t_i) = \pi_X^N(t_j)$. Then X is unrestricted by assumption, so from condition (iii) it follows that $X \subseteq \mathcal{X}_i \cap \mathcal{X}_j$. Thus, by definition of $\mathcal{X}_i, \mathcal{X}_j$, we have $Y \subseteq \mathcal{X}_i \cap \mathcal{X}_j$, which gives us $\pi_Y^N(t_i) = \pi_Y^N(t_j)$ by condition (i). It follows that $X \to Y$ holds on r.

Also by condition (i), the tuple t lies in $\bowtie r[\mathcal{D}]$. It remains to be shown that $t \notin r$. By assumption N_i is not a key of N, so $\mathcal{X}_i$ does not contain all units of N. Thus $t_i \neq t$ by condition (ii), which shows that $\mathcal{D}$ is not lossless.

Together with Lemma 7 this gives us an extension of Theorem 1 for complex-valued data bases.

Theorem 2. *Let N be a nested attribute with FDs Σ. If Σ is not LHS-restricted, then a dependency preserving decomposition of N is lossless if and only if it contains an extended subattribute which forms a key of N.*

Proof. Follows from Lemmas 7 and 10.

6   Conclusion

We have shown that in the relational model decompositions can be lossless without containing a key if domains are allowed to be finite. However, we found that this only occurs if attributes with finite domains appear in the left hand side of functional dependencies.

In complex-valued data models containing sets, finite domains occur naturally for subattributes such as $\{\lambda\}$, with $dom(\{\lambda\}) = \{\emptyset, \{ok\}\}$. It can be argued though that subattributes with finite domains rarely occur in the left hand side of functional dependencies. We were able to extend our results from the relational model, showing that when such LHS-restricted functional dependencies do not exist, then every lossless decomposition must contain a key (Lemma 10).

As a result, Theorem 2, which corresponds to Theorem 1 in the relational case, can usually be applied in our complex-valued data model as well. This simplifies the task of finding lossless (and dependency preserving) decompositions, since it is possible to apply methods similar to those used in the relational case, as e.g. in [3,10,18].

We wish to thank the anonymous reviewers of this paper for many helpful comments and suggestions.

References

1. Aho, A.V., Beeri, C., Ullman, J.D.: The theory of joins in relational databases. ACM Trans. Database Syst. 4(3), 297–314 (1979)
2. Arenas, M., Libkin, L.: An information-theoretic approach to normal forms for relational and XML data. In: PoDS, pp. 15–26 (2003)

3. Biskup, J., Dayal, U., Bernstein, P.A.: Synthesizing independent database schemas. In: SIGMOD Conference, pp. 143–151 (1979)
4. Codd, E.: Recent investigations in relational data base systems. In: IFIP Congress, pp. 1017–1021 (1974)
5. Hartmann, S., Link, S.: Deciding implication for functional dependencies in complex-value databases. Theor. Comput. Sci. 364(2), 212–240 (2006)
6. Hartmann, S., Link, S.: The nested list normal form for functional and multivalued dependencies. In: Dix, J., Hegner, S.J. (eds.) FoIKS 2006. LNCS, vol. 3861, pp. 137–158. Springer, Heidelberg (2006)
7. Hartmann, S., Link, S., Schewe, K.-D.: Axiomatisations of functional dependencies in the presence of records, lists, sets and multisets. Theor. Comput. Sci. 355(2), 167–196 (2006)
8. Hegner, S.J.: Characterization of desirable properties of general database decompositions. Ann. Math. Artif. Intell. 7(1–4), 129–195 (1993)
9. Hegner, S.J.: Unique complements and decomposition of database schemata. J. Comput. Syst. Sci. 48(1), 9–57 (1994)
10. Koehler, H.: Finding faithful Boyce-Codd normal form decompositions. In: Cheng, S.-W., Poon, C.K. (eds.) AAIM 2006. LNCS, vol. 4041, pp. 102–113. Springer, Heidelberg (2006)
11. Kolahi, S.: Dependency-preserving normalization of relational and XML data. J. Comput. Syst. Sci. 73(4), 636–647 (2007)
12. Levene, M., Loizou, G.: Semantics for null extended nested relations. ACM Trans. Database Syst. 18(3), 414–459 (1993)
13. Levene, M., Loizou, G.: A Guided Tour of Relational Databases and Beyond. Springer, Heidelberg (1999)
14. Link, S.: Dependencies in Complex-valued Databases. PhD Thesis (2004)
15. Maier, D.: The Theory of Relational Databases. Computer Science Press (1983)
16. Maier, D., Mendelzon, A.O., Sadri, F., Ullman, J.D.: Adequacy of decompositions of relational databases. J. Comput. Syst. Sci. 21(3), 368–379 (1980)
17. Mannila, H., Räihä, K.-J.: The Design of Relational Databases. Addison-Wesley, Reading (1987)
18. Osborn, S.L.: Testing for existence of a covering Boyce-Codd normal form. Information Processing Letters 8(1), 11–14 (1979)
19. Zaniolo, C.: A new normal form for the design of relational database schemata. ACM Trans. Database Syst. 7(3), 489–499 (1982)

SIM-PDT: A Similarity Based Possibilistic Decision Tree Approach

Ilyes Jenhani[1], Nahla Ben Amor[1], Salem Benferhat[2], and Zied Elouedi[1]

[1] LARODEC, Institut Supérieur de Gestion de Tunis, Tunisia
[2] CRIL, Université d'Artois, Lens, France
ilyes.j@lycos.com, nahla.benamor@gmx.fr, benferhat@cril.univ-artois.fr,
zied.elouedi@gmx.fr

Abstract. This paper investigates an extension of classification trees to deal with uncertain information where uncertainty is encoded in possibility theory framework. Class labels in data sets are no longer singletons but are given in the form of possibility distributions. Such situation may occur in many real-world problems and cannot be dealt with standard decision trees. We propose a new method for assessing the impurity of a set of possibility distributions representing instances's classes belonging to a given training partition. The proposed approach takes into account the mean similarity degree of each set of possibility distributions representing a given training partition. The so-called information closeness index is used to evaluate this similarity. Experimental results show good performance on well-known benchmarks.

Keywords: Data mining, Classification, Decision Trees, Possibility Theory, Uncertainty.

1 Introduction

Classification represents an important task in machine learning and data mining applications. It consists in 1) inducing a classifier from a set of historical examples (training set) with known class values and then 2) using the induced classifier to predict the class value (the category) of new objects on the basis of values of their attributes.

Classification tasks are ensured by several approaches such as: discriminant analysis, artificial neural networks, k-nearest neighbors, Bayesian networks and decision trees. The latter, namely, decision trees, is considered as one of the most popular classification techniques. They are able to represent knowledge in a flexible and easy form which justifies their use in decision support systems, intrusion detection systems, medical diagnosis, etc.

In many real-world problems, classes of examples in the training set may be partially defined and even missing. For example, for some instances, an expert may be unable to give the exact class value: A doctor who cannot specify the exact disease of a patient, a banker who cannot decide whether to give or not a

S. Hartmann and G. Kern-Isberner (Eds.): FoIKS 2008, LNCS 4932, pp. 348–364, 2008.

loan for a client, a network administrator who is not able to decide about the exact signature of a given connection, etc.

An interesting real example emphasizing the problem of having imprecise class labels is the one given in [4]. It consists in detecting certain transient phenomena (e.g. k-complexes and delta waves) in electroencephalogram (EEG) data. Such phenomena are usually difficult to detect, hence doctors are not always able to recognize them with full certainty. Consequently, it may be more easy for doctors to assess the possibility that certain phenomena are present in the data.

Hence, in these different examples, the expert can provide imprecise or uncertain classifications expressed in the form of a ranking on the possible classes. Obviously, rejecting these pieces of information in a learning process is not a good practice. A suitable theory dealing with such situations is possibility theory which is a non-classical theory of uncertainty proposed by [23] and developed by [5].

In order to deal with such uncertain information, standard classification techniques, such as decision trees [17], should be adequately adapted to such data representation (training instances whose class labels are given in the form of possibility distributions). Indeed, ignoring this uncertainty may affect classification results and even produce erroneous decisions. Some approaches have dealt with the induction of decision trees using other uncertainty formalisms: 1) from instances with vaguely defined linguistic attributes and classes [13,22] (fuzzy decision trees) and 2) from data with partially defined classes presented in the form of basic belief assignments (belief decision trees) [3,6].

In this paper, we will present an extension of a decision tree approach called Similarity based Possibilistic Decision Tree (Sim-PDT). By extension, we mean that if all class labels are precisely described, then our approach recovers standard decision trees.

It is important to mention that existing possibilistic decision trees do not deal with uncertainty in classes. For instance, the work proposed by Borgelt and al. [2] deals with crisp (standard) training sets. The authors encode the frequency distributions as possibility distributions (an interpretation which is based on the context model of possibility theory [2]) in order to define a possibilistic attribute selection measure. The possibilistic decision tree approach proposed by Hüllermeier [9] uses a possibilistic branching within the lazy decision tree technique. Again, this work does not deal with any uncertainty in the classes of the training objects.

Our approach allows the induction of decision trees from training instances whose class labels are given in the form of possibility distributions. We adapted the classical attribute selection measure, namely, the gain ratio criterion to the possibilistic setting by introducing the mean similarity measure of the possibility distributions belonging to each training partition when evaluating its entropy.

The rest of the paper is organized as follows: Section 2 describes some basics of the decision tree classification technique. Section 3 gives the necessary background concerning possibility theory. Section 4 describes our proposed similarity based possibilistic decision tree approach (Sim-PDT). Section 5 presents and analyzes experimental results carried out on modified versions of commonly used data sets from the U.C.I. repository [15]. Finally, Section 6 concludes the paper.

2 Decision Trees

A decision tree is a flow-chart-like hierarchical tree structure which is composed of three basic elements: decision nodes corresponding to attributes, edges or branches which correspond to the different possible attribute values. The third component consists of leaves including objects that typically belong to the same class or that are very similar. Such representation allows us to induce decision rules that will be used to classify new instances.

The majority of decision trees is made up of two major procedures, namely, the building (induction) and the classification (inference) procedures:

- **Building procedure:** Given a training set, building a decision tree is usually done by starting with an empty tree and selecting for each decision node the 'appropriate' test attribute using an attribute selection measure. The principle is to select the attribute that maximally diminish the mixture of classes between each training subset created by the test, thus, making easier the determination of object's classes. The process continues for each sub decision tree until reaching leaves and fixing their corresponding classes.

- **Classification procedure:** To classify a new instance, having only values of all its attributes, we start with the root of the constructed tree and follow the path corresponding to the observed value of the attribute in the interior node of the tree. This process is continued until a leaf is encountered. Finally, we use the associated label to obtain the predicted class value of the instance at hand.

Several algorithms for building decision trees have been developed. The most popular and applied ones are: **ID3** [16] and its successor **C4.5** "the state-of-the-art" algorithm developed by Quinlan [17]. These algorithms have many components to be defined:

a) **Attribute selection measure** generally based on information theory, serves as a criterion in choosing among a list of candidate attributes at each decision node, the attribute that generates partitions where objects are distributed less randomly, with the aim of constructing the smallest tree among those consistent with the data. The well-known measure used in the **C4.5** algorithm of Quinlan [17] is the gain ratio.

Given an attribute A_k, the information gain relative to A_k is defined as follows:

$$Gain(T, A_k) = E(T) - E_{A_k}(T) \tag{1}$$

where

$$E(T) = -\sum_{i=1}^{n} \frac{n(C_i, T)}{|T|} \ log_2 \ \frac{n(C_i, T)}{|T|} \tag{2}$$

and

$$E_{A_k}(T) = \sum_{v \in D(A_k)} \frac{|T_v^{A_k}|}{|T|} E(T_v^{A_k}) \tag{3}$$

$n(C_i, T)$ denotes the number of objects in the training set T belonging to the class C_i, $D(A_k)$ denotes the finite domain of the attribute A_k and $|T_v^{A_k}|$ denotes the cardinality of the set of objects for which the attribute A_k has the value v. Note that $\frac{n(C_i, T)}{|T|}$ corresponds to the probability of the class C_i in T. Thus, $E(T)$ corresponds to the *Shannon entropy* [19] of the set T. The gain ratio is given by:

$$Gr(T, A_k) = \frac{Gain(T, A_k)}{SplitInfo(T, A_k)} \tag{4}$$

where $SplitInfo(T, A_k)$ represents the potential information generated by dividing T into n subsets. It is given by:

$$SplitInfo(T, A_k) = - \sum_{v \in D(A_k)} \frac{|T_v^{A_k}|}{|T|} log_2 \frac{|T_v^{A_k}|}{|T|} \tag{5}$$

b) **Partitioning strategy** consisting in partitioning the training set according to all possible attribute values (for symbolic attributes) which leads to the generation of one partition for each possible value of the selected attribute. For continuous attributes, a discretization step is needed.

c) **Stopping criteria** stopping the partitioning process. Generally, we stop the partitioning if all the remaining objects belong to only one class, then the node is declared as a leaf labeled with this class value. We, also, stop growing the tree if there is no further attribute to test. In this case, we take the majority class as the leaf's label.

3   Possibility Theory

Possibility distribution

Given a universe of discourse $\Omega = \{\omega_1, \omega_2, ..., \omega_n\}$, a fundamental concept of possibility theory is the *possibility distribution* denoted by π. π corresponds to a function which associates to each element ω_i from the universe of discourse Ω a value from a bounded and linearly ordered valuation set $(L, <)$. This value is called a *possibility degree*: it encodes our knowledge on the real world. Note that, in possibility theory, the scale can be numerical (e.g. L=[0,1]): in this case we have numerical possibility degrees from the interval [0,1] and hence we are dealing with the quantitative setting of the theory.

In the qualitative setting, it is the ordering between the different possible values that is important. By convention, $\pi(\omega_i) = 1$ means that it is fully possible that ω_i is the real world, $\pi(\omega_i) = 0$ means that ω_i cannot be the real world (is impossible). Flexibility is modeled by allowing to give a possibility degree from $]0,1[$.

In possibility theory, extreme cases of knowledge are given by:

- *Complete knowledge:* $\exists \omega_i,\ \pi(\omega_i) = 1$ *and* $\forall\ \omega_j \neq \omega_i,\ \pi(\omega_j) = 0$.
- *Total ignorance:* $\forall\ \omega_i \in \Omega,\ \pi(\omega_i) = 1$ (all values in Ω are possible).

Possibility and Necessity measures

A Possibility measure is one of the fundamental concepts in possibility theory. From a possibility distribution, two dual measures can be derived: *Possibility* and *Necessity* measures. Given a possibility distribution π on the universe of discourse Ω, the corresponding possibility and necessity measures of any event $A \subseteq 2^\Omega$ are, respectively, determined by the formulas:

$$\Pi(A) = \max_{\omega \in A} \pi(\omega)$$

$$N(A) = \min_{\omega \notin A} (1 - \pi(\omega)) = 1 - \Pi(\overline{A})$$

$\Pi(A)$ evaluates at which level A is *consistent* with our knowledge represented by π while $N(A)$ evaluates at which level A is *certainly* implied by our knowledge represented by π.

Normalization

A possibility distribution π is said to be *normalized* if there exists at least one state $\omega_i \in \Omega$ which is totally possible (i.e. $\max_{\omega \in \Omega}\{\pi(\omega)\} = \pi(\omega_i)=1$). In the case of sub-normalized π,

$$Inc(\pi) = 1 - \max_{\omega \in \Omega}\{\pi(\omega)\} \tag{6}$$

is called the *inconsistency degree* of π. It is clear that, for normalized π, $\max_{\omega \in \Omega}\{\pi(\omega)\} = 1$, hence $Inc(\pi)=0$. The measure Inc is very useful in assessing the degree of conflict between two distributions π_1 and π_2 which is given by $Inc(\pi_1 \wedge \pi_2)$. We take the $\wedge$ as the minimum operator. Obviously, when $\pi_1 \wedge \pi_2$ gives a sub-normalized possibility distribution, it indicates that there is a conflict between π_1 and π_2 ($Inc(\pi_1 \wedge \pi_2) \in]0,1]$).

Non-specificity

An important concept that allows to compare possibility distributions is the principle of *minimum specificity*. A possibility distribution π_1 is said to be *more specific than* π_2 if and only if for each state of affairs $\omega_i \in \Omega$, $\pi_1(\omega_i) \leq \pi_2(\omega_i)$ [21]. Clearly, the more specific π, the more informative it is. The degree of information uncertainty of a possibility distribution is called non-specificity and it can be measured by the so-called *U-uncertainty* criterion [7].

Given a permutation of the degrees of a possibility distribution $\pi = \langle \pi_{(1)}, \pi_{(2)}, ..., \pi_{(n)} \rangle$ such that $\pi_{(1)} \geq \pi_{(2)} \geq ... \geq \pi_{(n)}$, the *U-uncertainty* of π, is given by the formula:

$$U(\pi) = \sum_{i=2}^{n}(\pi_{(i)} - \pi_{(i+1)})\ log_2\ i\ +\ (1 - \pi_{(1)})\ log_2\ n \tag{7}$$

where $\pi_{(n+1)} = 0$ by convention. Note that the range of U is $[0, log_2 n]$. $U(\pi) = 0$ is obtained for the case of complete knowledge (no uncertainty) and $U(\pi) = log_2 n$ is reached for instance in the case of total ignorance. Note also that the second term of the equation, i.e., $(1 - \pi_{(1)}) \, log_2 n$ generalizes U for sub-normalized π.

Information closeness: a possibilistic similarity measure
Comparing pieces of uncertain information given by several sources has attracted a lot of attention for a long time. This could be ensured by the use of similarity indexes. Few works have been done in this direction in the possibilistic framework. A deep study of existing similarity measures in possibility theory [8,11,18], shows that the *information closeness index* [8] is a well-established measure.

The information closeness index, denoted by G is an information variation based measure: function G is computed using the U-uncertainty measure [7] (Equation (7)) and it is applicable to any pair of normalized possibility distributions.

Definition 1. *Let π_1 and π_2 be two possibility distributions on the same universe of discourse Ω. The information closeness G between π_1 and π_2 is defined as:*

$$G(\pi_1, \pi_2) = g(\pi_1, \pi_1 \vee \pi_2) + g(\pi_2, \pi_1 \vee \pi_2) \tag{8}$$

where $g(\pi_i, \pi_j) = U(\pi_j) - U(\pi_i)$. $\vee$ is taken as the maximum operator and U is the non-specificity measure given by Equation (7). Consequently, function G can be written as:

$$G(\pi_1, \pi_2) = 2 * U(\pi_1 \vee \pi_2) - U(\pi_1) - U(\pi_2)$$

The less the value of G is, the more the information are similar (G behaves as a distance measure).

For sake of simplicity, in the rest of the paper, a possibility distribution π on a finite set $\Omega = \{\omega_1, \omega_2, ..., \omega_n\}$ will be denoted by $\pi[\pi(\omega_1), \pi(\omega_2), ..., \pi(\omega_n)]$.

Example 1. *Consider the following distributions π_1, π_2, π_3 and π_4 over $\Omega = \{\omega_1, \omega_2, \omega_3, \omega_4\}$: $\pi_1[1, \ 0.5, \ 0.3, \ 0.7]$, $\pi_2[1, 0, 0, 0]$, $\pi_3[0.9, 1, 0.3, 0.7]$, $\pi_4[0, 1, 0.3, 0.7]$. Let us try to find an order expressing which from the information given by π_2, π_3 and π_4 is closer to π_1. $G(\pi_1, \pi_2) = 1.12$, $G(\pi_1, \pi_3) = 0.52$, $G(\pi_1, \pi_4) = 1.08$. According to G, π_3 is the closest to π_1 and π_4 is closer to π_1 than π_2.*

4   Similarity Based Possibilistic Decision Trees

A similarity based possibilistic decision tree (Sim-PDT) is a decision tree with the same representation of an standard decision tree, i.e., it is composed of *decision nodes* for testing attributes, *branches* specifying attribute values and *leaves* dealing with classes of the training set.

In supervised learning, more specifically, in classification problems, we need a set of historical examples with known classes, called the training set, from which we will train a classifier (e.g. a decision tree). Then, this classifier will be used to predict the class value of each new object given known its attributes' values.

4.1 Imperfection in Classification Problems

As models of the real world, databases, or more specifically, training sets are often permeated with forms of imperfections, including imprecision and uncertainty. The topic of imperfect databases is gaining more and more attention the last years [12,14] since commercial database management systems are not able to deal with such kind of information. Now, we ask what is imperfect in a training set and why is it imperfect?

Imperfection in a training set may affect attribute values as well as class values, for instance, the *departure_time* of a flight, the *temperature* of a patient, the *property_value* of a client asking for a loan. Examples of imperfect class values include the exact type of an attack in an intrusion detection system, the exact cancer class of a patient in cancer diagnosis applications, the exact location or type of a detected aerial engine in military applications, etc.

These imperfections might result from using unreliable information sources, such as faulty reading instruments, or input forms that have been filled out incorrectly (intentionally or inadvertently). In other cases, imperfection is a result of system errors, including transmission noise, network latency for sensor networks applications, delays in processing update transactions, etc.

In a learning process, it is not appropriate to reject or ignore such information (by affecting the *null* value to such information) despite of its imperfection. On the contrary, we should benefit from the maximum amount of information which should be handled carefully else the learnt model could be inaccurate or even incorrect.

In this work, we only deal with imprecise class labels in the training set. Instead of rejecting instances having imprecise class labels or adding a *null* class value to such instances, we used a convenient mathematical model to deal with such kind of imperfection, namely possibility theory [5,23].

More formally, a possibility degree will be assigned to each possible class value indicating the possibility that the instance belongs to a given class [4,24]. These possibility degrees can be obtained from direct expert's elicitation, i.e., each expert is asked to quantify by a real number between 0 and 1 the possibility that a training instance belongs to each one of the different classes of the problem. The question that arises is: how to induce decision trees from training instances, classes of which are presented by means of possibility distributions?

To address this question, we propose to develop a new classifier, having the same structure as standard decision trees with the ability to treat uncertain classes. This new approach, named *similarity based possibilistic decision tree* and denoted by Sim-PDT is also based on a building and a classification procedure detailed in what follows.

4.2 Building Procedure

The main component of the building procedure of most decision tree algorithms is the *attribute selection measure*. Recall that, for each given node of the tree under construction, this measure tells us about the discriminative power of each attribute when partitioning the training set according to its different values. The attribute leading to more homogeneous partitions (having less randomly distributed classes) will be selected to be assigned to the node under study.

In the possibilistic setting, instances classes in the training set will be represented by possibility distributions over the different classes of the problem instead of exact classes. Consequently, each training partition will be characterized by a set of possibility distributions corresponding to the labels of the instances belonging to that partition. Hence, one must find a way for assessing homogeneity of such partitions. The idea consists in measuring the entropy of each partition weighted by the mean similarity degree of the possibility distributions in the corresponding partition. Let us define the basic components for the Sim-PDT approach:

a) Attribute selection measure

Given a training set T (the initial partition) containing n instances and given the set of attributes, let us denote by π_i the possibility distribution labeling the class of the instance i in T, we must find a way to assess the homogeneity of T, or say the homogeneity of the possibility distributions in T.

In standard decision trees, homogeneity of a partition is determined by the entropy of that partition. However, in our context, π_i's are most of the time very different, so it has no sense to directly compute their frequencies in order to determine the entropy of T (the entropy will be equal to 1). Moreover, one cannot simply view each π_i as a new class. First, because the number of classes will be exponential. Second, there are similar distributions that should be considered as globally expressing same or similar pieces of information. For instance, we cannot simply consider the distributions [1, 0.2] and [1, 0.21] as two different exclusive classes, but we will consider them as similar.

Hence, we need a finite set of Meta-Classes $MC_{j=1..m}$. Each MC_j corresponds to a meta-class which gathers together all possibility distributions similar to a predefined wrapper possibility distribution (say WD_j). More precisely, wrapper possibility distributions are binary possibility distributions (i.e., $\forall \omega \in \Omega, \pi(\omega) \in \{0, 1\}$) representing special cases of complete knowledge, partial ignorance and total ignorance representing the set of reference distributions.

Suppose that we have a problem with r classes (i.e., $\Omega = C = \{C_1, C_2, ..., C_r\}$). We define a degree of imprecision $dimp \in \{1, ..., r\}$ which will allow to determine the set of WD_j to be used. For instance, for a problem with three classes (r=3):

- If $dimp = 1$, we will only consider wrapper possibility distribution corresponding to complete knowledge (precise classes), i.e., $WD = \{[1, 0, 0], [0, 1, 0], [0, 0, 1]\}$.
  → $MC = \{MC_1, MC_2, MC_3\}$.

– If $dimp = 2$, we will also consider partial ignorance, i.e., $WD = \{[1, 0, 0],$
$[0, 1, 0], [0, 0, 1], [1, 1, 0], [1, 0, 1], [0, 1, 1]\}$.
→ $MC = \{MC_1, MC_2, MC_3, MC_4, MC_5, MC_6\}$.

– If $dimp = 3$, we will also consider total ignorance, i.e., $WD = \{[1, 0, 0], [0, 1, 0],$
$[0, 0, 1], [1, 1, 0], [1, 0, 1], [0, 1, 1], [1, 1, 1]\}$.
→ $MC = \{MC_1, MC_2, MC_3, MC_4, MC_5, MC_6, MC_7\}$.

For the rest of the paper, we will consider the general case ($dimp = r$). After specifying the set of Meta-Classes MC, we will assign to each possibility distribution π_i labeling an instance i a meta-class MC_j such that: $MC_j = \arg\min_{j=1}^{m}\{G(\pi_i, WD_j)\}$ where m is the total number of meta-classes and G corresponds to the information closeness index (Equation (8)). Note that, as in standard decision trees, ties are broken arbitrarily.

After mapping the different π_i's to their corresponding MC_j's, it becomes possible to assess the discriminative power of each attribute in partitioning a set into homogeneous subsets by extending the well-known gain ratio criterion [17]. First, we define the Similarity-Entropy Gain ($SGain$) of an attribute A_k by:

$$SGain(T, A_k) = SE(T) - SE_{A_k}(T) \qquad (9)$$

where

$$SE(T) = -\sum_{j=1}^{m}(1 + AvgG(MC_j)) * (\frac{|MC_j|}{|T|} log_2 \frac{|MC_j|}{|T|}) \qquad (10)$$

and

$$SE_{A_k}(T) = \sum_{v \in D(A_k)} \frac{|T_v^{A_k}|}{|T|} SE(T_v^{A_k}) \qquad (11)$$

where $|MC_j|$ in Equation (10) denotes the number of objects in the training set T belonging to the meta-class MC_j. Obviously, to compensate for the information loss resulting from grouping resemblant π_i's into their corresponding MC_j's, we have introduced the $AvgG(MC_j)$ factor which corresponds to the average distance between the original possibility distributions $\pi_{p=1..n}$ assigned to MC_j:

$$AvgG(MC_j) = \frac{\sum_{p=1}^{n-1} \sum_{q=p+1}^{n} G(\pi_p, \pi_q)}{\frac{n*(n-1)}{2}} \qquad (12)$$

Proposition 1. *When dealing with crisp training sets, i.e., with precise classes ($MC_j \equiv C_j$), we will always have $1 + AvgG(MC_j) = 1$ and $|MC_j|$ will correspond to number of instances labeled by the same class C_j, thus we recover the standard C4.5 approach.*

Second, the similarity-gain ratio is expressed in the same way as the classical gain ratio:

$$SGr(T, A_k) = \frac{SGain(T, A_k)}{SplitInfo(T, A_k)} \qquad (13)$$

where $SplitInfo(T, A_k) = -\sum_{v \in D(A_k)} \frac{|T_v^{A_k}|}{|T|} log_2 \frac{|T_v^{A_k}|}{|T|}.$

Obviously, the attribute maximizing SGr will be assigned to the decision node at hand.

Example 2. *Let us use a modified version of the golf data set [15] to illustrate the notion of wrapper distributions. Let T be the training set composed of fourteen instances $i_{=1..14}$ which are characterized by four attributes: **Outlook**, **Temp**, **Humidity** and **Wind**. Two classes are possible either, C_1 (play) or C_2 (don't play). A possibility distribution was given for each class of each instance of T. The training set T is given by Table 1.*

Table 1. Training set with imprecise class labels

	Outlook	Temp	Humidity	Wind	C_1	C_2
i_1	sunny	hot	high	weak	0.2	1
i_2	sunny	hot	high	strong	0.4	1
i_3	overcast	hot	high	weak	1	0.7
i_4	rainy	mild	high	weak	1	0
i_5	rainy	cool	normal	weak	1	0.8
i_6	rainy	cool	normal	strong	0.4	1
i_7	overcast	cool	normal	strong	1	0.9
i_8	sunny	mild	high	weak	0.3	1
i_9	sunny	cool	normal	weak	1	0.3
i_{10}	rainy	mild	normal	weak	1	0
i_{11}	sunny	mild	normal	strong	1	0.2
i_{12}	overcast	mild	high	strong	1	0
i_{13}	overcast	hot	normal	weak	1	0.3
i_{14}	rainy	mild	high	strong	0	1

The set of wrapper distributions relative to this example is $WD = \{[1,0], [0,1], [1,1]\}$. Consequently, $MC = \{MC_1, MC_2, MC_3\}$ such that $MC_1 = \{i_4, i_9, i_{10}, i_{11}, i_{12}, i_{13}\}$, $MC_2 = \{i_1, i_2, i_6, i_8, i_{14}\}$ and $MC_3 = \{i_3, i_5, i_7, \}$. The Similarity-Entropy of the set T is computed (using Equation (10)) as follows:

$$SE(T) = -1.173 * (\tfrac{6}{14} * log_2 \tfrac{6}{14}) - 1.2 * (\tfrac{5}{14} * log_2 \tfrac{5}{14}) - 1.133 * (\tfrac{3}{14} * log_2 \tfrac{3}{14}) = 1.791.$$

b) Partitioning strategy

Once an attribute is selected at a given decision node and since we only deal with nominal attributes, the partitioning strategy will be the same as with standard decision trees (see Section 2).

c) Stopping criteria

For the Sim-PDT approach, as for standard decision trees, we present the following cases for which we should stop the partitioning process for each generated training partition T_p and hence stop growing the tree:

1. There is no further attribute to test.
2. $SGain \leq 0$, i.e., no information is gained. In this case, continuing splitting produces less similar possibility distributions within the obtained partitions.
3. $|T_p|=0$, i.e., the generated partition does not contain any instance. In this case, we declare an empty leaf labeled by a randomly chosen wrapper possibility distribution from WD.

When stopping criterion 1 or 2 is satisfied for a training partition T_p containing n possibility distributions, we will declare a leaf labeled by the representative possibility distribution of that set (π_{Rep}), that is, the possibility distribution which corresponds to the closest distribution to all the remaining distributions in the set T_p:

$$\pi_{Rep} = \arg\min_{i=1}^{n}\{\frac{\sum_{j\neq i} G(\pi_i, \pi_j)}{(n-1)}\} \tag{14}$$

We have chosen to label each leaf of the tree by the representative possibility distribution and not by the corresponding wrapper distribution because using this latter will result in some information loss (loss=$U(\pi_{Rep})$-$U(WD_{\pi_{Rep}})$).

Note that in our context, fusion is not an appropriate tool to combine the possibility distributions of a training partition T_p. In fact, in the decision tree context, in each node, we have possibility distributions of distinct training instances reaching that node. These instances have some common attribute values (those values labeling edges of the path leading to that node) and the remaining attributes may have different values. So, it is clear that we cannot merge possibility distributions which are not dealing with the same "object": a necessary condition for information fusion problems.

Example 3. *Let us continue with Example 2 to illustrate the induction of a Sim-PDT:*

We should first compute the Similarity-Entropy of the set T:

$SE(T) = 1.791$ (see Example 2).

In this example, we will show a detailed computation of the similarity-gain ratio of only one attribute, namely, the "Humidity" attribute. So, let us compute $SE(T_{high}^{Humidity})$ and $SE(T_{normal}^{Humidity})$ using Equation (10):

$SE(T_{high}^{Humidity}) = -(1+0)*(\frac{1}{7}*log_2\frac{1}{7}) - (1+0.216)*(\frac{4}{7}*log_2\frac{4}{7}) - (1+0)* (\frac{2}{7}*log_2\frac{2}{7}) = 1.478.$

$SE(T_{normal}^{Humidity}) = -(1+0.166)*(\frac{4}{7}*log_2\frac{4}{7}) - (1+0.1)*(\frac{2}{7}*log_2\frac{2}{7}) - (1+ 0)*(\frac{1}{7}*log_2\frac{1}{7}) = 1.507.$

⇒ *Using Equation (11), we obtain: $SE_{Humidity}(T) = \frac{7}{14}*1.478 + \frac{7}{14}*1.507 = 1.492$.*

⇒ *Using Equation (9): $SGain(T, Humidity) = 1.791 - 1.492 = 0.299$.*

⇒ *Using Equation (5): $SplitInfo(T, Humidity) = -\frac{7}{14}*log_2\frac{7}{14} - \frac{7}{14}*log_2\frac{7}{14} = 1$.*

⇒ *Finally, using Equation (13): $SGr(T, Humidity) = \frac{0.299}{1} = 0.299$.*

Similarly, we obtain:

$SGr(T, Outlook) = 0.271$, $SGr(T, Temp) = 0.265$ and $SGr(T, Wind) = 0.093$

Hence, the attribute that will be assigned to the root node will be "Humidity" since it has the highest similarity-gain ratio among all the attributes.
We get the following Sim-PDT tree:

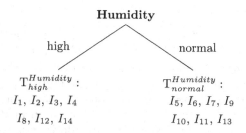

Fig. 1. First generated Sim-PDT tree

For the training subsets $T_{high}^{Humidity}$ and $T_{normal}^{Humidity}$, we apply the same process as we did for the training set T until one of the stopping criteria holds.
The final Sim-PDT tree induced by our algorithm is given by Fig. 2.

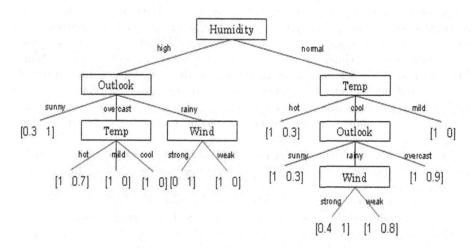

Fig. 2. Final Sim-PDT tree

Note that our approach also deals with what is known as ambiguous label classification (ALC) problems in which a simple set of disjunctive class values (with equal weights) are assigned to each instance [10]. In fact, a set S of disjunctive class values can be mapped into a binary possibility distribution assigning the degree 1 to the elements (class labels) belonging to S and the degree 0 to the remaining elements (that do not belong to S). Hence, we obtain a training set in which instances are labeled by special possibility distributions corresponding to partial ignorance, i.e., some classes appear as fully possible (with degree 1) and the remaining appear as impossible (with degree 0). Obviously, this can be dealt with the Sim-PDT approach which handles all kinds of possibility distributions (ranging from complete knowledge to total ignorance).

4.3 Classification Procedure

Once the Sim-PDT is constructed, we can classify any new object given its attributes values (see Section 2). As mentioned above, each leaf of our decision tree will be labeled by a possibility distribution over the different class values. Hence, to make a decision about the class of a given object, the decision maker can take the fully possible class label (i.e. the class having a possibility degree equal to 1). Moreover, in cases where there may be unequal predefined costs depending on several classes in classification, the decision maker could opt for a cost-sensitive classification. In our case, we take the more plausible class as the final decision.

5   Experimental Results

The purpose of our experimental study is to show that exploiting uncertain data for decision tree induction by using the proposed Sim-PDT approach is usually better than the obvious alternative, namely to ignore such data and learn with a standard decision tree algorithm from the remaining (exactly labeled) examples.

We split our experimental studies in two steps. First, we evaluate our Sim-PDT approach. Then, we compare our results with those of the C4.5 algorithm if we ignored uncertainty. Please note that we do not intend to compare Sim-PDT with C4.5 since this latter do not deal with uncertainty: the aim of the comparison is to show whether ignoring uncertainty in training data is a good practice or not.

For the evaluation of the similarity based possibilistic decision tree approach, we have developed programs using Matlab 6.5 implementing both of the building and classification procedures. The reported results of the C4.5 algorithm are obtained by using the well-known Weka data mining tool [20], especially, the J48 classifier without pruning since our approach does currently not involve a pruning step.

The experimental study is based on several data sets selected from the U.C.I repository [15]: (1) Wisconsin Breast cancer (699 instances, 8 attributes, 2 classes), (2) Voting (497, 16, 2), (3) Balance scale (625,4,3), (4) Solar Flare (1389, 10, 3), (5) Nursery (12960, 8, 5).

We have modified these data sets by transforming the original crisp classes by possibility distributions over the different classes. We used levels of uncertainty ($L\%$) when generating these possibilistic training sets. More precisely, for each training instance from the $L\%$ randomly chosen instances, we have assigned a possibility degree equal to 1 to the original class and a random possibility degree to the remainders in an uniform way. To each one of the remaining $(100 - L)\%$ instances of the original training set, we have assigned a completely sure possibility distribution corresponding to the original crisp instance's class. For our experiments, we have varied L from 0 (crisp training set) to 50 (half of the training instances has an uncertain class label).

In order to determine the accuracy of the induced trees, we have used two criteria, the first is relative to the percentage of correct classification:

$$PCC = \frac{number\ of\ well\ classified\ instances}{total\ number\ of\ classified\ instances} \times 100)$$

and the second corresponds to a similarity based criterion (Sim_crit) which we propose as a new criterion that is more appropriate to the possibilistic context:

$$Sim\_crit = [1 - (\frac{1}{NV} * \frac{\sum_{j=1}^{n} G(\pi_j^{res}, \pi_j)}{n})] * 100 \qquad (15)$$

Recall that within our possibilistic decision tree approach, the classification result is given in the form of a possibility distribution (π^{res}). Thus, the idea is to choose for each instance to classify the class having the highest possibility degree (equal to 1). If more than one class is obtained, then one of them is chosen randomly. The obtained class is considered as the class of the testing instance. Consequently, $number\ of\ well\ classified\ instances$ in PCC corresponds to the number of testing instances for which the resulting class obtained by the Sim-PDT approach is the same as their real more plausible class in the training set.

The limitation of this adaptation of the PCC criterion to the possibilistic setting, is that it chooses randomly one of the more plausible classes which may miss-classify some instances. Moreover, even when there is only one more plausible class, focusing on that class and ignoring the rest of the classes (classes with possibility degrees different from 1) is problematic. In fact, ignoring the rest of the degrees implies ignoring a part of the information given by the resulting possibility distribution (π^{res}).

Hence, we were inspired by the work in [1] to define the Sim_crit criterion which takes into account the mean similarity relative to all the classified testing instances: the average of the similarities between the resulting possibility distribution (π_j^{res}) and the real (completely sure) possibility distribution (π_j) of each classified instance j. Note that the value NV in Equation (15) ($NV = 2 * log2(r) - log2(r - 1)$) is used to obtain the required range $[0,1]$ where r stands for the number of classes of the problem. When Sim_crit is close to 100%, the classifier is good whereas when it falls to 0%, it is considered as a bad classifier.

Table 2 reports the different obtained results after varying the training sets' level of uncertainty $L\%$ from 0% to 50% for each database. Sim_crit values of the induced Sim-PDT trees are complemented by standard deviations after the use of a 10-fold cross validation testing process.

Table 2. Sim-PDT: Sim_crit and standard deviation

$L\%$	0%	10%	20%	30%	40%	50%
W.B.cancer	95.46(0.9)	94.56(1.2)	94.33(1.3)	93.29(2.1)	92.42(2.5)	91.64(2.8)
Voting	97.98 (1.8)	96.77(2.1)	96.29(1.9)	95.65(1.2)	95.16(1.2)	93.31(1.4)
Solar Flare	85.55(2.4)	85.28(2.2)	84.28(1.8)	83.70(1.7)	82.96(1.5)	82.95(1.5)
Balance	84.86 (1.4)	79.57(1.6)	76.93(2.3)	75.60 (1.2)	74.28(1.2)	73.94(1.1)
Nursery	98.96 (0.7)	97.80(1.1)	97.26(1.4)	97.03(1.3)	96.86(1.5)	96.05(1.6)

Note that high values of the *Sim_crit* criterion do not only imply that the induced trees are accurate but also imply that the possibility distributions provided by the induced Sim-PDT trees are of high quality and faithful to the original possibility distributions. From Table 2, we can see that *Sim_crit* values decrease when $L\%$ increases. This can be explained by the fact that the higher the level of uncertainty ($L\%$), the less informative the training set becomes (consequently, the harder the learning becomes), and therefore the less accurate the predictions are.

Now, let us see what happens when ignoring imprecisely labeled training instances when building decision trees. To respond to this question, we have conducted our experimentations as follows: for each training set and for each uncertainty level L (from 0% to 50%), we have induced a Sim-PDT tree. On the other hand, a C4.5 tree was induced from the corresponding training set, i.e., the standard training set from which we have discarded the $L\%$ instances to which we have assigned imprecise class labels since the C4.5 algorithm can not deal with such instances.

Then, both approaches are evaluated on the same testing sets: standard testing sets for C4.5 trees have been used and their corresponding testing sets (with completely sure possibility distributions on the original class labels) for Sim-PDT trees: this corresponds to one iteration of the 10-fold cross validation process used for the evaluation of the approach.

Table 3 reports the different obtained results after varying the training sets' level of uncertainty $L\%$ from 0% to 50% for each database. $MPCC$ denotes the mean PCC (complemented by standard deviation) of the induced decision trees (for both Sim-PDT and C4.5 approaches) for the 10-fold cross validation process.

Table 3 shows that the Sim-PDT approach gives interesting results when compared with the C4.5 algorithm. Again, we can see that classification accuracies of both approaches decrease when the level of uncertainty increases (for the same explanation provided above for Table 2).

Interestingly, in spite of this decrease in accuracy, we can see that the classification rate of Sim-PDT is always (even slightly) greater than the one of C4.5. Note that the aim of this comparison is not to directly compare the two approaches. In fact, the C4.5 is used only in certain environments: it is trained from reduced training sets (imprecisely labeled instances are omitted) while the Sim-PDT approach deals with both certain and uncertain environments: it is trained from complete training sets (including both precisely and imprecisely

Table 3. Results for C4.5 and Sim-PDT (MPCC and standard deviation)

Database	Method	$L = 0\%$	$L = 10\%$	$L = 20\%$	$L = 30\%$	$L = 40\%$	$L = 50\%$
W.B.cancer	C4.5	94.54(1.1)	93.86(1.4)	91.63(2.3)	91.05(2.5)	90.49(2.8)	90.11(3.2)
	Sim-PDT	94.54(1.1)	94.12(1.2)	93.85(2.1)	92.79(2.2)	92.53(2.4)	91.98(2.9)
Voting	C4.5	94.56(3.2)	93.42(3.2)	92.23(3.5)	90.15(3.8)	89.59(4.3)	87.27(4.6)
	Sim-PDT	94.56(3.2)	94.39(3.1)	93.77(3.3)	92.43(3.7)	92.04(4.0)	88.68(4.2)
Solar flare	C4.5	81.96(3.3)	80.38(3.5)	78.68(3.5)	77.03(3.7)	76.67(3.7)	74.37(3.9)
	Sim-PDT	81.96(3.3)	80.88(3.3)	79.53(3.4)	79.23(3.6)	78.71(3.4)	76.81(3.8)
Balance	C4.5	78.48(4.2)	77.12(4.3)	75.39(4.7)	74.78(5.3)	72.42(5.6)	70.38(5.7)
	Sim-PDT	78.48(4.2)	77.56(4.2)	77.12(4.3)	76.79(4.8)	76.32(5.3)	74.64(5.6)
Nursery	C4.5	98.78(0.8)	96.38(1.3)	95.27(1.4)	94.45(1.6)	93.73(2.3)	92.81(2.6)
	Sim-PDT	98.78(0.8)	98.42(0.9)	97.49(1.1)	96.88(1.1)	96.21(1.4)	94.67(2.3)

labeled instances). Besides, Table 3 confirms Proposition 1. In fact, our approach recovers the C4.5 one when dealing with crisp instances (with precise labels, i.e., $L\%=0$).

The principal result of this table is that, generally, rejecting training instances, classes of which are imprecisely defined, is not a good practice and reduces the accuracy of the induced classifier. This issue can be avoided and well handled by the use of the proposed Sim-PDT approach which can exploit the information contained in imprecise labels.

6 Conclusion

In this paper, we have developed a new approach so-called Similarity based possibilistic decision tree. This approach represents a generalization of the C4.5 approach to the imprecise setting. In fact, it has the advantage of allowing the induction of decision trees from training instances having possibilistic class labels. The Sim-PDT approach uses the information closeness index as an additional information when evaluating the entropy of a given training partition in the attribute selection step. Experiments have shown that rejecting training instances, classes of which are imprecisely defined, is not a good practice and reduces the accuracy of the induced classifier. We plan to automate the specification of the wrapper distributions, which is done quite ad-hoc here, by applying a clustering phase.

References

1. Ben Amor, N., Benferhat, S., Elouedi, Z.: Qualitative classification and evaluation in possibilistic decision trees. In: FUZZ-IEEE 2004. Proceedings of the IEEE International Conference on Fuzzy Systems, Budapest, Hungary, vol. 2, pp. 653–657 (2004)
2. Borgelt, C., Gebhardt, J., Kruse, R.: Concepts for Probabilistic and Possibilistic Induction of Decision Trees on Real World Data. In: EUFIT 1996. Proceedings of the 4th European Congress on Fuzzy and Intelligent Technologies, Aachen, pp. 1556–1560 (1996)

3. Denoeux, T., Bjanger, M.S.: Induction of decision trees from partially classified data. In: SMC 2000. Proceedings of the 2000 IEEE Int. Conf on Systems, Man and Cybernetics, Nashville, TN, pp. 2923–2928. IEEE, Los Alamitos (2000)
4. Denoeux, T., Zouhal, L.M.: Handling possibilistic labels in pattern classification using evidential reasoning. Fuzzy Sets and Systems 122(3), 47–62 (2001)
5. Dubois, D., Prade, H.: Possibility theory: An approach to computerized processing of uncertainty. Plenum Press, New York (1988)
6. Elouedi, Z., Mellouli, K., Smets, P.: Belief decision trees: Theoretical foundations. International Journal of Approximate Reasoning 28, 91–124 (2001)
7. Higashi, M., Klir, G.J.: Measures of uncertainty and information based on possibility distributions. International Journal of General Systems 9(1), 43–58 (1983)
8. Higashi, M., Klir, G.J.: On the notion of distance representing information closeness: Possibility and probability distributions. International Journal of General Systems 9, 103–115 (1983)
9. Hüllermeier, E.: Possibilistic Induction in decision tree learning. In: Elomaa, T., Mannila, H., Toivonen, H. (eds.) ECML 2002. LNCS (LNAI), vol. 2430, pp. 173–184. Springer, Heidelberg (2002)
10. Hüllermeier, E., Beringer, J.: Learning from Ambiguously Labeled Examples. Intelligent Data Analysis, 168–179 (2005)
11. Kroupa, T.: Measure of divergence of possibility measures. In: Proceedings of the 6th Workshop on Uncertainty Processing, Prague, 2003, pp. 173–181 (2003)
12. Kwan, S., Olken, F., Rotem, D.: Uncertain, incomplete, and inconsistent data in scientific and statistical databases. In: Motro, A., Smets, P. (eds.) Proceedings of the Workshop on Uncertainty Management in Information Systems: From Needs to Solutions, pp. 64–91 (1992)
13. Marsala, C.: Apprentissage inductif en présence de données imprécises: Construction et utilisation d'arbres de décision flous, PhD thesis, University P. et M. Curie, Paris, France (1998)
14. Motro, A.: Sources of Uncertainty, Imprecision and Inconsistency in Information Systems. In: Motro, A., Smets, P. (eds.) Uncertainty Management in Information Systems: From Needs to Solutions, pp. 9–34 (1996)
15. Murphy, P.M., Aha, D.W.: UCI repository of machine learning databases (1996)
16. Quinlan, J.R.: Induction of decision trees. Machine Learning 1, 81–106 (1986)
17. Quinlan, J.R.: C4.5: Programs for machine learning. Morgan Kaufmann, San Francisco (1993)
18. Sanguesa, R., Cabos, J., Cortes, U.: Possibilistic conditional independence: A similarity based measure and its application to causal network learning. International Journal of Approximate Reasoning (1997)
19. Shannon, C.E.: The mathematical theory of communication. The Bell system Technical Journal 27(3), 379–423 (1948)
20. Witten, I.H., Frank, E.: Data Mining: Practical machine learning tools and techniques, 2nd edn. Morgan Kaufmann publisher, San Francisco (2005)
21. Yager, R.R.: On the specificity of a possibility distribution. Fuzzy Sets and Systems 50, 279–292 (1992)
22. Yuan, Y., Shaw, M.J.: Induction of fuzzy decision trees. Fuzzy Sets and Systems 69, 125–139 (1995)
23. Zadeh, L.A.: Fuzzy sets as a basis for a theory of possibility. Fuzzy Sets ans Systems 1, 3–28 (1978)
24. Zemankova, M., Kandel, A.: Implementing imprecision in information systems. Information Sciences 37(1–3), 107–141 (1985)

Towards a Logic for Abstract MetaFinite State Machines

Qing Wang and Klaus-Dieter Schewe

Information Science Research Centre
Massey University, Private Bag 11222, Palmerston North, New Zealand
{q.q.wang,k.d.schewe}@massey.ac.nz

Abstract. The paper investigates the logic of database transformations based on abstract metafinite state machines. We first introduce a structure model that separates a metafinite state into database and algorithmic spaces with bridge functions providing a connection between them. Then abstract metafinite state machines are developed on top of metafinite states equipped with two kinds of updates: exclusive updates and aggregate updates. In order to characterize both static and dynamic aspects of abstract metafinite state machines, we present a logic which supports reasoning about the side effects of exclusive and aggregate updates in update multisets occurring over database transformations.

1 Introduction

In database theory, theoretical foundations for queries have been intensively studied based on a variety of paradigms [1]. However, a construction of the counterpart for updates appeared to be a very difficult task [1,4]. With the notion of database transformation, first defined in [5], database manipulations encompassing queries, updates, updatable views and database restructuring have been unified into a general framework. Although the search for a theory of database transformations under such a framework has received considerable research attention, only several classes of database transformations involving queries have been clearly characterized [2,19,20,3,18,17].

Recently, motivated by the methodologies of Abstract State Machines (ASMs) [11,7] which formalized different notions of "algorithm", a general computation model called Abstract Database Transformation Machine (ADTM) tailored for database transformations was developed in [21]. A database transformation is defined as a binary relation over a set of states comprising of not only the underlying database structures but also the relevant algorithmic structures. Furthermore, there are six postulates stipulated on such database transformations to deal with different concerns about runs, states, backgrounds, complexity, expressiveness and dynamic updates in the context of complex value databases. For this computation model, an immediate question arises as to its logical characterization. In this paper, we investigate the logic of database transformations

S. Hartmann and G. Kern-Isberner (Eds.): FoIKS 2008, LNCS 4932, pp. 365–380, 2008.
© Springer-Verlag Berlin Heidelberg 2008

on the basis of generalized ADTMs with metafinite states, which we will call *abstract metafinite state machines*.

Metafinite states are developed on the basis of metafinite model theory [10]. In classical database theory, database structures are treated as finite structures by ignoring elements of infinite domains that do not occur in relations. However, this approach gives rise to several concerns about infinity arising in database applications. More specifically, database computations over finite structures may generate new elements from countably infinite domains, outside of finite structures. For instance, counting queries that produce natural numbers even if no natural numbers occur in finite structures. Furthermore, database theory traditionally imposes the genericity principle on database transformations with the consequence that only structural properties are taken into considerations. However, a large number of database applications involving data interpretation can be found in practice. To remedy such deficiencies, we develop metafinite states to lay a ground for the structure model of database transformations. In contrast to ADTMs, abstract metafinite state machines enable interpretations of database structures by means of a set of bridge functions between database structures and mathematical structures.

A collection of updates is produced by executing transition rules of abstract metafinite state machines over metafinite states. Since in most situations database structures are essentially considered to be unordered, computations over database structures in parallel seem indispensable. The presence of parallelism unavoidably results in updates that may be duplicate but not trivial with respect to location contents. To deal with this, we will consider update multisets instead of update sets in abstract metafinite state machines. Two kinds of updates: exclusive updates and aggregate updates are specified. They provide different approaches to reflect side effects of updates into the underlying metafinite states.

The main goal of this paper is to investigate the logical characterization of abstract metafinite state machines. In the literature, many logics have been proposed for reasoning about ASMs from different perspectives [16,13,15,14]. We will follow the work of [16] and define a logic by introducing atomic formula for updates in an update multiset. Furthermore, a modal operator from dynamic logic is adopted for describing dynamic aspects of computations. As additional features of metafinite states, the set and multiset formulation rules facilitate mapping of generic structures from database space into mathematical structures from algorithmic space in metafinite states. A set of axioms and inference rules capturing properties of update multisets, l-operators and modal operators for programs is presented.

The reminder of the paper is organized as follows. In section 2, we formalize the notion of metafinite state and provide several illustrative examples. After that, Section 3 introduces abstract metafinite state machines by generalizing states of Abstract Database Transformation Machines and specifying two kinds of updates. The formal syntax and semantics of the logic characterizing abstract metafinite state machines are presented in Section 4. In addition, a set of axioms and interference rules are developed to capture static and dynamic features. Finally, we briefly discuss the completeness of the logic and conclude the paper in Section 5.

2   Metafinite States

In this section, we introduce the notion of *metafinite state* which is a variation of metafinite structures as suggested in metafinite model theory [10].

2.1   Basic Definitions

The essential idea of metafinite state is to separate a metafinite state into two independent spaces: *database space* and *algorithmic space*. Database space has finite, generic structures, while algorithmic space has infinite, mathematical structures. The links between two spaces are established by a set of *bridge functions*, which map generic structures of interest from database space into mathematical structures of interest from algorithmic space.

Definition 1. A *metafinite state* S is a triple $(\mathfrak{R}, \mathfrak{I}, \mathfrak{B})$ consisting of

- a finite structure $\mathfrak{R}$, called *database space*,
- an infinite structure $\mathfrak{I}$, called *algorithmic space*, and
- a finite set $\mathfrak{B}$ of *bridge functions*.

The *state vocabulary* V_S of a metafinite state S is constituted by function names defined over different components of S. More precisely, we have $V_S = V_{\mathfrak{R}} \cup V_{\mathfrak{I}} \cup V_{\mathfrak{B}}$, where $V_{\mathfrak{R}}$, $V_{\mathfrak{I}}$ and $V_{\mathfrak{B}}$ are pairwise disjoint, finite sets of database, algorithmic and bridge function names, respectively. Let $\mathbf{W} = \{O_i\}_{i \in [1,n]}$ be a fixed family of countably, finite or infinite domains and $\mathbb{W} = \bigcup_{i \in [1,n]} O_i$ be a universal domain. Then for a metafinite state $S = (\mathfrak{R}, \mathfrak{I}, \mathfrak{B})$, database space $\mathfrak{R}$ has a finite set $\mathbf{D} = \{D_1, ..., D_m\}$ of finite *database domains* that satisfy $\bigcup_{i \in [1,m]} D_i \subset \mathbb{W}$, and algorithmic space $\mathfrak{I}$ has a finite set $\mathbf{C}$ of *algorithmic domains* such that $\mathbf{C} \subseteq \mathbb{W}$. Some domains may be common to both, such as *Bool*. We assume $Bool \in \mathbf{D} \cap \mathbf{C}$ by default. The structure of database space is *generic* in the sense that elements in database domains are treated as being abstract and independent from specific data representations. We shall use $d_i (i \in [1,n])$ to represent abstract elements from database domains in the paper. Functions of a metafinite state are defined in the following.

Definition 2. Let $S = (\mathfrak{R}, \mathfrak{I}, \mathfrak{B})$ be a metafinite state with two finite sets $\mathbf{D}$ and $\mathbf{C}$ of database and algorithmic domains, respectively, then a function of S is a possibly partial mapping such that

- a *database function* $f_\Re$ with n-arity: $D_1 \times \cdots \times D_n \to D_0$ for $D_i \in \mathbf{D}$ ($i \in [0, n]$) and $f_\Re \in V_\Re$,
- an *algorithmic function* $f_\Im$ with n-arity: $C_1 \times \cdots \times C_n \to C_0$ for $C_i \in \mathbf{C}$ ($i \in [0, n]$) and $f_\Im \in V_\Im$, and
- a *bridge function* $f_\mathfrak{B} \colon D \to C$ for $D \in \mathbf{D}$, $C \in \mathbf{C}$ and $f_\mathfrak{B} \in V_\mathfrak{B}$.

Therefore, in terms of a metafinite state $S = (\Re, \Im, \mathfrak{B})$ with the state vocabulary $V_S = V_\Re \cup V_\Im \cup V_\mathfrak{B}$, the structure of $\Re$ is a set $\bigcup \mathbf{D}$ of database elements together with interpretations of database functions in $V_\Re$ over $\bigcup \mathbf{D}$, and the structure of $\Im$ is a set $\bigcup \mathbf{C}$ of algorithmic elements together with interpretations of algorithmic functions in $V_\Im$ over $\bigcup \mathbf{C}$. The following example illustrates some of the central concepts of metafinite states.

Example 1. Let $S = (\Re, \Im, \mathfrak{B})$ be a metafinite state associated with a set of domains $\mathbf{W} = \{D_1, D_2, String, Bool, \mathbb{N}\}$, then it is possible that

- $\Re$ has a function $person \colon D_1 \times D_2 \to Bool$ describing a set of persons with their names and ages,
- $\Im$ has a function $even \colon \mathbb{N} \to Bool$ meaning that $even$ returns true if a given natural number is a even number, and false otherwise,
- bridge functions $f_{\mathfrak{B}_1} \colon D_1 \to String$ and $f_{\mathfrak{B}_2} \colon D_2 \to \mathbb{N}$ are total and injective.

Over the above metafinite state S, the query "list all persons whose ages are even numbers" can be executed. The bridge function $f_{\mathfrak{B}_2}$ first interprets database domain D_2 of Age with algorithmic domain $\mathbb{N}$, and then the algorithmic function $even$ is applied over interpreted elements to figure out even numbers. Bridge functions can be partial. For example, suppose that $f_{\mathfrak{B}_2} \colon D_2 \to \mathbb{N}$ is a partial function with $d_1 \mapsto 4, d_4 \mapsto 35$, it means that only two database elements d_1 and d_4 from database domain D_2 in $\Re$ are interpreted as natural numbers in $\Im$.

Functions of a metafinite state are either dynamic or static, therefore, a metafinite state can be dynamically enlarged or shrink during database computations. More precisely, database functions are considered as being dynamic in order to reflect dynamic changes into database space. Algorithmic functions usually are static except for nullary functions, i.e., variables. For the purpose of supporting flexible interpretations, all bridge functions are dynamic.

Remark 1. Theoretically, structures of a database are regarded as finite structures in which only structural properties are of concerns. Nevertheless, most database computations in practice execute over elements with explicit interpretations. Furthermore, in order to properly capture structural properties of a database, it is indispensable to consider the use of abstract mathematical tools with mathematical structures and properties. For these reasons, we choose metafinite states as a general setting of structures for investigating database transformations.

2.2 Background Structures

Although we have presented the basic concepts of metafinite state in section 2.1, databases for the real world are usually much more complicated, due to the presence of complex values, such as sets, lists, tuples and even combinations of these. For clarity, let us look at the following example.

Example 2. Suppose that () and {} are two construct symbols denoting **tuple** and **set**, respectively, then the following relation ENROL that specifies papers enrolled by students in different years has complex values: tuples of student information and sets of paper numbers.

Student	PaperNo	StudyYear
(*Ellen Lee*, 003456)	{157100, 158289}	2006
(*Jane Wang*, 012582)	{175162, 158167, 157100}	2006
(*Nain Green*, 030567)	{157100}	2007

Following the idea of [8,6] in which various complex values are captured by means of flexible backgrounds, we augment metafinite states with backgrounds. Let V_K be a background vocabulary containing a finite set of construct symbols, $\mathbf{A}$ be a set of domains containing only atomic values and $\mathbb{N}$ be a set of natural numbers. Then a *background class* over $(\mathbf{A}, V_K)$ is a set $\widehat{\mathbf{A}}$ of *background domains* generated by applying the following rules:

- $\mathbf{A} \subseteq \widehat{\mathbf{A}}$.
- for every construct symbol $\llcorner\lrcorner \in V_K$ with unfixed arity, and every $m \in \mathbb{N}$, if $a_j \in \bigcup \widehat{\mathbf{A}}$ $(j \in [1, m])$, then $\llcorner a_1, ..., a_m \lrcorner \in \widehat{A}_{\llcorner\lrcorner}$ and $\widehat{A}_{\llcorner\lrcorner} \in \widehat{\mathbf{A}}$.
- for every construct symbol $\llcorner\lrcorner \in V_K$ with arity m, if $a_j \in \bigcup \widehat{\mathbf{A}}$ $(j \in [1, m])$, then $\llcorner a_1, ..., a_m \lrcorner \in \widehat{A}_{\llcorner\lrcorner}$ and $\widehat{A}_{\llcorner\lrcorner} \in \widehat{\mathbf{A}}$.

The database and algorithmic spaces of a metafinite state S may be associated with different background vocabularies. Let $\mathbf{D}$ and $\mathbf{C}$ be sets of database and algorithmic domains of S, respectively, and V_{DK} and V_{CK} be background vocabularies associated with database and algorithmic spaces of S, respectively, then a rich set of complex values can be obtained as $\bigcup \widehat{\mathbf{D}} \cup \bigcup \widehat{\mathbf{C}}$, where $\widehat{\mathbf{D}}$ is a set of finite subsets of background domains over $(\mathbf{D}, V_{DK})$ and $\widehat{\mathbf{C}}$ is a background class over $(\mathbf{C}, V_{CK})$, respectively. Specifically, for a metafinite state with backgrounds, database space has only finite elements from its background domains, however, algorithmic space has all of its background domains.

To incorporate backgrounds into a metafinite state, Definitions 1 and 2 need to be revised by extending $\mathbf{D}$ and $\mathbf{C}$ of database and algorithmic domains to $\widehat{\mathbf{D}}$ and $\widehat{\mathbf{C}}$ containing background domains. The permutation $\hbar$ over $\bigcup \widehat{\mathbf{D}}$ is obtained by extending the permutation h over $\bigcup \mathbf{D}$ in a canonical way such that

- for every $\llcorner a_1, ..., a_m \lrcorner \in \bigcup \widehat{\mathbf{D}}$, $\hbar(\llcorner a_1, ..., a_m \lrcorner) = \llcorner \hbar(a_1), ..., \hbar(a_m) \lrcorner$, and
- for $a \in \bigcup \mathbf{D}$, $\hbar(a) = h(a)$.

This approach of handling complex values simplifies our view of some common operations over complex value databases. For instance, set operations can be treated as binary functions over a background domain generated by applying the set construct, and similarly multiset operations are functions over a background domain generated by applying the multiset construct.

Example 3. Let S be a metafinite state with $\{Int\} \subseteq \mathbf{C}$, and $\{\}$ and $\{\!\{\}\!\}$ represent set and multiset constructs in V_{CK}, respectively, then we can obtain the background domains $\widehat{C}_{\{\}}$ and $\widehat{C}_{\{\!\{\}\!\}}$ such that

- aggregation operators, i.e., max, min, average and etc., are unary functions with $\widehat{C}_{\{\!\{\}\!\}} \to Int$, and
- set operations, i.e., union, difference and etc., are binary functions with $\widehat{C}_{\{\}} \times \widehat{C}_{\{\}} \to \widehat{C}_{\{\}}$.

3   Abstract Metafinite State Machines

In [21], Abstract Database Transformation Machines (ADTMs) were developed for investigating database transformations. In this section, we generalize ADTMs to abstract metafinite state machines.

Definition 3. An *abstract metafinite state machine* Λ with a fixed pair (Σ_1, Σ_2) of vocabularies consists of an input metafinite state over (Σ_1, Σ_2) and a transition program over (Σ_1, Σ_2).

A metafinite state over (Σ_1, Σ_2) refers to a metafinite state with state vocabulary Σ_1 and background vocabularry Σ_2 as described in the preceding section. In such a metafinite state, terms are defined by the induction on elements of background domains and function names. Formulae are defined over logical connectives and quantifiers as usual. A transition program over (Σ_1, Σ_2) is finitely and inductively defined by applying a set of transition rules over metafinite states of (Σ_1, Σ_2) such that

- *generalized update rule:* the value of term t_0 is assigned to the function f at the value of term t_1 by means of applying an operator op.
    $$f(t_1) \;=^{op}\; t_0$$
- *conditional rule:* if the value of term t is true, then rule r is executed.
    if t **then** r
- *forall rule:* for each value of variable $x \in X_{\Re}$ satisfying formula $\varphi(x)$, rule r is executed with that value for x in parallel.
    forall x **with** $\varphi(x)$ **do** r
- *choose rule:* if there are some values of variable $x \in X_{\Re}$ satisfying formula $\varphi(x)$, rule r is executed by arbitrarily choosing one of such values for x.
    choose x **with** $\varphi(x)$ **do** r
- *parallel rule:* two rules r_1 and r_2 are executed in parallel.
    par r_1 r_2 **par**
- *sequence rule:* two rules r_1 and r_2 are executed sequentially.
    seq r_1 r_2 **seq**

Let ζ be the structure of a metafinite state , $f \in V_S$ be a dynamic function name, a and b be elements of ζ, then an *update* is a triple $(f(a), b, p)$, where $f(a)$ is a location, b is an update value, p is an operator and $f^S(a)$ is a location content denoting the value of $f(a)$ in S. An *update set* is a set of updates. An *update multiset* is an unordered collection of updates, which allows duplicates. As discussed in [7,21], parallel subcomputations may be generated in a database computation. It gives rise to duplicate updates or collections of updates associated with different manipulations. Therefore, the notion of update multiset rather than update set appears to be more natural to describe updates yielded in the procedure of database computations.

We use the notation $\{\!\{\,\}\!\}$ for multiset. To deal with update multisets, we need to employ a kind of powerful operators, called *l-operators*, which are aggregate functions as defined in [9] adapting to metafinite states.

Definition 4. Let $D_1 \in \widehat{\mathbf{D}}$, $D_2 \in \widehat{\mathbf{D}} \cup \widehat{\mathbf{C}}$ and $\mathcal{M}(D_1)$ be the set of all non-empty multisets over D_1, then a *l-operator* $\rho = (\beta_1, \odot, \beta_2)$ is a function from $\mathcal{M}(D_1)$ to D_2, where

- $\beta_1 : D_1 \to D_2$ is a database or bridge function,
- $\odot : D_2 \times D_2 \to D_2$ is commutative and associative, and
- $\beta_2 : D_2 \to D_2$ is a terminating function.

Furthermore, for $m \in \mathcal{M}(D_1)$ and $m = \{\!\{b_1, ..., b_n\}\!\}$, $\rho(m)$ is defined as

$$\rho(m) = \beta_2(\beta_1(b_1) \odot \cdots \odot \beta_1(b_n))$$

Let $\mathcal{P}$ represent a set of l-operators, then an update either has a l-operator $p \in \mathcal{P}$ or an undefined operator $p = \bot$. Within an update multiset, multiple updates may be identical in the sense that all of them have the same location, update value and operator. In order to uniquely identify an update, we associate each update with an *update identifier*. Let $\mathcal{U}$ and $\mathcal{O}$ be a set of updates and update identifiers respectively, then there is an injective function from updates to update identifiers such that $\mathcal{U} \to \mathcal{O}$. Given that $\mathfrak{M}_1$ and $\mathfrak{M}_2$ are two update multisets, then

- the *union of two update multisets*, denoted by $\mathfrak{M}_1 \uplus \mathfrak{M}_2$, is an update multiset comprised of all the updates of $\mathfrak{M}_1$ together with all the updates of $\mathfrak{M}_2$.
- the *composition of two update multisets*, denoted by $\mathfrak{M}_1 \oslash \mathfrak{M}_2$, is an update multiset such that

$$\mathfrak{M}_1 \oslash \mathfrak{M}_2 :\equiv \mathfrak{M}_2 \uplus \{\!\{(f(a), b, p_1) \in \mathfrak{M}_1 | \neg \exists c, p_2(f(a), c, p_2) \in \mathfrak{M}_2\}\!\}.$$

Let $\ddot{\Delta}(r)$ denote an update multiset yielded by executing the rule r over the current metafinite state, $\|t\|$ refer to values of term t evaluated over the current metafinite state and $r|_{t_2 \leadsto t_1}$ represent that the value of term t_2 is bounded to term t_1 within rule r, then we have the following definition for update multisets yielded by transition rules.

$$- \ddot{\Delta}(f(t_1) =^{op} t_0) = \{\!\!\{(f(a_1), a_0, p\}\!\!\} \quad for \quad a_i \in [\![t_i]\!] (i \in [0, 1])$$

$$- \ddot{\Delta}(\text{if } t \text{ then } r) = \begin{cases} \ddot{\Delta}(r) & iff \ [\![t]\!] = \{true\} \\ \{\!\!\{\}\!\!\} & otherwise. \end{cases}$$

$$- \ddot{\Delta}(\textbf{forall } x \textbf{ with } \varphi(x) \textbf{ do } r) = \biguplus_{i \in [1,n]} \ddot{\Delta}(r|_{x \rightsquigarrow a_i}) \quad for \ \{x|\![\varphi(x)]\!] =$$

$$\{true\}\} = \{a_1, ..., a_n\}$$

$$- \ddot{\Delta}(\textbf{choose } x \textbf{ with } \varphi(x) \textbf{ do } r) = \ddot{\Delta}(r|_{x \rightsquigarrow a}) \quad for \ a \in \{x|\![\varphi(x)]\!] = $$

$$\{true\}\}$$

$$- \ddot{\Delta}(\textbf{par } r_1 \ r_2 \ \textbf{par}) = \ddot{\Delta}(r_1) \uplus \ddot{\Delta}(r_2)$$

$$- \ddot{\Delta}(\textbf{seq } r_1 \ r_2 \ \textbf{seq}) = \ddot{\Delta}(r_1) \oslash \ddot{\Delta}(r_2)$$

In abstract metafinite state machines, there are two kinds of updates that may occur over a metafinite state: *exclusive updates* and *aggregate updates*. They essentially differ in the way to update the structure. For an exclusive update, its operator is undefined and the location content is exclusively updated to its update value, whereas for an aggregate update, it has a l-operator and the location content is updated to a value calculated from all update values of that location in an update multiset. Therefore, in a consistent update multiset, each location can have at most one update value for exclusive updates, or a finite set of update values for aggregation updates.

Definition 5. An update multiset $\mathfrak{M}$ is *consistent* iff

- if $(f(a), b, p_1) \in \mathfrak{M}$ and $(f(a), c, p_2) \in \mathfrak{M}$, then $p_1 = p_2$, and
- if $(f(a), b, \bot) \in \mathfrak{M}$ and $(f(a), c, \bot) \in \mathfrak{M}$, then $b = c$.

Otherwise, $\mathfrak{M}$ is inconsistent.

We denote the resulting metafinite state after executing an update multiset $\mathfrak{M}$ over a metafinite state S as $S + \mathfrak{M}$. If $\mathfrak{M}$ is inconsistent, then $S + \mathfrak{M} = S$.

Definition 6. Let S be a metafinite state and $\mathfrak{M}$ be a consistent update multiset, then for all updates in $\mathfrak{M}$,

- if $(f(a), b, \bot) \in \mathfrak{M}$, then $f^{S+\mathfrak{M}}(a) = b$,
- if there are n updates $(f(a), b_1, \rho), ..., (f(a), b_n, \rho) \in \mathfrak{M}$ with the same location and $\rho \in \mathcal{P}$, then $f^{S+\mathfrak{M}}(a) = \rho(\{\!\!\{b_1, ..., b_n\}\!\!\})$, and
- if $\neg \exists p, b \ (f(a), b, p) \in \mathfrak{M}$, then $f^{S+\mathfrak{M}}(a) = f^S(a)$.

A *run* of an abstract metafinite state machine is a finite or infinite sequence of metafinite states such that it starts with the input metafinite state and goes to next states by applying update multisets generated by the transition program continuingly until two consecutive states are identical.

4   A Logical Formalization

In this section, we develop a logic characterization for abstract metafinite state machines. This logic provides a capability for reasoning about the consistency and side effects of update multisets occurring in abstract metafinite state machines.

4.1   Syntax and Semantics

The main concepts in the logic are term, formula and program. Let $V_S = (V_{\Re}, V_{\Im}, V_{\mathfrak{B}})$ be a fixed state vocabulary, $\widehat{\mathbf{D}}$ and $\widehat{\mathbf{C}}$ be two sets of database and algorithmic domains extended under their backgrounds, respectively, then a set Υ of terms, a set Ψ of formulae containing a subset Ψ^b of *basic formulae*, and a set Φ of programs are defined over $(V_S, \widehat{\mathbf{D}}, \widehat{\mathbf{C}})$. For convenience, we use $\mathrm{fr}(t)$ to denote a set of free variables occurring in a term t when necessary.

Definition 7. The set Υ of terms is defined by applying the following rules:

- Let $X_{\Re} \subseteq V_{\Re}$ be a set of nullary function names (i.e., variables) and $F_{\Re} \subseteq V_{\Re}$ be a set of non-nullary function names, then a set $\mathbb{T}_{\Re} \in \Upsilon$ of *database terms* is defined by
    - $X_{\Re} \subseteq \mathbb{T}_{\Re}$, and
    - $f(t_1, ...t_n) \in \mathbb{T}_{\Re}$ for $f \in F_{\Re}$ and $t_i \in \mathbb{T}_{\Re}$ $(i \in [1, n])$.
- Let $X_{\Im} \subseteq V_{\Im}$ be a set of nullary function names and $F_{\Im} \subseteq V_{\Im}$ be a set of non-nullary function names, then for a set $\mathbb{T}_{\Im} \in \Upsilon$ of *program terms*,
    - $\bigcup \widehat{\mathbf{C}} \subseteq \mathbb{T}_{\Im}$,
    - $X_{\Im} \subseteq \mathbb{T}_{\Im}$,
    - $f(t_1, ...t_n) \in \mathbb{T}_{\Im}$ for $f \in F_{\Im}$ and $t_i \in \mathbb{T}_{\Im}$ $(i \in [1, n])$, and
    - $f(t) \in \mathbb{T}_{\Im}$ for $f \in V_{\mathfrak{B}}$ and $t \in \mathbb{T}_{\Re}$.
- Terms are closed under the choice formulation rule: $\varepsilon x.\varphi(x) \in \Upsilon$ for $\varphi(x) \in \Psi^b$ and $x \in X_{\Re} \cup X_{\Im}$; moreover, $\mathrm{fr}(\varepsilon x.\varphi(x)) = \emptyset$.
- Terms are closed under the set formulation rule: $\{t(x, y) : \varphi(x)\} \in \Upsilon$ for $x \in X_{\Re}$, $\varphi(x) \in \Psi^b$ and $t(x, y) \in \Upsilon$; moreover, $\mathrm{fr}(\{t(x, y) : \varphi(x)\}) = \{y\}$ and $\mathrm{fr}(\varphi(x)) = \{x\}$.
- Terms are closed under the multiset formulation rule: $\{\!|t(x, y) : \varphi(x)|\!\} \in \Upsilon$ for $x \in X_{\Re}$, $\varphi(x) \in \Psi^b$ and $t(x, y) \in \Upsilon$; moreover, $\mathrm{fr}(\{\!|t(x, y) : \varphi(x)|\!\}) = \{y\}$ and $\mathrm{fr}(\varphi(x)) = \{x\}$.

Terms without variables are called ground terms, which can only be program terms; furthermore, terms may be built upon basic formulae (we will define them later in Definition 8) by applying the choice, set and multiset formulation rules. The value $vl_s\|t\|$ of a term t in a metafinite state S is recursively defined over the structure of S. For terms defined in the first two rules of Definition 7, their values are interpreted as usual. In the choice formulation rule, $vl_s\|\varepsilon x.\varphi(x)\|$ represents an arbitrarily nondeterministic element among the set of elements of S satisfying

$\varphi(x)$. If no element satisfies $\varphi(x)$, then $vl_s \llbracket \varepsilon x.\varphi(x) \rrbracket$ denotes the special symbol $\perp$. For the set and multiset formulation rules, $vl_s \llbracket \{t(x,y) : \varphi(x)\} \rrbracket$ is the set of all terms of $t(x,y)$ such that formula $\varphi(x)$ is satisfied over S, and similarly $vl_s \llbracket \{\!\{t(x,y) : \varphi(x)\}\!\} \rrbracket$ is the multiset of all terms of $t(x,y)$ such that formula $\varphi(x)$ is satisfied over S. Due to the restriction that x of $\varphi(x)$ is a variable in database space, the resulting sets or multisets after applying the set or multiset formulation rules always have a finite number of elements. Terms with boolean values are called boolean terms.

The intuition behind the set and multiset formulation rules is to capture properties of interest in a metafinite state's database space by employing powerful set and multiset structures. Two examples are presented in the following, where $\downarrow: D \to String$ is a bridge function such that $t^\downarrow \in String$ for $t \in D$.

Example 4. Suppose that the notation $t\lceil t_1, ..., t_n \rceil$ denotes a tree term, in which subtrees $t_1, ..., t_n$ have a common parent node t. Let us consider the tree (i) shown in Fig. 1. By using the term $\{\{x\lceil y\rceil : edge(x,y) \wedge color(y)^\downarrow = \text{"green"}\} : label(x)^\downarrow = \text{"A"}\}$, the following result can be obtained via the intermediate step $\{\{n_1\lceil y\rceil : edge(n_1, y) \wedge color(y)^\downarrow = \text{"green"}\}, \{n_6\lceil y\rceil : edge(n_6, y) \wedge color(y)^\downarrow = \text{"green"}\}, \{n_7\lceil y\rceil : edge(n_7, y) \wedge color(y)^\downarrow = \text{"green"}\}\},$

$$\{\{n_1\lceil n_3\rceil\}, \{n_6\lceil n_7\rceil, n_6\lceil n_9\rceil\}), \{n_7\lceil n_8\rceil\}\}.$$

Furthermore, assume that we have two tree algebraic operators defined over trees such that $\mathrm{II} : \{t_1, ..., t_n\} \to t$ refers to union trees $t_1, ..., t_n$ to be a tree t if $t_1, ..., t_n$ have a common root node, otherwise return an empty tree, and $\Omega^{t_0} : \{t_1, ..., t_n\} \to t_0\lceil t_1, ..., t_n\rceil$ refers to generate a common root node t_0 for all trees $t_1, ..., t_n$, respectively. By applying II and Ω on the above term such that,

- for $\Omega^{n_{10}} \{\mathrm{II}\{x\lceil y\rceil : edge(x,y) \wedge color(y)^\downarrow = \text{"green"}\} : label(x)^\downarrow = \text{"A"}\}$, we get the resulting tree as shown in Fig. 1(ii);
- for $\mathrm{II}\{\Omega^{n_{10}} \{x\lceil y\rceil : edge(x,y) \wedge color(y)^\downarrow = \text{"green"}\} : label(x)^\downarrow = \text{"A"}\}$, we get the resulting tree as shown in Fig. 1(iii).

Example 5. Consider the question "what is the maximal number of green nodes that have common parent node labeled as 'A'?" over Fig. 1 again. To deal with this question, we can write down a multiset term $max\{\!\{sum\{\!\{1 : edge(x,y) \wedge color(y)^\downarrow = \text{"green"}\}\!\} : label(x)^\downarrow = \text{"A"}\}\!\}$, which is evaluated to $max\{\!\{sum\{\!\{1\}\!\}, sum\{\!\{1,1\}\!\}, sum\{\!\{1\}\!\}\}\!\}$. Obviously, the result is 2 in this case.

Definition 8. The set Ψ of formulae is defined by the following grammar:

$$\varphi :\equiv bt \mid t_1 = t_2 \mid \varphi_1 \vee \varphi_2 \mid \varphi_1 \wedge \varphi_2 \mid \varphi_1 \Rightarrow \varphi_2 \mid \varphi_1 \Leftrightarrow \varphi_2 \mid \neg \varphi \mid \exists x \varphi \mid \forall x \varphi \mid$$

$$[\alpha]\varphi \mid upm(\alpha, f, y, z, p, o).$$

bt denotes a boolean term, t_1, t_2, y, z and $o \in \Upsilon$, $x \in X_\Re$, $f \in \Sigma_1$, $p \in \mathcal{P} \cup \{\perp\}$ and $\alpha \in \Phi$. The set Ψ^b of *basic formulae* contains all formulae of Ψ without any presence of $[\alpha]\varphi$ and $upm(\alpha, f, y, z, p, o)$. By introducing a modal operator $[\alpha]$

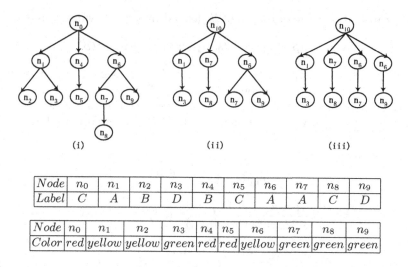

Node	n_0	n_1	n_2	n_3	n_4	n_5	n_6	n_7	n_8	n_9
Label	C	A	B	D	B	C	A	A	C	D

Node	n_0	n_1	n_2	n_3	n_4	n_5	n_6	n_7	n_8	n_9
Color	red	$yellow$	$yellow$	$green$	red	red	$yellow$	$green$	$green$	$green$

Fig. 1. labeled and colorful trees

for a program α, formula φ of $[\alpha]\varphi$ is indeed evaluated over a metafinite state after executing program α on the current metafinite state. A specific formula $\mathrm{upm}(\alpha, f, y, z, p, o)$ is introduced into the logic to characterize the dynamic aspects of computations in terms of update multisets. $\mathrm{upm}(\alpha, f, y, z, p, o)$ means that a program α generates an update with location $f(x)$, update value z, operator p and update identifier o. The semantics $el_s[\![\varphi]\!]$ of a formula φ over a metafinite state S is an evaluation function $el_s : \Psi \to Bool$ such that

$$el_s[\![bt]\!] = \begin{cases} \text{true} & \text{iff } vl_s[\![bt]\!] \text{ is true,} \\ \text{false} & \text{otherwise.} \end{cases}$$

$$el_s[\![t_1 = t_2]\!] = \begin{cases} \text{true} & \text{iff } vl_s[\![t_1]\!] = vl_s[\![t_2]\!], \\ \text{false} & \text{otherwise.} \end{cases}$$

$$el_s[\![\exists x \varphi]\!] = \begin{cases} \text{true} & \text{iff } \varphi|_{x \rightsquigarrow a} \text{ is true for some element } a \text{ of } \Re \text{ in } S, \\ \text{false} & \text{otherwise.} \end{cases}$$

$$el_s[\![\forall x \varphi]\!] = \begin{cases} \text{true} & \text{iff } \varphi|_{x \rightsquigarrow a} \text{ is true for every element } a \text{ of } \Re \text{ in } S, \\ \text{false} & \text{otherwise.} \end{cases}$$

$$el_s[\![[\alpha]\varphi]\!] = \begin{cases} \text{true} & \text{if program } \alpha \text{ generates an update multiset } \mathfrak{M} \text{ and} \\ & \quad \varphi \text{ is true in } S + \mathfrak{M}, \\ \text{false} & \text{otherwise.} \end{cases}$$

$$el_s[\![\mathrm{upm}(\alpha, f, y, z, p, o)]\!] = \begin{cases} \text{true} & \text{iff program } \alpha \text{ yields an update} \\ & \quad \text{multiset which has update} \\ & \quad (\mathrm{f}(vl_s[\![y]\!]), vl_s[\![z]\!], vl_s[\![p]\!]) \text{ with} \\ & \quad \text{update identifier } vl_s[\![o]\!], \\ \text{false} & \text{otherwise.} \end{cases}$$

We omit the semantics of formulae closed under logic connectives since they are evaluated in a standard manner.

Definition 9. The set Φ of programs is defined by the following grammar:

$$\alpha := \mathbf{skip} \mid f(t_1) =^{op} t_0 \mid \alpha_1 \; ; \; \alpha_2 \mid \alpha_1 \sqcup \alpha_2 \mid ?\varphi(\alpha),$$

$f \in \Sigma_1$, $op \in \mathcal{P} \cup \{\bot\}$, t_1 and $t_0 \in \Upsilon$, and $\varphi \in \Psi^b$. The semantics of a program can be described by an expression in form of $\frac{\mathfrak{P}}{\mathfrak{M} \to \mathfrak{M}'}(S, \alpha)$, which means that under a set $\mathfrak{P}$ of premises, a program α executing over a metafinite state S with an update multiset $\mathfrak{M}$ yields an update multiset $\mathfrak{M}'$. Therefore, we have

$$\frac{}{\mathfrak{M}_0 \to \mathfrak{M}_0}(S, \mathbf{skip}) \tag{1}$$

$$\frac{vl_s\llbracket t_1 \rrbracket = a \qquad vl_s\llbracket t_0 \rrbracket = b}{\mathfrak{M}_0 \to \mathfrak{M}_0 \uplus \{\!\{(f(a), b, op)\}\!\}}(S, f(t_1) =^{op} t_0) \tag{2}$$

$$\frac{\frac{}{\mathfrak{M}_0 \to \mathfrak{M}_0 \uplus \mathfrak{M}_i}(S, \alpha_i)\ (i \in [1,2])}{\mathfrak{M}_0 \to \biguplus_{j \in [0,2]} \mathfrak{M}_j}(S, \alpha_1 \sqcup \alpha_2) \tag{3}$$

$$\frac{\frac{}{\mathfrak{M}_0 \to \mathfrak{M}_0 \uplus \mathfrak{M}_1}(S, \alpha_1) \quad \frac{}{\{\!\{\}\!\} \to \mathfrak{M}_2}(S + (\mathfrak{M}_0 \uplus \mathfrak{M}_1), \alpha_2)}{\mathfrak{M}_0 \to (\mathfrak{M}_0 \uplus \mathfrak{M}_1) \oslash \mathfrak{M}_2}(S, \alpha_1 \; ; \; \alpha_2) \tag{4}$$

$$\text{if } \mathfrak{M}_0 \uplus \mathfrak{M}_1 \text{ is consistent.}$$

$$\frac{\frac{}{\mathfrak{M}_0 \to \mathfrak{M}_0 \uplus \mathfrak{M}_1}(S, \alpha_1)}{\mathfrak{M}_0 \to \mathfrak{M}_0 \uplus \mathfrak{M}_1}(S, \alpha_1 \; ; \; \alpha_2) \tag{5}$$

$$\text{if } \mathfrak{M}_0 \uplus \mathfrak{M}_1 \text{ is inconsistent.}$$

$$\frac{el_s\llbracket \varphi \rrbracket = true \quad \frac{}{\mathfrak{M}_0 \to \mathfrak{M}_1}(S, \alpha)}{\mathfrak{M}_0 \to \mathfrak{M}_1}(S, ?\varphi(\alpha)) \qquad \frac{el_s\llbracket \varphi \rrbracket = false}{\mathfrak{M}_0 \to \mathfrak{M}_0}(S, ?\varphi(\alpha)) \tag{6}$$

4.2   Axioms and Inference Rules

We formalize the properties of l-operators, update multisets and modal operator [ ] by presenting a set of axioms and inference rules. Inspired by the idea of [16], a formula $\mathrm{con}(\alpha)$ is used as an abbreviation for the following formula stating that $\mathrm{con}(\alpha)$ is true iff the update multiset produced by program α is consistent.

$$\mathrm{con}(\alpha) := \bigwedge_{f \in \Sigma_1} \forall y, z_1, z_2, p_1, p_2, o_1, o_2(\mathrm{ump}(\alpha, f, y, z_1, p_1, o_1) \wedge$$
$$\mathrm{ump}(\alpha, f, y, z_2, p_2, o_2)) \Rightarrow p_1 = p_2 \wedge (p_1 \neq \bot \vee z_1 = z_2)$$

With the use of modal operator [ ] for programs, the logic becomes a multi-modal logic. We write $\Gamma \vdash \varphi$ to represent that formula φ can be derived from a set Γ of formulae in all metafinite states of an abstract state machine by successively using the following axioms and inference rules.

- Axioms **m1-m5** assert that an update multiset for a program $\alpha \in \Phi$ is inductively generated. Compared with the work in [16], programs in our logic only have parallel, sequential and optional constructs in addition to atomic programs **skip** and $f(a) =^{op} b$, so firing all rules within a program can always terminate and result in an update multiset (possibly empty).

m1. $\mathrm{upm}(\mathbf{skip}, f, y, z, , p, o) \Leftrightarrow \mathrm{false}$
m2. $\mathrm{upm}(f(a) =^{op} b, f', a', b', p', o) \Leftrightarrow (f = f' \wedge a = a' \wedge b = b' \wedge p' = op)$
m3. $\mathrm{upm}(\alpha_1 ; \alpha_2, f, y, z, p, o) \Leftrightarrow ([\alpha_1]\mathrm{upm}(\alpha_2, f, y, z, p, o) \vee$
$(\mathrm{upm}(\alpha_1, f, y, z, p, o) \wedge \neg \exists z', p', o' [\alpha_1]\mathrm{upm}(\alpha_2, f, y, z', p', o')))$
m4. $\mathrm{upm}(\alpha_1 \sqcup \alpha_2, f, y, z, p, o) \Leftrightarrow (\mathrm{upm}(\alpha_1, f, y, z, p, o) \vee \mathrm{upm}(\alpha_2, f, y, z, p, o))$
m5. $\mathrm{upm}(?\varphi(\alpha), f, y, z, p, o) \Leftrightarrow (\varphi \wedge \mathrm{upm}(\alpha, f, y, z, p, o))$

- Axioms **l1-l3** state how an update multiset is applied over the underlying metafinite state. In the case of a consistent update multiset, axioms **l1** and **l2** deal with exclusive updates and aggregate updates, respectively. As for axiom **l3**, it says that no any side effect is made to the underlying metafinite state if applying an inconsistent update multiset.

l1. $\mathrm{upm}(\alpha, f, y, z, \bot, o) \wedge \mathrm{con}(\alpha) \Rightarrow [\alpha]f(y) = z$
l2. $\mathrm{upm}(\alpha, f, y, z, \rho, o) \wedge \mathrm{con}(\alpha) \wedge \rho \neq \bot \Rightarrow$
$\exists x (x = \{\!\!\{ z' | \exists o' \ \mathrm{upm}(\alpha, f, y, z', \rho, o') \}\!\!\} \wedge [\alpha]f(y) = \rho(x))$
l3. $\mathrm{upm}(\alpha, f, y, z, p, o) \wedge \neg \mathrm{con}(\alpha) \Rightarrow (\varphi \Leftrightarrow [\alpha]\varphi)$

- Axioms **p1-p3** and inference rules **gen** and **mp** are taken from dynamic logic [12]. Note that due to aggregate updates, the standard axiom of dynamic logic for the assignment rule: $[f(a) =^{op} b]\varphi \Leftrightarrow \varphi|_{f(a)\rightsquigarrow b}$ expressed by our notation cannot be adopted in our logic.

p1. $[\mathbf{skip}]\varphi \Leftrightarrow \varphi$
p2. $[\alpha_1 ; \alpha_2]\varphi \Leftrightarrow [\alpha_1][\alpha_2]\varphi$
p3. $[?\varphi_1(\alpha)]\varphi_2 \Leftrightarrow (\varphi_1 \wedge [\alpha]\varphi_2) \vee (\neg \varphi_1 \wedge \varphi_2)$
gen. $\varphi \vdash [\alpha]\varphi$
mp. $\varphi, \varphi \Rightarrow \psi \vdash \psi$

- As pointed out in the logic for Abstract State Machines (ASMs) [16] axioms and inference rules of the classical logic with equality and Barcan axiom can also be borrowed into our work. Such as, axioms **existential** and **universal** say that some element t in the metafinite state satisfying formula φ implies that there exists something satisfying formula φ, and formula φ satisfied by all elements in the metafinite state implies that there is some element t

satisfying formula φ, respectively. The **Barcan** axiom specifies that domains of metafinite states with respect to an abstract metafinite state machine are fixed.

existential. $\varphi|_{x \rightsquigarrow t} \Rightarrow \exists x \varphi$ for $t \in \widehat{\mathbf{D}} \cup \widehat{\mathbf{C}}$

universal. $\forall x \varphi \Rightarrow \varphi|_{x \rightsquigarrow t}$ for $t \in \widehat{\mathbf{D}} \cup \widehat{\mathbf{C}}$

Barcan. $\forall x [\alpha] \varphi \Rightarrow [\alpha] \forall x \varphi$

Since two parallel programs may produce aggregate updates, which are identical but all affect update values of locations, the following statement is not true in our logic.

$$[\alpha \sqcup \alpha] \varphi \Leftrightarrow [\alpha] \varphi$$

Example 6. Suppose that firing rules in a program α over a metafinite state S yields an update multiset $\{\!\!\{(f_1(a), 3, \Pi), (f_1(a), 3, \Pi), (f_2(a), 1, \bot)\}\!\!\}$, where $\Pi = (id, \times, id)$ for the identity function id and the binary multiply function $\times$, then

- program α over S with the yielded update multiset $\mathfrak{M}_1$ results in
  - $f_1^{S + \mathfrak{M}_1}(a) = \Pi(\{\!\!\{3, 3\}\!\!\}) = id(id(3) \times id(3)) = 9$, and
  - $f_2^{S + \mathfrak{M}_1}(a) = 1$.
- program $\alpha \sqcup \alpha$ over S with the yielded update multiset $\mathfrak{M}_2$ results in
  - $f_1^{S + \mathfrak{M}_2}(a) = \Pi(\{\!\!\{3, 3, 3, 3\}\!\!\}) = id(id(3) \times id(3) \times id(3) \times id(3)) = 81$, and
  - $f_2^{S + \mathfrak{M}_2}(a) = 1$.

In terms of an abstract metafinite state machine, we use $\Gamma \models \varphi$ to indicate that φ is true for every model that makes all of set Γ of sentences true.

Theorem 1. *If* $\Gamma \vdash \varphi$, *then* $\Gamma \models \varphi$.

5 Discussion and Conclusion

Since metafinite states with backgrounds have background domains extended over a fixed set of atomic domains, the structure of a metafinite state is many-sorted. Therefore, the logic discussed in this paper is indeed a many-sorted first-order logic built on top of metafinite states. Considering that many-sorted first-order logic enjoys the same properties as first-order logic, it is possible to prove the completeness of the logic by adapting the idea of [16] to the many-sorted case. Furthermore, as a logic proposed for database transformations, it would be more interesting to investigate the expressiveness of the logic under the case of metafinite states with finitely active structures ranging over different components. We leave these for future work.

In summary, we have focused on the following work in this paper:

- a structure model for database transformations was developed by specifying database and algorithmic structures with different properties,

- abstract metafinite state machines deal with updates in multiset semantics and distinguish them as two types: exclusive and aggregate updates, and
- a logic characterizing static and dynamic features of abstract metafinite state machines was proposed with a set of axioms and inference rules.

References

1. Abiteboul, S., Hull, R., Vianu, V. (eds.): Foundations of Databases: The Logical Level. Addison-Wesley Longman Publishing Co., Inc., Boston (1995)
2. Abiteboul, S., Kanellakis, P.C.: Object identity as a query language primitive. In: SIGMOD 1989. Proceedings of the 1989 ACM SIGMOD international conference on Management of data, pp. 159–173. ACM Press, New York (1989)
3. Abiteboul, S., Simon, E., Vianu, V.: Non-deterministic languages to express deterministic transformations. In: PODS 1990. Proceedings of the ninth ACM SIGACT-SIGMOD-SIGART symposium on Principles of database systems, pp. 218–229. ACM Press, New York (1990)
4. Abiteboul, S., Vianu, V.: A translation language complete for database update and specification. In: PODS 1987. Proceedings of the sixth ACM SIGACT-SIGMOD-SIGART symposium on Principles of database systems, pp. 260–268. ACM Press, New York (1987)
5. Abiteboul, S., Vianu, V.: Datalog extensions for database queries and updates. Tech. Rep. RR-0900, 09 (1988)
6. Blass, A., Gurevich, Y.: Background, reserve, and gandy machines. In: Proceedings of the 14th Annual Conference of the EACSL on Computer Science Logic, pp. 1–17. Springer, London (2000)
7. Blass, A., Gurevich, Y.: Abstract state machines capture parallel algorithms. ACM Trans. Comput. Logic 4(4), 578–651 (2003)
8. Blass, A., Gurevich, Y.: Background of computation. In: EATCS 1992. Bulletin of the European Association for Theoretical Computer Science (2007)
9. Cohen, S.: User-defined aggregate functions: bridging theory and practice. In: SIGMOD 2006. Proceedings of the 2006 ACM SIGMOD international conference on Management of data, pp. 49–60. ACM Press, New York (2006)
10. Grädel, E., Gurevich, Y.: Metafinite model theory. In: Leivant, D. (ed.) LCC 1994. LNCS, vol. 960, pp. 313–366. Springer, Heidelberg (1995)
11. Gurevich, Y.: Sequential abstract-state machines capture sequential algorithms. ACM Trans. Comput. Log. 1(1), 77–111 (2000)
12. Harel, D., Tiuryn, J., Kozen, D.: Dynamic Logic. MIT Press, Cambridge (2000)
13. Nanchen, S., Stärk, R.F.: A logic for secure memory access of abstract state machines. Theor. Comput. Sci. 336(2–3), 343–365 (2005)
14. Poetzsch-Heffter, A.: Deriving Partial Correctness Logics From Evolving Algebras. In: Pehrson, B., Simon, I. (eds.) IFIP 13th World Computer Congress, Technology/Foundations, pp. 434–439. Elsevier, Amsterdam, The Netherlands (1994)
15. Schönegge, A.: Extending dynamic logic for reasoning about evolving algebras
16. Stark, R., Nanchen, S.: A logic for abstract state machine. Journal of Universal Computer Science 7, 11 (2001)
17. Van den Bussche, J.: Formal Aspects of Object Identity in Database Manipulation. PhD thesis, University of Antwerp (1993)
18. Van den Bussche, J., Van Gucht, D.: Semi-determinism. In: PODS 1992. Proceedings of the eleventh ACM SIGACT-SIGMOD-SIGART symposium on Principles of database systems, pp. 191–201. ACM Press, New York (1992)

19. Van den Bussche, J., Van Gucht, D.: Non-deterministic aspects of object-creating database transformations. In: Selected Papers from the Fourth International Workshop on Foundations of Models and Languages for Data and Objects, pp. 3–16. Springer, London, UK (1993)
20. Van Den Bussche, J., Van Gucht, D., Andries, M., Gyssens, M.: On the completeness of object-creating database transformation languages. J. ACM 44(2), 272–319 (1997)
21. Wang, Q., Schewe, K.-D.: Axiomatization of database transformations. In: Proceedings of the ASM'07: The 14th International ASM Workshop (2007)

Towards a Fuzzy Logic for Automated Multi-issue Negotiation

Azzurra Ragone[1], Umberto Straccia[2], Tommaso Di Noia[1], Eugenio Di Sciascio[1], and Francesco M. Donini[3]

[1] SisInfLab, Politecnico di Bari, Bari, Italy
{a.ragone,t.dinoia,disciascio}@poliba.it
[2] ISTI-CNR, Pisa, Italy
straccia@isti.cnr.it
[3] Università della Tuscia, Viterbo, Italy
donini@unitus.it

Abstract. We present a novel logic-based approach to automate multi-issue bilateral negotiation in e-marketplaces. In such frameworks issues to negotiate on can be multiple, interrelated, and may not be fixed in advance. We use logic to model relations among issues and to allow agents express their preferences on them. In particular, we introduce the logic $\mathcal{P}(\mathcal{N})$, a fuzzy propositional logic extended with concrete domains in order to handle numerical, as well as non numerical features, and to deal with vagueness in buyer/seller preferences. Hence, agents can express preferences as *e.g., I am searching for a passenger car costing about 25000€ yet if the car has a GPS system and more than two-year warranty I can spend up to 28000€.*

We illustrate the theoretical framework, the logical language, the protocol we adopt and show that using a mediator with a proactive behavior we can compute Pareto-efficient agreements.

1 Introduction

Parsons et al. [16] define negotiation as "the process by which a group of agents communicate with one other to try and come to a mutually acceptable agreement on some matter." Several negotiation mechanisms have been proposed in literature to model different scenarios, as each scenario has its own peculiarities and issues. In this paper we refer to mechanisms to automate multi-issue bilateral negotiation in peer-to-peer (P2P) e-marketplaces [25], where products (cars, houses, Personal Computers, etc.) or services (travel booking, wedding service, etc.) can be, at the same time, provided by suppliers or searched by potential customers who are endowed of peer opportunitites as they enter the marketplace. Automated negotiation mechanisms in such e-marketplaces need to represent, in a machine understandable way, the product characteristics, the request/offer descriptions and the preferences of the users entering the markeplace. In fact, differently from e-marketplaces dealing with undifferentiated products (*e.g.*, oil, commodities) in *e.g.*, an automotive e-marketplace price cannot be the only issue to negotiate on, but also other features as warranty, delivery time, as well as model, color,

S. Hartmann and G. Kern-Isberner (Eds.): FoIKS 2008, LNCS 4932, pp. 381–396, 2008.

optionals, have to be taken into account. Moreover such issues may not be established in advanced, as it is a common assupmtion in many other negotiation scenarios (task and resource allocation, auctions). Therefore there is a need for Knowledge Representation languages able to model relations among issues and to allow agents share a common protocol during the negotiation.

We propose here a *fuzzy propositional logic* endowed with *concrete domains* to model relations among issues and as a communication language between agents. We may represent facts such as that a *Ferrari is an Italian car* (Ferrari⇒ ItalianMaker), or that a *Sedan is a type of Passenger Car* (Sedan ⇒ PassengerCar), or the fact that a car cannot have at the same time a *Diesel and a GAS engine* (Diesel ⇒ ¬Gasoline). Such kind of relations can be expressed in a Theory (from now on an Ontology) $\mathcal{T}$. Furthermore, we may represent *preferences*, such as *e.g.*, a seller can state that *"If you want an embedded alarm system you'll have to wait more than one month"* (AlarmSystem⇒ deliverytime $\geq$ 30), as well as a buyer can state that *"I would like a passenger car with an alarm system if it costs more than 25000€"*(PassengerCar ∧ (price $\geq$ 25000 ⇒ AlarmSystem)). In our proposal, concrete domains allow to deal with numerical features, which are mixed, in preferences, with non numerical ones.

We note that in the negotiation scenario we model a buyer request, as well as a seller supply, can be split into two parts: one involving issues that have to be necessarily satisfied in order to accept a final agreement, which we call *hard constraints*, and another one involving issues buyer and seller are willing to negotiate on, we call these *soft constraints*. Among *soft constraints* there can be also *fuzzy constraints*, which are preferences involving numerical features. *Fuzzy constraints* are represented in our approach using fuzzy membership functions, see Section 3, therefore while a simple *soft constraint* can or cannot be satisfied, a *fuzzy constraints* can also be satisfied to a "certain degree". For example, a buyer can state, among soft constraints, that *if a GPS system is mounted on the car she can spend up to 25000 for a sedan*; if the price in the proposed agreement is equal to 25500 we should not simply say that the preference is not satisfied at all, but rather that is satisfied to a certain degree, as will be better described later on (see Section 4).

We note that in our framework it will be possible to model *positive* and *negative* preferences (I would like a car black or gray, but not red), as well as *conditional preferences* (I would like leather seats if the car is black) involving both numerical features and non numerical ones (If you want a car with GPS system you have to wait at least one month) or only numerical ones (I accept to pay more than 25000€ only if there is more than a two-year warranty).

Besides we model *quantitative* preferences; thanks to the weight assigned to each preference it is possible to determine a relative importance among them, rather than only a total order between them. Obviously, the whole approach holds also if the user does not specify a weight for each preference, but only a global order on preferences. However, in that case, the relative importance among preferences is missed.

The rest of the paper is structured as follows: next section discusses the assumptions we make and the negotiation mechanism we adopt. In Section 3 we illustrate the modeling of issues through our logical language and then we define the multi-issue negotiation problem and how to compute Pareto agreements. In Section 6 the whole negotiation process is highlighted with the aid of a simple example. Related Work and discussion close the paper.

2   Negotiation Mechanism

We start outlining the scenario and assumptions that characterize the proposed negotiation mechanism. Following [22] we define the *Space of possible deals*, the *Negotiation Protocol* and the *Negotiation Strategy*. The *Space of possible deals* is the set of all possible agreements, in our framework we define an agreement as a model for the theory and the set of *hard constraints* (see Section 4). Furthermore we are not only interested in a *feasible* agreement, but in agreements which are *Pareto efficient*[1]. In order to ensure that agents reached agreements which are Pareto efficients a *protocol* and *strategies*, suitable for such a protocol, have to be defined. We adopt a *one-shot* protocol with the intervention of a *mediator* with a proactive behavior. Differently from the classical Single-shot bargaining [20], where one player proposes a deal and the other player may only accept or refuse it [2], in our framework we hypothesize the presence of an electronic mediator, that may automatically explore the negotiation space and discover Pareto-efficient agreements to be proposed to both parties. As pointed out in [21, p.311], usually bargainers are obviously reluctant to disclose their true preferences or utilities to the other party, but they are more willing to reveal these information to a trusted – automated – mediator, helping negotiating parties to achieve efficient and equitable outcomes. The presence of a mediator and the one-shot protocol is an incentive for the two parties to reveal the true preferences, as they can trust the fairness of the mediator and they have a single possibility to reach the agreement with that counterpart. Thanks to the presence of a mediator we can model a negotiation with *incomplete information*, where agents do not know anything about their counterparts, neither preferences nor worth of them. For what concerns *strategy*, the bargainers reveal their preferences to the mediator and then, once it has computed a solution, they can accept or refuse the agreement proposed to them; if one of them or both refuse the agreement proposed by the mediator the negotiation ends with a *conflict deal*. Bargainers may refuse if they think possible to reach a better agreement looking for another partner or for a different set of bidding rules.

3   Representation of Issues

We divide issues involved in a negotiation in two categories. Some issues may simple express properties that are true or false, like *e.g.*, in an automotive domain, Sedan, DriverInsurance. We represent such issues as propositional atoms $A_1, A_2, \ldots$ from

---

[1] An agreement is Pareto-efficient if there is no other agreement that will make at least one participant better off without making at least one other participant worse off. If a negotiation outcome is not Pareto-efficient, then there is another outcome that will make at least one participant happier while keeping everyone else at least as happy [9].

a finite set $\mathcal{A}$. While we represent issues involving numerical features as variables $f_1, f_2, \ldots$, each one taking values in its specific domain $D_{f_1}, D_{f_2}, \ldots$, such as $[0, 96]$ (months) for month_warranty, or $[1,000, 50,000]$ (euros), for price.

The variables representing numerical features are either involved in *hard constraints* or *soft constraints*. In hard constraints, the variables are always constrained by comparing them to some constant, like price $< 20,000$, or month_warranty ≥ 60, and such constraints can be combined into complex propositional requirements – also involving propositional issues – *e.g.*, Sedan $\wedge$ (price $\leq 25,000$) $\wedge$ (deliverytime < 30) (representing a sedan, costing no more than 25,000 euros, delivered in less than 30 days), or AlarmSystem $\wedge$ (price $> 26,000$) (expressing the seller's requirement "if you want an alarm system mounted you'll have to spend more than 26,000 euros"). Vice-versa when numerical features are involved in *soft constraints*, also called *fuzzy constraints*, the variables representing numerical features are constrained by so-called fuzzy membership functions, as shown in Figure 1. For instance, price $ls(18000, 22000)$ dictates that given a price it returns the degree of truth to which the constraint is satisfied. Essentially, price $ls(18000, 22000)$ states that if the price is no higher than 18000 then the constraint is definitely satisfied, while if the price is higher than 22000 then the constraint is definitely not satisfied. In between 18000 and 22000, we use linear interpolation, given a price, to evaluate the satisfaction degree of the constraint.

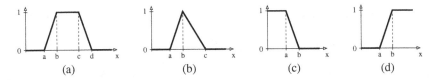

Fig. 1. (a) Trapezoidal function $trz(a, b, c, d)$, (b) triangular function $tri(a, b, c)$, (c) left shoulder function $ls(a, b)$, and (d) right shoulder function $rs(a, b)$

We now give precise definitions for the above intuitions, borrowing from a previous formalization of so-called *concrete domains* [1] from Knowledge Representation languages.

Definition 1 (Concrete Domains, [1]). *A concrete domain D consists of a non-empty set $\Delta_c(D)$ of numerical values, and a set of predicates $C(D)$ expressing numerical constraints on D.*

For our numerical features, predicates will always be of the form $\geq k, \leq k, = k$ (k is a value in D) if *hard constraints*, e.g., price ≤ 26000, or of the form $ls(a, b), rs(a, b)$, $tri(a, b, c), trz(a, b, c, d)$ if *soft constraints*, e.g., deliverytime $ls(30, 40)$. The former predicate expresses a strict constraint (*hard*), *my budget is such that I cannot spend more than 26000 euros*, the latter one expresses a *soft* constraint since I *would prefer not to wait more than one month for a car, but I can be disposal to wait until 40 days, furthermore in such a case I will be less satisfied.*

Once we have defined a concrete domain and constraints, we can formally extend propositional logic in order to handle numerical features. We call this language $\mathcal{P}(\mathcal{N})$.

Definition 2 (The language $\mathcal{P}(\mathcal{N})$). *Let $\mathcal{A}$ be a set of propositional atoms, and F a set of pairs $\langle f, D_f \rangle$ each made of a feature name and an associated concrete domain D_f, and let k be a value in D_f. Then the following formulas are in $\mathcal{P}(\mathcal{N})$:*

1. *every atom $A \in \mathcal{A}$ is a formula in $\mathcal{P}(\mathcal{N})$*
2. *if $\langle f, D_f \rangle \in F$, $k \in D_f$, and $c \in \{\geq, \leq, =\}$ then $(f\ c\ k)$ is a formula in $\mathcal{P}(\mathcal{N})$*
3. *if $\langle f, D_f \rangle \in F$ and c is of the form $ls(a, b), rs(a, b), tri(a, b, c), trz(a, b, c, d)$ then $(f\ c)$ is a formula in $\mathcal{P}(\mathcal{N})$*
4. *if ψ and φ are formulas in $\mathcal{P}(\mathcal{N})$ and $n \in [0, 1]$ then so are $\neg\psi$, $\psi \wedge \varphi$, $\psi \vee \varphi$, $\psi \Rightarrow \varphi$ and $n \cdot \psi$. We use $\psi \Leftrightarrow \varphi$ in place of $(\psi \Rightarrow \varphi) \wedge (\varphi \Rightarrow \psi)$.*

In order to define a formal semantics of $\mathcal{P}(\mathcal{N})$ formulas, we consider interpretation functions $\mathcal{I}$ that map propositional atoms into $[0, 1]$, feature names into values in their domain, and define the truth of composite formulas as follows. Let $\neg: [0, 1] \rightarrow [0, 1]$, $\wedge: [0, 1] \times [0, 1] \rightarrow [0, 1]$, $\vee: [0, 1] \times [0, 1] \rightarrow [0, 1]$ and $\Rightarrow: [0, 1] \times [0, 1] \rightarrow [0, 1]$ be functions to be used to interpret negation, conjunction (a t-norm), disjunction (s-norm) and implication, respectively [8]. Note that we use the the same symbol, e.g., $\neg$ for both to identify the negation of a formula as well as the negation of a truth value. The meaning will always be determined by the signature of the symbol. The choice of them is not arbitrary, but is restricted, as usual, by the conditions described in Figure 2. Some specific choices instead are described in Figure 3, while Figure 4 highlights some salient properties of them. It is important to note that we can never enforce that a choice of the interpretation of the connectors satisfies all properties of Figure 4, because then the logic will collapse to classical boolean propositional logic.

Axiom Name	T-norm	S-norm
Tautology / Contradiction	$a \wedge 0 = 0$	$a \vee 1 = 1$
Identity	$a \wedge 1 = a$	$a \vee 0 = a$
Commutativity	$a \wedge b = b \wedge a$	$a \vee b = b \vee a$
Associativity	$(a \wedge b) \wedge c = a \wedge (b \wedge c)$	$(a \vee b) \vee c = a \vee (b \vee c)$
Monotonicity	if $b \leq c$, then $a \wedge b \leq a \wedge c$	if $b \leq c$, then $a \vee b \leq a \vee c$

Axiom Name	Implication Function	Negation Function
Tautology / Contradiction	$0 \rightarrow b = 1$	$\neg 0 = 1, \neg 1 = 0$
	$a \rightarrow 1 = 1$	
Antitonicity	if $a \leq b$, then $a \rightarrow c \geq b \rightarrow c$	if $a \leq b$, then $\neg a \geq \neg b$
Monotonicity	if $b \leq c$, then $a \rightarrow b \leq a \rightarrow c$	

Usually,

$$a \rightarrow b = \sup\{c : a \wedge c \leq b\}$$

is used and is called *r-implication* and depends on the t-norm only.

Fig. 2. Conditions on norms

Definition 3 (Interpretation and models). *An interpretation $\mathcal{I}$ for $\mathcal{P}(\mathcal{N})$ is a function (denoted as a superscript $\cdot^{\mathcal{I}}$ on its argument) that maps each atom in $\mathcal{A}$ into a truth value $A^{\mathcal{I}} \in [0, 1]$, each feature name f into a value $f^{\mathcal{I}} \in D_f$, and assigns truth values in $[0, 1]$ to formulas as follows:*

	Łukasiewicz Logic	Gödel Logic	Product Logic	Zadeh
$\neg x$	$1 - x$	if $x = 0$ then 1 else 0	if $x = 0$ then 1 else 0	$1 - x$
$x \wedge y$	$\max(x + y - 1, 0)$	$\min(x, y)$	$x \cdot y$	$\min(x, y)$
$x \vee y$	$\min(x + y, 1)$	$\max(x, y)$	$x + y - x \cdot y$	$\max(x, y)$
$x \Rightarrow y$	if $x \leq y$ then 1 else $1 - x + y$	if $x \leq y$ then 1 else y	if $x \leq y$ then 1 else y/x	$\max(1 - x, y)$

Fig. 3. Typical norms

Property	Łukasiewicz Logic	Gödel Logic	Product Logic	Zadeh Logic
$x \wedge \neg x = 0$	•	•	•	
$x \vee \neg x = 1$	•			
$x \wedge x = x$		•		•
$x \vee x = x$		•		•
$\neg \neg x = x$	•			•
$x \rightarrow y = \neg x \vee y$	•			•
$\neg(x \rightarrow y) = x \wedge \neg y$	•			•
$\neg(x \wedge y) = \neg x \vee \neg y$	•	•	•	•
$\neg(x \vee y) = \neg x \wedge \neg y$	•	•	•	•

Fig. 4. Some properties of norms

- *for hard constraints, $(f\ c\ k)^{\mathcal{I}} = 1$ iff the relation $f^{\mathcal{I}}\ c\ k$ is true in D_f, $(f\ c\ k)^{\mathcal{I}} = 0$ otherwise*
- *for soft constraints, $(f\ c)^{\mathcal{I}} = c(f^{\mathcal{I}})$, i.e., the result of evaluating the fuzzy membership function c on the value $f^{\mathcal{I}}$*
- $(\neg\psi)^{\mathcal{I}} = \neg\psi^{\mathcal{I}}, (\psi \wedge \varphi)^{\mathcal{I}} = \psi^{\mathcal{I}} \wedge \varphi^{\mathcal{I}}, (\psi \vee \varphi)^{\mathcal{I}} = \psi^{\mathcal{I}} \vee \varphi^{\mathcal{I}}, (n \cdot \psi)^{\mathcal{I}} = n \cdot \psi^{\mathcal{I}}$ *and* $(\psi \Rightarrow \varphi)^{\mathcal{I}} = \psi^{\mathcal{I}} \Rightarrow \varphi^{\mathcal{I}}.$

Given a formula φ in $\mathcal{P}(\mathcal{N})$, we denote with $\mathcal{I} \models \varphi$ the fact that $\mathcal{I}$ assigns 1 to φ. If $\mathcal{I} \models \varphi$ we say $\mathcal{I}$ is a model *for φ, and $\mathcal{I}$ is a model for a set of formulas when it is a model for each formula.*

Clearly, an interpretation $\mathcal{I}$ is completely defined by the values it assigns to propositional atoms and numerical features.

Example 1. Let $\mathcal{A} = \{\texttt{PassengerCar}, \texttt{Diesel}\}$ be a set of propositional atoms, $D_{\texttt{price}} = \{0, \ldots, 60000\}$ and $D_{\texttt{year\_warranty}} = \{0, 1, \ldots, 5\}$ be two concrete domains for the features price, year_warranty, respectively. A model $\mathcal{I}$ for both formulas:

$$\left\{ \begin{array}{l} \texttt{PassengerCar} \wedge (\texttt{Diesel} \Rightarrow (\texttt{year\_warranty} \geq 1)), \\ (\texttt{price} \leq 5{,}000) \end{array} \right\}$$

is $\texttt{PassengerCar}^{\mathcal{I}} = 1, \texttt{Diesel}^{\mathcal{I}} = 0, \texttt{year\_warranty}^{\mathcal{I}} = 0, \texttt{price}^{\mathcal{I}} = 4{,}500.$

Definition 4 (Łukasiewicz Logic as intended semantics). *For the sake of our purpose, for the remainder of the paper we will use Łukasiewicz Logic as the specific interpretation of the connectives.*

The reason for this choice is due to the nice logical and computational properties of Łukasiewicz Logic. Furthermore, note that $x \wedge_G y = \min(x, y)$ and $x \vee_G y = \max(x, y)$ can also be defined in it by means of $x \wedge (x \rightarrow y)$ and $\neg(\neg x \wedge_G \neg y)$, respectively.

Given a set of formulas $\mathcal{T}$ in $\mathcal{P}(\mathcal{N})$ (representing an ontology), we denote with $\mathcal{I} \models \mathcal{T}$ that $\mathcal{I}$ is a *model* for $\mathcal{T}$. An ontology is *satisfiable* if it has a model. $\mathcal{T}$ logically implies a formula φ, denoted by $\mathcal{T} \models \varphi$ iff $\varphi^{\mathcal{I}} = 1$ in all models of $\mathcal{T}$. We denote with $\mathcal{M}_{\mathcal{T}}$, the set of all models for $\mathcal{T}$, and omit the subscript when no confusion arises. We also denote with

$$|\phi|_{\mathcal{T}} = \inf_{\mathcal{I} \models \mathcal{T}} \phi^{\mathcal{I}} \, ,$$

the lower bound of ϕ's truth degree over all models $\mathcal{T}$, while with

$$\langle \phi \rangle_{\mathcal{T}} = \sup_{\mathcal{I} \models \mathcal{T}} \phi^{\mathcal{I}} \, ,$$

we denote the maximal truth degree of ϕ over all models $\mathcal{T}$.

Example 2. Consider $\mathcal{T}$ as the set of formulae

$$\left\{ \begin{array}{l} \texttt{PassengerCar} \wedge (\texttt{Diesel} \Rightarrow (\texttt{year\_warranty} \geq 1)), \\ (\texttt{price} \leq 5{,}000), \end{array} \right\}$$

Then, $|(\texttt{price rs}(3000, 6000))|_{\mathcal{T}} = 0$, while $\langle (\texttt{price rs}(3000, 6000)) \rangle_{\mathcal{T}} = 1/3$.

4   Multi Issue Bilateral Negotiation in $\mathcal{P}(\mathcal{N})$

Following [18], we use logic formulas in $\mathcal{P}(\mathcal{N})$ to model the buyer's demand and the seller's supply. Relations among issues, both propositional and numerical, are represented by a set $\mathcal{T}$ – for Theory – of $\mathcal{P}(\mathcal{N})$ formulas.

As we have stated before, in a typical bilateral negotiation scenario, issues within both the buyer's request and the seller's offer can be split into *hard constraints* and *soft constraints*. In the rest of the paper we call *hard constraints*, issues that have to be necessarily satisfied in the final agreement, *demand/supply*. *Soft constraints*, denoting issues they are willing to negotiate on, *preferences*.

Example 3. Suppose to have a buyer's request like: "I am searching for a Passenger Car provided with Diesel engine. I need the car as soon as possible, and I can not wait more than one month. Preferably I would like to pay less than 22,000 € furthermore I am willing to pay up to 24,000 € if warranty is greater than 160000 km. (I won't pay more than 27,000 €)". In this example it is possible to distinguish between *hard constraints* (**demand**) and *soft constraints* (**preferences**).

demand: I want a Passenger Car provided with Diesel feeding. I can not wait more than one month. I won't pay more than 27,000 € .
preferences: I would like to pay less than 22,000 € furthermore I am willing to pay up to 24,000 € if warranty is greater than 160000 km.

Definition 5 (Demand, Supply, Agreement). *Given an ontology $\mathcal{T}$ represented as a set of formulas in $\mathcal{P}(\mathcal{N})$ representing the knowledge on a marketplace domain*

– *a buyer's* demand *is a formula* β *(for Buyer) in* $\mathcal{P}(\mathcal{N})$ *such that* $\mathcal{T} \cup \{\beta\}$ *is satisfiable.*
– *a seller's* supply *is a formula* σ *(for Seller) in* $\mathcal{P}(\mathcal{N})$ *such that* $\mathcal{T} \cup \{\sigma\}$ *is satisfiable.*
– $\mathcal{I}$ *is a* possible deal *between* β *and* σ *iff* $\mathcal{I} \models \mathcal{T} \cup \{\sigma, \beta\}$, *that is,* $\mathcal{I}$ *is a model for* $\mathcal{T}$, σ, *and* β. *We also call* $\mathcal{I}$ *an* agreement.

The seller and the buyer model in σ and β the minimal requirements they accept for the negotiation. On the other hand, if seller and buyer have set *hard constaints* that are in conflict with each other, that is $\mathcal{T} \cup \{\sigma, \beta\}$ has no models, then the negotiation ends immediately because, it is impossible to reach an agreement. If the participants are willing to avoid the *conflict deal* [22], and continue the negotiation, it will be necessary they revise their *hard constaints*.

In the negotiation process both the buyer and the seller express some preferences on attributes, or their combination in terms of weighted formulae. While there may be many different ways to define preferences, for the sake of our work we define:

Definition 6 (Preferences). *The buyer's negotiation preference* $\mathcal{B}$ *is a formula of the form* $n_1 \cdot \beta_1 \vee \ldots \vee n_k \cdot \beta_k$, *where each* β_i *represents the subject of a buyer's preference, and* n_i *is the utility associated to it. We assume that* $\Sigma_i n_i = 1$. *Analogously, the seller's negotiation preference* $\mathcal{S}$ *is a formula of the form* $m_1 \cdot \sigma_1 \vee \ldots \vee m_h \cdot \sigma_h$, *where each* σ_i *represents the subject of a seller's preference, and* m_i *is the utility associated to it. We assume that* $\Sigma_i m_i = 1$.

Note that a formula of the form $n_1 \cdot \beta_1 \vee \ldots \vee n_k \cdot \beta_k$ is under Łukasiewicz logic the weighted sum of the degree of truth of the β_i.

For instance, the Buyer's request in Example 3 is formalized as:

$$\beta = \texttt{PassengerCar} \wedge \texttt{Diesel} \wedge (\texttt{price} \leq 27,000) \wedge$$
$$(\texttt{deliverytime} \leq 30)$$
$$\beta_1 = (\texttt{price}, \texttt{ls}(22000, 25000))$$
$$\beta_2 = (\texttt{km\_warranty}, \texttt{rs}(140000, 160000)) \Rightarrow (\texttt{price}, \texttt{ls}(24000, 27000))$$

As usual, both agents' utilities are normalized to 1 to eliminate outliers, and make them comparable. Since we assumed that utilities are additive, the utility function, that we call *preference utility*, is just a weighted sum of the utilities of preferences satisfied in the agreement.

Definition 7 (Preference Utilities). *Let* $\mathcal{B}$ *and* $\mathcal{S}$ *be respectively the buyer's and seller's preference, and* $\mathcal{M}_{\mathcal{T} \cup \{\alpha, \beta\}}$ *be their agreements set. The preference utility of an agreement* $\mathcal{I} \in \mathcal{M}_{\mathcal{T} \cup \{\alpha, \beta\}}$ *for a buyer and a seller, respectively, are defined as:*

$$u_{\beta, \mathcal{P}(\mathcal{N})}(\mathcal{I}) \doteq \mathcal{B}^{\mathcal{I}}$$
$$u_{\sigma, \mathcal{P}(\mathcal{N})}(\mathcal{I}) \doteq \mathcal{S}^{\mathcal{I}}.$$

Where $\mathcal{B}^{\mathcal{I}}$, as well as $\mathcal{S}^{\mathcal{I}}$, is a weighted sum of the degree of truth of the β_i under Łukasiewicz logic.

Notice that if one agent *e.g.*, the buyer, does not specify *soft constraints*, but only *hard constraints*, it is as $\beta_1 = \top$ and $\mathcal{B}^\mathcal{I} = 1$, which reflects the fact that an agent accepts whatever agreement not in conflict with its *hard constraints*.

From the formulas related to Example 3, we note that while considering numerical features, it is still possible to express hard and soft constraints on them. A *hard constraint* expresses on a numerical feature is surely the ***reservation value*** [21]. In Example 3 the buyer expresses two reservation values, one on price *"more than 27,000 €"* and the other on delivery time *"less than 1 month"*.

Both buyer and seller can express a reservation values on numerical feature involved in the negotiation process. It is the maximum (or minimum) value in the range of possible feature values to reach an agreement, *e.g.*, the maximum price the buyer wants to pay for a car or the minimum warranty required, as well as, from the seller's perspective the minimum price he will accept to sell the car or the minimum delivery time. Usually, each participant knows its own reservation value and ignores the opponent's one. Referring to price and the two corresponding reservation values $r_{\beta,price}$ and $r_{\sigma,price}$ for the buyer and the seller respectively, if the buyer expresses $price \leq r_{\beta,price}$ and the seller $price \geq r_{\sigma,price}$, in case $r_{\sigma,price} \leq r_{\beta,price}$ we have $[r_{\sigma,price}, r_{\beta,price}]$ as a **Z**one **O**f **P**ossible **A**greement — $ZOPA(price)$, otherwise no agreement is possible [21]. More formally, given an agreement $\mathcal{I}$ and a feature f, $f^\mathcal{I} \in ZOPA(f)$ must hold.

Keeping the price example, let us suppose that the maximum price the buyer is willing to pay is 25,000, while the seller minimum allowable price is 20,000, then we can set the two reservation values: $r_{\beta,price} = 25,000$ and $r_{\sigma,price} = 20,000$, so the *agreement price* will be in the interval $ZOPA(price) = [20000, 25000]$.

Obviously, the reservation value is considered as private information and will not be revealed to the other party, but will be taken into account by the mediator when the agreement will be computed. Since setting a reservation value on a numerical feature is equivalent to set a strict requirement, then, once the buyer and the seller express their strict requirements, reservation values constraints have to be added to them (see Example 3).

In order to formally define a Multi-issue Bilateral Negotiation problem in $\mathcal{P}(\mathcal{N})$, the only other elements we still need to introduce are the *disagreement thresholds*, also called disagreement payoffs, t_β, t_σ. They are the minimum utility that each agent requires to pursue a deal. Minimum utilities may incorporate an agent's attitude toward concluding the transaction, but also overhead costs involved in the transaction itself, *e.g.*, fixed taxes.

Definition 8 (MBN-$\mathcal{P}(\mathcal{N})$). *Given a $\mathcal{P}(\mathcal{N})$ set of axioms $\mathcal{T}$, a demand β and a set of buyer's preferences $\mathcal{B}$ with utility function $\mathcal{B}^\mathcal{I}$ and a disagreement threshold t_β, a supply σ and a set of seller's preferences $\mathcal{S}$ with utility function $\mathcal{S}^\mathcal{I}$ and a disagreement threshold t_σ, a **Multi-issue Bilateral Negotiation problem (MBN)** is finding a model $\mathcal{I}$ (agreement) such that all the following conditions hold:*

$$\mathcal{I} \models \mathcal{T} \cup \{\sigma, \beta\} \tag{1}$$
$$\mathcal{B}^\mathcal{I} \geq t_\beta \tag{2}$$
$$\mathcal{S}^\mathcal{I} \geq t_\sigma \tag{3}$$

Note that not every agreement $\mathcal{I}$ is a solution of an MBN, if either $\mathcal{B}^{\mathcal{I}} < t_\sigma$ or $\mathcal{S}^{\mathcal{I}} < t_\beta$. Such an agreement represents a deal which, although satisfying strict requirements, is not worth the transaction effort. Also notice that, since reservation values on numerical features are modeled in β and σ as strict requirements, for each feature f, the condition $f^{\mathcal{I}} \in ZOPA(f)$ always holds by condition (1).

5 Computing Pareto Agreements in $\mathcal{P}(\mathcal{N})$

Among all possible agreements that we can compute, given a theory $\mathcal{T}$ as constraint, we are interested in agreements that are Pareto-efficient and *fair* for both the participants, in order to make them equally, and as much as possible, satisfied. Formally, let ψ, φ be two formulae, and let $*$ be a connective interpreted as the product t-norm (see Figure 2) Then a *Pareto agreement* is defined as follows. Let $\mathcal{T}$ be an ontology, let β be the buyer's demand, let σ be the seller's supply, let $\mathcal{B}$ and $\mathcal{S}$ be respectively the buyer's and seller's preferences. Let r be a rational in $[0, 1]$ and let us assume that we admit formulae of the form $r \Rightarrow \psi$ and $\psi \Rightarrow r$. We define $\mathcal{I} \models r \Rightarrow \psi$ iff $\psi^{\mathcal{I}} \geq r$ (the truth degree of ψ is equal or greater than r), while $\mathcal{I} \models \psi \Rightarrow r$ iff $\psi^{\mathcal{I}} \leq r$ (the truth degree of ψ is equal or less than r). Furthermore, let $\bar{\mathcal{T}}$ be the ontology

$$\bar{\mathcal{T}} = \mathcal{T} \cup \{\beta, \sigma\} \cup \{\texttt{buy} \Leftrightarrow \mathcal{B}, \texttt{sell} \Leftrightarrow \mathcal{S}, t_\beta \Rightarrow \mathcal{B}, t_\sigma \Rightarrow \mathcal{S}\} .$$

Then a *Pareto agreement* is an interpretation $\bar{\mathcal{I}}$ such that

$$\bar{\mathcal{I}} = \arg \max_{\mathcal{I} \models \bar{\mathcal{T}}} (\texttt{buy})^{\mathcal{I}} * (\texttt{sell})^{\mathcal{I}} .$$

It is not difficult to see that Pareto agreements can be also characterized as the set of all models $\mathcal{I}$ of $\bar{\mathcal{T}}$, such that

$$(\texttt{buy})^{\mathcal{I}} * (\texttt{sell})^{\mathcal{I}} = \langle \texttt{buy} * \texttt{sell} \rangle_{\bar{\mathcal{T}}} .$$

The value $\langle \texttt{buy} * \texttt{sell} \rangle_{\bar{\mathcal{T}}}$ is called the *Pareto agreement value*. It is easily verified that while the Pareto agreement value is unique, there may be many different Pareto agreements (*i.e.*, interpretations) with the same Pareto agreement value.

Computing a Pareto agreement is in fact easy, using Quadratic Mixed Integer Linear Programming. We start with replacing any formula ψ in $\bar{\mathcal{T}}$ with the formula $1 \Rightarrow \psi$ (ψ is true to degree 1). Now, it is not difficult to see that we can recursively associate to any formula $1 \Rightarrow \psi \in \bar{\mathcal{T}}$ a set of linear in-equations $Eq(1 \Rightarrow \psi)$, by assigning to any propositional letter p a variable x_p (see, *e.g.*, [8,14], see also Figure 5).

Then, we solve the Quadratic Mixed Integer Linear Programming problem[2]

$$\max x_{\texttt{buy}} \cdot x_{\texttt{sell}}$$

$$\bigcup_{1 \Rightarrow \psi \in \bar{\mathcal{T}}} Eq(1 \Rightarrow \psi)$$

Any assignment to the variables in the optimal solution corresponds to a Pareto agreement.

---

[2] Note that the fuzzy membership functions in Figure 1 are combination of linear functions and, thus, can be mapped into a set of linear in-equations as well, which we do not report here, – see [10,23].

$$\varphi \lor \psi \qquad\qquad \mapsto \neg(\neg\varphi \land \neg\psi)$$
$$r \to p \qquad\qquad \mapsto x_p \geq r, x_p \in [0, 1]$$
$$p \to r \qquad\qquad \mapsto x_p \leq r, x_p \in [0, 1]$$
$$r \to \neg\varphi \qquad\qquad \mapsto \varphi \to (1 - r)$$
$$\neg\varphi \to r \qquad\qquad \mapsto (1 - r) \to \varphi$$
$$r \to (\varphi \land \psi) \mapsto x_1 \to \varphi, x_2 \to \psi, y \leq 1 - r, x_i \leq 1 - y, x_1 + x_2 = r + 1 - y,$$
$$x_i \in [0, 1], y \in \{0, 1\}$$
$$(\varphi \land \psi) \to r \mapsto x_1 \to \neg\varphi, x_2 \to \neg\psi, x_1 + x_2 = 1 - r, x_i \in [0, 1]$$
$$r \to (\varphi \to \psi) \mapsto \varphi \to x_1, x_2 \to \psi, r + x_1 - x_2 = 1, x_i \in [0, 1]$$
$$(\varphi \to \psi) \to r \mapsto x_1 \to \varphi, \psi \to x_2, y - r \leq 0, y + x_1 \leq 1, y \leq x_2, y + r + x_1 - x_2 = 1,$$
$$x_i \in [0, 1], y \in \{0, 1\}$$
$$r \to n\varphi \qquad\qquad \mapsto r/n \to \varphi$$
$$n\varphi \to r \qquad\qquad \mapsto \varphi \to r/n$$

Fig. 5. Transformation rules, where φ, ψ are formulae, p is a propositional letter and r, n are rationals in $[0, 1]$

6 The Bargaining Process

Summing up, the negotiation process covers the following steps:

Pre-negotiation Phase. The buyer defines *hard constraints* β and preferences (*soft constraints*) $\mathcal{B}$ with corresponding weigths for each preference $n_1, n_2, ..., n_k$, as well as the threshold t_β, and similarly the seller σ, $\mathcal{S}$, m_h and t_σ. Here we are not interested in how to compute t_β, t_σ, n_i and m_i; we assume they are determined in advance by means of either direct assignment methods (Ordering, Simple Assessing or Ratio Comparison) or pairwise comparison methods (like AHP and Geometric Mean) [17]. After the previous elements have been set, both agents inform the mediator about these specifications and the theory $\mathcal{T}$ they refer to. Notice that for numerical features involved in the negotiation process, both in β and σ their respective reservation values are set either in the form $f \leq r_f$ or in the form $f \geq r_f$.

Negotiation-Core phase. Once the mediator have collected the sets of *hard* and *soft* *constraints*, the theory $\mathcal{T}$ they refer to, the weights n_i and m_i and the thresholds t_β, t_σ from the bargainers, it exploits such an information in order to compute Pareto agreements (see Section 5). With respect to the set of constraints represented by the theory $\mathcal{T}$, the *hard constraints* β, σ and the thresholds t_β, t_σ the mediator solves an optimization problem, trying to maximizing the utility of both buyer and seller, *i.e.*, trying to maximizing the number of satisfied preferences of both players in the final agreement. The returned solution to the optimization problem is the agreement proposed to the buyer and the seller. The solution proposed by the mediator is not only a Pareto-optimal one, as it is also a *fair* solution [21]. In fact, among all the Pareto-optimal solutions we take the one maximizing the product of utilities of the players.

From this point on, it is a *take-it-or-leave-it* offer: the bargainers can either accept or reject the proposed agreement [9]. If both players accept then an agreement is reached, otherwise, the negotiation ends in a *conflict deal*.

Let us present a tiny example in order to better clarify the approach. Given the toy ontology $\mathcal{T}$,

$$\mathcal{T} = \begin{cases} \texttt{Sedan} \Rightarrow \texttt{PassengerCar} \\ \texttt{ExternalColorBlack} \Rightarrow \neg\texttt{ExternalColorGray} \\ \texttt{SatelliteAlarm} \Rightarrow \texttt{AlarmSystem} \\ \texttt{InsurancePlus} \Leftrightarrow \texttt{DriverInsurance} \wedge \texttt{TheftInsurance} \\ \texttt{NavigatorPack} \Leftrightarrow \texttt{SatelliteAlarm} \wedge \texttt{GPS\_system} \end{cases}$$

The buyer and the seller specify their *hard* and *soft constraints*. For each numerical feature involved in *soft constraints* we associate a fuzzy function. If the bargainer has stated a reservation value on that feature, it will be used in the definition of the fuzzy function, otherwise a default value will be used.

$\beta\ = \texttt{PassengerCar} \wedge \texttt{price} \le 26000$
$\beta_1 = (\texttt{price}, \texttt{rs}(23000, 26000)) \wedge \texttt{AlarmSystem}$
$\beta_2 = \texttt{DriverInsurance} \wedge (\texttt{TheftInsurance} \vee \texttt{FireInsurance})$
$\beta_3 = \texttt{AirConditioning} \wedge (\texttt{ExternalColorBlack} \vee \texttt{ExternalColorGray})$
$\beta_4 = (\texttt{price}, \texttt{ls}(22000, 24000)) \vee (\texttt{km\_warranty}, \texttt{rs}(140000, 160000))$
$\mathcal{B}\ = 0.1 \cdot \beta_1 \vee 0.2 \cdot \beta_2 \vee 0.3 \cdot \beta_3 \vee 0.4 \cdot \beta_4$
$t_\beta = 0.7$

$\sigma\ = \texttt{Sedan} \wedge \texttt{price} \ge 24000$
$\sigma_1 = \texttt{NavigatorPack} \wedge (\texttt{price}, \texttt{rs}(24000, 26000))$
$\sigma_2 = \texttt{InsurancePlus}$
$\sigma_3 = (\texttt{km\_warranty}, \texttt{ls}(150000, 170000))$
$\sigma_4 = \texttt{ExternalColorBlack} \wedge \texttt{AirConditioning}$
$\mathcal{S}\ = 0.3 \cdot \sigma_1 \vee 0.1 \cdot \sigma_2 \vee 0.4 \cdot \sigma_3 \vee 0.2 \cdot \sigma_4$
$t_\sigma\ = 0.6$

Let

$$\bar{\mathcal{T}} = \mathcal{T} \cup \{\beta, \sigma\} \cup \{\texttt{buy} \Leftrightarrow \mathcal{B}, \texttt{sell} \Leftrightarrow \mathcal{S}, t_\beta \Rightarrow \mathcal{B}, t_\sigma \Rightarrow \mathcal{S}\}$$

Then, by definition, an interpretation $\bar{\mathcal{I}}$ such that

$$\bar{\mathcal{I}} = \arg\max_{\mathcal{I} \models \bar{\mathcal{T}}} (\texttt{buy})^{\mathcal{I}} * (\texttt{sell})^{\mathcal{I}} .$$

is a *Pareto agreement* which is equivalent to solve

$$\max x_{\texttt{buy}} \cdot x_{\texttt{sell}}$$

$$\bigcup_{1 \Rightarrow \psi \in \bar{\mathcal{T}}} E_{1 \Rightarrow \psi} .$$

It turns out that an optimal $\bar{\mathcal{I}}$ is such that

$$\max x_{\texttt{buy}} \cdot x_{\texttt{sell}} = 0.933 \cdot 0.7 = 0.651 ,$$

that is, $0.651 = \texttt{buy}^{\bar{\mathcal{I}}} * \texttt{sell}^{\bar{\mathcal{I}}} = 0.933 \cdot 0.7$. Furthermore, all $\beta_i^{\bar{\mathcal{I}}} = 1$ except $\beta_1^{\bar{\mathcal{I}}} = 0.333$ and $\sigma_i^{\bar{\mathcal{I}}} = 1$, except $\sigma_1^{\bar{\mathcal{I}}} = 0.0$.

In particular, the final agreement is:

$\text{Sedan}^{\bar{\mathcal{I}}} = 1.0,$

$\text{PassengerCar}^{\bar{\mathcal{I}}} = 1.0,$

$\text{InsurancePlus}^{\bar{\mathcal{I}}} = 1.0,$

$\text{AlarmSystem}^{\bar{\mathcal{I}}} = 1.0,$

$\text{DriverInsurance}^{\bar{\mathcal{I}}} = 1.0,$

$\text{AirConditioning}^{\bar{\mathcal{I}}} = 1.0,$

$\text{NavigatorPack}^{\bar{\mathcal{I}}} = 1.0,$

$(\text{km\_warranty ls}(150000, 170000))^{\bar{\mathcal{I}}} = 0.5,$ *i.e.*, $\text{km\_warranty}^{\bar{\mathcal{I}}} = 160000,$

$(\text{price}, \text{ls}(23000, 26000))^{\bar{\mathcal{I}}} = 0.33,$ *i.e.*, $\text{price}^{\bar{\mathcal{I}}} = 24000,$

$\text{TheftInsurance}^{\bar{\mathcal{I}}} = 1.0,$

$\text{FireInsurance}^{\bar{\mathcal{I}}} = 1.0,$

$\text{ExternalColorBlack}^{\bar{\mathcal{I}}} = 1.0,$

$\text{ExternalColorGray}^{\bar{\mathcal{I}}} = 0.0.$

Notice that $\beta_1^{\bar{\mathcal{I}}} = 0.333$, as the preference is not *fully* satisfied — the price is equal to 24000. Furthermore, thanks to its fuzzy representation, it is possible to say that it is satisfied with a *certain degree*.

7 Related Work and Discussion

Automated bilateral negotiation has been widely investigated, both in artificial intelligence and in microeconomics research communities, so this section is necessarily far from complete.

AI-oriented research has usually focused on automated negotiation among agents, and on designing high-level protocols for agent interaction [13]. Agents can play different roles: act on behalf of a buyer or seller, but also play the role of a mediator or facilitator. Depending on the presence of a mediator we can distinguish between *centralized* and *distributed* approaches. In the former, agents elicit their preferences and then a mediator, or some central entity, selects the most suitable deal based on them. In the latter, agents negotiate through various negotiation steps reaching the final deal by means of intermediate deals, without any external help [5]. Distributed approaches do not allow the presence of a mediator because – as stated in [12, p.25] – agents cannot agree on any entity, so they do not want to disclose their preferences to a third party, that, missing any relevant information, could not help agents. In dynamic systems a predefined conflict resolution cannot be allowed, so the presence of a mediator is discouraged. On the other hand the presence of a mediator can be extremely useful in designing negotiation mechanisms and in practical important commerce settings. As stated in [15], negotiation mechanisms often involve the presence of a mediator [3], which collects information from bargainers and exploits them in order to propose an efficient negotiation outcome. Various recent proposals adopt a mediator, including [6,11,7]. In

---

[3] The most well known –and running– example of mediator is eBay site, where a mediator receives and validates bids, as well as presenting the current highest bid and finally determining the auction winner [15].

[6] an extended alternating-offers protocol is presented, with the presence of a media-tor, which improves the utility of both agents. No inter-dependent issues are taken into account. In [11] a mediated-negotiation approach is proposed for complex contracts, where inter-dependency among issues is investigated. The agreement is a vector of is-sues, having value 0 or 1 depending on the presence or absence of a given contract clauses. Only binary dependencies between issues are considered: the agent's utility is computed through an influence matrix, where each cell represents the utility of a given pair of issues. However in this approach no semantic relations among issues are investigated.

Several recent logic-based approaches to negotiation are based on propositional logic. In [3], Weighted Propositional Formulas (WPF) are used to express agents pref-erences in the allocation of indivisible goods, but no common knowledge (as our on-tology) is present. The use of an ontology allows *e.g.*, to catch inconsistencies between demand and supply or find out if an agent preference is implied by a preference of its opponent, which is fundamental to model an e-marketplace. Utility functions expressed through WPF are classified in [4] according to the properties of the utility function (sub/super-additive, monotone, etc.). We used the most expressive functions according to that classification, namely, weights over unrestricted propositional formulas.

The work presented in [27] adopts a kind of propositional knowledge base arbitra-tion to choose a fair negotiation outcome. However, *common knowledge* is considered as just more entrenched preferences, that could be even dropped in some deals. Instead, the logical constraints in our ontology $\mathcal{T}$ must *always* be enforced in the negotiation outcomes, and we introduce a fuzzy propositional logic with concrete domains. Finally we devised a *protocol* which the agents should adhere to while negotiating; in contrast, in [27] a game-theoretic approach is taken, presenting no protocol at all, since commu-nication between agents is not considered.

We borrow from [26] the definition of agreement as a model for a set of formulas from both agents. However, in [26] only multiple-rounds protocols are studied, and the approach leaves the burden to reach an agreement to the agents themselves, although they can follow a protocol. The approach does not take preferences into account, so that it is not possible to guarantee the reached agreement is Pareto-efficient. Our ap-proach, instead, aims at giving an *automated* support to negotiating agents to reach, in one shot, Pareto agreements. The work presented here builds on [19], where a basic propositional logic framework endowed of a logical theory was proposed. In [18] the approach was extended and generalized and complexity issues were discussed. In this paper we further extended the framework, introducing the extended logic $\mathcal{P}(\mathcal{N})$, thus effectively handling numerical features involved in *fuzzy constraints*, and showed we are able to compute Pareto-efficient agreements, by solving an optimization problem and adopting a one-shot negotiation protocol. We are aware that there is no universal approach to automate negotiation fitting every scenario, but rather several frameworks suitable for different scenarios, depending on the assumptions made about the domains and agents involved in the interaction. Here, we have proposed a logic-based frame-work to automate multi-issue bilateral negotiation in P2P e-marketplaces, where agents communicate using the logic $\mathcal{P}(\mathcal{N})$, which allows to handle both numerical features and non numerical ones. Modeling issues in a $\mathcal{P}(\mathcal{N})$ ontology it is possible to catch inconsistency between preferences and then reach consistent agreements, as well as to discover implicit relations (such as implication) among preferences that do not immediately appear at the syntactic level. Moreover, thanks to fuzzy representation it has been possible to model *fuzzy constraints* on numerical features. Exploiting a mediator the proposed approach allows to deal with the problem of incomplete information about opponent's preferences. We adopted a one-shot protocol, using a mediator to solve an optimization problem that ensures the Pareto-efficiency of the outcomes.

In the near future we plan to extend the approach using more expressive logics, namely, Fuzzy Description Logics [24], to increase the expressiveness of supply/demand descriptions. We are also investigating other negotiation protocols, without the presence of a mediator, allowing to reach an agreement in a reasonable amount of communication rounds.

References

1. Baader, F., Hanschke, P.: A schema for integrating concrete domains into concept languages. In: Proc. of IJCAI 1991, pp. 452–457 (1991)
2. Binmore, K.: Fun and Games. A Text on Game Theory. D.C. Heath and Company (1992)
3. Bouveret, S., Lemaitre, M., Fargier, H., Lang, J.: Allocation of indivisible goods: A general model and some complexity results. In: Proc. of AAMAS 2005, pp. 1309–1310 (2005)
4. Chevaleyre, Y., Endriss, U., Lang, J.: Expressive power of weighted propositional formulas for cardinal preference modeling. In: Proc. of KR 2006, pp. 145–152 (2006)
5. Chevaleyre, Y., Endriss, U., Lang, J., Maudet, N.: Negotiating over small bundles of resources. In: Proc. of AAMAS 2005, pp. 296–302 (2005)
6. Fatima, S., Wooldridge, M., Jennings, N.R.: Optimal agendas for multi-issue negotiation. In: Proc. of AAMAS 2003, pp. 129–136 (2003)
7. Gatti, N., Amigoni, F.: A decentralized bargaining protocol on dependent continuous multi-issue for approximate pareto optimal outcomes. In: Proc. of AAMAS 2005, pp. 1213–1214 (2005)
8. Hájek, P.: Metamathematics of Fuzzy Logic. Kluwer, Dordrecht (1998)
9. Jennings, N.R., Faratin, P., Lomuscio, A.R., Parsons, S., Wooldridge, M.J., Sierra, C.: Automated negotiation: Prospects, methods and challenges. Int. J. of Group Decision and Negotiation 10(2), 199–215 (2001)
10. Jeroslow, R.G.: Logic-based Decision Support. Mixed Integer Model Formulation. Elsevier, Amsterdam (1989)
11. Klein, M., et al.: Negotiating complex contracts. In: Proc. of AAMAS 2002, pp. 753–757 (2002)
12. Kraus, S.: Strategic Negotiation in Multiagent Environments. The MIT Press, Cambridge (2001)
13. Lomuscio, A.R., Wooldridge, M., Jennings, N.R.: A classification scheme for negotiation in electronic commerce. Int. Journal of Group Decision and Negotiation 12(1), 31–56 (2003)
14. Lukasiewicz, T., Straccia, U.: Tutorial: Managing uncertainty and vagueness in semantic web languages. In: AAAI 2007. Twenty-Second Conference on Artificial Intelligence (2007)
15. MacKie-Mason, J.K., Wellman, M.P.: Automated markets and trading agents. In: Handbook of Computational Economics. North-Holland, Amsterdam (2006)
16. Parsons, S., Sierra, C., Jennings, N.: Agents that reason and negotiate by arguing. Journal of Logic and Computation 8(3), 261–292 (1998)
17. Pomerol, J.C., Barba-Romero, S.: Multicriterion Decision Making in Management. In: Kluwer Series in Operation Research. Kluwer Academic, Dordrecht (2000)

18. Ragone, A., Di Noia, T., Di Sciascio, E., Donini, F.M.: A logic-based framework to compute pareto agreements in one-shot bilateral negotiation. In: Proc. of ECAI 2006, pp. 230–234 (2006)
19. Ragone, A., Di Noia, T., Di Sciascio, E., Donini, F.M.: Propositional- logic approach to one-shot multi issue bilateral negotiation. ACM SIGecom Exchanges 5(5), 11–21 (2006)
20. Raiffa, H.: The Art and Science of Negotiation. Harvard University Press, Cambridge (1982)
21. Raiffa, H., Richardson, J., Metcalfe, D.: Negotiation Analysis - The Science and Art of Collaborative Decision Making. The Belknap Press of Harvard University Press, Cambridge (2002)
22. Rosenschein, J.S., Zlotkin, G.: Rules of Encounter. MIT Press, Cambridge (1994)
23. Straccia, U.: Description logics with fuzzy concrete domains. In: Bachus, F., Jaakkola, T. (eds.) UAI 2005. 21st Conference on Uncertainty in Artificial Intelligence, pp. 559–567. AUAI Press, Edinburgh, Scotland (2005)
24. Straccia, U.: A fuzzy description logic for the semantic web. In: Sanchez, E. (ed.) Fuzzy Logic and the Semantic Web, Capturing Intelligence, ch. 4, pp. 73–90. Elsevier, Amsterdam (2006)
25. Wellman, M.P.: Online marketplaces. In: Practical Handbook of Internet Computing. CRC Press, Boca Raton (2004)
26. Wooldridge, M., Parsons, S.: Languages for negotiation. In: Proc. of ECAI 2004, pp. 393–400 (2000)
27. Zhang, D., Zhang, Y.: A computational model of logic-based negotiation. In: Proc. of the AAAI 2006, pp. 728–733 (2006)

Author Index

Lecture Notes in Computer Science

Sublibrary 3: Information Systems and Application, incl. Internet/Web and HCI

For information about Vols. 1– 4526
please contact your bookseller or Springer

Vol. 4744: Y. de Kort, W. IJsselsteijn, C. Midden, B. Eggen, B.J. Fogg (Eds.), Persuasive Technology. XIV, 316 pages. 2007.

Vol. 4740: L. Ma, M. Rauterberg, R. Nakatsu (Eds.), Entertainment Computing – ICEC 2007. XXX, 480 pages. 2007.

Vol. 4730: C. Peters, P. Clough, F.C. Gey, J. Karlgren, B. Magnini, D.W. Oard, M. de Rijke, M. Stempfhuber (Eds.), Evaluation of Multilingual and Multi-modal Information Retrieval. XXIV, 998 pages. 2007.

Vol. 4723: M. R. Berthold, J. Shawe-Taylor, N. Lavrač (Eds.), Advances in Intelligent Data Analysis VII. XIV, 380 pages. 2007.

Vol. 4721: W. Jonker, M. Petković (Eds.), Secure Data Management. X, 213 pages. 2007.

Vol. 4718: J. Hightower, B. Schiele, T. Strang (Eds.), Location- and Context-Awareness. X, 297 pages. 2007.

Vol. 4717: J. Krumm, G.D. Abowd, A. Seneviratne, T. Strang (Eds.), UbiComp 2007: Ubiquitous Computing. XIX, 520 pages. 2007.

Vol. 4715: J.M. Haake, S.F. Ochoa, A. Cechich (Eds.), Groupware: Design, Implementation, and Use. XIII, 355 pages. 2007.

Vol. 4714: G. Alonso, P. Dadam, M. Rosemann (Eds.), Business Process Management. XIII, 418 pages. 2007.

Vol. 4704: D. Barbosa, A. Bonifati, Z. Bellahsène, E. Hunt, R. Unland (Eds.), Database and XML Technologies. X, 141 pages. 2007.

Vol. 4690: Y. Ioannidis, B. Novikov, B. Rachev (Eds.), Advances in Databases and Information Systems. XIII, 377 pages. 2007.

Vol. 4675: L. Kovács, N. Fuhr, C. Meghini (Eds.), Research and Advanced Technology for Digital Libraries. XVII, 585 pages. 2007.

Vol. 4674: Y. Luo (Ed.), Cooperative Design, Visualization, and Engineering. XIII, 431 pages. 2007.

Vol. 4663: C. Baranauskas, P. Palanque, J. Abascal, S.D.J. Barbosa (Eds.), Human-Computer Interaction – INTERACT 2007, Part II. XXXIII, 735 pages. 2007.

Vol. 4662: C. Baranauskas, P. Palanque, J. Abascal, S.D.J. Barbosa (Eds.), Human-Computer Interaction – INTERACT 2007, Part I. XXXIII, 637 pages. 2007.

Vol. 4658: T. Enokido, L. Barolli, M. Takizawa (Eds.), Network-Based Information Systems. XIII, 544 pages. 2007.

Vol. 4656: M.A. Wimmer, J. Scholl, Å. Grönlund (Eds.), Electronic Government. XIV, 450 pages. 2007.

Vol. 4655: G. Psaila, R. Wagner (Eds.), E-Commerce and Web Technologies. VII, 229 pages. 2007.

Vol. 4654: I.-Y. Song, J. Eder, T.M. Nguyen (Eds.), Data Warehousing and Knowledge Discovery. XVI, 482 pages. 2007.

Vol. 4653: R. Wagner, N. Revell, G. Pernul (Eds.), Database and Expert Systems Applications. XXII, 907 pages. 2007.

Vol. 4636: G. Antoniou, U. Aßmann, C. Baroglio, S. Decker, N. Henze, P.-L. Patranjan, R. Tolksdorf (Eds.), Reasoning Web. IX, 345 pages. 2007.

Vol. 4611: J. Indulska, J. Ma, L.T. Yang, T. Ungerer, J. Cao (Eds.), Ubiquitous Intelligence and Computing. XXIII, 1257 pages. 2007.

Vol. 4607: L. Baresi, P. Fraternali, G.-J. Houben (Eds.), Web Engineering. XVI, 576 pages. 2007.

Vol. 4606: A. Pras, M. van Sinderen (Eds.), Dependable and Adaptable Networks and Services. XIV, 149 pages. 2007.

Vol. 4605: D. Papadias, D. Zhang, G. Kollios (Eds.), Advances in Spatial and Temporal Databases. X, 479 pages. 2007.

Vol. 4602: S. Barker, G.-J. Ahn (Eds.), Data and Applications Security XXI. X, 291 pages. 2007.

Vol. 4601: S. Spaccapietra, P. Atzeni, F. Fages, M.-S. Hacid, M. Kifer, J. Mylopoulos, B. Pernici, P. Shvaiko, J. Trujillo, I. Zaihrayeu (Eds.), Journal on Data Semantics IX. XV, 197 pages. 2007.

Vol. 4592: Z. Kedad, N. Lammari, E. Métais, F. Meziane, Y. Rezgui (Eds.), Natural Language Processing and Information Systems. XIV, 442 pages. 2007.

Vol. 4587: R. Cooper, J. Kennedy (Eds.), Data Management. XIII, 259 pages. 2007.

Vol. 4577: N. Sebe, Y. Liu, Y.-t. Zhuang, T.S. Huang (Eds.), Multimedia Content Analysis and Mining. XIII, 513 pages. 2007.

Vol. 4568: T. Ishida, S. R. Fussell, P. T. J. M. Vossen (Eds.), Intercultural Collaboration. XIII, 395 pages. 2007.

Vol. 4566: M.J. Dainoff (Ed.), Ergonomics and Health Aspects of Work with Computers. XVIII, 390 pages. 2007.

Vol. 4564: D. Schuler (Ed.), Online Communities and Social Computing. XVII, 520 pages. 2007.

Vol. 4563: R. Shumaker (Ed.), Virtual Reality. XXII, 762 pages. 2007.

Vol. 4561: V.G. Duffy (Ed.), Digital Human Modeling. XXIII, 1068 pages. 2007.

Vol. 4560: N. Aykin (Ed.), Usability and Internationalization, Part II. XVIII, 576 pages. 2007.

Vol. 4559: N. Aykin (Ed.), Usability and Internationalization, Part I. XVIII, 661 pages. 2007.

Vol. 4558: M.J. Smith, G. Salvendy (Eds.), Human Interface and the Management of Information, Part II. XXIII, 1162 pages. 2007.

Vol. 4557: M.J. Smith, G. Salvendy (Eds.), Human Interface and the Management of Information, Part I. XXII, 1030 pages. 2007.

Vol. 4541: T. Okadome, T. Yamazaki, M. Makhtari (Eds.), Pervasive Computing for Quality of Life Enhancement. IX, 248 pages. 2007.

Vol. 4537: K.C.-C. Chang, W. Wang, L. Chen, C.A. Ellis, C.-H. Hsu, A.C. Tsoi, H. Wang (Eds.), Advances in Web and Network Technologies, and Information Management. XXIII, 707 pages. 2007.

Vol. 4531: J. Indulska, K. Raymond (Eds.), Distributed Applications and Interoperable Systems. XI, 337 pages. 2007.